Rate Processes of Extractive Metallurgy

Rate Processes of Extractive Metallurgy

Edited by
Hong Yong Sohn
and
Milton E. Wadsworth
University of Utah
Salt Lake City, Utah

Plenum Press · New York and London

Library of Congress Cataloging in Publication Data

Main entry under title:

Rate processes of extractive metallurgy.

Includes bibliographical references and index.
1. Metallurgy. I. Sohn, Hong Yong. II. Wadsworth, Milton E.
TN665.R27 669 78-15941
ISBN 978-1-4684-9119-7 ISBN 978-1-4684-9117-3 (eBook)
DOI 10.1007/978-1-4684-9117-3

A Divison of Plenum Publishing Corporation
227 West 17th Street, New York, N.Y. 10011
Softcover reprint of the hardcover 1st edition 1979

Contributors

Robert W. Bartlett, Stanford Research Institute, Menlo Park, California; formerly Kennecott Research Center, Salt Lake City, Utah

Ivan B. Cutler, Department of Materials Science and Engineering, University of Utah, Salt Lake City, Utah

James W. Evans, Department of Materials Science and Mineral Engineering, University of California, Berkeley, California

John A. Herbst, Department of Metallurgy and Metallurgical Engineering, University of Utah, Salt Lake City, Utah

C.-H. Koo, Department of Materials Science and Mineral Engineering, University of California, Berkeley, California

Sanaa E. Khalafalla, Twin Cities Metallurgy Research Center, U.S. Bureau of Mines, Twin Cities, Minnesota

Jan D. Miller, Department of Metallurgy and Metallurgical Engineering, University of Utah, Salt Lake City, Utah

Charles H. Pitt, Department of Metallurgy and Metallurgical Engineering, University of Utah, Salt Lake City, Utah

Hong Yong Sohn, Department of Metallurgy and Metallurgical Engineering, University of Utah, Salt Lake City, Utah

Julian Szekely, Department of Materials Science and Engineering, Massachusetts Institute of Technology, Cambridge, Massachusetts

E. T. Turkdogan, Research Laboratory, U.S. Steel Corporation, Monroeville, Pennsylvania

Milton E. Wadsworth, Department of Metallurgy and Metallurgical Engineering, University of Utah, Salt Lake City, Utah

Preface

Computer technology in the past fifteen years has essentially revolutionized engineering education. Complex systems involving coupled mass transport and flow have yielded to numerical analysis even for relatively complex geometries. The application of such technology together with advances in applied physical chemistry have justified a general updating of the field of heterogeneous kinetics in extractive metallurgy. This book is an attempt to cover significant areas of extractive metallurgy from the viewpoint of heterogeneous kinetics.

Kinetic studies serve to elucidate fundamental mechanisms of reactions and to provide data for engineering applications, including improved ability to scale processes up from bench to pilot plant. The general theme of this book is the latter—the scale-up. The practicing engineer is faced with problems of changes of order of magnitude in reactor size. We hope that the fundamentals of heterogeneous kinetics will provide increasing ability for such scale-up efforts. Although thermodynamics is important in defining potential reaction paths and the end products, kinetic limitations involving molecular reactions, mass transport, or heat flow normally influence ultimate rates of production. For this reason, rate processes in the general field of extractive metallurgy have been emphasized in this book.

This book is an edited version of the notes for an intensive short course of the same title held at the University of Utah in 1975. The section on oxide reduction was added subsequently. Although various chapters and sections are contributions by different authors, we selected and organized the topics so that coherence and continuity were preserved as far as possible. There are some slight differences in the notation in the various sections, but in such cases the symbols are defined where they first appear in a given section.

We wish to thank all those who have helped in making the short course possible and in the preparation of this book. While we cannot possibly name all of them, we would like to mention the following: We gratefully acknowledge the support of the Kennecott Copper Corporation, the U.S. Steel Corporation, and the U.S. Bureau of Mines, and the encouragement and assistance of Dr. K. J. Richards of Kennecott Research Center. Professor Henry Eyring of the University of Utah, a pioneer in the field of chemical kinetics, delivered a keynote lecture on the subject of "Reaction-Rate Theory and Some Applications." Professor Julian Szekely of the Massachusetts Institute of Technology deserves thanks for introducing one of us (HYS) to extractive and process metallurgy and for instilling in him an "erroneous" idea, through his many short courses and conferences, that a short course of this type is an easy venture. Professors R. D. Pehlke (University of Michigan) and T. J. O'Keefe (University of Missouri–Rolla) offered encouragement and support of our effort and made valuable suggestions regarding the contents of the book.

The inception of the short course was a result of the combined efforts by our colleagues in the Department of Metallurgy and Metallurgical Engineering, University of Utah. We also thank the participants of the short course for providing lively discussions during the sessions and helpful comments afterward. Our warmest thanks go to our secretarial staff for typing much of the manuscript and assisting us in many ways with its preparation.

H. Y. Sohn
M. E. Wadsworth

Contents

2. Rate Processes in Multiparticle Metallurgical Systems

J. A. Herbst

2A. Chemical Process Analysis and Design for Multiparticle Systems

Robert W. Bartlett

3. Hydrometallurgical Processes

Sec. 3.1. Milton E. Wadsworth
Sec. 3.2. Milton E. Wadsworth
Sec. 3.3. J. D. Miller

4. *Pyrometallurgical Processes*

Sec. 4.1. S. E. Khalafalla
Sec. 4.2. J. W. Evans and C.-H. Koo
Sec. 4.3. H. Y. Sohn and E. T. Turkdogan
Sec. 4.4. I. B. Cutler
Sec. 4.5. E. T. Turkdogan
Sec. 4.6. C. H. Pitt

1

Fundamentals of the Kinetics of Heterogeneous Reaction Systems in Extractive Metallurgy

H. Y. Sohn

Changes that must be effected to extract metal values from mineral ores generally occur through heterogeneous processes. Although most reactions involve a rather complex set of steps and may require individual treatments, there are certain elementary aspects of the overall reaction that are common to a wide range of reactions. Such aspects are amenable to systematic treatment. In this chapter we shall examine these elementary steps and their application to a number of individual reaction systems, with an emphasis on noncatalytic *fluid–solid* reactions. Understanding these individual systems is essential for the analyses of more complex systems; thus this chapter lays a foundation for analyzing the rates of actual processes in extractive metallurgy to be discussed in subsequent chapters.

The following discussion will include the elementary steps of fluid–solid reactions, the reaction of a single nonporous solid particle, the reaction of a single porous particle, and reactions between solids that occur with the aid of gaseous intermediates.

1.1. Elementary Steps

Let us consider a fluid–solid reaction of the following type:

$$\mathrm{A(fluid)} + b\mathrm{B(solid)} = c\mathrm{C(fluid)} + d\mathrm{D(solid)} \qquad (1.1\text{-}1)$$

Hong Yong Sohn • Department of Metallurgy and Metallurgical Engineering, University of Utah, Salt Lake City, Utah

Examples of reactions of this type include:

$$Fe_2O_3(s) + CO(g) = 2\,FeO + CO_2(g)$$

$$CuO(s) + H_2(g) = Cu(s) + H_2O(g)$$

$$CuO \cdot SiO_2 \cdot 2\,H_2O(s) + 2\,H^+(liq) = Cu^{2+}(liq) + SiO_2 \cdot nH_2O(s) + (3-n)H_2O(liq)$$

The overall reaction process may involve the following individual steps, as sketched in Figure 1.1-1:

1. mass transfer of reactants and products between the bulk of the fluid and the external surface of the solid particle;
2. diffusion of reactants and products within the pores of the solid;
3. chemical reaction between the reactants in the fluid and in the solid.

Only recently it has been recognized that the rate-controlling step can change depending upon reaction conditions, and thus rate information obtained under a given set of conditions may not be applicable under another set of conditions. Furthermore, frequently there may not be a single rate-controlling step because several steps may have more or less equal effects on determining the overall rate. The relative importance of these steps could also change in the course of reaction. Therefore, understanding how the individual reaction steps interact with each other is important in determining not only the rate-controlling step under given reaction conditions but also whether more than a single step must be considered in expressing the overall rate.

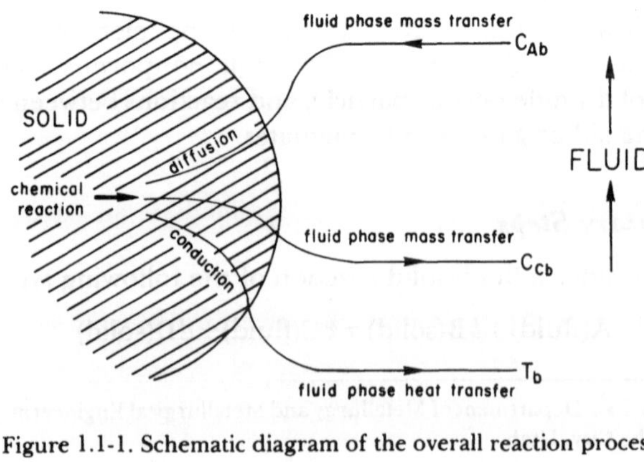

Figure 1.1-1. Schematic diagram of the overall reaction process.

In addition to the above steps involving the chemical change of species, there are two other processes that may have significant influences on the overall rate, i.e., heat transfer and changes in the structure of the solid during reaction. Many fluid–solid reactions either generate or consume heat. The heat of reaction must be transferred from the surrounding to where the reaction takes place, or vice versa. Heat transfer involves (1) convection and/or radiation between the surrounding and the solid surface, and (2) conduction within the solid. The chemical reaction and heat may cause sintering or other changes in the pore structure, which in turn could have significant effects on the overall reaction rate.

In the following we shall review the elementary steps which make up fluid–solid reaction systems. The discussion will be rather brief because many of these steps are not unique for fluid–solid reactions, and there are more comprehensive treatments available in the literature.[1–5]

1.1.1. Mass Transfer between a Solid Surface and a Fluid

This "external" mass transfer step has been extensively studied, and thus is perhaps the best understood of the overall reaction sequence. Although it is possible in certain cases to calculate the rate of mass transfer between a moving fluid stream and a solid surface by the solution of the appropriate flow and diffusion equation,[1,6] here we shall adopt a practical approach of obtaining an average mass transfer coefficient from empirical correlations.

Let us consider a situation where species A is being transferred from a solid surface into a moving fluid stream. The concentration of A is C_{As} and C_{Ab} near the solid surface and in the bulk of the fluid, respectively. Then the rate at which mass is transferred per unit surface area of the solid is given by

$$n_A = k_m(C_{Ab} - C_{As}) \tag{1.1-2}$$

where k_m is the mass transfer coefficient, and n_A is the flux of A.

The correlation for estimating k_m is usually expressed in terms of the Sherwood number as a function of other dimensionless groups such as the Reynolds and the Schmidt numbers. There are numerous such correlations in various systems depending on the mode of fluid–solid contacting, such as a single particle in a large amount of fluid, a packed bed, or a fluidized bed. The reader is referred to other sources[3,5,7] for actual correlations and detailed discussions on the subject.

Although the appropriate choice of correlations must be made considering the nature of individual systems, the following are some examples most frequently used in estimating the mass transfer coefficient:

For a single particle in a large amount of fluid,

$$\mathrm{Sh} = 2.0 + 0.6\,\mathrm{Re}^{1/2} \cdot \mathrm{Sc}^{1/3} \tag{1.1-3}$$

where

$$\mathrm{Sh} \equiv \frac{k_m d_p}{D}$$

$$\mathrm{Re} \equiv \frac{d_p \rho u}{\mu}$$

and

$$\mathrm{Sc} \equiv \frac{\mu}{\rho D}$$

For a particle in a packed bed,[64,65]

$$\varepsilon j_D = \frac{0.357}{\mathrm{Re}^{0.359}} \qquad \text{gases, } 3 < \mathrm{Re} < 2000 \tag{1.1-4a}$$

$$\varepsilon j_D = \frac{0.250}{\mathrm{Re}^{0.31}} \qquad \text{liquids, } 55 < \mathrm{Re} < 1500 \tag{1.1-4b}$$

$$\varepsilon j_D = \frac{1.09}{\mathrm{Re}^{2/3}} \qquad \text{liquids, } 0.0016 < \mathrm{Re} < 55 \tag{1.1-4c}$$

where

$$j_D \equiv \frac{k_m \rho}{G} Sc^{2/3}, \qquad \mathrm{Re} \equiv \frac{d_p G}{\mu}$$

To calculate the mass transfer coefficient from the above correlations, one needs information on the diffusivity and the viscosity of the fluid. The reader is referred to other sources[1,8] for the estimation methods for these properties.

1.1.2. Diffusion of Fluid Species through the Pores of a Solid

When the reactant solid is porous, diffusion through the pores is important in that a much greater portion of the solid other than the external surface becomes potentially available for reaction. It may also be important in the reactions of nonporous solids which produce porous solid products, because diffusion occurs much more rapidly through pores than through a solid phase.

Pore diffusion is much more complex than diffusion in a fluid phase. Complications arise because in a porous solid the diffusion paths are not simple but tortuous, and the existence of a tortuosity that depends on the pore structure is difficult to estimate. Furthermore, when the pores are smaller than the mean free path of the various species, the laws of molecular diffusion no longer apply, but Knudsen diffusion becomes important. Under certain circumstances, significant pressure gradients may be established within the solid, and mass transfer due to a pressure gradient may also have to be taken into consideration.

Pore diffusion is of considerable importance in many other fields of science and engineering, including heterogeneous catalysis, isotope separation, and drying. Useful discussions on the subject are available in the literature.[3,5,9–11]

Diffusion in porous media is described by Fick's law in the following form:

$$n_A = -D_e \nabla C_A \tag{1.1-5}$$

where D_e is the effective diffusivity of species A in the porous medium; D_e takes into consideration the portion of cross-sectional area occupied by the solid (and thus not available for diffusion) and also the fact that the pores are not straight, which has the effect of increasing the distance of diffusion. ∇C_A is the concentration gradient with C_A being the actual concentration of A in the pore space. In general the effective diffusivity of species A in a porous medium is estimated using the following Bosanquet formula:

$$\frac{1}{D_{Ae}} = \frac{\tau}{\varepsilon}\left(\frac{1}{D_{AK}} + \frac{1}{D_{AB}}\right) \tag{1.1-6}$$

where

D_{Ae} is the effective diffusivity of A,
D_{AK} is the "Knudsen diffusivity" (a single-capillary coefficient),
D_{AB} is the molecular diffusivity of A in the mixture,
τ is the tortuosity of the solid, and
ε is the porosity of the solid.

The tortuosity factor, which is a measure of the increased diffusion distance in the porous medium, cannot be satisfactorily estimated *a priori*. It must be determined experimentally. Its numerical value usually lies in the range of 2 to 10. The porosity ε in the above equation takes care of the cross-sectional area occupied by the solid and thus not available for diffusion.

1.1.3. Intrinsic Kinetics of Heterogeneous Reactions on Solid Surfaces

As noted above, certain generalizations can be made for mass transfer and pore diffusion of the fluid species. The adsorption and chemical reaction, however, are highly specific to the nature of the substances involved. In most work on fluid–solid reactions, relatively little attention has been paid to the detailed reaction mechanisms (particularly the concentration dependence) of the intrinsic kinetics. A simple first-order rate expression has often been assumed, for mathematical simplicity rather than for valid reasons. In many practical cases, fluid–solid reactions can indeed be approximated as first-order reactions. In general, however, the concentration dependence is much more complex.

There are two types of adsorption of fluid species on a solid surface, namely physical adsorption and chemisorption. Chemisorption, which takes place as a result of much stronger interaction than physical adsorption, is mainly responsible for fluid–solid reactions as well as catalytic reactions. Again, a detailed discussion of the subject, which can be found elsewhere,[5,12–14] is beyond the scope of this article. It will suffice here to state that the chemical kinetics of fluid–solid reactions may be broken down into individual steps, namely the adsorption of reactants, the reaction between the adsorbed species and the solid at the surface, and the desorption of products. The rate expression obtained by considering the kinetics of each of these steps may usually be simplified by applying the steady-state approximation and the knowledge of the reaction equilibria. The resultant rate expressions are generally of the following form and the constants are determined by which of the steps is rate-controlling:

$$\text{Rate} = \frac{k(C_R^n - C_P^m/K_E)}{1 + K_R C_R^r + K_p C_P^p} \tag{1.1-7}$$

where C_R denotes the reactant concentration and C_P the product concentration in the fluid phase. When the concentrations are low and $n = m = 1$, the rate becomes first order, which is why first-order kinetics is frequently used for fluid–solid reactions, in addition to its mathematical simplicity.

1.1.4. Heat Transfer between a Solid Surface and a Moving Fluid Stream

The "external" heat transfer step is essentially analogous to the "external" mass transfer discussed above. The rate of heat transfer between a solid surface and a fluid flowing by it may be expressed as follows:

$$q = h(T_s - T_b) \tag{1.1-8}$$

where q is the rate of heat transfer per unit surface area, h is the heat transfer coefficient, and T_s and T_b are the surface and bulk-fluid temperatures, respectively. The close analogy with the equation for the external mass transfer is readily apparent. As in the case of the mass transfer coefficient, we generally rely on empirical correlations for estimating the heat transfer coefficient. In most practical systems the correlation for the heat transfer coefficient may be obtained from those previously given for mass transfer by replacing the Sherwood and Schmidt numbers, respectively, with the Nusselt and Prandtl numbers.[1,5]

At high temperatures, thermal radiation may become an important mechanism for heat transfer between a solid particle and its surroundings. The net radiant flux received by a particle is given by

$$q_r = \sigma(\varepsilon_a T_a^4 - \varepsilon_s T_s^4) \tag{1.1-9}$$

where

ε is the emissivity of the surfaces,
σ is the Boltzmann constant, and
T_a is the temperature of the surroundings.

1.1.5. Conduction of Heat in Porous Solids

The conduction of heat in a porous medium is described by Fourier's law in the following form:

$$q = -\lambda_e \nabla T \tag{1.1-10}$$

where λ_e is the effective thermal conductivity, and ∇T is the temperature gradient. Here the gas in the pore space is assumed to be in thermal equilibrium with the solid.

The thermal conductivity of a consolidated solid with closed pores may be estimated by the following[15]:

$$\lambda_e = \lambda_s(1-\varepsilon) \tag{1.1-11}$$

where λ_s is the thermal conductivity of the solid itself. The effective thermal conductivity of a porous medium which is an aggregate of fine particles is more difficult to estimate. An extensive discussion on the subject can be found in the literature.[3,16]

1.1.6. Summary

In this section we have discussed the elements that are involved in the reaction between a fluid and a solid particle. It is seen that in many instances even these individual steps themselves are quite complex and inadequately understood, which precludes the development of an entirely general model for fluid–solid reactions.

The reactions which are well understood tend to be controlled by one of the elementary steps described above, which renders the interpretation much simpler. Nonetheless, it is important to understand the principal mechanisms that constitute the overall reaction, because different steps may become the controlling step under different conditions.

1.2. Reaction of a Single Nonporous Particle

If the reactant solid is initially nonporous, the reaction occurs at a sharp interface between two phases, i.e., either fluid–solid or solid–solid, depending on whether or not a solid product is formed and if so, whether it is porous or nonporous.

If no solid product is formed, as in the gasification or dissolution, or the solid product is removed from the surface as it is formed, the solid reactant will always be in contact with the bulk fluid and the size of the particle will diminish as the reaction progresses. If, on the other hand, a coherent layer of solid product is formed around the reactant solid, the reaction will occur at the interface between the unreacted and the completely reacted zones. If the solid product is porous, the fluid reactant can reach the reaction interface by diffusing through the pores of the product solid. If the product is nonporous, either the fluid species must diffuse into the solid by solid-state diffusion, or a constituent species of the solid reactant must diffuse to the surface to react with fluid reactants. The overall size of the solid will depend on whether the solid product has a greater or smaller volume than the reactant solid.

In the reaction of a nonporous solid and a fluid, the chemical reaction and mass transport are connected in series, making the analysis much easier than in the case of a porous solid.

In certain fluid–solid reactions, nucleation presents an important step. The growth of nuclei is a rather complex phenomenon. As the solid size becomes larger or the reaction temperature is raised, the time within which nucleation is important becomes a small portion of the total reaction time and thus nucleation becomes less important. We shall limit our discussion here to reactions occurring at a sharp boundary which advances in parallel to the external surface of the solid.

1.2.1. Reactions in Which No Solid Product Layer Is Formed

Examples of such reactions are the dissolution of metal in acids, the formation of nickel carbonyl, the chlorination of metals, and the roasting of cinnabar (HgS). This type of a reaction may in general be

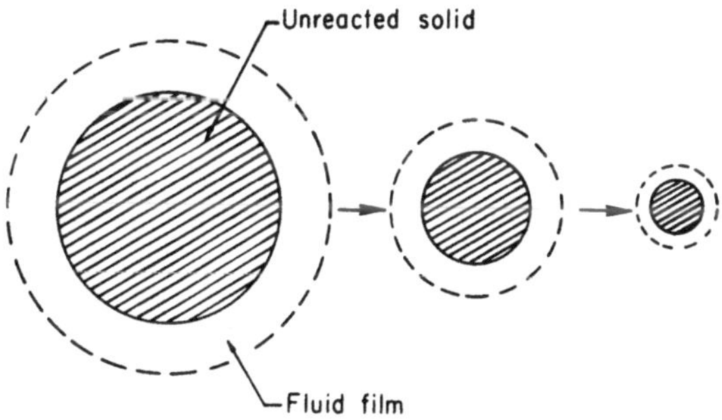

Figure 1.2-1. Schematic diagram of a shrinking particle while reacting with the surrounding fluid.

described by the following scheme:

$$\text{A(fluid)} + b\text{B(solid)} = c\text{C(fluid)} + d\text{D(removable solid)} \qquad (1.2\text{-}1)$$

where b, c, and d are stoichiometric coefficients.

Let us consider a spherical particle of a nonporous solid reacting with a fluid without forming a solid product, as illustrated in Figure 1.2-1. The rate of consumption of the fluid species A at the solid surface by reaction is given as follows:

$$n_A = kf(C_{As}) \qquad (1.2\text{-}2)$$

where n_A is the rate per unit surface area, k is the reaction rate constant, and f designates the dependence of the rate on concentration. Neglecting accumulation in the boundary layer surrounding the solid, the rate of chemical reaction must equal the rate at which fluid species are transferred between the surface and the bulk fluid. As discussed earlier, the rate of external mass transport is described by

$$n_A = k_m(C_{Ab} - C_{As}) \qquad (1.2\text{-}3)$$

Thus, combining equations (1.2-2) and (1.2-3), we obtain

$$kf(C_{As}) = h_m(C_{Ab} - C_{As}) \qquad (1.2\text{-}4)$$

The overall rate can be determined by solving this equation for the unknown C_{As}, and substituting it back into either equation (1.2-2) or (1.2-3). Before obtaining the general solution including the effects of both chemical kinetics and external mass transfer, it is instructive to examine asymptotic cases first.

When $k \ll k_m$, equation (1.2-4) yields $C_{As} \simeq C_{Ab}$. This is the case where external mass transfer offers little resistance, and thus chemical reaction controls the overall rate of reaction. The rate is then given by

$$n_A = kf(C_{Ab}) \qquad (1.2\text{-}5)$$

On the other hand, when $k \gg k_m$, $f(C_{As})$ tends to zero; this occurs when the concentration of A approaches its equilibrium concentration under the conditions prevailing at the surface of the solid, C^*_{As}. Thus, in this case chemical reaction offers little resistance and external mass transfer controls the overall rate. Substituting C^*_{As} for C_{As} in equation (1.2-3), we obtain

$$n_A = k_m(C_{Ab} - C^*_{As}) \tag{1.2-6}$$

In the intermediate regime where both chemical kinetics and mass transfer offer significant resistances, the analysis will be illustrated with a first-order irreversible reaction. (The case of first-order, reversible reaction is described in Ref. 5.) Thus, we have

$$n_A = kC_{As} \tag{1.2-7}$$

and

$$C^*_{As} = 0 \tag{1.2-8}$$

Solving equations (1.2-3) and (1.2-7) simultaneously to eliminate C_{As}, we obtain

$$n_A = \frac{C_{Ab}}{1/k + 1/k_m} \tag{1.2-9}$$

Equation (1.2-9) indicates the familiar result for first-order processes coupled in series; the resistances are additive. It is also noted that equation (1.2-9) reduces to either equation (1.2-6) or (1.2-7) under appropriate conditions.

In order to obtain the overall conversion vs. time, the rate of disappearance of A, n_A, must be equated with the rate of consumption of the solid B. From the stoichiometry of the reaction (1.2-1) we have

$$n_A = -\frac{\rho_B}{b}\frac{dr_c}{dt} \tag{1.2-10}$$

where ρ_B is the molar concentration of solid B and r_c is the radius of the solid at any time. Therefore, from equations (1.2-9) and (1.2-10), we get

$$\frac{dr_c}{dt} = \frac{bC_{Ab}/\rho_B}{1/k + 1/k_m} \tag{1.2-11}$$

All parameters, except k_m on the right-hand side of equation (1.2-11), are independent of r_c. If k_m could be considered to be independent of r_c, integration of equation (1.2-11) would be straightforward. In reality, however, k_m will vary with r_c with the relationship given by one of the correlations discussed in Section 1.1.1. Given such a relationship, one can perform the integration to obtain r_c and hence the conversion as a function of time. An example of the procedure can be found in Ref. 5. The general procedure can be illustrated as follows by rearranging equation (1.2-11) and integrating,

$$t = \frac{\rho_B}{bc_{Ab}}\left[\frac{r_0 - r_c}{k} + \int_{r_c}^{r_0} \frac{dr_c}{k_m(r)_c}\right] \tag{1.2-12}$$

which gives the relationship between r_c and time. Conversion X is related to r_c by the following:

$$X = 1 - \left(\frac{r_c}{r_0}\right)^3 \tag{1.2-13}$$

where r_0 is the original radius of the solid. Equation (1.2-12) also shows that the time necessary to attain a certain r_c (and hence a certain conversion) is the sum of the time to attain the same r_c in the absence of mass transfer resistance and the time to reach the same r_c under mass transfer control. This important result applies to any reaction system made up of first-order rate processes coupled in series.

When the reaction accompanies a significant enthalpy change, considerations must be given to the transfer of heat as well as of mass. The complete equations involving heat conduction inside the solid as well as external heat transfer would be rather difficult to solve.

The formulation of the problem is in many ways similar to that in the case of a fluid-solid reaction forming a solid product. The effect of heat of reaction is much more pronounced in the latter case. Therefore, the detailed discussion of nonisothermal effect will be postponed until later when the system of shrinking unreacted core is described. Interested readers may consult other articles dealing with this subject.[17,18]

1.2.2. Reactions in Which a Product Layer Is Formed

This type of a reaction is frequently encountered in extractive metallurgy, some typical examples being the leaching of minerals from ores, the reduction of metal oxides, the oxidation of metals, and the roasting of sulfide ores. This group of reactions may in general be described by the following:

$$\text{A(fluid)} + b\text{B(solid)} = c\text{C(fluid)} + d\text{D(solid)} \tag{1.2-14}$$

Figure 1.2-2 illustrates how the reaction progresses in this type of system. The overall process can be divided into three steps: external mass transfer, diffusion through the product layer, and the chemical reaction at the interface between the unreacted and completely reacted zones. In the following we will formulate equations including all these steps and show conditions under which one of these may become the

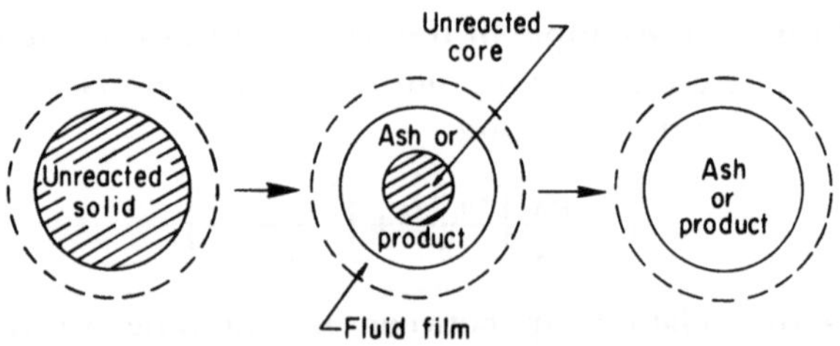

Figure 1.2-2. Schematic diagram of a shrinking unreacted core system.

controlling step. The criteria for these asymptotic regimes will also be developed. The analysis will be made for an isothermal system on a first-order, irreversible reaction occurring in an initially nonporous spherical particle. Generalization for non-first-order or reversible reactions and other geometries can be found in Refs. 5 and 19.

1.2.2.1. *Mathematical Formulation*

At steady state the interfacial chemical reaction and mass transfer processes must occur at the same rate. Thus, in terms of species A the following three rates are equal:

Interfacial chemical reaction

$$-N_A = 4\pi r_c^2 k C_{Ac} \tag{1.2-15}$$

Diffusion through the product layer

$$-N_A = 4\pi r^2 D_e \frac{dC_A}{dr} \tag{1.2-16}$$

External mass transport

$$-N_A = 4\pi r_p^2 k_m (C_{Ab} - C_{As}) \tag{1.2-17}$$

where

$-N_A$ is the total rate of transport of A into the sphere,
r_c is the radius of the unreacted core,
C_{Ac} is the concentration of A at r_c,
D_e is the effective diffusivity of A in the product layer, and
r_p is the radius of the sphere at any time.

Equation (1.2-16) assumes that the diffusion of gaseous species is either equimolar counterdiffusion or occurs at low concentrations of diffusing species. The solution is obtained by applying the pseudo-steady-state approximation, that is, the movement of r_c is much slower compared with the time scale for establishing the concentration profile of A. Thus, as far as the diffusion of A is concerned, the position of r_c appears to be stationary and N_A is independent of position. Then, equation (1.2-16) can be integrated for constant N_A with equations (1.2-15) and (1.2-17) as boundary conditions to give the concentration profile as a function of r_c. From the concentration profile thus obtained, N_A may be calculated using any of equations (1.2-15)–(1.2-17). As before, the consumption of A can be related to that of solid B through the following relationship:

$$N_A = \frac{4\pi r_c^2 \rho_B}{b} \frac{dr_c}{dt} \tag{1.2-18}$$

If the volume of solid product is different from the volume of initial solid reactant, r_p will change as the reaction progresses. In many fluid–solid reactions the change is negligible. Therefore, we will discuss the case of constant r_p. The system with variable r_p will be discussed later.

The procedure for solving equations (1.2-15)–(1.2-17) together with equation (1.2-18) is straightforward and may be found in Refs. 5 and 20. For constant r_p, the result can be expressed in terms of r_c as follows:

$$\frac{bkC_{Ab}}{\rho_B r_p} t = 1 - \frac{r_c}{r_p} + \frac{kr_p}{6D_e}\left\{1 - 3\left(\frac{r_c}{r_p}\right)^2 + 2\left(\frac{r_c}{r_p}\right)^3 + \frac{2D_e}{k_m r_p}\left[1 - \left(\frac{r_c}{r_p}\right)^2\right]\right\} \tag{1.2-19}$$

It should be noted that, in the presence of an inert solid mixed with solid B, ρ_B represents only the number of moles of species B per unit volume of the entire solid mixture.

The following expression has been derived[5] to systematically describe the conversion-vs.-time relationship for an isothermal first-order reaction of a nonporous solid with a fluid in which the solid may be an infinite slab, an infinite cylinder, or a sphere:

$$t^* = g_{F_p}(X) + \sigma_s^2\left[p_{F_p}(X) + \frac{2X}{\mathrm{Sh}^*}\right] \tag{1.2-20}$$

where

$$t^* \equiv \frac{bkC_{Ab}}{\rho_B}\left(\frac{A_p}{F_p V_p}\right) t \tag{1.2-21}$$

$$\sigma_s^2 \equiv \frac{k}{2D_e}\left(\frac{V_p}{A_p}\right) \tag{1.2-22}$$

$$\mathrm{Sh}^* \equiv \frac{k_m}{D_e}\left(\frac{F_p V_p}{A_p}\right) \quad \text{(modified Sherwood number)} \tag{1.2-23}$$

and A_p and V_p are the external surface area and the volume of the particle, respectively, and F_p is the particle shape factor which takes the value of 1, 2, or 3 for an infinite slab, an infinite cylinder, or a sphere, respectively. We note that $F_p V_p / A_p$ is the half-thickness of an infinite slab, and the radius of an infinite cylinder or a sphere. Other quantities in equation (1.2-20) are defined as follows:

$$g_{F_p}(X) \equiv 1 - (1-X)^{1/F_p} \tag{1.2-24}$$

$$\begin{aligned} p_{F_p}(X) &\equiv X^2 && \text{for } F_p = 1 \\ &\equiv X + (1-X)\ln(1-X) && \text{for } F_p = 2 \\ &\equiv 1 - 3(1-X)^{2/3} + 2(1-X) && \text{for } F_p = 3 \end{aligned} \tag{1.2-25}$$

and we have also used the relationship

$$X = 1 - \left(\frac{r_c}{r_p}\right)^{F_p} = 1 - \left(\frac{A_p r_c}{F_p V_p}\right)^{F_p} \tag{1.2-26}$$

1.2.2.2. The Case of Changing Particle Size

The case in which particle size changes during the reaction may be solved[5] by using the following relationship in integrating equation (1.2-16):

$$(r_p)^{F_p}_{\text{at any time}} = Z(r_p)^{F_p}_{\text{original}} + (1-Z)r_c^{F_p} \tag{1.2-27}$$

where Z is the volume of the product solid formed from a unit volume of the reactant solid. If we neglect the change of k_m with particle size, the solution is of the same form as equation (1.2-20) except that the following definition of $p_{F_p}(X)$ should be used:

$$\begin{aligned} p_{F_p}(X) &\equiv ZX^2 && \text{for } F_p = 1 \\ &\equiv \frac{[Z+(1-Z)(1-X)]\ln[Z+(1-Z)(1-X)]}{Z-1} \\ &\quad + (1-X)\ln(1-X) && \text{for } F_p = 2 \\ &\equiv 3\left\{\frac{Z-[Z+(1-Z)(1-X)]^{2/3}}{Z-1} - (1-X)^{2/3}\right\} && \text{for } F_p = 3 \end{aligned} \tag{1.2-28}$$

We note that, as Z approaches unity, equations (1.2-28) reduce to equations (1.2-25).

1.2.2.3. Additivity of Reaction Times

It is seen in equation (1.2-20) that the time required to attain a certain conversion is the sum of the times to reach the same conversion under the control of the three separate steps, the first term being that of chemical reaction, the second that of product-layer diffusion, and the third that of external mass transport. This is analogous to the result obtained for reactions in which no solid product layer is formed, given by equation (1.2-12).

1.2.2.4. The Importance of σ_s^2

It is noted that σ_s^2 provides the numerical criteria for establishing the respective regimes of chemical reaction and diffusion controls. For porous solids, Sh* is usually quite large due to the fact that D_e is about an order of magnitude smaller than the molecular diffusivity. Thus, external mass transport plays only a secondary role to the diffusion through the product layer. Therefore, we will examine the effect of the shrinking-core reaction modulus, $\hat{\sigma}_s^2$, for $\text{Sh}^* = \infty$. Figure 1.2-3 shows the conversion function vs. time for various σ_s^2. When σ_s^2 is less than 0.1, the $\sigma_s^2 = 0$ asymptote is approached. When σ_s^2 is larger than 10, the system may be assumed to have reached the condition of infinite σ_s^2. When σ_s^2 approaches ∞, equation (1.2-20) can be rearranged to give

$$t^{\dagger} \equiv \frac{2bF_pD_eC_{Ab}}{\rho_B}\left(\frac{A_p}{F_pV_p}\right)^2 t = p_{F_p}(X) + \frac{2X}{\text{Sh}^*} \tag{1.2-29}$$

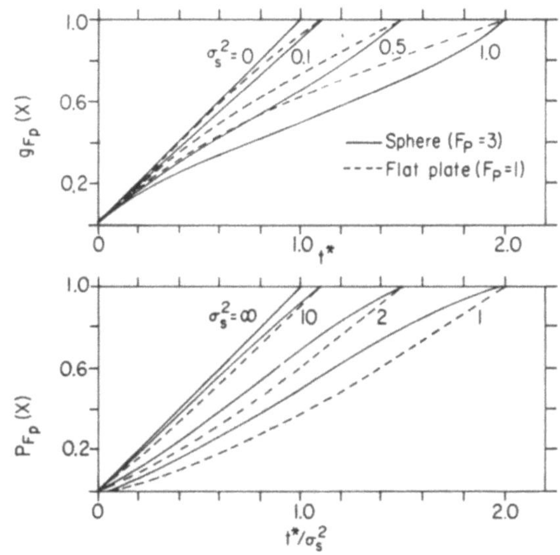

Figure 1.2-3. Effect of σ_s^2 on conversion-vs.-time relationship for the reaction of a nonporous solid with a fluid. [Adapted from J. Szekely, J. W. Evans, and H. Y. Sohn, *Gas–Solid Reactions*, Academic Press, New York (1976).]

Thus, we have established important criteria in terms of σ_s^2, as follows: When $\sigma_s^2 \leq 0.1$, chemical reaction controls, and when $\sigma_s^2 \geq 10$, diffusion through product layer and external mass transfer controls the overall rate. The criteria are general in that its numerical values for defining the asymptotic regimes are the same for all three geometries considered.

We can determine k and D_e from experimental data obtained in the regimes controlled by chemical reaction and diffusion, respectively.

1.2.2.3. Comments Regarding the Shape Factor F_p

An advantage of equation (1.2-20) and the definitions of dimensionless quantities using F_p, A_p, and V_p (for example, using $F_p V_p/A_p$ rather than a characteristic dimension such as radius) is that the equation can be expected to hold approximately valid for geometries other than the three basic ones used in obtaining the result, provided that A_p and V_p are known and F_p can be estimated suitably. For example, equation (1.2-20) is expected to be valid for a cube or a cylinder whose diameter is equal to its length if $F_p = 3$ is used. A cylinder whose length is finite but greater than its diameter will have an F_p between 2 and 3. For a particle which has dimensions that are "more or less" equal in three directions, F_p may be assumed to be 3. A particle with two more or less equal dimensions but a long third dimension can be assumed to have F_p equal to 2.

1.2.2.4. Further Remarks

It is seen in Figure 1.2-3 that the relationship between $g_{F_p}(X)$ and time is approximately linear for σ_s^2 up to 0.5 where significant diffusional resistances are present. This shows that one cannot assume chemical reaction control based only on the linear relationship between $g_{F_p}(X)$ and time. Such mistakes have often been made in the past. On the other hand, $p_{F_p}(X)$ vs. t^*/σ_s^2 is seen to be approximately linear for σ_s^2 as small as 1 where chemical reaction and diffusion are of equal importance. Again, the existence of a linear relationship between $p_{F_p}(X)$ and time does not guarantee control by product-layer diffusion. A more reliable method would be to vary the particle size and test if the reaction time is proportional to the size of the particle or its square.

Remarks may be made at this point regarding the Jander equation[(21)] which has been used to describe a diffusion-controlled gas–solid reaction in a spherical solid[(22,23)]:

$$[1-(1-X)^{1/3}]^2 = \frac{2bD_e C_{Ab}}{\rho_B r_p^2} t \tag{1.2-30}$$

This equation was obtained assuming that the product layer around the spherical particle is flat. It has been shown[5] that, when the conversion is less than about 40%, this equation shows a reasonable agreement with the exact solution of equation (1.2-29) with $Sh^* = \infty$. As conversion increases, the assumption of flat product layer becomes inapplicable and the Jander equation becomes grossly erroneous. In fact, as can be seen by comparing equations (1.2-29) and (1.2-30), the Jander equation does not even give the correct time for complete conversion. Furthermore, Carter[24] has shown, using the data on the oxidation of nickel particles, that the Jander equation becomes clearly inapplicable at conversions higher than 60%. Thus, the exact relationship given by equation (1.2-29) should be used instead of the Jander equation.

1.2.3. Nonisothermal Reactions in Shrinking-Unreacted-Core Systems

When the reaction involves a substantial enthalpy change, temperature gradients will develop within the particle. The reaction rate may be significantly influenced by this temperature difference. With an exothermic reaction, the increased temperature in the solid will enhance the reaction rate. The rate, however, will not increase indefinitely because, when the chemical reaction rate is very fast, diffusion through the product layer will control the overall rate. The rate of diffusion is relatively insensitive to further increases in temperature. Another interesting aspect of exothermic gas–solid reactions is the possible existence of multiple steady states and thermal instabilities.

1.2.3.1. Mathematical Formulation

For a nonisothermal reaction system, energy balance is needed in addition to mass balance over the solid particle. To facilitate mathematics, we will develop the equations for a first-order, irreversible reaction. We will also assume that the effective diffusivity, the thermal conductivity, and total concentration are constant within the temperature ranges. The energy balance equations are analogous to mass balance equations given by equations (1.2-15)–(1.2-17). We will develop the equations for solids with small heat capacities. Then, the energy equations in a spherical shrinking-core system are given as follows:

At the reaction interface

$$Q = 4\pi r_c^2 k(T_c) C_{Ac}(-\Delta H) \qquad (1.2\text{-}31)$$

where T_c is the temperature at the reaction interface;

Conduction through the product layer

$$Q = 4\pi r^2 \lambda_e \frac{dT}{dr} \tag{1.2-32}$$

where λ_e is the effective thermal conductivity of product layer;

External heat transfer

$$Q = 4\pi r_p^2 h(T_s - T_b) \tag{1.2-33}$$

where T_s and T_b are the external surface and the bulk temperatures, respectively.

The solution can be obtained by determining C_{Ac} and T_c from equations (1.2-15)–(1.2-17) and (1.2-31)–(1.2-33) with k in (1.2-31) evaluated at T_c. In keeping with our assumptions, D_e, λ_e, k_m, and h are kept constant. Equations (1.2-31)–(1.2-33) are solved using a pseudo-steady-state assumption, that is, at any time Q is constant at any r. The results are as follows[5]:

$$\frac{C_{Ab}}{C_{Ac}} - 1 - [\sigma_s(T_b)]^2 \left[\frac{2/\text{Sh}^* + p'_{F_p}(X)}{g'_{F_p}(X)}\right] \exp\left[\gamma\left(1 - \frac{T_b}{T_c}\right)\right] = 0 \tag{1.2-34}$$

and

$$\frac{T_c}{T_b} - 1 - \beta[\sigma_s(T_b)]^2 \frac{C_{Ac}}{C_{Ab}} \left[\frac{2/\text{Nu}^* + p'_{F_p}(X)}{g'_{F_p}(X)}\right] \exp\left[\gamma\left(1 - \frac{T_b}{T_c}\right)\right] = 0 \tag{1.2-35}$$

where

$$\gamma \equiv E/RT_b \tag{1.2-36}$$

$$\beta \equiv \frac{(-\Delta H) D_e C_{Ab}}{\lambda_e T_b} \tag{1.2-37}$$

$$\text{Nu}^* \equiv \frac{h}{\lambda_e}\left(\frac{F_p V_p}{A_p}\right) \tag{1.2-38}$$

and $g'_{F_p}(X)$ and $p'_{F_p}(X)$ represent the derivatives with respect to X of $g_{F_p}(X)$ and $p_{F_p}(X)$, respectively. C_{Ac} and T_c can now be determined for each conversion X by solving equations (1.2-34) and (1.2-35) simultaneously. The conversion-vs.-time relationship is obtained by substituting C_{Ac} and T_c into equation (1.2-15) and combining it with equation (1.2-18) together with equation (1.2-26).

1.2.3.2. *Multiple Steady States and Thermal Instability*

These phenomena occur due to the fact that for an exothermic reaction equations (1.2-34) and (1.2-35) may have more than one solution at a fixed conversion. This can be best explained by comparing the rates of heat generation at and conduction away from the reaction interface when the interface is at a given r_c.

Thus, we obtain from equation (1.2-31)

$$Q_{\text{generation}} = 4\pi r_c^2 k(T_c) C_{Ac}(-\Delta H) \tag{1.2-39}$$

and from equations (1.2-32) and (1.2-33)

$$Q_{\text{conduction}} = \lambda_e (A_p/V_p)(T_c - T_b)\left[\frac{g'_{F_p}(X)}{1/\text{Nu}^* + p'_{F_p}(X)/2}\right] \tag{1.2-40}$$

In deriving equation (1.2-40), we again made use of equation (1.2-26). The heat generation is typically an S-shaped function of T_c: At a low T_c the reaction is chemically controlled and the reaction rate increases exponentially with T_c. As T_c increases, chemical kinetics becomes fast and the overall reaction is controlled by the product-layer diffusion which increases very slowly with temperature. The conduction term given by equation (1.2-40) is essentially a straight line with respect to T_c. Thus, there exist the possibility of three solutions for T_c and hence C_{Ac}, as shown in Figure 1.2-4. Of the three solutions, the middle solution is unstable and cannot exist in reality. The upper solution is usually controlled by diffusion, whereas the lower solution is controlled by chemical reaction due to the low temperature and hence slow reaction rate. Figure 1.2-4 represents the possible relative positions of $Q_{\text{generation}}$ and $Q_{\text{conduction}}$ at a certain r_c. During the reaction r_c changes with time and the position of these curves will also change. These changes may occur such that no sudden change in T_c is experienced. But under

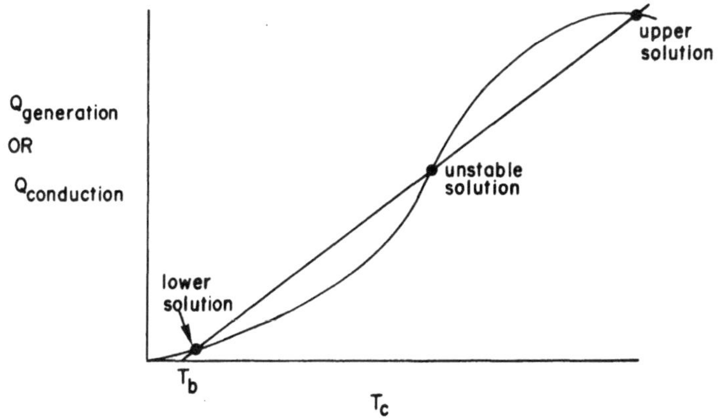

Figure 1.2-4. Schematic diagram of the possible existence of multiple steady state.

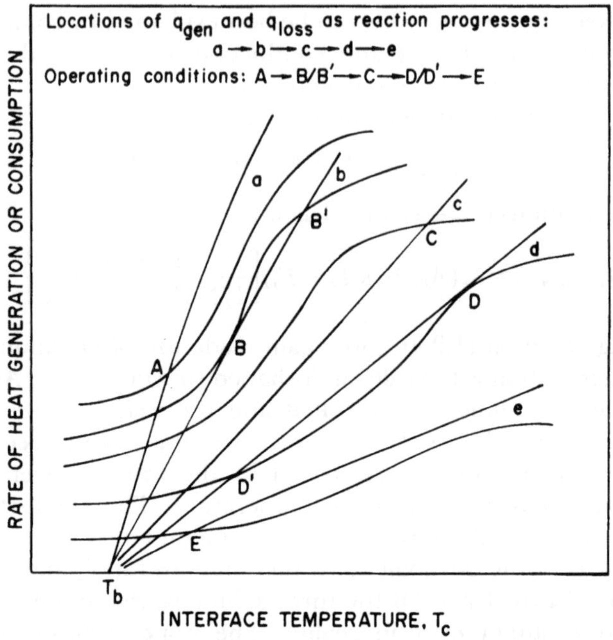

Figure 1.2-5. Thermal balance at the reaction interface. [Adapted from J. Szekely, J. W. Evans, and H. Y. Sohn, *Gas–Solid Reactions*, Academic Press, New York (1976).]

conditions depicted in Figure 1.2-5 a sudden transition from the lower to the upper operating condition will occur, resulting in a thermal instability. In this system, the bulk gas is maintained at temperature T_b. The reaction starts at the external surface of the solid where heat transfer is rapid and hence the reaction temperature is not much higher than T_b, as shown by point A. At this low temperature, chemical reaction is likely to control the overall rate. As reaction progresses into the interior of the particle, heat transfer becomes slower and point B is reached followed by a rapid transition to point B′. Thus, the system is "ignited" to the upper reaction regime, and will be in general operated at a state represented by point C. As the reaction progresses further (or in a system starting at point C), the system may go through point D at which a sudden jump to point D′ occurs causing an "extinction" of the reaction.

Criteria have been developed for the possibility of a thermal instability in gas–solid reactions.[20,25,26] In these studies the heat capacity of the solid is assumed to be negligible. Wen and Wang[27] have discussed the effect of solid heat capacity in a nonisothermal gas–solid reaction.

1.2.3.4. Maximum Temperature Rise in Diffusion-Controlled Gas–Solid Reactions

Excessive temperatures occurring in gas–solid reactions may cause severe structural changes such as serious sintering which may close pores, thus hindering further reaction. It is, therefore, of practical importance to be able to predict the magnitude of the maximum temperature that the solid may encounter in the course of reaction. A rigorous solution could be found from T_c obtained by solving equations (1.2-34) and (1.2-35) for various conversion values. This procedure, however, is rather tedious. An alternative procedure is possible by recognizing that, if an appreciable temperature rise is attained within the solid, the reaction is likely to be controlled by diffusion in the product layer, as discussed earlier. Luss and Amundson[28] obtained a numerical solution for the maximum temperature rise for diffusion-controlled gas–solid reactions occurring in a spherical particle. It is noted here that, if the reaction is controlled by diffusion, the solution is identical whether the solid is initially nonporous or porous. Sohn[29] has obtained an analytical solution which is exact for an infinite slab and is approximately correct for a particle of other geometries. He also obtained, from the analytical solution, useful asymptotic solutions for the prediction of maximum temperature rise. Only the final solutions will be presented below. Interested readers are referred to the original article for detailed derivations. Sohn showed[29] that when the reaction front is at a certain position, the temperature rise in the solid is given by

$$\theta = 2Ae^{-x^2}[D(x) - D(\alpha)] \tag{1.2-41}$$

where

$$\theta \equiv \frac{h(T_s - T_b)}{b(-\Delta H)D_e C_{Ab}} \left(\frac{F_p V_p}{A_p}\right) \tag{1.2-42}$$

$$A \equiv \left(\frac{h}{\rho_s c_s}\right)\left(\frac{\rho_B F_p}{b D_e C_{Ab}}\right)\left(\frac{F_p V_p}{A_p}\right) \tag{1.2-43}$$

(here, ρ_B is the molar concentration of only the reactant B in the solid which might contain other inert solids, and ρ_s is the density of the solid including the inerts if any), and

$$x \equiv \frac{A^{1/2}}{2}\left(\delta + \frac{1}{\mathrm{Sh}^*}\right); \qquad \delta \equiv 1 - \frac{r_c}{r_p} \tag{1.2-44}$$

$$\alpha \equiv \frac{A^{1/2}}{2\mathrm{Sh}^*} \tag{1.2-45}$$

and $D(x)$ is defined by the Dawson's integral

$$e^{-w^2}D(w) \equiv e^{-w^2}\int_0^w e^{t^2}\,dt \tag{1.2-46}$$

which is tabulated in the literature.[29–31]

The maximum temperature rise can be obtained from the maximum value of θ at a corresponding position r_c. However, the maximum temperature rise can be explicitly determined from the following asymptotic and approximate solutions of equation (1.2-41).[29] For a large Sh* (thus small α), equation (1.2-41) reduces to

$$\theta = (2A)^{1/2} e^{-x^2} D(x) \tag{1.2-47}$$

From the maximum value of the Dawson's integral,[30] we get

$$\theta_{\max,\mathrm{Sh}^* \to \infty} = 0.765A^{1/2} \tag{1.2-48}$$

which occurs at

$$\delta = 1.31/A^{1/2} \tag{1.2-49}$$

This large Sh* asymptote is recommended for $\alpha \leq 0.1$.[30] For large values of α $(1 \leq \alpha)$, the asymptotic solution is

$$\theta_{\max,\alpha \to \infty} = \mathrm{Sh}^* \left(\frac{1}{1 + 1/a^2} \right) \tag{1.2-50}$$

For $0.1 < \alpha < 1$,

$$\theta_{\max} = (2A)^{1/2} \exp\left\{ -\left[\frac{\alpha + (\alpha^2 + 2)^{1/2}}{2} \right]^2 \right\} \left\{ D\left[\frac{\alpha + (\alpha^2 + 2)^{1/2}}{2} \right] - D(\alpha) \right\} \tag{1.2-51}$$

The asymptotic and approximate solutions are shown in Figure 1.2-6 with the exact solution obtained from equation (1.2-41).

1.2.4. Summary and Comments

The reaction of a nonporous solid with a fluid has been discussed and solutions to the governing equations have been derived. We have attempted a systematic generalization at the sacrifice of some less important details. As a result, we were able to establish an important criterion in σ_s^2 for determining the asymptotic regimes of chemical-reaction-controlled or diffusion-controlled reactions. Another consequence of this approach is the generalization for various geometries. Thus, the numerical values of σ_s^2 characterizing the asymptotic regimes are the same for all geometries.

The shrinking-unreacted-core model is attractive for its simplicity. But it should be noted that this is valid only for the reaction of a

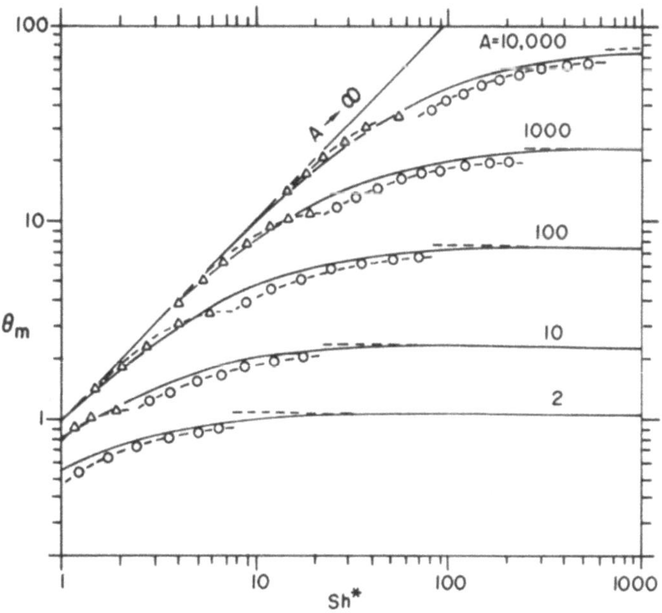

Figure 1.2-6. Asymptotic and approximate analytical solutions for the maximum temperature rise. Key: —— exact solution; - - - asymptotic solution ($\alpha \leq 0.1$); -O-O- approximate solution ($0.1 < \alpha < 1$); -△-△- approximate solution ($1 < \alpha$). [Adapted from H. Y. Sohn, *AIChE J.* **19**, 191 (1973).]

nonporous solid occurring at a well-defined sharp reaction interface. Various previous investigators have applied this model, mainly for its simplicity, to the reaction of a porous solid where chemical reaction occurs in a diffuse zone rather than at a sharp interface. This is permissible only for diffusion-controlled reactions for which mathematical expressions are identical. In general, the application of the shrinking-core model to the reaction of porous solids results in erroneous analyses of experimental data, namely, incorrect dependence of the reaction rate on different reaction conditions (for example, activation energy) and various physical parameters (for example, the dependence of rate on particle size). The proper analysis of the reaction of porous solids with fluids will be discussed in the following section.

1.3. Reaction of a Single Porous Particle

When the reactant solid is initially porous, the fluid reactant will diffuse into the solid while reacting with it on its path. Thus, chemical reaction and diffusion occur in parallel over a diffuse zone rather than at a sharp boundary. The reaction of a porous solid has not been studied as extensively as that of a nonporous solid. Only recently has it received much attention from a number of research groups. As in the case of a nonporous solid, it is important to understand the relative importance of chemical kinetics and transport phenomena. The resistance to mass and heat transport may significantly influence the apparent activation energy, the apparent reaction order, and the dependence of overall rate on the size and other structures of the particle.

1.3.1. Reactions in Which No Solid Product Is Formed

Examples of such reactions are the combustion of porous carbon, the formation of nickel carbonyls from relatively pure nickel, the dissolution of porous minerals, and the Boudouard (or solution-loss) reaction between porous carbon and carbon dioxide. This type of reaction may in general be described by

$$\text{A(fluid)} + b\text{B(solid)} \rightarrow \text{fluid products} \tag{1.3-1}$$

In the case of nonporous solids reacting without forming a solid product layer, it was shown that the overall reaction may be controlled by chemical reaction or by external mass transfer. In the case of porous solids, the diffusion of fluid reactants within the pores of the solid creates an additional regime where the overall reaction is *strongly influenced* by the pore diffusion (but *not controlled* by it).

1.3.1.1. Uniformly Reacting Porous Particles

At low temperatures where intrinsic kinetics is slow, the fluid species can diffuse deeply into the interior of the solid, and reaction occurs throughout the solid under a uniform concentration of fluid A at its bulk value. Thus all the kinetic measurements will yield intrinsic values.

One of the first models on the gasification of solids, taking into account the changes in pore structure by the consumption of solid, is due to Petersen,[32] who assumed that the solid contains uniform cylindrical pores with random intersections. The result of the model will be summarized without going into the detailed mathematical derivation. When the diffusion through the pores is rapid and hence offers little resistance, the concentration of the reactant is uniform throughout the particle and the rate of reaction per unit volume of solid is given by

$$\text{rate per volume} = k\left(\frac{\varepsilon_0}{r_0}\right)\frac{(2G-3\xi)\xi}{G-1}C_A^n \tag{1.3-2}$$

where

k = gas–solid reaction rate constant,
ε_0 = initial porosity of the solid,
r_0 = initial radius of pores,
$\xi = r/r_0$ (r = pore radius at anytime),
C_A = concentration of reactant A, and
G = the root of

$$\tfrac{4}{27}\varepsilon_0 G^3 - G + 1 = 0 \tag{1.3-3}$$

The variation of ξ with time is obtained by integrating

$$\frac{dr}{dt} = \frac{bk}{\rho_s} C_A^n \tag{1.3-4}$$

to give

$$\xi = 1 + t/\tau_c \tag{1.3-5}$$

where

$$\tau_c \equiv \frac{r_0 \rho_s}{bkC_A^n} \tag{1.3-6}$$

The relationship between the conversion X and time[5] is given by

$$X = \frac{\varepsilon_0}{1-\varepsilon_0}\left[\left(1+\frac{t}{\tau_c}\right)^2\left(\frac{G-1-t/\tau_c}{G-1}\right)-1\right] \tag{1.3-7}$$

It is emphasized that this expression is valid only when diffusion offers little resistance and hence the fluid concentration is uniform within the solid. When diffusion offers a significant resistance, the concentration and reaction rate will vary with the position inside the particle. This problem has been treated by Petersen[32] and, more recently, by Hashimoto and Silveston.[33]

Applying the model to the reaction of porous graphite with carbon dioxide, Petersen[32] found a reasonable agreement, despite the assumptions that the pores are cylindrical and uniform in size and the neglect of the coalescence of adjacent pores as they grow. Furthermore, this model is one of the few models that contain the realistic feature that the surface area may increase with reaction, go through a maximum and then decrease.

1.3.1.2. Reactions Occurring Under the Strong Limitation of Pore Diffusion or Under the Control of External Mass Transfer

As the reaction becomes faster, the fluid reactants cannot penetrate deeply into the solid without reacting. Under these conditions, the

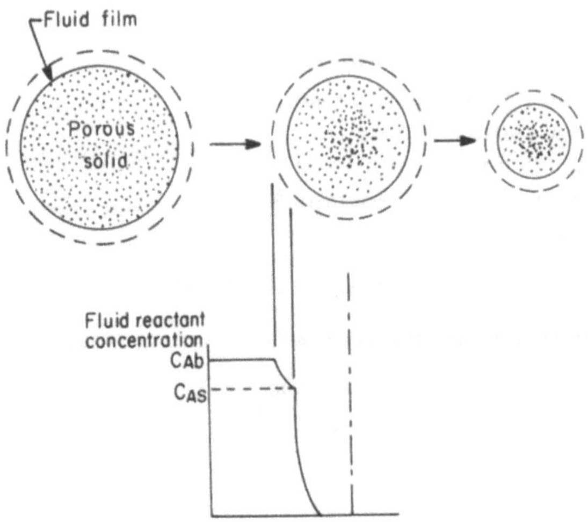

Figure 1.3-1. Reaction of a porous particle without forming a solid product layer and shrinking in overall size.

reaction will mostly occur in a narrow region near the external surface consuming the solid from the surface into the center, as described in Figure 1.3-1.

We will illustrate the quantitative aspects of such a reaction for an irreversible reaction. Since the reaction mainly occurs in a narrow layer near the external surface, the reaction zone may be considered flat regardless of the actual overall geometry of the solid. The mass balance, therefore, may be written in one direction only:

$$D_e \frac{d^2 C_A}{dx^2} - k S_v C_A^n = 0 \tag{1.3-8}$$

where x is the distance normal to the external surface and S_v is the surface area per unit volume. Here, we made the assumption that the bulk flux is negligible, which is valid for equimolar counterdiffusion or when A is present at low concentrations. The case of substantial volume change due to reaction will be examined later. The boundary conditions are

$$C_A = \begin{cases} C_{As} & \text{at } x = 0 \text{ (outer surface of the solid)} & (1.3\text{-}9) \\ dC_A/dx = 0 & \text{as } x \to \infty & (1.3\text{-}10) \end{cases}$$

The first boundary condition assumes a negligible resistance due to external mass transfer. The effect of external mass transfer can be easily incorporated as shown later. The second boundary condition is justified when it is recognized that the pellet dimension is much larger than the thin layer in which the fluid reactants are completely reacted. This condition is satisfied when the following inequality is valid[2]:

$$\frac{V_p}{A_p}\left[\left(\frac{n+1}{2}\right)\frac{kS_v C_{As}^{n-1}}{D_e}\right]^{1/2} > 3 \tag{1.3-11}$$

the solution of equation (1.3-8) is given by†

$$\frac{dC_A}{dx} = -\left[\left(\frac{2}{n+1}\right)\frac{kS_v}{D_e} C_A^{n+1}\right]^{1/2} \tag{1.3-12}$$

From the concentration gradient at the external surface ($X = 0$), the overall rate per unit external surface area can be obtained as follows:

$$n_A = -D_e\left(\frac{dC_A}{dx}\right)_{\text{at external surface}} = \left(\frac{2}{n+1} kS_v D_e\right)^{1/2} C_{As}^{(n+1)/2} \tag{1.3-13}$$

Note that the rate is proportional to $(kD_e)^{1/2}$ which means that the apparent activation energy is the arithmetic average of the activation energies of intrinsic reaction and diffusion:

$$E_{\text{app}} = \tfrac{1}{2}(E_r + E_d) \tag{1.3-14}$$

For the diffusion of gaseous reactant, $E_d \simeq 0$ and the apparent activation energy is one-half of the intrinsic value. The apparent reaction order is also changed:

$$n_{\text{app}} = \tfrac{1}{2}(n_{\text{intrinsic}} + 1) \tag{1.3-15}$$

Equation (1.3-13) shows why diffusion does not control the overall rate, that is, even when chemical reaction is fast (k large), a larger k still increases the overall rate. This is because chemical reaction and diffusion occurs in parallel, not in series. The overall conversion of the solid may be described using n_A given by equation (1.3-13). The procedure is entirely analogous to that for a nonporous particle described previously.

Combining equations (1.2-10) and (1.3-13), integrating, and substituting equation (1.2-26), we get

$$\frac{b\left(\dfrac{2}{n+1} kS_v D_e\right)^{1/2} C_{As}^{(n+1)/2}}{\rho_B}\left(\frac{A_p}{F_p V_p}\right)t = 1 - (1-X)^{1/F_p} \equiv g_{F_p}(X) \tag{1.3-16}$$

assuming that the external mass transfer offers little resistance. We shall examine the effect of external mass transfer subsequently. It is seen that the X-vs.-t relationship is analogous for porous and nonporous solids.

† Substitute $p \equiv dC_A/dx$ and thus $d^2C_A/dx^2 = p\,dp/dC_A$ and solve for p with C_A as the independent variable.

Thus, frequently no distinction has been made in analyzing this type of a reaction whether the solid is porous or not. On closer inspection, however, the kinetics of the reaction of a porous solid is falsified by the influence of diffusion. Instead of intrinsic kinetic parameters, one obtains the apparent values of activation energy and reaction order. The kinetics is also strongly affected by the physical structure of the porous solid such as the specific surface area and the effective diffusivity.

The above analysis was made for a power-law kinetic expression. Many fluid–solid reactions may be more accurately described by the Langmuir–Hinshelwood type expression

$$\text{rate} = \frac{kC_A}{1 + KC_A} \tag{1.3-17}$$

which can be shown[34] to also represent, using appropriate transformation of variables, a class of more complex mechanisms accounting for a reversible surface chemical reaction and the adsorption of other species. When this kinetic expression is incorporated in equation (1.3-8), the result is

$$n_A = (2kS_v D_e)^{1/2} \frac{[KC_{As} - \ln(1 + KC_{As})]^{1/2}}{KC_{As}} C_{As} \tag{1.3-18}$$

When there is a substantial change in fluid volume upon reaction, which is expected more likely in gas–solid reactions, the bulk flux of fluid species in the pores must be taken into account. If the volume increases upon reaction, as in $C + O_2 \rightarrow 2\,CO$, the bulk outward flow of CO hinders the diffusion of O_2 into the solid. A reduction in volume has the opposite effect.

Thiele[35] first showed the effect of volume change on the rate of a catalytic reaction in a porous catalyst pellet. His work was extended later by Weekman and Gorring.[36] We can apply a similar analysis to the fluid–solid noncatalytic reactions we are concerned with here. The mass balance equation including the bulk flux term at constant pressure is as follows:

$$D_e \frac{d^2 C_A}{dx^2} - \frac{\theta / C_{As}}{1 + \theta C_A / C_{As}} \left(\frac{dC_A}{dx} \right)^2 - kS_v C_A^n \left(1 + \frac{\theta C_A}{C_{As}} \right) = 0 \tag{1.3-19}$$

where $\theta \equiv (\nu - 1) y_{As}$ and ν is the volume of fluid species produced per unit volume of fluid species reacted, and y_{As} is the mole fraction of gas A at the pellet surface. Solution of equation (1.3-19) with the boundary conditions given in equations (1.3-9) and (1.3-10) gives the overall rate of reaction per unit area of external surface as follows:

For a zeroth-order reaction $(n=0)$,

$$n_A=(2kS_vD_e)^{1/2}\left[\frac{1}{\theta}\ln(1+\theta)\right]^{1/2}C_{As}^{1/2} \tag{1.3-20}$$

For a first-order reaction $(n=1)$,

$$n_A=(2kS_vD_e)^{1/2}\left[\frac{1}{\theta}-\frac{1}{\theta^2}\ln(1+\theta)\right]^{1/2}C_{As} \tag{1.3-21}$$

For a second-order reaction $(n=2)$,

$$n_A=(2kS_vD_e)^{1/2}\left[\frac{1}{2\theta}-\frac{1}{\theta^2}+\frac{1}{\theta^3}\ln(1+\theta)\right]^{1/2}C_{As}^{3/2} \tag{1.3-22}$$

Up to now, we have implicitly assumed that the area of the external surface is negligible compared with S_v. This may not be valid for solids of very low porosity or very fast reaction. In such a case, the reaction occurring at the external surface may contribute significantly to the overall rate.[37] The reaction rate at the external surface may be written as

$$n_{A,\,\text{external surface}}=kfC_{As}^n \tag{1.3-23}$$

where f is the roughness factor for the external surface defined as the ratio of true area to the projected area of external surface. This term is added to equations (1.3-13), (1.3-18), and (1.3-20)–(1.3-22) to obtain the total reaction rate, which must be used to determine the conversion.

For a first-order reaction without appreciable volume change, the overall rate, including the resistance due to the external mass transfer, is

$$n_A=\frac{C_{Ab}}{\{[2kS_vD_e/(n+1)]^{1/2}+kf\}^{-1}+1/k_m} \tag{1.3-24}$$

It can be readily verified that the external mass transfer controls the overall rate when $k \gg k_m$. When the external mass transfer offers little resistance $(k \ll k_m)$, the overall rate becomes

$$n_A=\{[2kS_vD_e/(n+1)]^{1/2}+kf\}C_{Ab} \tag{1.3-25}$$

When the contribution of the second term is small compared with that of the first, the expression reduces to that of equation (1.3-13) with C_{As} equal to C_{Ab}.

Nonisothermal behavior in reaction systems under consideration has been studied,[5] based on the analysis by Petersen[2,38] for a nonisothermal catalytic reaction in a porous catalyst pellet. Only the result for a first-order reaction will be presented here:

$$n_A = (2k_s S_v D_e)^{1/2} \frac{(e^{\delta} - 1 - \delta)^{1/2}}{\delta} C_{As} \tag{1.3-26}$$

where

$$k_s = k(T_s) \tag{1.3-27}$$

the reaction-rate constant evaluated at external surface temperature, and

$$\delta \equiv \gamma\beta \tag{1.3-28}$$

$$\gamma \equiv E/RT_s \tag{1.3-29}$$

$$\beta \equiv \frac{(-\Delta H) D_e C_{As}}{\lambda_e T_s} = \left(\frac{T - T_s}{T_s}\right)_{\max} \tag{1.3-30}$$

It has been assumed in deriving equation (1.3-26) that $\beta \ll 1$, for which the solution is valid for values of $|\delta|$ up to at least 5. For other values of β the solution involves numerical integration, and the reader is referred to Refs. 5 and 39.

1.3.2. Reactions in Which a Product Layer Is Formed

This type of a reaction is similar to those discussed in Section 1.2.2, except that the initial reactant solid is porous. The reaction can again be described by the following:

$$\text{A(fluid)} + b\text{B(solid)} = c\text{C(fluid)} + d\text{D(solid)} \tag{1.3-31}$$

In a porous solid, the reaction occurs in a diffuse zone rather than a sharp interface. There is a gradual change in conversion of solid over the pellet. In general, the external layer will be completely reacted first and the thickness of the completely reacted layer will grow towards the interior of the porous solid. Figure 1.3-2 illustrates the reaction of a porous solid in which a solid product layer is formed.

When pore diffusion is fast compared with the rate of chemical reaction, the concentration of fluid reactant is uniform throughout the pellet, and the reaction occurs at a uniform rate. If chemical kinetics is much faster than the rate of diffusion, however, the reaction occurs in a narrow region between the unreacted and the completely reacted zones. This latter situation is identical to the diffusion-controlled shrinking-core reaction of a nonporous solid discussed previously. We will formulate equations including both intrinsic chemical kinetics and diffusion, and derive the criteria for asymptotic regimes where a particular step controls the overall rate. The analysis will be made for an

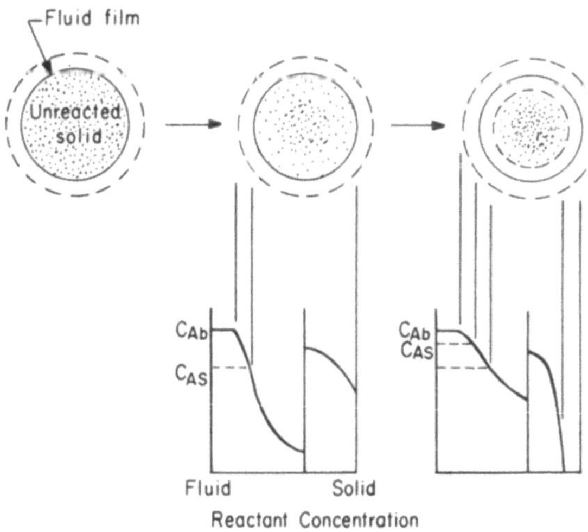

Figure 1.3-2. Reaction of a porous particle forming a solid product layer.

isothermal system of a first-order, irreversible reaction. (Extension to a reversible reaction can be found in Ref. 5.) We also neglect the structural changes that occur during the reaction, although certain aspects of such changes can be incorporated into the analysis as discussed by Szekely, Evans, and Sohn.[5]

The discussion to follow is based on the work of Sohn and Szekely.[40–43] Unlike those of many other investigators,[44–48] their analysis is general in terms of solid geometry and gives approximate closed-form solutions valid for the cases that must otherwise be solved numerically. They also derived important criteria for asymptotic regimes for which simple, exact solutions are possible.

1.3.2.1. Mathematical Formulation

Let us assume the porous solid to be an aggregate of fine grains. In order to facilitate the visualization and description of the model, we will assume that the grains have the shape of flat plates, long cylinders, or spheres.† The external shape of the pellet may also be approximated by one of these geometries.

We will follow the procedure described in Ref. 40 in developing the mathematical equations for the generalized grain model. Additional

† The assumed shape of particles is to facilitate the visualization of the system. The real requirement is to know the variation in the rate of reaction with solid conversion in the absence of resistance due to intrapellet diffusion. This must always be determined by experiments.

assumptions to be made are:

1. The pseudo-steady-state approximation is valid for determining the concentration profile of the fluid reactant within the pellet.[49,50]

2. The resistance due to external mass transfer is negligible. (This assumption will be relaxed later and the effect of external mass transfer will be studied.)

3. Intrapellet diffusion is either equimolar counterdiffusion or occurs at low concentrations of diffusing species.

4. Diffusivities are constant throughout the pellet.

5. Diffusion through the product layer around the individual grain is fast. (This assumption will also be relaxed later.)

The conservation of fluid reactant may be described by

$$D_e \nabla^2 C_A - v_A = 0 \tag{1.3-32}$$

where v_A is the local rate of consumption of the fluid reactant A, in moles per time per unit volume of pellet.

Within each grain, the conservation of solid reactant may be described by

$$-\rho_B \frac{\partial r_c}{\partial t} = bkC_A \tag{1.3-33}$$

where r_c is the distance from the center of symmetry to the reaction interface.

An expression for v_A may be obtained by determining the surface area for reaction available per unit volume of the pellet:

$$v_A = \alpha_B k \left(\frac{A_g}{V_g}\right)\left(\frac{A_g r_c}{F_g V_g}\right)^{F_g - 1} C_A \tag{1.3-34}$$

where α_B is the volume fraction of the pellet occupied by solid B.

Equations (1.3-32) and (1.3-33) may be expressed in dimensionless forms by introducing the following dimensionless variables:

$$\psi \equiv \frac{C_A}{C_{Ab}} \tag{1.3-35}$$

$$\xi \equiv \frac{A_g r_c}{F_g V_g} = \frac{r_c}{r_g} \tag{1.3-36}$$

$$t^* \equiv \left(\frac{bkC_{Ab}}{\rho_B} \frac{A_g}{F_g V_g}\right) t \tag{1.3-37}$$

$$\eta \equiv \frac{A_p R}{F_p V_p} = \frac{R}{R_p} \tag{1.3-38}$$

$$\sigma \equiv \frac{F_p V_p}{A_p}\left(\frac{\alpha_B k}{D_e}\frac{A_g}{V_g}\right)^{1/2} \tag{1.3-39}$$

In the above, R is the distance from the center of symmetry in the pellet. The dimensionless forms of (1.3-32) and (1.3-33) are

$$\nabla^{*2}\psi - \sigma^2 \psi \xi^{F_g - 1} = 0 \tag{1.3-40}$$

and

$$\frac{\partial \xi}{\partial t^*} = -\psi \tag{1.3-41}$$

where ∇^{*2} is the Laplacian operator with η as the position coordinate.

The initial and boundary conditions for equations (1.3-40) and (1.3-41) are

$$\xi = 1, \qquad \text{at } t^* = 0 \tag{1.3-42}$$

$$\psi = 1, \qquad \text{at } \eta = 1$$

$$\frac{d\psi}{d\eta} = 0, \qquad \text{at } \eta = 0 \tag{1.3-43}$$

For most practical purposes the desired results are in terms of the fraction of the solid reacted, rather than ψ or ξ, as a function of time. This is obtained as follows:

$$X = \frac{\int_0^1 \eta^{F_p - 1}(1 - \xi^{F_g})\, d\eta}{\int_0^1 \eta^{F_p - 1}\, d\eta} \tag{1.3-44}$$

It is noted that the dimensionless representation of the governing equations indicates that the dependent variables ξ and ψ, and thus X, are related to t^* and η through a single parameter σ. This quantity σ is a measure of the ratio of the capacities for the system to react chemically and for it to diffuse reactants into the pellet. The modulus σ contains both structural and kinetic parameters and is very useful in characterizing the behavior of the system.

1.3.2.2. Asymptotic Behaviors

When σ approaches zero, the overall rate is controlled by chemical kinetics, the diffusion being rapid compared with the rate of chemical reaction. Thus, the reactant concentration is uniform within the pellet and equal to that in the bulk ($\psi = 1$). This can be directly obtained from equation (1.3-40) with $\sigma = 0$. The variable ξ is then independent of η and equation (1.3-41) is readily integrated to give:

$$\begin{aligned} \xi &= 1 - t^*, \qquad &&\text{for } 0 < t^* \le 1 \\ \xi &= 0, \qquad &&\text{for } t^* \ge 1 \end{aligned} \tag{1.3-45}$$

Using equation (1.3-44) we obtain the following relationship between X and t^*:

$$t^* = 1-(1-X)^{1/F_g} \equiv g_{F_g}(X) \qquad (1.3\text{-}46)$$

which is identical to equation (1.2-20) with $\sigma_s = 0$ for chemical reaction control. Equation (1.3-46) provides a convenient means of determining the reaction rate constant from experimental data obtained under conditions where diffusional resistance is absent. *When σ approaches infinity,* the overall rate is controlled entirely by the diffusion of the gaseous reactant within the pellet. This case is identical to the diffusion-controlled reaction of nonporous solids discussed in Section 1.2.2. The result can be expressed as follows using equation (1.2-29):

$$p_{F_p}(X) = \frac{2F_p b D_e C_{Ab}}{\alpha_B \rho_B}\left(\frac{A_p}{F_p V_p}\right)^2 t \qquad (1.3\text{-}47)$$

$$= \frac{2F_g F_p}{\sigma^2} t^* = \frac{t^*}{\hat{\sigma}^2} \qquad (1.3\text{-}48)$$

where $p_{F_p}(X)$ has been defined in equations (1.2-25). Equation (1.3-47) provides a convenient means of determining the effective diffusivity by plotting experimental data obtained under diffusion control according to this equation. Equation (1.3-48) suggests the following generalized modulus:

$$\hat{\sigma} = \frac{\sigma}{(2F_g F_p)^{1/2}} = \frac{V_p}{A_p}\left[\frac{\alpha_B k F_p}{2D_e}\left(\frac{A_g}{F_g V_g}\right)\right]^{1/2} \qquad (1.3\text{-}49)$$

As will be seen subsequently, this generalized modulus enables us to determine criteria for the chemically controlled and diffusion-controlled asymptotic regimes. Furthermore, the numerical values of $\hat{\sigma}$ defining the respective asymptotic regimes will be the same for all combinations of grain and pellet geometries.

1.3.2.3. Complete Solutions

Analytical solutions are possible for $F_g = 1$. The case of $F_p = 3$ and $F_g = 1$ has been solved by Ishida and Wen,[48] and the solution for the case of $F_p = F_g = 1$ is also available.[40] In other cases the solution must be obtained numerically. Results of numerical computation have been reported by Sohn and Szekely,[40] and are shown in Figures 1.3-3 and 1.3-4.

Figure 1.3-3 shows a plot of $g_{F_g}(X)$ vs. t^* for small values of $\hat{\sigma}$. The ordinate was chosen to allow the convenient presentation of the appropriate asymptotic solution of $g_{F_g}(X) = t^*$. It is apparent from this figure

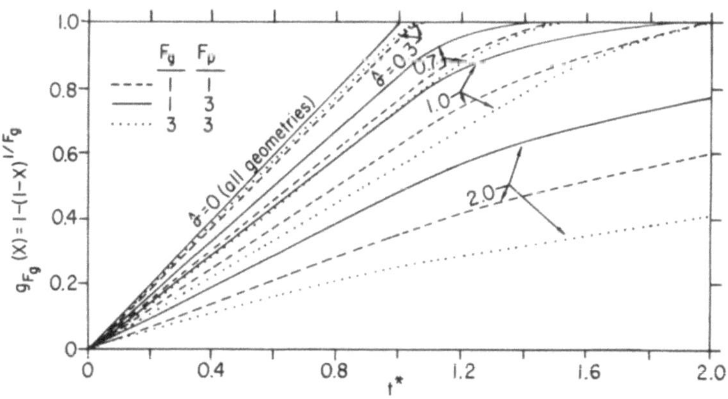

Figure 1.3-3. Conversion function vs. reduced time for small values of $\hat{\sigma}$. [Adapted from H. Y. Sohn and J. Szekely, *Chem. Eng. Sci.* **27**, 763 (1972).]

that the $\hat{\sigma} = 0$ asymptote is valid for $\hat{\sigma} < 0.3$ (or $\hat{\sigma}^2 < 0.1$), *regardless of the geometry*. Thus, when this criterion is met, the system is under the control of the reaction of individual grains and the solution is given by equation (1.3-46), which eliminates the need for numerical computation.

Figure 1.3-4 shows a plot of $p_{F_p}(X)$ vs. $t^*/\hat{\sigma}^2$ for large values of $\hat{\sigma}$. The choice of the ordinate was dictated by the form of the asymptotic solution of equation (1.3-48). It is seen that the $\hat{\sigma} \to \infty$ asymptote is approached when $\hat{\sigma} > 3.0$ (or $\hat{\sigma}^2 > 10$), again *regardless of the geometry.*

The solution for $0.1 < \hat{\sigma}^2 < 10$ depends on the geometries of the grain and the pellet. Sohn and Szekely[40] obtained the following

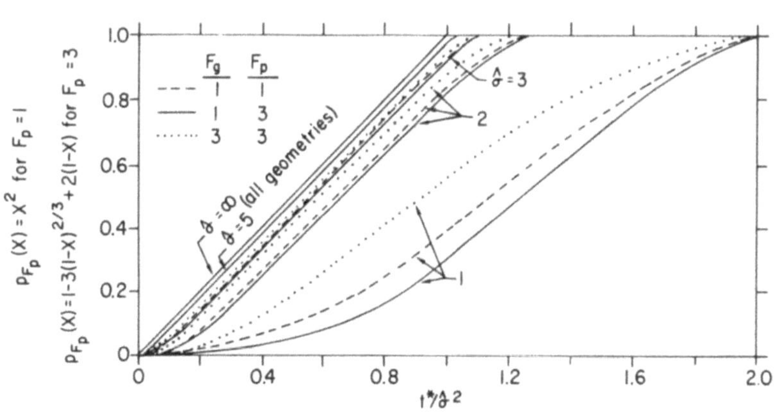

Figure 1.3-4. Conversion function vs. reduced time for large values of $\hat{\sigma}$. [Adapted from H. Y. Sohn and J. Szekely, *Chem. Eng. Sci.* **27**, 763 (1972).]

closed-form approximate solution that is applicable to any combination of geometries:

$$t^* \simeq g_{F_g}(X) + \hat{\sigma}^2 p_{F_p}(X) \tag{1.3-50}$$

The detailed derivation of this relation can be found in the original article.

Equation (1.3-50) provides not only a simple and easy-to-use relationship between X and t^*, but also a greater insight into the problem by clearly showing the relative importance of chemical reaction and diffusion through a parameter $\hat{\sigma}$. Another important feature of equation (1.3-50) is that the time required to attain a certain conversion is shown to be the sum of the time to reach the same conversion under chemical reaction control and that under the diffusion control. This can be recognized by comparing equation (1.3-50) with equations (1.3-46) and (1.3-48).

Comparison of equation (1.3-50) with an exact solution is shown in Figure 1.3-5. The comparison is made for intermediate values of $\hat{\sigma}$ for which the difference is largest. The approximate solution is seen to be a satisfactory representation of the exact solution. For smaller and larger values of $\hat{\sigma}$, agreement is better than shown in this figure. In fact, the approximation is asymptotically correct as $\hat{\sigma} \to 0$ or $\hat{\sigma} \to \infty$. Furthermore, it has been shown[40] that equation (1.3-50) is exact at $X = 1$ for all the combinations of geometries (F_g and F_p). Thus

$$t^*_{X=1} = 1 + \hat{\sigma}^2 \tag{1.3-51}$$

The above analyses have been successfully applied to the reduction of nickel oxide pellets with hydrogen.[42]

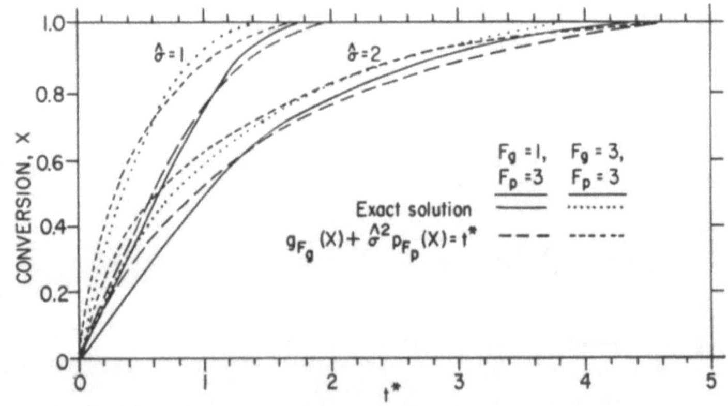

Figure 1.3-5. Comparison of approximate solution with exact solution for $\hat{\sigma} = 1$ and 2. [Adapted from H. Y. Sohn and J. Szekely, *Chem. Eng. Sci.* **27**, 763 (1972).]

1.3.2.4. Further Comments

The effect of external mass transfer has also been studied.[40] The appropriate relationship was determined to be

$$t^* \simeq g_{F_g}(X) + \hat{\sigma}^2\left[p_{F_p}(X) + \frac{2X}{\text{Sh}^*}\right] \tag{1.3-52}$$

The case where the solid product has different effective diffusivity from that of the initial reactant solid has been studied by Ishida and Wen[48] and Szekely and Sohn.[43] The results for incorporating the Langmuir–Hinghelwood type kinetics, rather than a first-order kinetics, have also been reported.[41]

All of the above analyses assume that the diffusion through the product layer around each grain offers negligible resistance. When this intragranular diffusion is important, the approximate relationship between X and t^* has been shown[51] to be

$$t^* \simeq g_{F_g}(X) + \hat{\sigma}_g^2 \cdot p_{F_g}(X) + \hat{\sigma}^2\left[p_{F_p}(X) + \frac{2X}{\text{Sh}^*}\right] \tag{1.3-53}$$

where

$$\hat{\sigma}_g^2 = \frac{k}{2D_g}\left(\frac{V_g}{A_g}\right)$$

Nonisothermal reaction between a porous solid and a gas is generally very complex, although a few simple asymptotic cases can be solved systematically. For example, the maximum temperature rise in a diffusion-controlled system discussed earlier for nonporous solid applies directly to that for porous solid. The effect of nonisothermal reaction on the rate of reaction and the problem of multiple steady state and instability has been discussed in the literature.[5,52] The reader is referred to these articles for further details, because a detailed discussion of a nonisothermal reaction in a porous solid is quite involved and thus beyond the scope of this discussion.

1.3.3. Concluding Remarks

The reaction of a porous solid with a fluid involves chemical reaction and intrapellet diffusion occurring in parallel. Thus, the analysis of such a reaction system is generally more complicated than that of the reaction of nonporous solids. The *grain model* which was discussed above is a rather recent development in this area, and represents one of the distributed models proposed in the last few years. These distributed models have been tested against experiments and found to describe the reaction of porous solids reasonably well.[42,46,53–55] In the text, the grain model was described in terms of dimensionless equations enabling us to establish an important parameter $\hat{\sigma}$ that characterizes the behavior of the reaction system and also gives us numerical criteria for asymptotic regimes. Furthermore, the closed-form approximate solution obtained in the text is very useful, especially when we are confronted with analyzing multiparticle systems. Without such a closed-form solution, both the process within the individual particles and that over the whole system must be solved numerically. Even with the availability of impressive capacities of modern computers, this presents a formidable problem. The application of the approximate solution to multiparticle systems has been discussed in the literature.[56,57]

1.4. Reactions between Two Solids Proceeding through Gaseous Intermediates

A number of reactions between solids are of considerable importance in pyrometallurgical processes. Some of these reactions are true solid–solid reactions which take place in the solid state between two species in contact with each other. Many other important reactions, however, occur through gaseous intermediates. These latter reactions may be considered as coupled gas–solid reactions, and can thus be analyzed in light of the mathematical analyses developed in the previous sections. In this section, we will restrict our discussion to those reactions proceeding through gaseous intermediates.

It is generally recognized that the carbothermic reduction of most metal oxides occurs through the gaseous intermediates of CO and CO_2. The reaction mechanism is

$$\begin{array}{c} Me_xO_y(s) + CO(g) = Me_xO_{y-1}(s) + CO_2(g) \\ z\ CO_2(g) + z\ C(s) = 2z\ CO(g) \\ \hline Me_xO_y(s) + z\ C(s) = Me_xO_{y-1}(s) + (2z-1)CO(g) + (1-z)CO_2(g) \end{array} \tag{1.4-1}$$

Some examples of this type of reaction are:

Reduction of iron oxides[58–60]

$$\begin{array}{c} Fe_xO_y + y\ CO = x\ Fe + y\ CO_2 \\ CO_2 + C = 2\ CO \end{array} \tag{1.4-2}$$

Reaction between ilmenite and carbon[61]

$$FeTiO_3 + CO = Fe + TiO_2 + CO_2$$
$$CO_2 + C = 2\,CO \tag{1.4-3}$$

Reaction between chromium oxide and chromium carbide[62]

$$\tfrac{1}{6}\,Cr_{23}C_6 + CO_2 = \tfrac{23}{6}\,Cr + 2\,CO$$
$$\tfrac{1}{3}\,Cr_2O_3 + CO = \tfrac{2}{3}\,Cr + CO_2 \tag{1.4-4}$$

In these reactions, there is a net generation of gaseous species CO and CO_2 resulting in a bulk flow of the gas mixture from the reaction zone. There are other groups of reactions between solids proceeding without a net generation of gases. The preparation of metal carbides from metals and carbon is an example:

$$C(s) + 2\,H_2(g) = CH_4(g)$$
$$Me(s) + CH_4(g) = MeC(s) + 2\,H_2(s)$$
$$\overline{\qquad\qquad\qquad\qquad\qquad\qquad\qquad\qquad}$$
$$Me(s) + C(s) = MeC(s) \tag{1.4-5}$$

In this case there is no net consumption or generation of the gaseous intermediates and the gaseous species do not appear in the overall stoichiometry. The gaseous species simply act as carriers of the solid reactant (carbon). In this respect, the mechanism for this type of reactions might be termed *catalysis by gases.*

In spite of the practical importance of the reactions between solids, there have been few mathematical descriptions developed for these systems which would provide suitable bases for prediction and interpretation of their behaviors. Until recently, the majority of investigators reported the kinetic data as were obtained from experiments without appropriate reaction models. In more recent studies the possibility of applying the analysis of gas–solid reactions to solid–solid reactions has been recognized.

In early theoretical work the kinetics of solid–solid reactions were analyzed for specific conditions of one of the gas–solid reactions controlling the overall reaction. Thus, the rate of reaction could be represented from the consideration of the rate of this single rate-controlling step. Under such conditions, the analysis of the reaction of nonporous solids presented in Section 1.2 would apply directly. The criterion for which the reaction is rate controlling under what conditions was not established. Rather, experimental observation was relied upon for the determination of such a controlling step.

The studies on the carbothermic reduction of hematite by Rao,[60] on the reaction between ilmenite and solid carbon by El-Guindy and Davenport,[61] and on the reaction between chromium oxide and chromium carbide by Maru *et al.*[62] make use of the assumption of such a single controlling reaction. These studies represent significant advances in presenting the rates of solid–solid reactions quantitatively. It would, however, be of interest to have a theoretical basis for establishing a criterion for the controlling reaction. This would be very important especially for reactions whose controlling step might change under different reaction conditions. A first step in the direction has been made by Sohn and Szekely[63] who considered reactions between solids proceeding through gaseous intermediates controlled by chemical kinetics.

1.4.1. Formulation of Model

A reaction between solids B and D which proceeds through the gaseous intermediates of A and C can in general be expressed as follows:

$$A(g) + bB(s) = cC(g) + eE(s) \tag{1.4-6}$$

$$C(g) + dD(s) = aA(g) + fF(s) \tag{1.4-7}$$

In order for this reaction to be self-sustaining without requiring external supply of gaseous species, the condition $a \times c \geq 1$ must be satisfied.

Let us consider a porous pellet made up of uniformly mixed grains of B and D as sketched in Figure 1.4-1. Solids B and D may have different shapes and sizes. The particle size of either solid is assumed to be uniform. We shall consider an isothermal system, which reacts under a uniform concentration of gaseous species. This latter condition will be valid when (1) the pore diffusion is fast and the external gas concen-

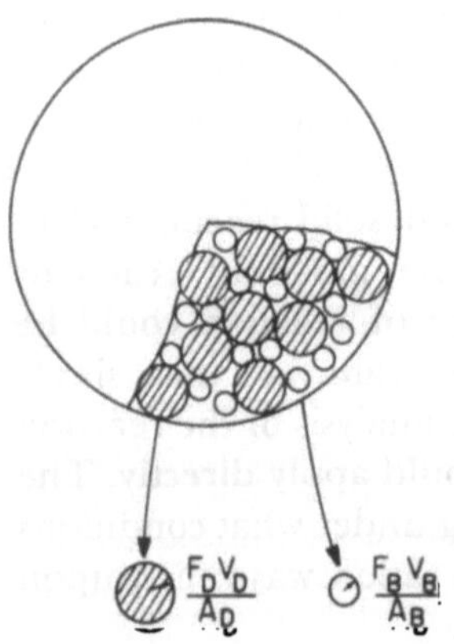

Figure 1.4-1. Schematic representation of uniformly mixed grains of solid reactants. [Adapted from H. Y. Sohn and J. Szekely, *Chem. Eng. Sci.* **28**, 1789 (1973).]

tration is kept constant, and (2) the net generation of gaseous intermediate is sufficiently fast so as to prevent any ambient gas diffusing into the interior of the pellet.[59] The first case is rather trivial and the reaction of each solid depends on the given gas concentration. In the following we will discuss the case where the second condition is satisfied.

The rates at which the reactions proceeds, in terms of the generation of gaseous intermediates, are given as follows:

$$\frac{dn_A}{dt} = -v_1 + av_2 \tag{1.4-8}$$

$$\frac{dn_C}{dt} = cv_1 - v_2 \tag{1.4-9}$$

where v_1 and v_2 are the net forward rates of reactions (1.4-8) and (1.4-9), respectively, per unit volume of the pellet.

If we assume that each grain reacts according to the scheme of the shrinking unreacted core, the reaction rates of the solids may be expressed as

$$-\rho_B \frac{dr_B}{dt} = bk_1 C_A \tag{1.4-10}$$

$$-\rho_D \frac{dr_D}{dt} = dk_2 C_C \tag{1.4-11}$$

Here we assumed that the reactions are irreversible and of first order. It is possible to incorporate other rate expressions.[63] We can also express v_1 and v_2 as follows:

$$v_1 = \alpha_B \left(\frac{A_B}{V_B}\right)\left(\frac{A_B r_B}{F_B V_B}\right)^{F_B - 1} k_1 C_A \tag{1.4-12}$$

$$v_2 = \alpha_D \left(\frac{A_D}{V_D}\right)\left(\frac{A_D r_D}{F_D V_D}\right)^{F_D - 1} k_2 C_C \tag{1.4-13}$$

where α_B and α_D are volumes occupied by solids B and D, respectively, per unit volume of the pellet. These expressions are analogous to equation (1.3-34).

If the total pressure of the system is maintained constant, the following relationships hold:

$$\frac{dn_A}{dt} = \frac{C_A}{V_p}\frac{dV}{dt} \tag{1.4-14}$$

$$\frac{dn_C}{dt} = \frac{C_C}{V_p}\frac{dV}{dt} \tag{1.4-15}$$

where dV/dt is the rate of generation of the gas mixture and V_p is the volume of the pellet. We have made a pseudo-steady-state assumption that the gas-phase concentrations at any time are at the steady-state values corresponding to the amounts and sizes of the solids at that time, namely $C\,dV/dt \gg V\,dC/dt$.

We also have the condition that

$$C_A + C_C = C_T \tag{1.4-16}$$

Equations (1.4-8)–(1.4-16) may be solved for C_A and C_C. Using these values, the rates of reaction of the solids may then be obtained from equations (1.4-10) and (1.4-11). It would be worthwhile to determine the criteria for the regimes where one of the pair of gas–solid reactions controls the overall rate. The derivation is facilitated if we rearrange the governing equations in the following dimensionless forms:

$$\psi_A(\psi_C + c\psi_A) = \gamma\beta\left(\frac{F_D \xi_D^{F_D-1}}{F_B \xi_B^{F_B-1}}\right)\psi_C(\psi_A + a\psi_C) \tag{1.4-17}$$

which is obtained by combining equations (1.4-8), (1.4-9), (1.4-12), and (1.4-13) with (1.4-14) and (1.4-15);

$$\frac{d\xi_B}{dt^*} = -\psi_A \tag{1.4-18}$$

$$\frac{d\xi_D}{dt^*} = -\beta(1-\psi_A) \tag{1.4-19}$$

$$\psi_A + \psi_C = 1 \tag{1.4-20}$$

where

$$\psi_A \equiv \frac{C_A}{C_T}, \qquad \psi_C \equiv \frac{C_C}{C_T} \tag{1.4-21}$$

$$\xi_B \equiv \frac{A_B r_B}{F_B V_B}, \qquad \xi_D \equiv \frac{A_D r_D}{F_D V_D} \tag{1.4-22}$$

$$\gamma \equiv \left(\frac{b}{d}\right)\left(\frac{\alpha_D}{\alpha_B}\right)\left(\frac{\rho_D}{\rho_B}\right) \tag{1.4-23}$$

$$\beta \equiv \left(\frac{d}{b}\right)\left(\frac{\rho_B}{\rho_D}\right)\left(\frac{F_B V_B/A_B}{F_D V_D/A_D}\right)\left(\frac{k_2}{k_1}\right) \tag{1.4-24}$$

$$t^* \equiv \frac{b k_1 C_T}{\rho_B}\left(\frac{A_B}{F_B V_B}\right)t \tag{1.4-25}$$

The conversion of the solid reactants may then be calculated as follows:

$$X_B = 1 - \xi_B^{F_B} \tag{1.4-26}$$

$$X_D = 1 - \xi_D^{F_D} \tag{1.4-27}$$

It is noted that the conversion and the rate of reaction are related to time (t^*) through two parameters, γ and β. The parameter γ represents the relative molar quantities of the two solids, and β corresponds to the ratio of reactivities of the two solids. As will be seen subsequently, the quantities γ and β allow us to define asymptotic regimes and hence the rate-controlling step.

On combining equations (1.4-17) and (1.4-20), we get

$$\psi_A = \frac{v - (v^2 - 4uw)^{1/2}}{2u} \tag{1.4-28}$$

where

$$u \equiv (a-1)\gamma\beta \frac{F_D \xi_D^{F_D - 1}}{F_B \xi_B^{F_B - 1}} - (c-1) \tag{1.4-29}$$

$$v \equiv (2a-1)\gamma\beta \frac{F_D \xi_D^{F_D - 1}}{F_B \xi_B^{F_B - 1}} + 1 \tag{1.4-30}$$

$$w \equiv a\gamma\beta \frac{F_D \xi_D^{F_D - 1}}{F_B \xi_B^{F_B - 1}} \tag{1.4-31}$$

We can now obtain the relationship between ξ_B and ξ_D and t^* by substituting equation (1.4-28) into equations (1.4-18) and (1.4-19) and integrating with the following initial conditions:

$$\xi_B = \xi_D = 1 \qquad \text{at } t^* = 0 \tag{1.4-32}$$

1.4.2. Results

In general the integration of equations (1.4-18) and (1.4-19) must be done numerically. For the case of $F_B = F_D = 1$, analytical solution is possible because u, v, and w and hence ψ_A become constant. We shall discuss only the latter case here. For the cases of other values of F_B and F_D, the reader is referred to the original paper.[63]

When $F_B = F_P = 1$, that is, the surface areas within the grains available for reaction do not change with conversion (zero-order reaction with respect to the solid), equation (1.4-28) gives

$$\psi_A = \frac{(2a-1)\gamma\beta + 1 - [(\gamma\beta)^2 + 2(2ac-1)\gamma\beta + 1]^{1/2}}{2[(a-1)\gamma\beta - c + 1]} \tag{1.4-33}$$

This expression reduces to $\psi_A = \gamma\beta/(1+\gamma\beta)$ when $a = c = 1$, which can also be obtained directly from equation (1.4-17).

1.4.2.1. Asymptotic Behavior

When $\gamma\beta$ approaches zero, reaction (1.4-7) is much slower than reaction (1.4-6), and hence controls the overall rate. The gaseous intermediates consist mainly of C and the concentration of gas A is small. [In the case of reversible reactions, the gaseous intermediates corresponds to the equilibrium gas mixture for reaction (1.4-6)]. This expected behavor is readily demonstrated by the asymptotic behavior of the solution.

When $\gamma\beta \to 0$, equation (1.4-33) reduces to

$$\psi_A \cong a\gamma\beta \tag{1.4-34}$$

Equations (1.4-18) and (1.4-19) become

$$\frac{d\xi_B}{dt^*} = -a\gamma\beta \tag{1.4-35}$$

$$\frac{d\xi_D}{dt^*} = -\beta \tag{1.4-36}$$

Integration of these equations gives

$$\xi_B = 1 - a\gamma\beta t^* \tag{1.4-37}$$

$$\xi_D = 1 - \beta t^* \tag{1.4-38}$$

The latter asymptote corresponds to the reaction of flat nonporous particles of solid D in a constant gas composition, namely pure C in the present case and the equilibrium mixture for reaction (1.4-6) for reversible reaction. The reaction of B is then determined from the following stoichiometric relationship obtained from equations (1.4-37) and (1.4-38):

$$X_B = a\gamma X_D \tag{1.4-39}$$

As shown above, the attainment of this asymptotic regime is characterized by the numerical value of $\gamma\beta$. It will be shown subsequently that $\gamma\beta < 10^{-2}$ adequately defines the $\gamma\beta = 0$ asymptote, where the kinetics of reaction of D is independent of γ or the kinetics of reaction of B. It then follows that the intrinsic kinetics of reaction of D can be determined from the conversion data obtained under the conditions of $\gamma\beta \to 0$, where the gas composition corresponds to that of the equilibrium mixture for reaction (1.4-6).

When $\gamma\beta$ is large, reaction (1.4-6) controls the overall rate. Mathematical development readily follows the procedure described for small $\gamma\beta$. Thus, when $\gamma\beta \to \infty$, equation (1.4-33) gives

$$1 - \psi_A = \psi_C = c/\gamma\beta \tag{1.4-40}$$

$$\frac{d\xi_B}{dt^*} = -1 \tag{1.4-41}$$

$$\frac{d\xi_D}{dt^*} = -\frac{c}{\gamma} \tag{1.4-42}$$

and

$$\xi_B = 1 - t^* \tag{1.4-43}$$

$$\xi_D = 1 - \frac{c}{\gamma} t^* \tag{1.4-44}$$

The relationship between X_B and X_D obtained from equations (1.4-43) and (1.4-44) is

$$X_D = \frac{c}{\gamma} X_B \tag{1.4-45}$$

It is seen in this case of large $\gamma\beta$ that the kinetics of reaction of B is independent of γ or β. Thus, the intrinsic kinetics of the reaction of B can be obtained from the conversion data obtained under the condition of $\gamma\beta \to \infty$, where the gas composition is at the equilibrium concentration for reaction (1.4-7). It will be shown below that the $\gamma\beta \to \infty$ asymptote is approached when $\gamma\beta > 10^2$.

For intermediate values of $\gamma\beta$, the composition of the gas is given by equation (1.4-33). The conversions of the solids are given by equations (1.4-18) and (1.4-19). One of the more important reactions between two solids proceeding through gaseous intermediates is the carbothermic reduction of metal oxides described in equations (1.4-1)–(1.4-3). For these reactions, $a = 2$ and $c = 1$. The solution to be presented below will be for these values of a and c.

Figure 1.4-2 shows the rates of reactions and the concentrations of the gaseous intermediates as functions of γ for various values of β. Since the rates are constant [see equations (1.4-18), (1.4-19), and (1.4-33)] until either one of the solids is completely reacted (when either B or D is all consumed the reaction stops), the behavior of a given system is represented by a point corresponding to given γ and β. When both F_B and F_D are not unity, the rate changes with conversion.[63]

It was shown in equations (1.4-36) and (1.4-41) that when $\gamma\beta \to 0$ or ∞, $d\xi_D/dt^*$ or $d\xi_B/dt^*$, respectively, is independent of γ. Figure 1.4-2

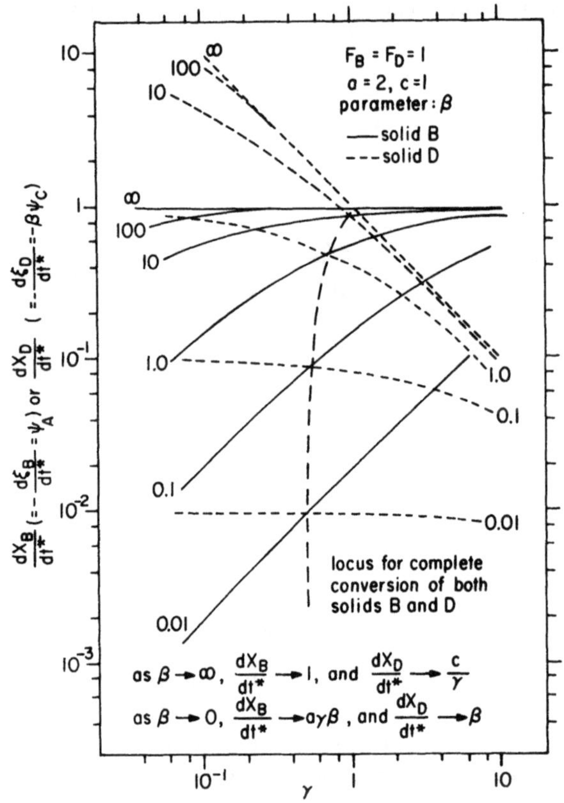

Figure 1.4-2. Rate of reaction and concentration vs. γ for $F_B = F_D = 1$. [Adapted from H. Y. Sohn and J. Szekely, *Chem. Eng. Sci.* **28**, 1789 (1973).]

shows that these asymptotes are valid when $\gamma\beta < 10^{-2}$ and $\gamma\beta > 10^{2}$, respectively. These criteria are valid also for other values of F_B and F_D.[63]

The intersection of the curves for B and D for the same value of β provides the locus for the complete conversion of B and D simultaneously. It is of interest to note that the relative amount (γ) of B and D for simultaneous complete conversion, i.e., the "stoichiometric" number of moles, vary with the relative kinetics (β). In the extreme regimes of $\gamma\beta \to 0$ and $\gamma\beta \to \infty$, the values of γ for the simultaneous complete conversion of both B and D may be obtained from equations (1.4-39) and (1.4-45), respectively:

$$\gamma = \begin{cases} 1/a & \text{when } \gamma\beta \to 0 \qquad (1.4\text{-}46) \\ c & \text{when } \gamma\beta \to \infty \qquad (1.4\text{-}47) \end{cases}$$

These values can be seen in Figure 1.4-2 (note that $a = 2$ and $c = 1$ in this figure). For intermediate values of $\gamma\beta$, the value of γ for the simultaneous complete conversion of B and D depends on the relative kinetics (β).

1.4.3. Discussion

A quantitative analysis has been described for the reactions between solids proceeding through gaseous intermediates when the effect of intrapellet diffusion is negligible. The criteria for the negligible diffusional effects have been established elsewhere.[63] Many systems fall within this category.

Few systematic experimental data are available that may be compared with the model. The work of Otsuka and Kunii[59] and Rao[60] on the reduction of hematite with carbon comes close to the analysis presented here. Although there is qualitative agreement between these experimental results and the calculated results of the model, a quantitative comparison cannot be made because of the difficulties arising from the two-step reaction in the hematite–carbon system and the limited range of experimental parameters. Furthermore, the above analysis made use of a number of simplifying assumptions to facilitate the mathematical development. Although this was useful in expressing the underlying concept of the model, few systems can be expected to satisfy all the simplifying assumptions used above. For example, the kinetics of the component gas–solid reactions used in equations (1.4-10) and (1.4-11) may be of different expressions such as the Langmuir–Hishelwood type, or the reactions of solids B and D may be controlled by the intragrain diffusion, i.e., diffusion of the gaseous reactant through the product layer around each grain. These cases can be treated by writing appropriate expressions for equations (1.4-10)–(1.4-13).

It is noted that the analysis of a system with reversible, first-order reactions follows directly, with minor adjustments, that for an irreversible reaction described in this Section.

It is hoped that the model will not only aid the analysis of experimental results but also stimulate further research on this interesting area of solid–solid reaction systems.

1.5. Notation

a stoichiometric coefficient in reaction (1.4-7)

A dimensionless quantity defined by equation (1.2-43)

A_B, A_D surface area of a grain of solids B and D, respectively

A_g, A_p external surface area of an individual grain and the pellet, respectively

b number of moles of solid B reacted by one mole of fluid reactant A

c stoichiometric coefficient in reaction (1.4-6)

c_s specific heat of the particle including the inert solid, if any

C molar concentration of fluid species

C^* equilibrium molar concentration of fluid species

D_e effective diffusivity in porous solid

D_g effective diffusivity in the product layer around a grain

f roughness factor defined as the ratio of the true external surface area to the apparent surface area

F shape factor (=1, 2, and 3 for flat plates, long cylinders, and spheres, respectively)

$g(X)$ conversion function defined by equation (1.2-24) or (1.3-46)

G parameter defined by equation (1.3-3) or mass velocity

h external heat transfer coefficient

$(-\Delta H)$ molar heat of reaction

k reaction-rate constant

k_m external mass transfer coefficient

Nu^* the modified Nusselt number defined by equation (1.2-38)

$p(X)$ conversion function defined by equation (1.2-25)

r distance from the center of symmetry in a nonporous particle or a grain

R distance from the center of symmetry in a porous pellet

Re the Reynolds number defined in equation (1.1-3)

Sc the Schmidt number defined in equation (1.1-3)

Sh, Sh^* the Sherwood and modified Sherwood numbers defined by equations (1.1-3) and (1.2-23), respectively

S_v surface area per unit volume of the pellet

t time

t^+, t^* dimensionless times defined by equations (1.2-29), and (1.2-21) or (1.3-37), respectively

T temperature

V volume

x parameter defined by equation (1.2-44), or distance coordinate in equation (1.3-8)

X fractional conversion of the solid

Z volume of solid product formed from unit volume of initial solid

Greek Symbols

α dimensionless quantity defined by equation (1.2-45), or fraction of volume of pellet occupied by reactant solid

β dimensionless quantity defined by equation (1.2-37) or (1.3-30), or the relative reactivities of solids defined by equation (1.4-24)

γ dimensionless quantity defined by equation (1.2-36) or (1.3-29), or the relative molar quantities of solids defined by equation (1.4-23)

δ dimensionless variable defined by equation (1.2-44), or quantity defined by equation (1.3-28)

ε emissivity or porosity

η dimensionless distance defined by equation (1.3-38)

θ dimensionless temperature defined by equation (1.2-42) or just after equation (1.3-19)

λ_e, λ_s effective thermal conductivity of a porous solid, and the true thermal conductivity of solid, respectively

ξ dimensionless position of the reaction front in the grain, defined by equation (1.3-36) or (1.4-22)

ρ molar concentration of solid reactants

ρ_s density of the particle including the inert solid, if any

σ dimensionless parameter defined by equation (1.3-39)

$\hat{\sigma}$ generalized gas–solid reaction modulus defined by equation (1.3-49)

$\hat{\sigma}_g$ shrinking-core reaction modulus defined after equation (1.3-53)

σ_s shrinking-core reaction modulus defined by equation (1.2-22)

τ_c parameter defined by equation (1.3-6)

ψ dimensionless concentration defined by equation (1.3-35)

Subscripts

A gas A

b bulk property

B solid B

c value at reaction interface

C gas C

D solid D

g grain

o original value

p particle or pellet

s value at external surface, or for entire solid including inert solid, if any

References

1. R. B. Bird, W. E. Stewart, and E. N. Lightfoot, *Transport Phenomena*, Wiley, New York (1960).
2. E. E. Petersen, *Chemical Reaction Analysis*, Prentice-Hall, Englewood Cliffs, N.J. (1965).
3. C. N. Satterfield, *Mass Transfer in Heterogeneous Catalysis*, MIT Press, Cambridge, Mass. (1970).
4. J. Szekely and N. J. Themelis, *Rate Phenomena in Process Metallurgy*, Wiley, New York (1971).
5. J. Szekely, J. W. Evans, and H. Y. Sohn, *Gas–Solid Reactions*, Academic Press, New York (1976).
6. V. G. Levich, *Physicochemical Hydrodynamics*, Prentice-Hall, Englewood Cliffs, N.J. (1962).
7. D. Kunii and O. Levenspiel, *Fluidization Engineering*, Wiley, New York (1969).
8. R. C. Reid and T. K. Sherwood, *Properties of Gases and Liquids*, 2nd ed., McGraw-Hill, New York (1966).
9. J. M. Smith, *Chemical Engineering Kinetics*, 2nd ed., McGraw-Hill, New York (1970).
10. E. A. Mason and T. R. Marrero, *Adv. At. Mol. Phys.* **6**, 155–232 (1970).
11. T. R. Marrero and E. A. Mason, *J. Phys. Chem. Ref. Data* **1**, 1–118 (1972).
12. J. M. Thomas and W. J. Thomas, *Introduction to the Principles of Heterogeneous Catalysis*, Academic Press, New York (1967).
13. O. A. Hougen and K. M. Watson, *Chemical Process Principles*, Part 3, Kinetics and Catalysis, Wiley, New York (1947).
14. K. J. Laidler, in *Catalysis*, P. H. Emmett, ed., Vol. 1, Reinhold, New York (1954), pp. 75, 119, 195.
15. J. Francl and W. D. Kingery, *J. Am. Ceram. Soc.* **37**, 99 (1954).
16. A. V. Luikov, A. G. Shashkov, L. L. Vasillies, and Y. E. Fraimon, *Int. J. Heat Mass Transfer* **11**, 117 (1968).

17. D. A. Frank-Kamenetskii, *Diffusion and Heat Transfer in Chemical Kinetics*, 2nd ed., Plenum, New York (1969).
18. M. F. R. Mulcahy and I. W. Smith, *Rev. Pure Appl. Chem.* **19**, 81 (1969).
19. H. Y. Sohn and J. Szekely, *Can. J. Chem. Eng.* **50**, 674 (1972).
20. J. Shen and J. M. Smith, *Ind. Eng. Chem. Fundam.* **4**, 293 (1965).
21. W. Jander, *Z. Anorg. Allg. Chem.* **163**, 1 (1927).
22. F. Habashi, *Extractive Metallurgy*, Vol. 1, General Principles, Chap. 8, Gordon and Breach, New York (1969).
23. P. P. Budnikov and A. M. Ginstling, *Principles of Solid State Chemistry, Reaction in Solids*, K. Shaw, transl. ed., Chap. 5, McLaren and Sons, London (1968).
24. R. E. Carter, in *Ultrafine Particles*, W. E. Kuhn, ed., Wiley, New York (1963), p. 491; *J. Chem. Phys.* **35**, 1137 (1961).
25. R. Aris, *Ind. Eng. Chem. Fundam.* **6**, 316 (1967).
26. G. S. G. Beveridge and P. J. Goldie, *Chem. Eng. Sci.* **23**, 913 (1968).
27. C. Y. Wen and S. C. Wang, *Ind. Eng. Chem.* **62**(8), 30 (1970).
28. D. Luss and N. R. Amundson, *AIChE J.* **15**, 194 (1969).
29. H. Y. Sohn, *AIChE J.* **19**, 191 (1973); **20**, 416 (1974).
30. M. Abramowitz and I. A. Stegun, *Handbook of Mathematical Functions*, Dover Publications Inc., New York (1965).
31. J. B. Rosser, *Theory and Application of*

$$\int_0^x e^{-x^2}\,dx \quad \textit{and} \quad \int_0^x e^{-p^2y^2}\,dy \int_0^y e^{-x^2}\,dx$$

Mapleton House, New York (1948).
32. E. E. Petersen, *AIChE J.* **3**, 443 (1957).
33. K. Hashimoto and P. L. Silveston, *AIChE J.* **19**, 268 (1973).
34. P. Schneider and P. Mitschka, *Chem. Eng. Sci.* **21**, 455 (1966).
35. E. W. Thiele, *Ind. Eng. Chem.* **31**, 916 (1939).
36. V. W. Weekman, Jr. and R. L. Gorring, *J. Catal.* **4**, 260 (1965).
37. P. L. Walker, Jr., F. Rusinko, Jr., and L. G. Austin, *Adv. Catal.* **11**, 133 (1959).
38. E. E. Petersen, *Chem. Eng. Sci.* **17**, 987 (1962).
39. P. B. Weisz and J. S. Hicks, *Chem. Eng. Sci.* **17**, 265 (1962).
40. H. Y. Sohn and J. Szekely, *Chem. Eng. Sci.* **27**, 763 (1972).
41. H. Y. Sohn and J. Szekely, *Chem. Eng. Sci.* **28**, 1169 (1973).
42. J. Szekely, C. I. Lin, and H. Y. Sohn, *Chem. Eng. Sci.* **28**, 1975 (1973).
43. J. Szekely and H. Y. Sohn, *Trans. Inst. Min. Met.* **82**, C92 (1973).
44. J. Szekely and J. W. Evans, *Chem. Eng. Sci.* **25**, 1019 (1970).
45. J. Szekely and J. W. Evans, *Chem. Eng. Sci.* **26**, 1901 (1971).
46. R. H. Tien and E. T. Turkdogan, *Met. Trans.* **3**, 2039 (1972).
47. D. Papanastassiou and G. Bitsianes, *Met. Trans.* **4**, 477 (1973).
48. M. Ishida and C. Y. Wen, *AIChE J.* **14**, 311 (1968).
49. K. B. Bischoff, *Chem., Eng. Sci.* **18**, 711 (1963); **20**, 783 (1965).
50. D. Luss, Can. *J. Chem. Eng.* **46**, 154 (1968).
51. H. Y. Sohn and J. Szekely, *Chem. Eng. Sci.* **29**, 630 (1974).
52. A. Calvelo and J. M. Smith, *Proceedings of Chemeca 70*, Paper 3.1, Butterworths, Australia (1971).
53. J. Szekely and J. W. Evans, *Met. Trans.* **2**, 1699 (1971).
54. A. K. Lahiri and V. Seshadri, *J. Iron Steel Inst. London* **206**, 1118 (1968).
55. S. Strijbos, *Chem. Eng. Sci.* **28**, 205 (1973).
56. J. W. Evans and S. Song, *Met. Trans.* **4**, 170 (1973).
57. J. W. Evans and S. Song, *I/EC Proc. Design Develop.* **13**, 146 (1974).
58. B. G. Baldwin, *J. Iron Steel Inst.* (*London*) **179**, 30 (1955).

59. K. Otsuka and D. Kunii, *J. Chem. Eng. Japan* **2**(1), 46 (1969).
60. Y. K. Rao, *Met. Trans.* **2**, 1439 (1971).
61. M. I. El-Guindy and W. G. Davenport, *Met. Trans.* **1**, 1729 (1970).
62. Y. Maru, Y. Kuramasu, Y. Awakura, and Y. Kondo, *Met. Trans.* **4**, 2591 (1973).
63. H. Y. Sohn and J. Szekely, *Chem. Eng. Sci.* **28**, 1789 (1973).
64. L. J. Petrovic and G. Thodos, *Ind. Eng. Chem. Fundam.* **7**, 274, (1968).
65. E. J. Wilson and C. J. Geankoplis, *Ind. Eng. Chem. Fundam.* **5**, 9 (1966).

2

Rate Processes in Multiparticle Metallurgical Systems

J. A. Herbst

2.1. General Approach to Describing Rate Processes in Multiparticle Systems

2.1.1. Introduction

In the preceding chapter, the kinetic behavior of individual particles isolated in an infinite fluid medium has been discussed. The various steps or combinations of steps which can control the rate of heterogeneous reactions were identified and appropriate types of mathematical models to describe individual particle behavior were reviewed. In the present chapter, consideration is extended to the behavior of multiparticle systems. By definition, a *multiparticle system* consists of an assembly of particles that make up the *disperse phase* plus the environment surrounding the particles that makes up the *continuous phase* in a processing vessel. Virtually all particulate assemblages encountered in extractive metallurgical practice are polydisperse in nature, i.e., the particles being processed have a broad distribution of properties such as size, mineralogical composition, etc., which contribute to the overall behavior of the system. In addition, in practical systems the particles often interact with one another and/or with the fluid environment. If one wishes to accurately design a reactor, optimize an existing operation, or specify an effective automatic control strategy for an extractive metallurgical process, it is necessary to be able to describe, in

John A. Herbst • Department of Metallurgy and Metallurgical Engineering, University of Utah, Salt Lake City, Utah

quantitative terms, the influence of material property distributions and particle–particle or particle–fluid interactions on the overall reaction behavior of the system.

Early approaches to the quantification of behavior of multiparticle systems were quite primitive. In some instances the particulate nature of the system was completely ignored, whereas in others the distribution of material properties was suppressed by representing an entire assembly of particles by an "average" particle. A rigorous approach to the modeling of such systems that recognized the distributed nature of particle properties did not emerge until the early 1960's.[1,2] This latter approach is sufficiently general to allow the treatment of most systems of importance in extractive metallurgy and is the approach adopted in this chapter.

In the analysis that follows the kinetics of chemical and physical change for individual particles provide the basic building blocks for the description of the behavior of an entire assembly of particles. The influence of property distributions and interactions on assembly behavior are treated by a special type of particle-accounting procedure termed *population balance*. Using this framework, it is possible to keep track of the behavior of each particle "type" in an assembly and its influence on all other particles. In this chapter, the magnitude of the effects of property distributions and particle interactions on assembly behavior is explored first (Section 2.1.2). Next, particle characteristics of importance in extractive metallurgy are reviewed along with methods of experimental measurement (Section 2.1.3), and the mathematical representation of property distributions for particle assemblies is discussed (Section 2.1.4). Following this introductory material, the general population-balance framework for representing change in multiparticle systems will be presented (Section 2.1.5), and selected example applications of "population-balance" models to the description of batch systems, and continuous systems (Section 2.1.6) of importance in extractive metallurgy will be discussed.

2.1.2. Motivation

Before discussing general aspects of multiparticle system behavior it is instructive to examine the behavior of a specific system. The purpose here is threefold: (1) to make it clear that multiparticle system kinetics cannot in general be analyzed in the same manner as individual particle kinetics; (2) to provide the reader with a feel for the types of variables important in multiparticle systems; and (3) to examine the magnitude of effects of material property distributions and particle–fluid interactions on the overall reaction kinetics of an assembly of particles.

The specific system chosen for discussion is the oxidation leaching of copper sulfide minerals in an agitated vessel. It is known that the oxygen–acid leaching of pure chalcopyrite at high temperatures and oxygen pressures follows the reaction

$$CuFeS_2 + \tfrac{17}{4}\,O_2 + H^+ \rightarrow Cu^{2+} + Fe^{3+} + 2\,SO_4^{2-} + \tfrac{1}{2}\,H_2O \qquad (2.1\text{-}1)$$

and that the kinetics of leaching of individual particles is controlled by surface reaction.[3] Thus, according to the discussion presented in Chapter 1, a plot of the integrated rate expression for batch leaching, $1-(1-\alpha)^{1/3}$, vs. time for nearly spherical monosize particles leached in a dilute suspension should yield a straight line with a slope that is inversely proportional to particle size. Figure 2.1-1a shows that individual particle kinetics for narrow-size fraction feeds (48×65 and 65×100 mesh) leached in dilute suspension do follow the "$1-(1-\alpha)^{1/3}$ law." Figure 2.1-1b shows the same type of plot for the leaching of a chalcopyrite feed having a *wide distribution of particle sizes* (a Gaudin–Schuhmann distribution[4] with size modulus, $d_{max} = 121\ \mu m$, and distribution modulus, $m = 0.7$). Notice in this case, even though the kinetics of individual particles are controlled by surface-reaction rate, the overall reaction behavior, shown in Figure 2.1-1b deviates strongly from linearity demonstrating that the *overall size distribution kinetics* do not follow the "$1-(1-\alpha)^{1/3}$ law." Each of the particle sizes continues to react at a characteristic rate of the form

$$1-(1-\alpha)^{1/3} = (K/d_0)t \qquad (2.1\text{-}2)$$

but the overall rate for the assembly is a compound rate. Its form is determined by the nature of the size distribution.

Size distributions are essentially infinitely variable in form and magnitude, and the detailed structure of the distribution depends upon the material which makes up the particles and the method by which the size distribution is produced.[5] Two important parameters used to characterize a size distribution are the mean size, or central tendency of the distribution, μ, and the coefficient of variation or the relative spread of sizes around the mean, CV. Figure 2.1-2a shows a log–log representation of a series of Gaudin–Schuhmann distributions for which the coefficient of variation is constant at 0.73 and the mean size varies from 12.5 to 200 μm. Figure 2.1-2b shows a series of distributions for which the mean is constant at 50 μm and the coefficient of variation ranges from 0 to 1.20. Figure 2.1-2c shows "fraction-reacted-vs.-time plots" for chalcopyrite feeds with the same coefficient of variation but different mean sizes. It is evident that the mean size of the distribution has a large effect on the overall kinetics. As one would expect, the smaller the

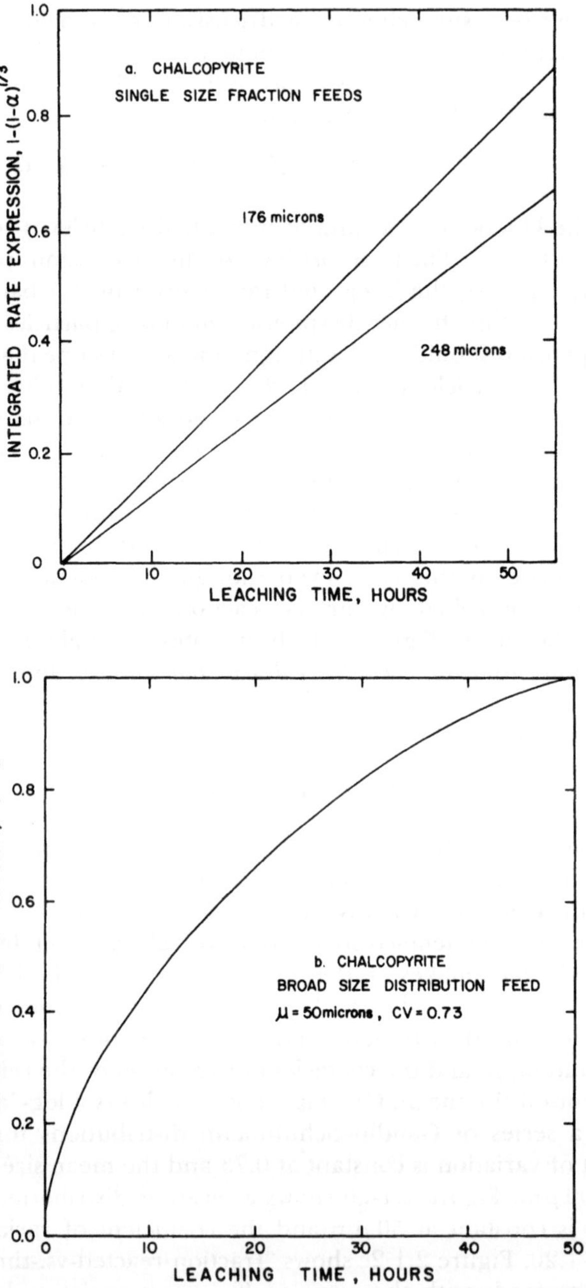

Figure 2.1-1. Comparison of integrated expression plots for chalocopyrite leaching kinetics with single size fraction feeds (a) and a broad size distribution feed (b).

average particle size in the feed, the faster the overall reaction rate is for the assembly.

The influence of the spread of the particle size distribution for the fixed mean size is shown in Figure 2.1-2d. The effect is not as dramatic as for changes in the mean, but is still significant, indicating that the average particle size of the feed alone is not sufficient to completely characterize the leaching response. Note that the curves in Figure 2.1-2d cross over one another; those for small CV values have low initial rates and relatively high rates for extended leaching, and those for high CV values have high initial rates and low extended leaching rates. This occurs because increasing values for the coefficient of variation result in an increasing abundance of both small and large diameters (relative to the mean) in the population. The small particles leach at a high rate giving rise to the initial shape of the curves, whereas the large particles leach at a slow rate giving rise to the shape of the curves after extended leaching.

Figure 2.1-3 illustrates the effect of another type of property distribution on the overall kinetics of reaction of an assembly. In this case, a 50–50 mixture of narrow-size fractions of chalcopyrite and a second copper sulfide phase is being leached in a batch reactor. As with chalcopyrite, the leaching kinetics of the second phase are controlled by surface reaction, but the surface-reaction-rate constant is approximately four times larger than that for chalcopyrite. In this figure, the individual responses of the two minerals are shown, as well as the composite response of the mixture. Once again the composite does not follow the "$1-(1-\alpha)^{1/3}$ law," which each of the individual components follows. Analysis of the latter data as a feed consisting of a single mineralogical component would clearly be inappropriate in this case.

The nature of the interaction between particles and the fluid phase in which they are reacting is illustrated in Figure 2.1-4. The previous examples were for very dilute suspensions in which the reactant concentration that the particles "see" at their surface does not change significantly with extent of reaction. In more concentrated suspensions, the number of moles of reactant consumed from the fluid phase by the reacting particles may be large enough to significantly reduce the concentration driving force for a reaction. The degree to which a reaction

$$a\,\mathrm{A}_{\text{solid}} + b\,\mathrm{B}_{\text{fluid}} \rightarrow c\,\mathrm{C}_{\text{solid}} + d\,\mathrm{D}_{\text{fluid products}} \tag{2.1-3}$$

is slowed down as a result of this interaction can be characterized by a dimensionless parameter η defined by

$$\eta = \frac{a}{b}\,\frac{VC_{\mathrm{B}_0}}{M_{\mathrm{A}_0}} \tag{2.1-4}$$

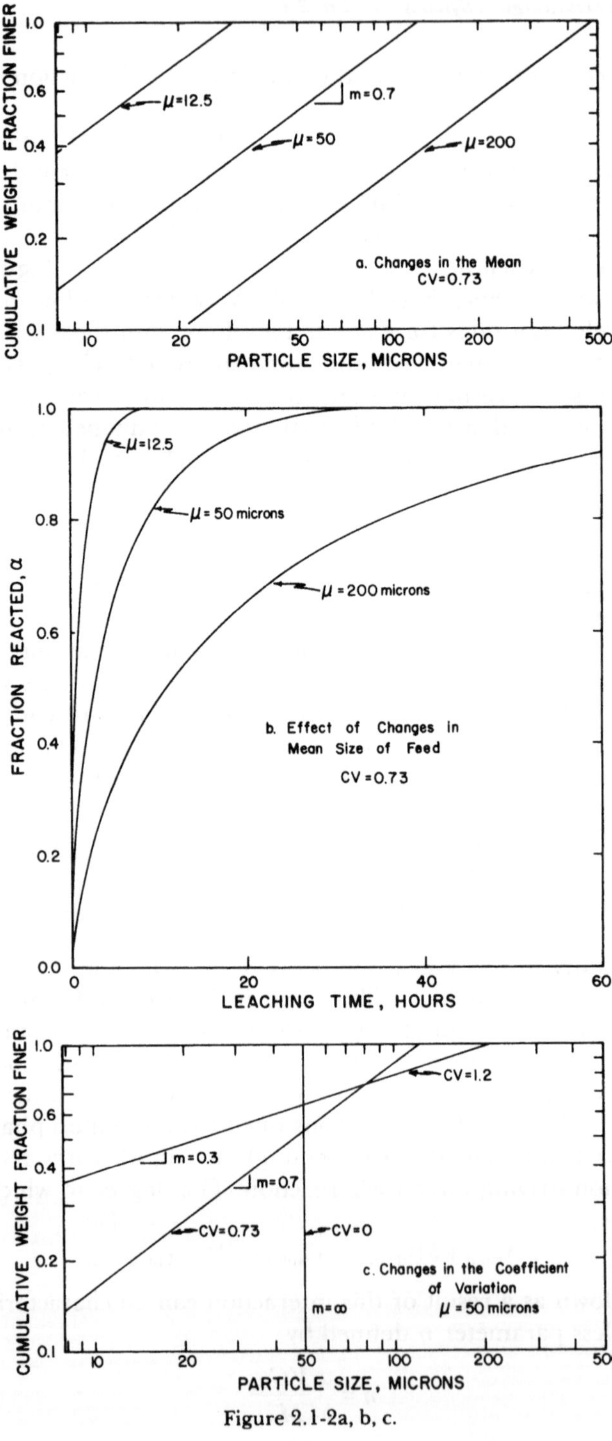

Figure 2.1-2a, b, c.

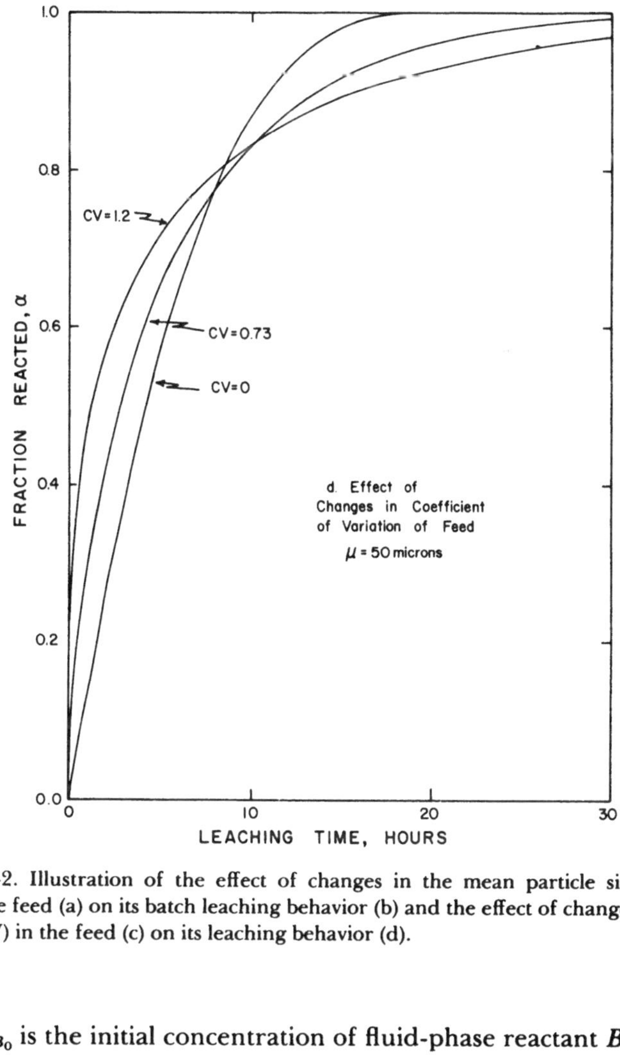

Figure 2.1-2. Illustration of the effect of changes in the mean particle size (μ) of a chalcopyrite feed (a) on its batch leaching behavior (b) and the effect of changes in spread of sizes (CV) in the feed (c) on its leaching behavior (d).

where C_{B_0} is the initial concentration of fluid-phase reactant B, V is the initial volume of the fluid phase, and M_{A_0} is the initial number of moles of the solid phase A present in the system. Physically, η represents the number of moles of A which could be converted by the amount of B initially present divided by the actual number of moles of A initially present. Large values of η are equivalent to dilute suspensions, small values to concentrated suspensions. Figure 2.1-4 shows the influence of the parameter η on the reaction progress for the batch leaching of monosize chalcopyrite particles. Notice that the initial kinetics of reaction are the same for all cases because all particles see the same initial concentration of reactant. It is apparent from these plots, however, that

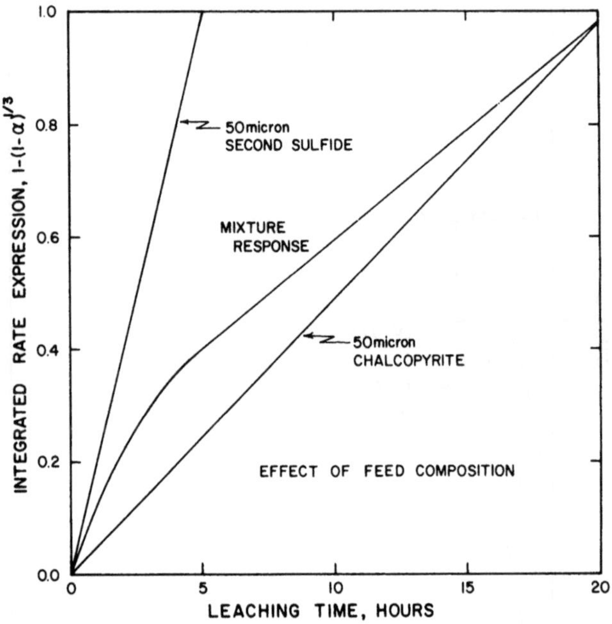

Figure 2.1-3. Illustration of the effect of feed composition for the leaching of single size fraction copper sulfide feeds.

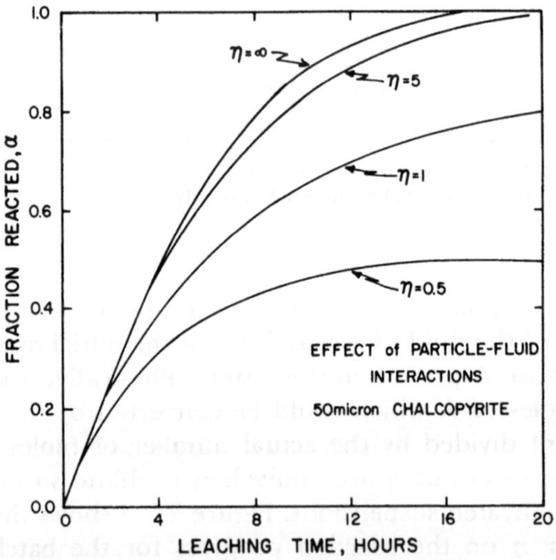

Figure 2.1-4. Illustration of the effect of particle–fluid interactions on the batch leaching behavior of chalcopyrite ($\eta \rightarrow \infty$ corresponds to dilute suspensions, $\eta \rightarrow 0$ corresponds to concentration suspensions).

the interaction of the particles with the fluid phase cannot be ignored for small values of η, particularly at high conversions.

The preceding discussion has attempted to illustrate the effect of property distributions and particle–fluid interactions on the kinetics of reaction of a multiparticle system. The example used was acid–O_2 leaching of copper sulfide minerals in a batch system. However, similar trends would have been observed had an example been chosen from solvent extraction, cementation, roasting, etc., in either a batch or continuous-processing mode. The magnitude of the effects observed here demonstrates the need for proper characterization of material property distributions and the need for a general framework within which to analyze the kinetics of multiparticle systems.

2.1.3. Particle Characterization

As pointed out in the preceding section, proper particle characterization is a necessary prerequisite for quantifying the behavior of a particulate system. Proper characterization, in this case, entails identifying those particle properties of importance for the specific system of interest and subsequently measuring these properties for individual particles experimentally. In some instances, the properties considered will be only those which are themselves undergoing change in the process of interest, e.g., particle size in a grinding or an agglomerating system. In other instances, it will also be necessary to consider those properties which have an effect on the process but do not themselves undergo change, e.g., particle size in a leaching system in which a change in size does not accompany the leaching reaction. Table 2.1-1 lists several particle properties which are often important in one or both of the above categories for mineral processing and chemical metallurgy systems. The identification of the *minimum* set of properties necessary to describe the behavior of a system requires a thorough understanding of individual particle kinetics for both physical and chemical change.

Once the properties of importance for a specific system have been identified, experimental methods for these measurements must be chosen. Some of the more commonly used measurement methods are summarized in Table 2.1-1. Other methods plus detailed descriptions of experimental procedures are available in several texts.[6–8] Upon examining Table 2.1-1, the reader may be puzzled because in general there is no practical method for *uniquely* determining particle size, shape, composition, etc., for irregularly shaped heterogeneous particles. Table 2.1-1 shows that particle "size" can be measured microscopically using statistical diameters; by determining the terminal velocity of the particle falling through a fluid and calculating an equivalent hydro-

Table 2.1-1. Particle Characteristics and Their Measurement

Property	Method
Size	Microscopic, sedimentation, sieving
Shape	Size (linear dimension), volume, and area measurement
Mineralogical composition	Microscopic, X-ray diffraction
Density	Picnometer
Surface area	
external	Permeametry
external and internal	Gas adsorption, chromatography
Porosity and pore size	Porosimetry
Surface charge	Electrophoresis, streaming potential
Magnetic susceptibility, electrical conductivity	Standard methods of physics
Wettability	Contact angle, microcalorimetry
Strength	Compression, drop weight, grindability tests, etc.

dynamic sphere diameter; or by determining the aperture size of a sieve through which the particle just passes. Each method, in general, yields a slightly different particle "size." Similarly, a unique description of particle shape requires a common description of particle topology. Commonly used measures of shape include the volume–shape factor, $C_3 =$ volume of the particle divided by the particle "diameter" cubed, and area–shape factor, $C_2 =$ external area of the particle divided by the particle "diameter" squared. These quantities can be related to regular geometrical shapes, e.g., spheres, cubes, tetrahedra, disks, etc. One of the most commonly used measures of a particle's shape is its sphericity, $\phi =$ surface area of a sphere with the same volume as the particle divided by the actual surface area of the particle. In terms of C_2 and C_3, particle sphericity is given by $\phi = 4.84\ C_3^{2/3}/C_2$. It is apparent from this expression that an "infinite" number of combinations of the shape factors C_2 and C_3 can yield the same values of ϕ.

Measurements of particle composition are subject to similar uniqueness problems. Microscopic measurements of the areas of different phases exposed in a particle thin section will not, in general, yield the same mineralogical composition as X-ray diffraction. In fact, the determination of virtually all of the characteristics listed in Table 2.1-1 suffer from similar uniqueness problems.

In spite of these problems, it is generally true that the more that is known about the characteristics of the particles to be processed, the better will be the understanding of their behavior, and ultimately the better will be the system analysis and process design. For a given application, one should decide which particle characteristics must be measured and choose a measurement method for each. These characterization methods will in fact define what is meant by size, shape, etc., for the system under consideration. The measured characteristics themselves can be correlated with the kinetic response of the system. This procedure will lead to a set of self-consistent particle characteristics which is useful in describing the system behavior.

2.1.4. Representation of Property Distributions

Once the properties of importance in an assembly have been identified and characterization methods chosen, it is important to be able to describe, in quantitative terms, the "amount" of particles with a given set of properties in an assembly. If it is known that particles of a given size and composition react in a certain manner, then we must know what fraction of the particle assembly has this size and composition to predict the overall assembly behavior. The required "fractions" are specified mathematically through the use of *density functions* and *distribution functions* similar in form to probability distributions.[9–11] The distribution of each of the properties of interest in an assembly can be represented individually or collectively by such functions. Figures 2.1-5a and b show these two functions for a typical distribution of particle sizes in an assembly. The quantity $f(d)$ is termed the particle-size density function for the assembly. Physically $f(d)\,d(d)$ is equal to the fraction of the particles in a population that lie in a differential size interval d to $d+d(d)$. Geometrically this fraction can be represented by the shaded area under the density function between the differential limits d to $d+d(d)$ shown in Figure 2.1-5a. In order to find the fraction of particles smaller than some size d', one must add the fractions of particles $f(d)\,d(d)$ from the minimum size in the population, $d_{\min}$, to the size of interest, d'. This summation is accomplished by integration of the density function, i.e.,

$$F(d') = \sum_{\substack{d_{\min} \\ \text{all } d(d)}}^{d'} f(d)\,d(d) = \int_{d_{\min}}^{d'} f(d)\,d(d) \tag{2.1-5}$$

The function $F(d')$, termed the *distribution function*, gives the fraction of the population with size less than d'. It is apparent from equation (2.1-5) that $F(d')$, depicted in Figure 2.1-5b, equals the area under the density function curve between $d_{\min}$ and d' (cross-hatched area in Figure 2.1-5a). It is easily shown that the fraction of particles between any two sizes d_a and d_b ($d_b > d_a$) is given by $F(d_b) - F(d_a)$, that $F(d_{\max}) = 1.0$, and that $f(d_a) = [dF(d)/d(d)]_{d=d_a}$. It is important to recognize that once $f(d)$ *or* $F(d)$

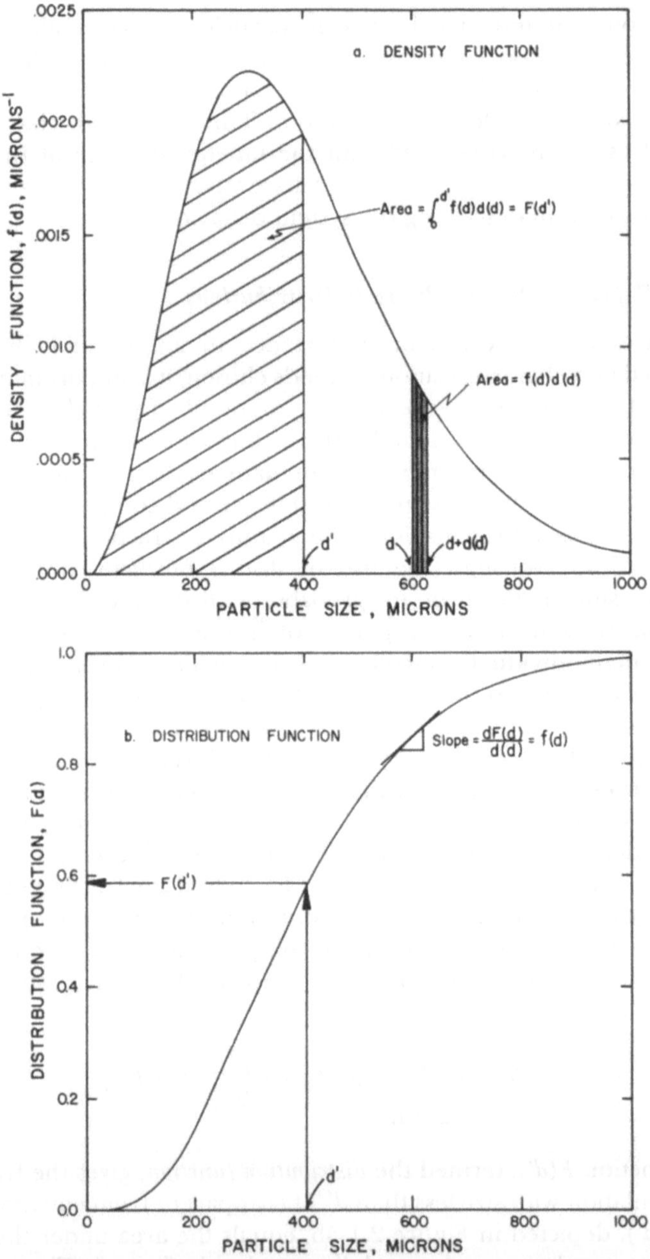

Figure 2.1-5. Plot of a typical density function (a) and distribution function (b) for particle size in an assembly of particles. The equivalence of the distribution function and the area under the density function curve and the density function and the slope of the distribution function curve are illustrated.

is known, everything about the distribution of particle sizes in the assembly is known.

In many instances it is not necessary, nor is it possible experimentally, to determine the complete density function or distribution function for a certain property. In such instances it may suffice to determine *selected characteristics* of the distribution. One such set of characteristics is the fraction of particles in a series of discrete property intervals. If, for example, the entire size range of particles d_{min} to d_{max} is broken up into a series of n discrete subintervals, then the fraction in the ith interval (bounded by d_i above and d_{i+1} below) is given by

$$f_i = \int_{d_{i+1}}^{d_i} f(d)\,d(d) = F(d_i) - F(d_{i+1}), \qquad i = 1, 2, 3, \ldots, n \quad (2.1\text{-}6)$$

Information of this type might be generated experimentally by sizing an assembly using a series of n sieves. The set of f_i values determined in this manner does not completely characterize the distribution (since no information is contained in f_i concerning the distribution within the interval d_{i+1} to d_i), but approximations to the complete density function and distribution function can be obtained from these data. It is apparent from equation (2.1-6) that the following equality holds for the distribution function at the discrete points d_i, $i = 1, 2, \ldots, n$:

$$F(d_i) = \sum_{j=1}^{n} f_j \qquad (2.1\text{-}7)$$

From the mean-value theorems of calculus the density function can be approximated by

$$f(d_i^*) \simeq \frac{F(d_i) - F(d_{i+1})}{d_i - d_{i+1}} = \frac{f_i}{d_i - d_{i+1}} \qquad (2.1\text{-}8)$$

where d_i^* is an average value of d in the interval d_i to d_{i+1}. Commonly used values for d_i^* include the arithmetic average $(d_i + d_{i+1})/2$, the geometric average $(d_i d_{i+1})^{1/2}$, and the harmonic average $2d_i d_{i+1}/(d_i + d_{i+1})$. Figure 2.1-6 shows the density and distribution functions of Figures 2.1-5a and b along with the discrete approximations obtained for ten equispaced size intervals, $d_i^* = (d_i + d_{i+1})/2$. The discrete approximations are seen to reproduce the essential characteristics of the continuous functions; such approximations become even better as the number of size intervals is increased.

In certain instances, only one or two parameters are used to characterize a distribution. The mean and variance of the distribution are often used in such instances. The mean or average size of a population is

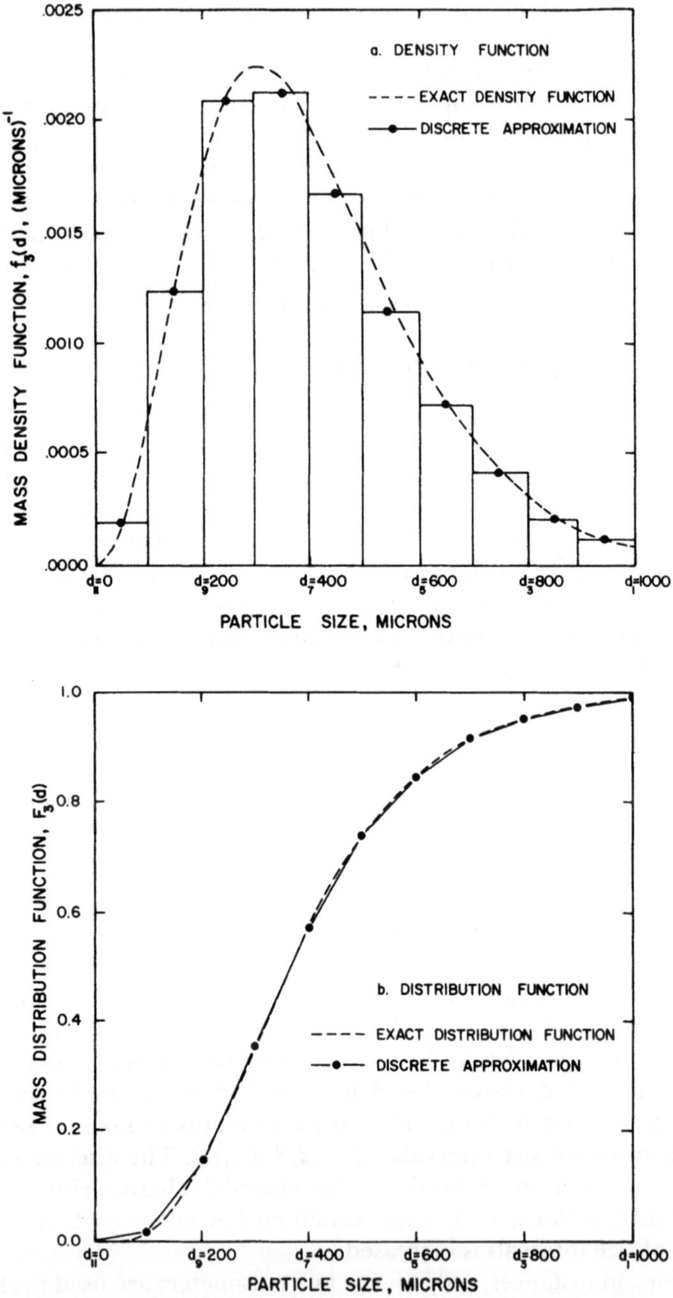

Figure 2.1-6. Plot illustrating ten size fraction discrete approximations to a density function (a) and distribution function (b).

determined from the density function using the defining equation

$$\mu = \int_{d_{\min}}^{d_{\max}} df(d)\, d(d) \tag{2.1-9}$$

Its corresponding approximation obtained from a set of f_i values is

$$\hat{\mu} = \sum_{i=1}^{n} d_i^* f(d_i^*)\, \Delta d_i = \sum_{i=1}^{n} d_i^* f_i \tag{2.1-10}$$

The variance or spread of sizes (around the mean) is given by

$$\sigma^2 = \int_{d_{\min}}^{d_{\max}} (d-\mu)^2 f(d)\, d(d) \tag{2.1-11}$$

or approximated by

$$\hat{\sigma}^2 = \sum_{i=1}^{n} (d_i^* - \hat{\mu})^2 f_i \tag{2.1-12}$$

The normalized spread of sizes is characterized by the coefficient of variation

$$\mathrm{CV} = \sigma/\mu \tag{2.1-13}$$

Figures 2.1-7a and b illustrate the way in which these characteristics change with changes in the density function, $f(d)$. Figure 2.1-7a shows distributions with the same coefficient of variation but different mean sizes and Figure 2.1-7b shows distributions with the same mean size but different CV values.

Several commonly used *empirical* size distributions are listed in Table 2.1-2 along with the corresponding mean sizes, variances, and coefficients of variation. These two parameter distributions can be fitted to experimental-size distribution data by plotting or moment matching.[12] The best fit values of the adjustable parameters for each of the distributions can be determined directly from plots or from computed values of the mean and variance.

In the preceding discussion of $f(d)$ and $F(d')$ it was not specified whether these quantities referred to fractions of the total number of particles or fractions of the total mass or volume of particles in population. This is an important distinction and the ability to convert from one type of size distribution (e.g., fraction by number) to another (e.g., fraction by volume or mass) is required in many applications. Number distributions, denoted by $f_0(d)$ and $F_0(d)$, are generally considered to be the most fundamental form of distribution. This type of distribution forms the basis for the development of population balance models in the next section. On the other hand, we often determine distributions

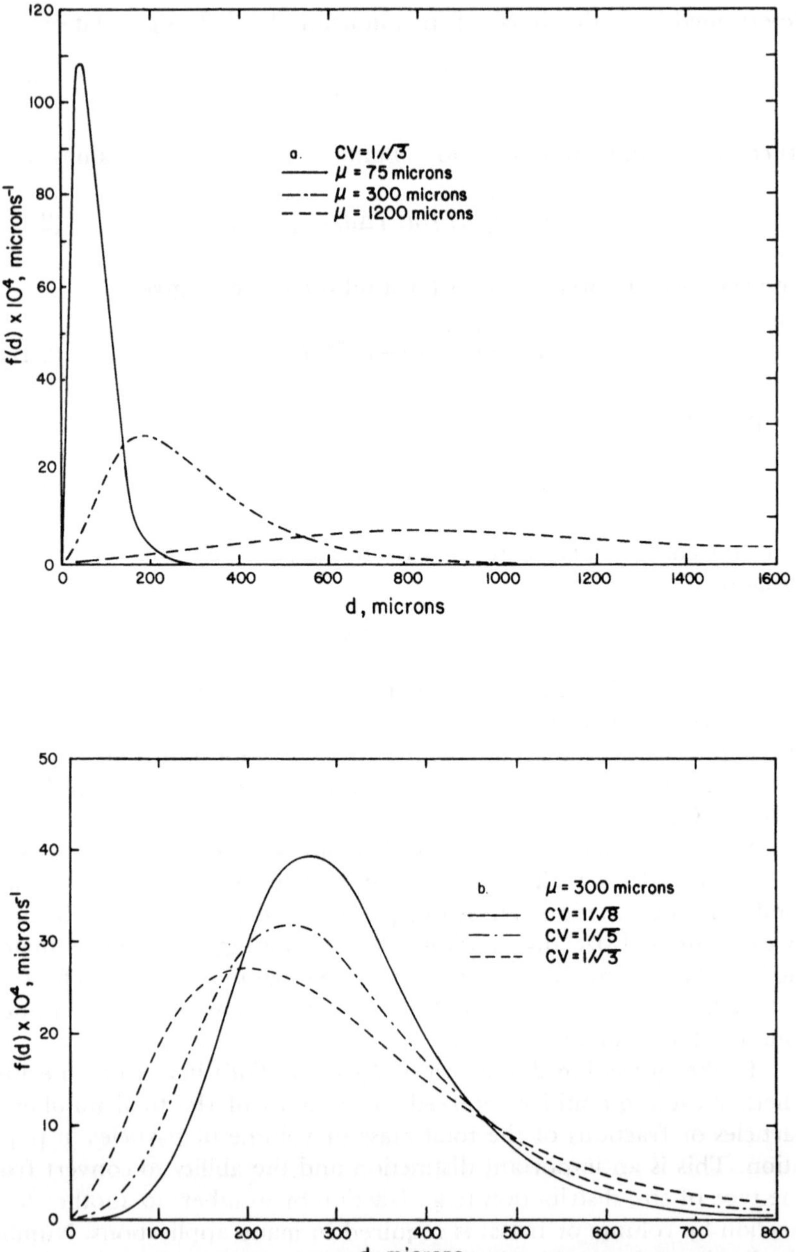

Figure 2.1-7. Illustration of density function changes resulting from changes in the mean for a fixed coefficient of variation (a) and changes in the coefficient of variation for a fixed mean (b).

Table 2.1-2. Commonly Used Empirical Size Distributions

Name	Gaudin–Schuhmann	Rosin–Rammler	Log-Normal	Gamma[a]
Adjustable parameters	m, d_{max}	m, l	$\log\sigma, l'$	b, p
Density function, $f(d)$	$m\dfrac{d^{m-1}}{d_{max}^m}$	$\dfrac{m}{l^m}d^{m-1}\exp\left[-\left(\dfrac{d}{l}\right)^m\right]$	$\dfrac{1}{(2\pi)^{1/2}d\log\sigma}\exp\left[-\dfrac{\log^2(d/l')}{2\log^2\sigma}\right]$	$\dfrac{b^p d^{p-1}e^{-bd}}{\Gamma(p)}$
Distribution function, $F(d)$	$\left(\dfrac{d}{d_{max}}\right)^m$	$1-\exp\left[-\left(\dfrac{d}{l}\right)^m\right]$	numerical integration of density function or standard tabulations	
Range, $[d_{min}, d_{max}]$	$[0, d_{max}]$	$[0, \infty]$	$[0, \infty]$	$[0, \infty]$
Mean, μ	$\dfrac{m}{m+1}d_{max}$	$l\Gamma\left(\dfrac{m+1}{m}\right)$	$l'\exp(\frac{1}{2}\log^2\sigma)$	$\dfrac{p}{b}$
Variance, σ	$\left[\dfrac{m}{(m+2)}-\dfrac{m^2}{(m+1)^2}\right]d_{max}^2$	$l^2\left[\Gamma\left(\dfrac{m+2}{m}\right)+\Gamma^2\left(\dfrac{m+1}{m}\right)\right]$	$l'^2[\exp(2\log^2\sigma)-\exp(\log^2\sigma)]$	$\dfrac{p}{b^2}$
Coefficient of variation, CV	$\dfrac{\dfrac{m}{m+2}-\dfrac{m^2}{(m+1)^2}}{\dfrac{m}{(m+1)}}$	$\dfrac{\left[\Gamma\left(\dfrac{m+2}{m}\right)-\Gamma^2\left(\dfrac{m+1}{m}\right)\right]^{1/2}}{\Gamma\left(\dfrac{m+1}{m}\right)}$	$\dfrac{[\exp(2\log^2\sigma)-\exp(\log^2\sigma)]^{1/2}}{\exp(\frac{1}{2}\log^2\sigma)}$	$\dfrac{1}{p^{1/2}}$

[a] By definition the complete gamma function is given by $\Gamma(p)=\int_0^\infty z^{p-1}e^{-z}\,dz$, $\Gamma(p+1)=p\Gamma(p)$. For p an integer, the incomplete gamma function is given by $F(d)=1-\{b^p e^{-bd}\sum_{r=0}^{p-1}d^{p-1-r}/[(p-1-r)\cdot b^{r+1}]\}$.

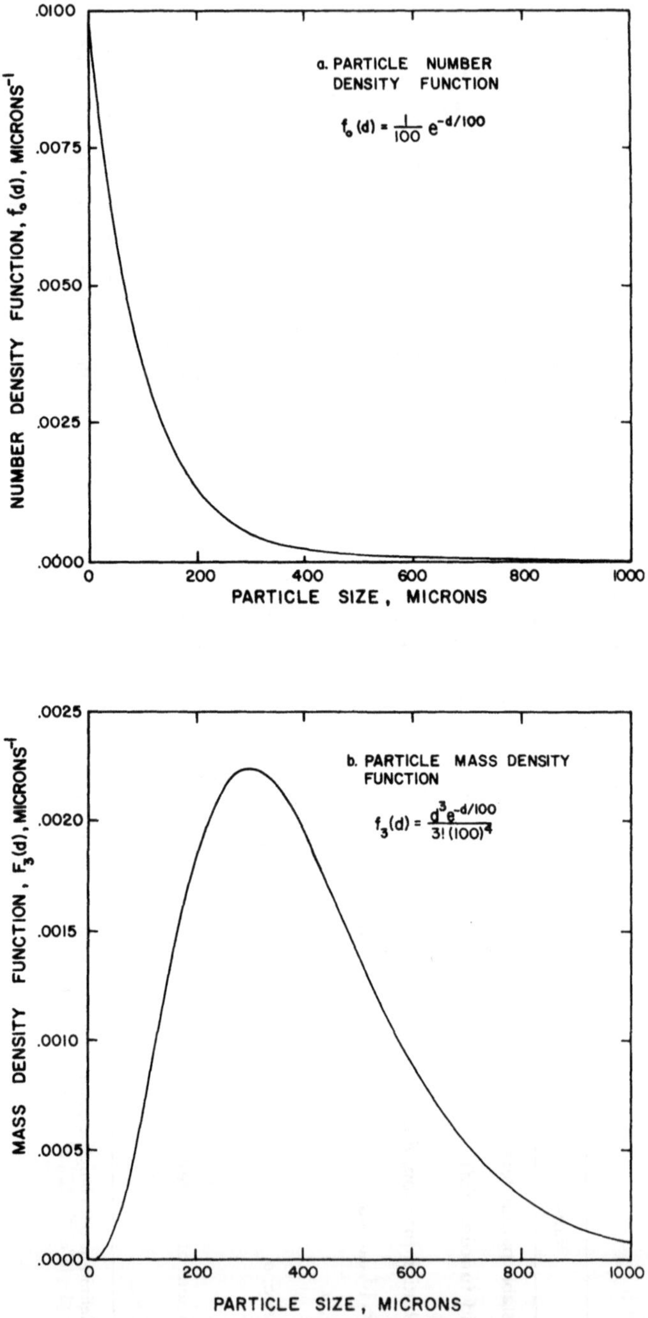

Figure 2.1-8. Illustration of the transformation of

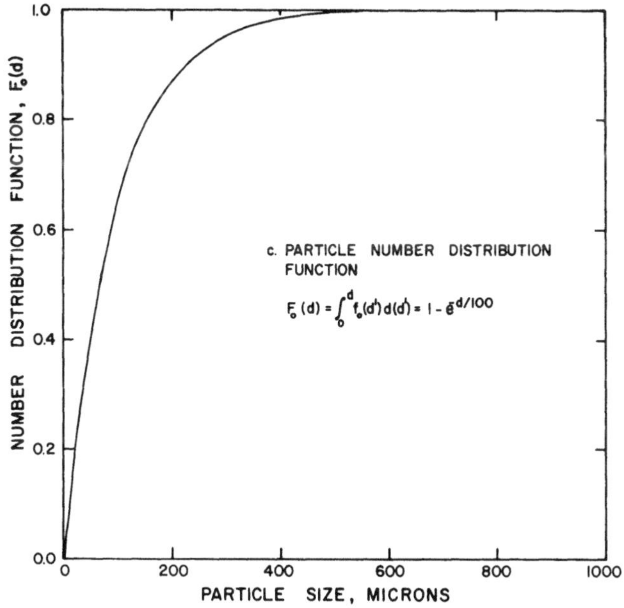

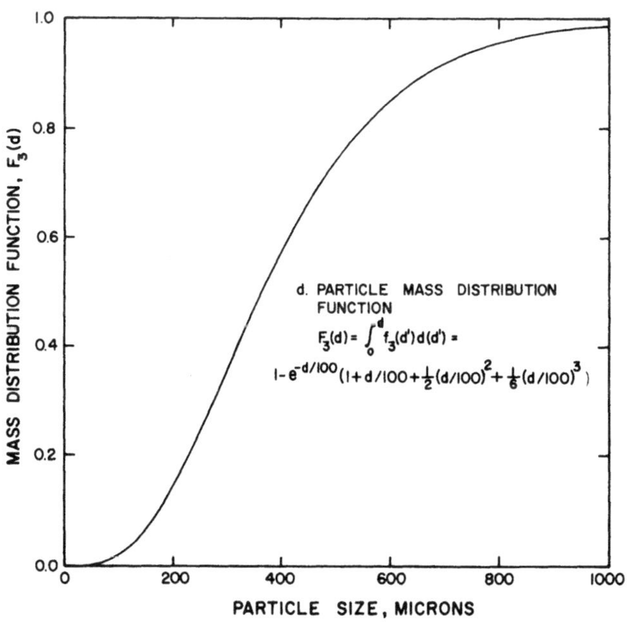

a size distribution from a number to a mass basis.

experimentally on the basis of mass (e.g., sieving and weighing). Hence the mass distribution, denoted by $f_3(d)$ and $F_3(d')$, is important in many instances. Figure 2.1-8 shows a comparison of a typical number distribution and the associated mass distribution for a particle assembly. The general relationship between a number distribution and a mass distribution can be obtained from the following equality for particles of size d to $d+d(d)$ in any assembly:

$$\begin{bmatrix}\text{mass of particles of}\\ \text{size } d\end{bmatrix} = \begin{bmatrix}\text{mass of particles of}\\ \text{size } d\end{bmatrix}$$

$$\begin{bmatrix}\text{total mass of}\\ \text{particles}\end{bmatrix}\begin{bmatrix}\text{mass fraction}\\ \text{of particles}\\ \text{of size } d\end{bmatrix} = \begin{bmatrix}\text{mass of a}\\ \text{particle}\\ \text{of size } d\end{bmatrix}\begin{bmatrix}\text{number of}\\ \text{particles}\\ \text{of size } d\end{bmatrix}$$

$$Wf_3(d)\,d(d) = \rho C_3 d^3 N f_0(d)\,d(d) \tag{2.1-14}$$

where N is the total *number* of particles in the population, $f_0(d)\,d(d)$ is the *number fraction* of the particles with size d to $d+d(d)$, W is the total *mass* of particles in the population, $f_3(d)\,d(d)$ is the *mass fraction* of particles with size d to $d+d(d)$, ρ is the solid density, and C_3 is the volume–shape factor. If the solid density and volume–shape factor are independent of size, then equation (2.1-14) can be used in conjunction with the equations $\int f_0(d)\,d(d) = 1$ and $\int f_3(d)\,d(d) = 1$ for integration over all values of d to show that

$$f_3(d) = \frac{d^3 f_0(d)}{\int_{d_{\min}}^{d_{\max}} d^3 f_0(d)\,d(d)} \tag{2.1-15a}$$

$$f_0(d) = \frac{d^{-3} f_3(d)}{\int_{d_{\min}}^{d_{\max}} d^{-3} f_0(d)\,d(d)} \tag{2.1-15b}$$

The last two equations permit the transformation from one type of size distribution to the other. The reader should use the above expressions to confirm the transformation $f_0(d) \to f_3(d)$ depicted in Figure 2.1-8. It is also instructive to compute the mean and variance of particle size for the number and mass distributions shown in the figure using equations (2.1-9) and (2.1-11). Note that, in general, the mean and variance are different for the number size distribution and the mass size distribution.

The preceding discussion of single property distributions has centered around particle size distributions. Clearly, the same type of treatment can be made for any property of interest in a particle assembly. Thus for a general property Z, (e.g., composition, strength, etc.) we can define $f(\zeta)\,d\zeta =$ the fraction of particles in the assembly with property Z between ζ and $\zeta + d\zeta$, and $F(\zeta) =$ the fraction of particles with property $Z \leq \zeta$. From these definitions we can find the fractions in a set of discrete property intervals [ζ_i to ζ_{i+1}] and can compute means and variances of the distribution in an analogous fashion to equations (2.1-6)–(2.1-13). In the general case, transformations from number to mass distributions can only be made if information is available concerning the relationship between the number and mass of particles in a given property interval.

Up to this point we have treated the case in which only one property of the assembly is of interest. The representation of the distribution of two or more properties simultaneously is a logical extension of what has been done above. Let us say, for example, that we are interested in the distribution of size, d, *and* the mass of some valuable constituent, m_A, in a particle assembly. In this case we can denote the number density function by $f_0(d, m_A)$. This joint density function may look something like that shown in Figure 2.1-9. From the joint density function $f_0(d, m_A)$ we can find the number fraction of particles with size between d and $d+d(d)$ and mass of valuable between m_A and $m_A+d(m_A)$ from the expression

$$f_0(d, m_A)\, d(d)\, d(m_A) \tag{2.1-16}$$

Geometrically this fraction can be represented by the volume under the density function curve between the differential limits shown in the

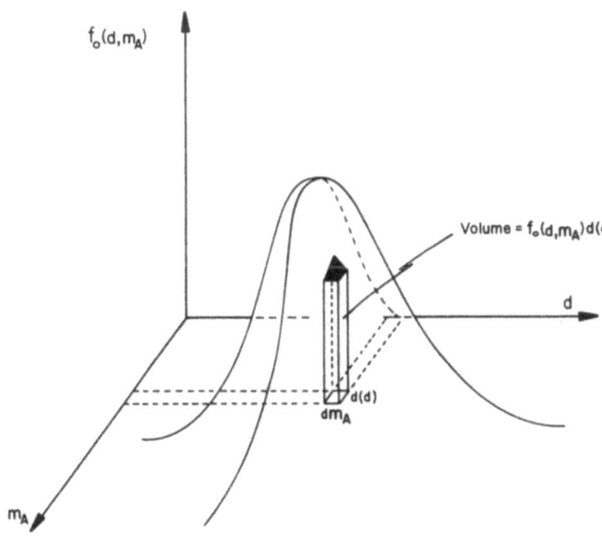

Figure 2.1-9. Graphical representation of a two-property (size and mass of valuable) density function.

figure. The distribution function is given by

$$F_0(d', m'_A) = \int_{m_{A_{\min}}}^{m'_A} \int_{d_{\min}}^{d'} f_0(d, m_A)\, d(d)\, d(m_A) \qquad (2.1\text{-}17)$$

which gives the number fraction of particles with size less than d' and mass of valuable less than m'_A. Clearly, $F_0(d_{\max}, m_{A_{\max}}) = 1.0$ (the sum of all fractions in the population).

The number size distribution (irrespective of the mass of valuable) can be recovered from $f_0(d, m_A)$ by summing over all m_A values:

$$f_0(d) = \int_{m_{A_{\min}}}^{m_{A_{\max}}} f_0(d, m_A)\, d(m_A) \qquad (2.1\text{-}18)$$

and similarly the distribution of valuable (irrespective of particle size) can be found by summing over all sizes:

$$f_0(m_A) = \int_{d_{\min}}^{d_{\max}} f_0(d, m_A)\, d(d) \qquad (2.1\text{-}19)$$

The two latter quantities are termed the *marginal* density functions for d and m_A, respectively. The mean, variance, and coefficient of variation for size and mass of valuable can be computed using equations (2.1-18) and (2.1-19) in defining equations (2.1-9), (2.1-11), and (2.1-12). The transformation from number to mass distribution can be obtained from equation (2.1-14) written for particles of size d with mass of valuable m_A.

The above procedure for representing property distributions can be extended to any number of properties of interest $(\zeta_1, \zeta_2, \ldots, \zeta_J)$, using the general joint density function $f(\zeta_1, \zeta_2, \ldots, \zeta_J)$. However, in practice it is rare to consider more than two properties simultaneously because of the complexity of the distributions.

2.1.5. Population-Balance Framework

In the preceding section, the quantitative representation of particle property distributions for single and multiple properties was discussed. In this section, a general accounting procedure will be presented which can be used to *predict* changes in property distributions for kinetic processes involving particles. We shall see that this procedure provides the basis for predicting changes in particle-size distribution resulting from grinding or agglomeration, the changes in the distribution of valuable constituents during leaching or roasting, and changes in many other distributions of importance in extractive metallurgy.

In 1964, two groups of investigators, Hulburt and Katz[1] and Randolph and Larson,[2,13] observed that many problems involving change in particulate systems could not be solved using the usual continuity of mass and rate expressions. These investigators recognized that particulate processes are unique in that the disperse phase is made up of a countable number of entities, and these entities typically possess a distribution of properties. They proposed the use of an equation for the continuity of particulate numbers, termed ***population balance***, as a basis for describing the behavior of such systems. This number balance is developed from the general conservation equation

$$\text{accumulation} = \text{input} - \text{output} + \text{net generation} \tag{2.1-20}$$

applied to particles having a specified set of properties $\{\zeta_1, \zeta_2, \ldots, \zeta_J\}$. In this balance, input and output terms represent changes in the number of particles in the specified property intervals resulting from convective flow (influx and efflux), while the generation term accounts for particles entering and leaving the specified property intervals as a result of individual particle kinetics such as chemical reaction, particle breakage, etc. Using this framework, the authors developed two forms of the population balance: a microscopic form, which accounts for changes in a particle population in an infinitesimal volume at any geometrical position, x, y, z, as a function of time, and a macroscopic form, which accounts for *average* changes in a particle population within an entire process vessel (spatial dependence suppressed). The microscopic form requires detailed vessel flow-pattern information (particle velocity profiles) for its implementation. The macroscopic form is applicable when the vessel is to a reasonable first approximation well mixed or when residence time distribution information is available. The extractive metallurgy examples which follow in the next section involve well mixed vessels, hence, only the macroscopic population balance model will be considered here.

For a rigorous derivation of the macroscopic population-balance model, the reader is referred to the original work of Hulbert and Katz[1] and Randolph and Larson.[2,13] The development that follows is a heuristic formulation which should be useful for understanding the basis of this model and its applications. Referring to Figure 2.1-10, let us assume that inside a "reactor" (grinding mill, leaching vessel, pelletizer, etc.) at time t we have a total of $\bar{n}(t)$ particles per unit volume† and that $f_0(\zeta_1, \zeta_2, \zeta_3, \ldots, \zeta_J; t)\, d\zeta_1\, d\zeta_2 \cdots d\zeta_J$ is the number fraction of these particles with property Z_1 between ζ_1 and $\zeta_1 + d\zeta_1$, property Z_2 between ζ_2 and $d\zeta_2$, etc. In abbreviated notation

$$\bar{\psi}\, dZ = \bar{n}(t) f_0(\zeta_1, \zeta_2, \ldots, \zeta_J; t)\, d\zeta_1\, d\zeta_2 \cdots d\zeta_J \tag{2.1-21}$$

† The pertinent volume can be taken as particle-free volume or volume of particles plus environment as long as volume flow rates and kinetic terms are defined in a consistent fashion.[13] Examples presented in this chapter will be based on the latter.

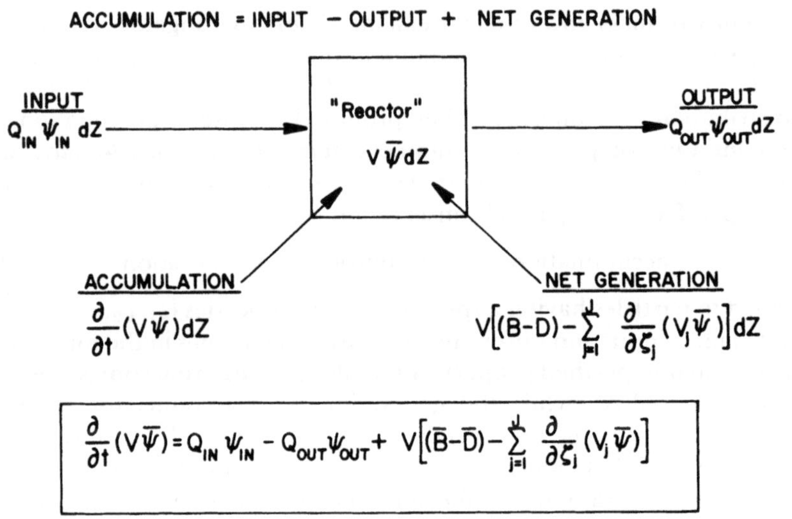

Figure 2.1-10. Pictorial representation of the macroscopic population-balance model.

is the number of particles per unit volume in the "reactor" with properties in the unspecified intervals. The accumulation of particles with properties in these intervals inside a "reactor" of volume V is, by definition, the time rate of change of particle number

$$\text{accumulation} = \frac{\partial(V\bar{\psi})}{\partial t}\, dZ \tag{2.1-22}$$

The rate at which particles in the specified property intervals are entering the "reactor" is equal to the total number of particles per unit volume in the inlet stream, n_{in}, times the volume flow rate, Q_{in}, times the fraction of particles in the inlet with properties in the specified intervals, $f_{0_{\text{in}}}(\zeta_1, \zeta_2, \ldots, \zeta_J; t)\, d\zeta_1 \cdots d\zeta_J$:

$$\text{input} = n_{\text{in}} Q_{\text{in}} f_{0_{\text{in}}}(\zeta_1, \zeta_2, \ldots, \zeta_J; t)\, d\zeta_1 \cdots d\zeta_J = Q_{\text{in}} \psi_{\text{in}}\, dZ \tag{2.1-23}$$

Similarly, the rate at which particles in the specified property intervals leave the "reactor" is

$$\text{output} = n_{\text{out}} Q_{\text{out}} f_{0_{\text{out}}}(\zeta_1, \zeta_2, \ldots, \zeta_J; t)\, d\zeta_1 \cdots d\zeta_J = Q_{\text{out}} \psi_{\text{out}}\, dZ \tag{2.1-24}$$

The net generation of particles in the specified property intervals within the reactor consists of two parts: (1) generation because of *discrete changes* in which a particle enters or leaves the property intervals as a result of a catastrophic event (e.g., particle breakage or coalescence), and

(2) generation because of *continuous changes* in which a particle enters or leaves the property intervals because of a systematic "drift" in particle properties (e.g., the drift in particle composition which occurs as a result of a leaching reaction). Net generation because of discrete changes can be defined in terms of the average rate of birth for particles in the intervals of interest, $V\bar{B}\,dZ$, minus the average death rate for particles in the intervals, $V\bar{D}\,dZ$, where V is the volume of the vessel, and $\bar{B}\,dZ$ and $\bar{D}\,dZ$ are the average birth and death rate (number per unit time per unit volume of vessel), respectively. The contribution because of continuous changes in particle properties can be represented in terms of the number flux for each property, $-(\partial/\partial\zeta_j)(v_j\bar{\psi})$; $j = 1, 2, \ldots, J$, where v_j is the time rate of change of property ζ_j for particles in the specified intervals (i.e., $v_j = d\zeta_j/dt$) which is termed the *velocity* of property j. Thus the overall net *generation* of particles in the specified property intervals in the vessel is given by

$$\text{net generation} = V\left[(\bar{B} - \bar{D}) - \sum_{j=1}^{J} \frac{\partial}{\partial\zeta_j}(v_j\bar{\psi})\right] dZ \qquad (2.1\text{-}25)$$

Substitution of equations (2.1-22)–(2.1-25) into equation (2.1-20) yields the general macroscopic population-balance model:

$$\frac{\partial}{\partial t}(V\bar{\psi}) = Q_{\text{in}}\psi_{\text{in}} - Q_{\text{out}}\psi_{\text{out}} + V\left[(\bar{B} - \bar{D}) - \sum_{j=1}^{J} \frac{\partial}{\partial\zeta_j}(v_j\bar{\psi})\right] \qquad (2.1\text{-}26)$$

In principle, equation (2.1-26) can be used to find the change in particle-property distribution in a vessel for *any* particle processing system for which the individual particle kinetics are known. In order to apply this equation, one must be able to supply expressions for $\bar{D}$, $\bar{B}$, and the $d\zeta_j/dt$, specify the relationship between $\bar{\psi}$ and ψ_{out}, and solve the resulting differential equation.

Once $\bar{\psi}$ has been obtained from equation (2.1-26), the total number of particles per unit volume of vessel and the property density function for the process of interest can be recovered by integrating equation (2.1-21) over all ζ's with the recognition that the integral of the density function over all values is unity,

$$\bar{n}(t) = \int_{\zeta_{1_{\min}}}^{\zeta_{1_{\max}}} \cdots \int_{\zeta_{J_{\min}}}^{\zeta_{J_{\max}}} \bar{\psi}\,dZ \qquad (2.1\text{-}27)$$

and substituting this result into the defining equation, equation (2.1-21), to obtain

$$f_0(\zeta_1, \zeta_2, \ldots, \zeta_J; t) = \bar{\psi}\Big/\int_{\zeta_{1_{\min}}}^{\zeta_{1_{\max}}} \cdots \int_{\zeta_{J_{\min}}}^{\zeta_{J_{\max}}} \bar{\psi}\,dZ \qquad (2.1\text{-}28)$$

If necessary, the number density function can be transformed to the corresponding mass density function and selected characteristics of the property distribution such as means and variances can be obtained by using the relationships given in Section 2.1.4.

The death term, $\bar{D}$, birth term, $\bar{B}$, and property velocity terms, v_j, required to solve the equation must be obtained from fundamental considerations concerning individual particle kinetics for the process of interest. In some instances, particularly those involving heterogeneous reaction, these terms depend directly on the concentration of one or more consumable species in the continuous phase. Thus, in leaching or roasting reactions where the time rate of change of particle composition (composition velocity) depends on the concentration of lixiviant or oxidant in the fluid phase, one must know how the fluid phase composition varies with extent of reaction. This information can be obtained from a species mass balance for the continuous phase. If one uses the general conservation equation,

$$\text{accumulation} = \text{input} - \text{output} + \text{generation}$$

to write a macroscopic mass balance for species i in the continuous phase the result is[14]

$$\frac{d(VC_i)}{dt} = Q_{\text{in}}C_{i\,\text{in}} - Q_{\text{out}}C_{i\,\text{out}} + r_i V \qquad (2.1\text{-}29)$$

where C_i, $C_{i\,\text{in}}$, and $C_{i\,\text{out}}$ are the mass concentrations of continuous-phase species i inside the vessel, and in the vessel inlet and outlet, respectively. V is the vessel volume and r_i is the rate of generation of continuous-phase species i per unit volume of vessel. Changes in the continuous phase are linked directly to the disperse phase through r_i; thus in general the continuous-phase mass balance, equation (2.1-29), must be solved simultaneously with the disperse-phase population balance, equation (2.1-26), to obtain a complete solution to a particle-processing problem.

One of the first problems solved using the macroscopic population balance approach was the prediction of the crystal size distribution and mass yield of a well-mixed crystallizer operated at steady state.[13] Crystallization problems of this type are important in such extractive metallurgical processes as alumina production by the Bayer process, selective crystallization of KCl from brine solutions, and various metal precipitation processes. This section ends with the population-balance solution to the crystallizer problem along with experimental verification of this solution. This first example of the application of population balance is presented at this point to introduce a step-by-step approach for applying this equation. This same approach will be adopted in the next section of this chapter which discusses the general application of equations (2.1-26) and (2.1-29) to batch and continuous systems with several additional examples.

Example 1. *Description of the Steady-State Behavior of a Well Mixed Crystallizer*

Problem. Derive an equation(s) that can be used to predict the mass yield, M_T, and crystal size distribution, ψ_{out}, from a well mixed crystallizer of volume V operated at steady state (see Figure 2.1-11). The feed to the crystallizer is to be a supersaturated solution of species i containing no crystals or nuclei. The nucleation rate within the crystallizer can be assumed to have a power-law dependence on supersaturation, whereas the growth rate of crystals can be assumed to follow the McCabe Δd law.[13]

Solution. The formal solution to the problem consists of four parts: (1) writing the general multiparticle model equations; (2) simplifying the general equations to suit the crystallizer problem; (3) specifying the form of property velocity (growth) and birth terms (nucleation); and (4) solving the resulting descriptive differential equations subject to appropriate side conditions to find $\psi_{out}(d)$ and M_T.

1. Write the general macroscopic population-balance (PBM) and mass-balance models (MBM).

$$\text{PBM:}\quad \frac{1}{V}\frac{\partial}{\partial t}(v\bar{\psi})+\sum_{j=1}^{J}\frac{\partial}{\partial \zeta_j}(v_j\bar{\psi})+\bar{D}-\bar{B}+\frac{1}{V}(Q_{in}\psi_{in}-Q_{out}\psi_{out}) \tag{1}$$

$$\text{MBM:}\quad \frac{1}{V}\frac{\partial}{\partial t}(C_iV)+r_i=\frac{1}{V}(Q_{in}C_{i\,in}-Q_{out}C_{i\,out}) \tag{2}$$

2. Simplify equations (1) and (2) to suit problem.
   a. The only property of interest in this case is the linear dimension (size) of a crystal, $\zeta_1 = d$.
   b. Since the crystallizer is to be operated at steady state, it follows that

$$\frac{d(V\bar{\psi})}{dt}=\frac{d(VC_i)}{dt}=0$$

   c. Assume that the volume flow rate of solution "in" is equal to the volume flow rate of slurry "out," i.e., $Q_{in} = Q_{out}$.
   d. Assume that no crystal breakage occurs in the crystallizer, i.e., $\bar{D} = 0$.

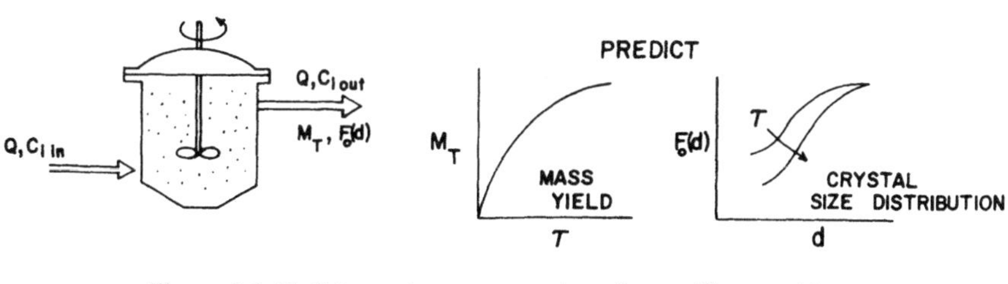

Figure 2.1-11. Schematic representation of crystallizer problem.

e. From the problem statement, assume that there are no crystals or nuclei in feed, therefore $\psi_{in} = 0$.
f. Since both the continuous and disperse phase are well mixed without preferential product removal, it follows that $\psi_{out} = \bar{\psi}$ and $C_{i\,out} = C_i$.
g. For this system, the mass rate of generation of species i, $r_i V$, is equal but opposite in sign to the mass rate of production of crystals in the crystallizer, whereas the mass rate of production of crystals is given by the volume flow rate times the mass of crystals per unit volume in the outlet. Therefore, $r_i V = -Q \int_0^\infty \rho C_3 d^3 \bar{\psi}\, d(d)$.

Considering the simplification given by a–g, equations (1) and (2) become

$$\frac{\partial}{\partial d}(v_i \bar{\psi}) - \bar{B} = -\frac{Q_{out}}{V}\bar{\psi} \tag{1a}$$

$$C_{i\,in} - C_i = \int_0^\infty \rho C_3 d^3 \bar{\psi}\, d(d) \tag{2a}$$

3. Specify the form of property velocities and birth and death rates.

Property velocities. If crystal growth occurs according to the "McCabe Δd law"[13] (rate of change of the mass of a crystal is proportional to its surface area) with a first-order dependence on the supersaturation of species i, then the rate of change of crystal size is given by

$$v_1 = \frac{d(d)}{dt} = g(C_i - C_{i\,sat}) = G \tag{3}$$

where C_i is the concentration of solute in the bulk phase, $C_{i\,sat}$ is the saturation value, and g is a growth-rate constant. Note this velocity is independent of crystal size.

Birth and death rates: In d. above it was assumed that the death rate for crystals is zero. The only particles which are born in the system *are nuclei* of infinitesimal size, $d = 0^+$. If it is assumed that the rate of crystal nuclei generation is proportional to the nth power of the solution supersaturation,[13] then the birth rate, B, for particles of size d is given by

$$\bar{B} = b(C_i - C_{i\,sat})^n \cdot \delta(d - 0^+) = B_0\, \delta(d - 0^+) \tag{4}$$

where $\delta(d - 0^+)$ is the unit impulse† at $d = 0^+$.

4. Solve equations subject to side conditions.

Substitution of equations (3) and (4) into equation (1a) gives the first-order nonhomogeneous differential equation

$$G\frac{\partial \bar{\psi}}{\partial d} + \frac{Q_{out}}{V}\bar{\psi} - B_0\, \delta(d - 0^+) = 0 \tag{1b}$$

This equation can be solved by the Laplace transformation method or with the integrating factor method with the factor exp $(d/G\tau)$,[15] subject to the condition that there can be crystals with $d < 0$,

$$\bar{\psi}(d) = \left(\frac{B_0}{G}\right) \exp\left[-\frac{d}{G(Q/V)}\right] \tag{5}$$

† It is important to recognize the following properties of the unit impulse at $d = d'$: $\delta(d - d') = 0$ for $d \neq d'$; $\int_a^b \delta(d - d')\, d(d) = 1$, and $\int_a^b \delta(d - d') f(d)\, d(d) = f(d')$ for $d' \in [a, b]$.[15]

By applying equations (2.1-27) and (2.1-28) to equation (5) for a mean retention time, $\tau = V/Q$, it follows that the total number of crystals per unit volume in the outlet is given by

$$\bar{n} = B_0\tau \tag{6}$$

and the number density function for crystal size is given by an exponential distribution†

$$f_0(d) = \frac{1}{G\tau}\exp\left[-\frac{d}{G\tau}\right] \tag{7}$$

Since both B_0 and G depend on the supersaturation $(C_i - C_{i\,\text{sat}})$, it is apparent that the mass balance [equation (2a)] must also be solved to complete the solution. Substitution of equation (5) into equation (2a) and performing the indicated integration yields

$$C_{i\,\text{in}} - C_i = 6\rho C_3 b g^3 \tau^4 (C_i - C_{i\,\text{sat}})^{n+3} \tag{8}$$

Knowing the values of solid density, ρ, volume shape factor for the crystals, C_3, the saturation concentration for the solute, $C_{i\,\text{sat}}$, and the nucleation and growth-rate constants, b, n, and g (see Randolph and Larson[13] for methods of estimating these constants), this nonlinear algebraic equation can be solved for a given retention time, τ, to obtain C_i. Evaluation of $G = g(C_i - C_{i\,\text{sat}})$ allows the prediction of the crystal size distribution according to equation (7) and the mass yield can be obtained from:

$$M_T = Q(C_{i\,\text{in}} - C_i) \tag{9}$$

Experimental Verification. Figures 2.1-12a and b show two tests of equation (5), NaCl and $(NH_4)_2SO_4$ crystallizations, respectively; for different mean residence times in a well mixed crystallizer.[16] Note that in each case the size distribution data conform quite closely to the straight line predicted by equation (5). Randolph and Larson[13] present additional conformation of this crystallizer model and give several examples of the use of population-balance models for the steady-state and dynamic simulation of crystallizer performance.

2.1.6. *Application of Population Balance to Extractive Metallurgical Systems*

The purpose of this section is to demonstrate the power and versatility of the population-balance approach in describing the behavior of extractive metallurgical systems. This is done through a series of example applications to diverse types of systems. In each application the problem is stated, the population-balance and mass-balance models are simplified to suit the problem, appropriate expressions for individual particle kinetics are introduced, and the resulting equations are solved. Wherever possible, experimental verification of model predictions is presented. Detailed mathematical and physical arguments are for the most part omitted; for these the reader is referred to the cited literature.

† See Table 2.1-2, gamma distribution, $p = 1$

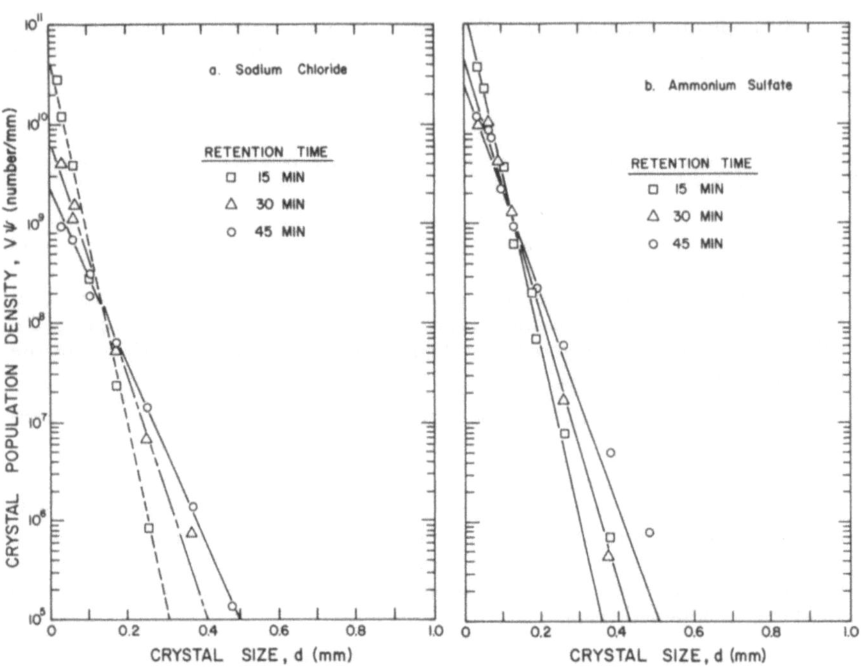

Figure 2.1-12. Experimental verification of model predictions of crystal size distributions in a continuous crystallizer[16] for (a) NaCl and (b) $NH_4)_2SO_4$.

Batch Processes

For batch processes there is no flow of material into or out of the "reaction" vessel ($Q_{\text{in}} = Q_{\text{out}} = 0$), hence, the general macroscopic population-balance model, equation (2.1-26), reduces to

$$\frac{1}{V}\frac{\partial(V\bar{\psi})}{\partial t} + \sum_{j=1}^{J} \frac{\partial}{\partial \zeta_j}(v_j\bar{\psi}) + \bar{D} - \bar{B} = 0 \qquad (2.1\text{-}30)$$

Substitution of appropriate expressions for the kinetics of property changes, v_j, and the birth and death terms, $\bar{B}$ and $\bar{D}$, into equation (2.1-30) and solving the resulting partial differential equation yields an expression for the evolution of any initial property distribution, $\bar{\psi}_0$, with time in a batch system. In instances where the expressions for v_j or $\bar{B}$ and $\bar{D}$ depend upon the concentration of some consumable "reactant" in the fluid phase, a mass balance for the consumable species must be solved simultaneously with the PBM. For batch conditions owing to the no-flow-in/no-flow-out condition, the mass-balance model, equation (2.1-29), reduces to

$$\frac{\partial(VC_i)}{\partial t} = Vr_i \qquad (2.1\text{-}31)$$

Three examples of the application of equations (2.1-30) and (2.1-31) to metallurgical systems are given below. In the first, it is desired to predict the size distribution of particles undergoing size reduction in a batch ball mill. In this case, there are no continuous changes in size (e.g., no attrition) hence $v_j = 0$, but the form of birth and death terms associated with particle breakage must be specified to solve the problem. In the second example, it is desired to predict the equilibrium size distribution of droplets in a solvent extraction mixer; once again there are no continuous changes in particle size and it is only necessary to specify the form of the birth and death terms for droplet coalescence and breakage. This second example demonstrates the importance of particle–particle interactions in some systems. In the third example, the general problem of predicting batch heterogeneous reaction behavior is discussed. Here, continuous changes in the amount of solid reactant remaining in the particles are of prime importance and the velocity associated with such changes must be specified. In this example, a mass balance for the fluid phase reactant is required to treat particle–fluid interactions.

Example 2. *Description of Batch Grinding in a Ball Mill*

Problem. Derive an equation that can be used to predict the evolution of the size distribution of a ground product produced in a batch ball mill (see Figure 2.1-13). The feed to the ball mill is to have an arbitrary size distribution, $f_3(d, t)|_{t=0}$. Assume that the material in the ball charge is well mixed and that the kinetics of breakage are linear.[17]

Solution.

1. Write the batch macroscopic population-balance and mass-balance equations, equations (2.1-30) and (2.1-31).

$$\text{PBM:}\quad \frac{\partial}{\partial t}(V\bar{\psi}) = V\bar{B} - V\bar{D} - V\sum_{j=1}^{J}\frac{\partial}{\partial \zeta_j}(v_j\bar{\psi}) \tag{1}$$

MBM: not required

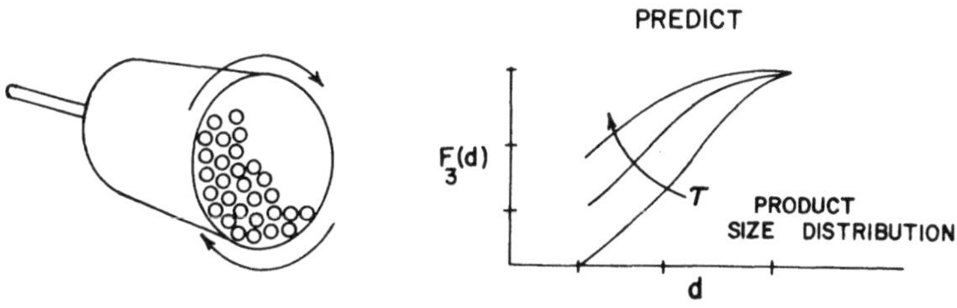

Figure 2.1-13. Schematic representation of a batch ball-milling problem.

2. Simplify equation (1) to suit the problem.
   a. Only one property of interest, i.e., particle size d.
   b. No continuous change in size $d(d)/dt = 0$ (no attrition).

Considering a and b, equation (1) becomes

$$\frac{\partial}{\partial t}[V\bar{\psi}(d, t)] = V\bar{B}(d) - V\bar{D}(d) \tag{1a}$$

3. Specify the form of property velocities and birth and death rates.

Birth and death rates

a. If the kinetics of breakage are linear, the breakage rate (death rate) for particles of size d is proportional to the number of that size present[18,19]; therefore

$$V\bar{D}(d)\, d(d) = S(d)\, V\bar{\psi}(d, t)\, d(d) \tag{2}$$

where the quantity $S(d)$ is the fractional rate of breakage for particles of size d.
b. The production rate (birth rate) arises from the production of fragments of size d from the breakage of larger particles. Since each breakage event involving a particle d' ($d' > d$) produces $b^0(d', d)\, d(d')$ fragments of size d, and from equation (2) there are $S(d')\, V\bar{\psi}(d', t)\, d(d')$ such events occurring per unit time and we must sum over breakage events for all particles $d' > d$, it follows that the birth rate for particles of size d can be expressed as

$$V\bar{B}(d)\, d(d) = \int_d^{d_{max}} b^0(d, d') S(d')\, V\bar{\psi}(d', t)\, d(d')\, d(d) \tag{3}$$

4. Solve equations subject to side conditions. Substitution of equations (2) and (3) into equation (1a) yields the integrodifferential equation

$$\frac{\partial}{\partial t}\bar{\psi}(d, t) = -S(d)\bar{\psi}(d, t) + \int_d^{d_{max}} b^0(d, d') S(d') \bar{\psi}(d', t)\, d(d') \tag{4}$$

Since size distributions in grinding are normally expressed as mass distributions rather than number distributions, by recognizing that $V\bar{\psi}(d, t) = Nf_0(d, t)$ and applying the transformation given by equation (2.1-14) to equation (4), one obtains

$$\frac{\partial f_3(d, t)}{\partial t} = -S(d) f_3(d, t) + \int_d^{d_{max}} b(d, d') S(d) f_3(d', t)\, d(d') \tag{5}$$

where $b(d, d') = (d/d')^3 b^0(d, d')$ is the mass density function for the primary breakage of particles of size d'.

Several methods are available for solving equation (5) subject to the initial condition $f_3(d, t)|_{t=0} = f_3(d, 0)$. An especially simple form of the solution arises for the special case $S(d')b(d, d') = \alpha k_0 d^{\alpha-1}$. Substitution of this expression in equation (5) yields a cumulative distribution of the form

$$F_3(d, t) = 1 - [1 - F_3(d, 0)] \exp[-k_0 d^{\alpha} t] \tag{6}$$

It has been shown that this solution reduces to several of the standard empirical energy-size reduction models for specific values of α and $F_3(d, 0)$.[20,21] This two-parameter model (α and k_0) is not as accurate as the general solution to equation (5) but often provides a reasonable first approximation for predicting size distribution changes in a batch mill.

One of the most efficient general solutions comes from discretizing the size variable d (into Tyler $2^{1/2}$ sieve intervals) to produce a set of ordinary differential equations, one for each size fraction (f_i, $i = 1, \ldots, n$) in the assembly

$$\frac{df_i(t)}{dt} = -S_i f_i(t) + \sum_{j=1}^{i-1} b_{ij} S_j f_j(t) \tag{7}$$

In this case, quantities S_i and b_{ij} (termed the *size discretized selection* and *breakage functions*) are interval-averaged values of $S(d)$ and $b(d, d')$.[21-23] When S_i and b_{ij} are independent of size distribution changes in the mill, then equation (6) can be solved subject to the initial condition, $f_i(0) = f_{i0}$ to obtain[24]

$$f_i(t) = \sum_{j=1}^{i} A_{ij} e^{-S_j t} \tag{8}$$

where

$$A_{ij} = \begin{cases} 0 & i < j \\ \displaystyle\sum_{k=j}^{i-1} \frac{b_{ik} S_k}{S_i - S_j} A_{kj} & i > j \\ \displaystyle f_{i0} - \sum_{k=1}^{i-1} A_{ik} & i = j \end{cases}$$

The parameters S_i and b_{ij} can be measured directly from narrow size fraction feed experiments, tracer tests, or from standard grinding tests using nonlinear regression.[25]

Experimental Verification. Grinding models are the most highly developed form of population-balance models in use in mineral processing and extractive metallurgy today. The literature is replete with data verifying the population-balance grinding models. Figure 2.1-14a–c shows examples in which S_i and b_{ij} values obtained by a simple experimental scheme[26] have been used in conjunction with equation (6) to *predict* the size distribution produced for the grinding of various feed size distributions of dolomite in a 10 in. by 11.5 in. batch ball mill. Note that in each case the predictions are in good agreement with experimentally observed product size distributions.

Example 3. *Description of Droplet Size Distributions Produced in an SX Mixer*

Problem. Extraction efficiency in a solvent extraction mixer–settler unit is strongly dependent on the droplet size distribution of the disperse phase (see Figure 2.1-15). Derive an equation which can be used to predict the *equilibrium* droplet size distribution, $f_0(\nu)$, produced in a batch mixer. In this derivation assume that the kinetics of droplet break-up are linear and that droplet coalescence occurs in a strictly random fashion. In addition, assume that the vessel contains an arbitrary volume fraction of disperse phase, δ, and that the phases are well mixed throughout the vessel volume.

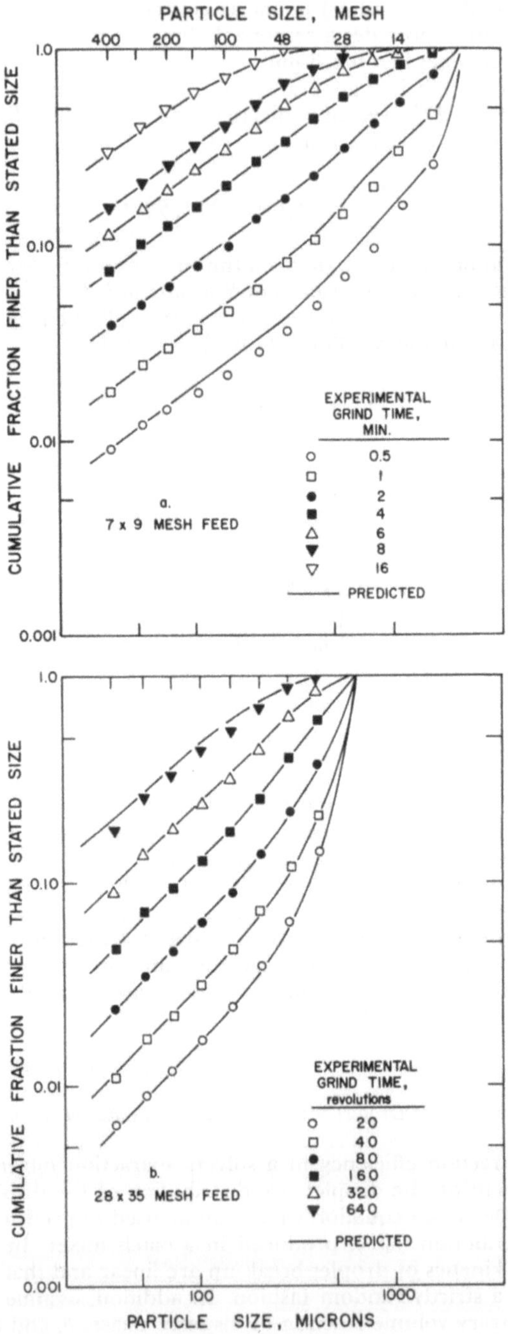

Figure 2.1-14a, b.

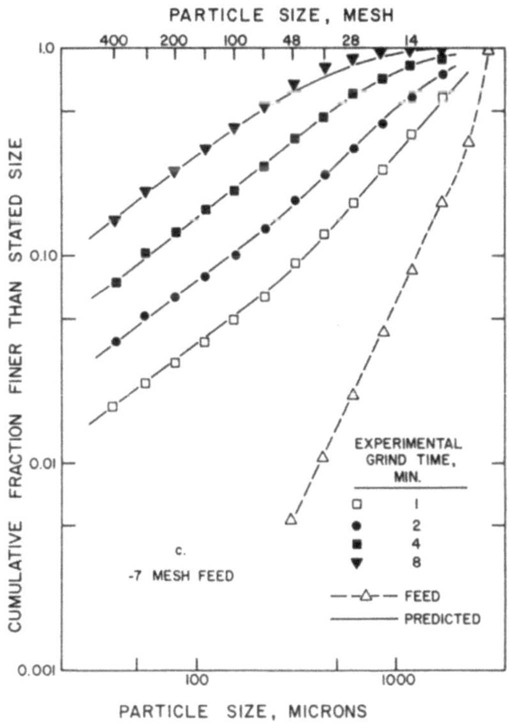

Figure 2.1-14. Experimental verification of model predictions of product size distributions produced from a variety of dolomite feeds in a batch ball[26]: (a) 7×9 mesh feed; (b) 28×35 mesh feed; (c) −7 mesh feed.

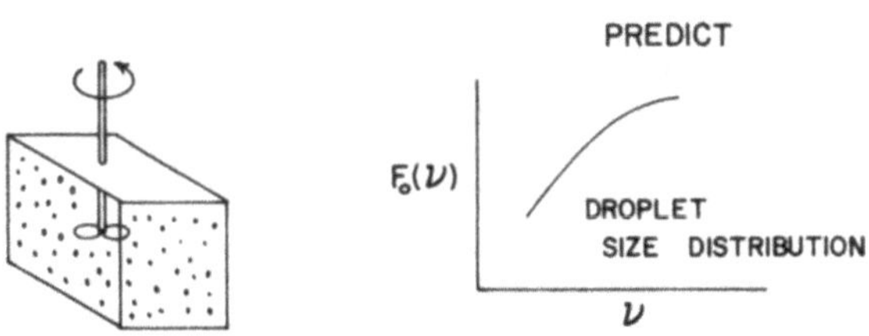

Figure 2.1-15. Schematic representation of batch reactor problem.

Solution

1. Write the general batch population-balance and mass-balance equations, equations (2.1-30) and (2.1-31).

$$\text{PBM:}\quad \frac{1}{V}\frac{\partial}{\partial t}(V\bar{\psi})+\sum_{j=1}^{J}\frac{\partial}{\partial \zeta_j}(v_j\bar{\psi})+\bar{D}-\bar{B}=0 \tag{1}$$

MBM: not required

2. Simplify equation (1) to suit problem.
   a. One property of droplets of interest is the droplet volume, $J=1$, and $\zeta_1=\nu$.
   b. Equilibrium conditions prevail in the vessel, $\partial(V\bar{\psi})/\partial t=0$.
   c. No continuous changes in droplet volumes occur, $\dot{\zeta}_1=d\nu_1/dt=0$.
   d. Death events for droplets of volume ν occur as a result of both breakage and coalescence events, $\bar{D}(\nu)=\bar{D}_B(\nu)+\bar{D}_C(\nu)$; similarly, birth events occur as a result of both breakage and coalescence, $\bar{B}(\nu)=\bar{B}_B(\nu)+\bar{B}_C(\nu)$.

Considering the simplifications a–d, equation (1) becomes

$$\bar{D}_B^{(\nu)}+\bar{D}_C^{(\nu)}-\bar{B}_B^{(\nu)}-\bar{B}_C^{(\nu)}=0 \tag{1a}$$

3. Specify the form of property velocities and birth and death rates.

Birth and death rates. For linear breakage kinetics the rate of breakage of drops of volume ν is proportional to the number of drops of that size present. It follows from Example 2 that the disappearance and appearance rates for droplets of volume ν resulting from droplet break-up are

$$\bar{D}_B^{(\nu)}=S(\nu)\bar{\psi}(\nu)$$

$$\bar{B}_B^{(\nu)}=\int_{\nu}^{\infty}S(\nu')b^0(\nu,\nu')\bar{\psi}(\nu')\,d(\nu')$$

For random coalescence the rate of coalescence between two sizes of drop is proportional to the number of each size present but independent of their sizes. In this case, the disappearance and appearance rates for droplets of volume ν resulting from droplet coalescence are given by[27,28]

$$\bar{D}_C^{(\nu)}=\lambda\bar{\psi}(\nu)$$

$$\bar{B}_C^{(\nu)}=\frac{\lambda}{2}\frac{\int_0^{\nu}\bar{\psi}(\nu-\nu')\bar{\psi}(\nu')\,d\nu'}{\int_0^{\infty}\bar{\psi}(\nu)\,d\nu}$$

It should be noted that the coalescence process, in contrast to the breakage process, is inherently nonlinear. The convolution integral (the birth term) clearly shows that a strong particle–particle interaction exists in this system since the rate at which drops of a given volume, v, are produced by coalescence depends on the product of two density functions, $\bar{\psi}(\nu-\nu')$ and $\bar{\psi}(\nu')$.

Substitution of the above expressions for death and birth rates into equation (1a) gives the nonlinear integral equation

$$[S(\nu)+\lambda]\bar{\psi}(\nu)-\int_{\nu}^{\infty}S(\nu')b^0(\nu,\nu')\bar{\psi}(\nu')\,d\nu'-\frac{\lambda}{2}\frac{\int_0^{\nu}\bar{\psi}(\nu-\nu')\bar{\psi}(\nu')\,d\nu'}{\int_0^{\infty}\bar{\psi}(\nu)\,d\nu}=0 \tag{2}$$

4. Solve equation (2) subject to side conditions.

A general solution to equation (2) has not been identified. However, if one assumes certain specific forms for $S(v)$ and $b^0(v, v')$, a solution can be found that agrees with experimental observations. In particular, if the fractional rate of drop break up for a given size drop is proportional to its volume to the nth power,[29,30] i.e.,

$$S(\nu) = k\nu^n, \qquad \nu \geq 0 \tag{3}$$

and the daughter droplets produced from the breakage of a droplet of volume ν' obey a uniform distribution, i.e.,

$$b^0(\nu, \nu') = 2/\nu', \qquad 0 \leq \nu \leq \nu' \tag{4}$$

then the solution to equation (2) for $n = 1$ is

$$\bar{\psi}(\nu) = \delta(2k^2/\lambda)e^{-(2k/\lambda)\nu} \tag{5}$$

where δ is the volume fraction of disperse phase in the system, i.e.,

$$\delta = \frac{\text{total volume of drops}}{\text{total volume of system}} = \int_0^\infty \nu\bar{\psi}(\nu)\, d\nu \tag{6}$$

The reader should confirm that equation (5) is a solution by direct substitution into equation (2). The number density function, $f_0(v)$, can be found from equations (1) and (2.1-28) to be an exponential distribution,

$$f_0(\nu) = (1/\nu_0)e^{-\nu/\nu_0} \tag{7}$$

with a mean droplet volume equal to $\nu_0 = \lambda/2k$. Equation (7) can, in turn, be used to calculate the equilibrium specific surface area of the disperse phase, $\bar{a}$, in terms of the ratio of the breakage rate parameter, k, to the coalescence rate parameter, λ, as follows:

$$\bar{a} = \frac{\text{surface area}}{\text{volume of disperse phase}} = \int_0^\infty a_\nu f_0(\nu)\, d\nu = 4.37\left(\frac{2k}{\lambda}\right)^{1/3}$$

Solutions to equation (2) for values of n different from unity, equation 3, must in general be obtained numerically.

Experimental Verification. Figure 2.1-16 shows two sets of droplet size distribution data obtained by Brown and Pitt[31] for a kerosene–water system agitated with a flat blade impeller at a speed of 250 rpm. Also shown in Figure 2.1-16 is the number density function predicted by equation (7) with $v_0 = 4 \times 10^{-6}\ \text{cm}^3$ (corresponding to a mean diameter $d_0 = 0.0197$ cm). Considering the rather large spread in the data for small sizes, the agreement between the experimental data and equation (7) is quite satisfactory.

Example 4. *Description of Batch Heterogeneous Reaction Systems*

Problem. Derive an equation(s) that can be used to predict the overall extent of conversion of A in an assembly, $\alpha(t)$, of particles undergoing the reaction

$$a\,\mathrm{A}_{\text{solid}} + b\,\mathrm{B}_{\text{fluid}} \rightarrow c\,\mathrm{C}_{\text{solid}} + \text{fluid products}$$

in a well mixed, isothermal batch reactor. The batch reactor is charged with an arbitrary feed size distribution and arbitrary amounts of solid- and fluid-phase

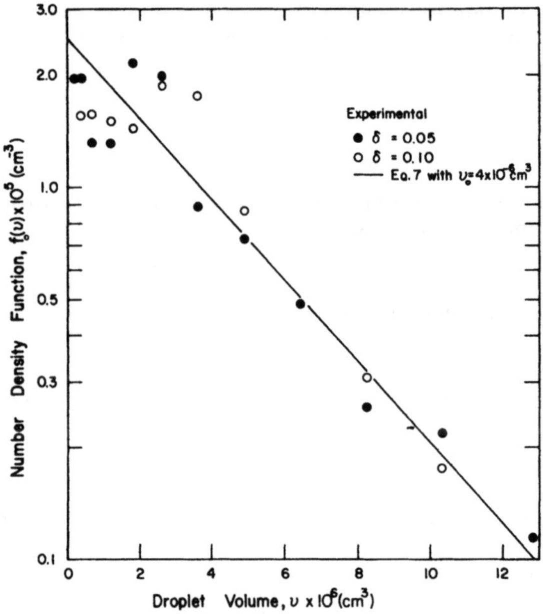

Figure 2.1-16. Experimental verification of model predictions of equilibrium droplet size distribution produced in a batch mixer.[31]

reactants (see Figure 2.1-17). Assume that the intrinsic reaction kinetics for individual particles are known, including the size and fluid-phase reactant concentration dependence.

Solution.

1. Write the batch macroscopic population-balance and mass-balance models, equations (2.1-30) and (2.1-31):

$$\text{PBM:} \quad \frac{\partial}{\partial t}(V\bar{\psi}) = V\bar{B} - V\bar{D} - V\sum_{j=1}^{J}\frac{\partial}{\partial \zeta_j}(v_j\bar{\psi}) \tag{1}$$

$$\text{MBM:} \quad \frac{\partial(VC_i)}{\partial t} = Vr_i \tag{2}$$

2. Simplify equations (1) and (2) to suit the problem.
   a. Only one property of interest, mass of A remaining in a particle, m_A, thus $\bar{\psi} = \bar{\psi}(m_A, t)$ and $v_1 = dm_A/dt$.
   b. No breakage or coalescence of particles occurs, $\bar{B} = \bar{D} = 0$.

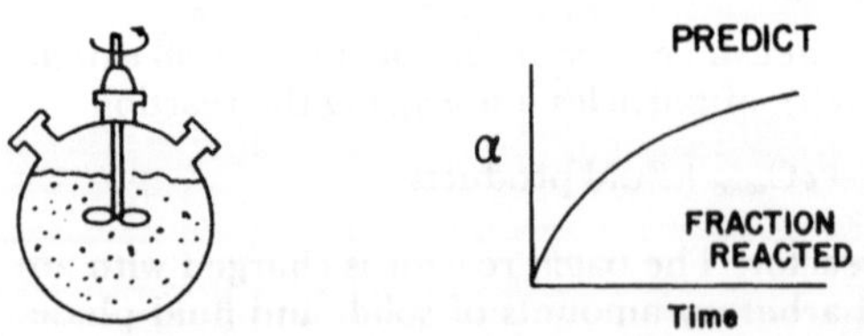

Figure 2.1-17. Schematic representation of two phase mixer problem.

c. Species B is consumed from fluid phase at a rate that is equal to (b/a) times the rate of consumption of solid phase reactant A, i.e.,

$$Vr_i = -\left(\frac{b}{a}\right) V \frac{\partial}{\partial t}\left[\int_0^{m_{A_{\max}}} m_A \bar{\psi}(m_A, t)\, dm_A\right]$$

Considering the simplifications a–c, equations (1) and (2) become

$$\frac{\partial}{\partial t}[V\bar{\psi}(m_A, t)] = -V \frac{\partial}{\partial m_A}[v_1 \bar{\psi}(m_A, t)] \tag{1a}$$

$$\frac{\partial (VC_i)}{\partial t} = -\left(\frac{b}{a}\right) V \frac{\partial}{\partial t}\left[\int_0^{m_{A_{\max}}} m_A \bar{\psi}(m_A, t)\, dm_A\right] \tag{1b}$$

3. Specify the form of property velocities and birth and death rates.

Property velocities: The time rate of change of m_A in a particle of initial size d is

$$v_1 = dm_A/dt$$

This velocity can be expressed in terms of the fraction reacted for *a particle* of initial size d at time t when exposed to a fluid-phase reactant concentration, C_B, $\alpha(d, C_B, t)$, by recognizing that

$$m_A = [1 - \alpha(d, C_B, t)]\rho_A C_3 d^3 \tag{3}$$

Specific expressions for intrinsic kinetics, $\alpha(d, C_B, t)$, can be obtained by the methods described in Chapter 1.

4. Solve equations subject to side conditions. The formal solution to the first-order partial differential equation (PDE) (equation la) for an initial distribution $\bar{\psi}(m_A, 0)$ is:

$$\bar{\psi}(m_A, t)\, dm_A = \bar{\psi}(m_{A_0}, 0)\, dm_{A_0} \tag{4}$$

where $m_{A_0} = m_A - \int_0^t v_1\, dt$ and

$$VC_B(t) - VC_B(0) = -(b/a)V\left[\int m_A \bar{\psi}(m_A, t)\, dm_A - \int m_A \bar{\psi}(m_A, 0)\, dm_A\right] \tag{5}$$

Equations (4) and (5) can be solved simultaneously to obtain $C_B(t)$ and the number distribution of solid reactant A in the particles $\bar{\psi}(m_A, t)$. The problem asks for the overall conversion of A in the assembly of particles as a function of time, i.e.,

$$\alpha^{(t)}_{\text{overall}} = \frac{V\int_{m_{A_{\min}}}^{m_{A_{\max}}} m_A \bar{\psi}(m_A, 0)\, dm_A - V\int_0^{m_{A_{\max}}} m_A \bar{\psi}(m_A, t)\, dm_A}{v\int_{m_{A_{\min}}}^{m_{A_{\max}}} m_A \bar{\psi}(m_A, 0)\, dm_A} \tag{6}$$

If we substitute equation (4) into equation (6) along with equation (3) for m_A and transform from a number distribution to a mass distribution via equation (2.1–14), then equation (6) becomes simply

$$\alpha_{\text{overall}} = \int_{d_{\min}}^{d_{\max}} \alpha(d_0, C_B, t) f_3(d, 0)\, d(d) \tag{7}$$

Hence the overall conversion of A for material with feed size distribution $f_3(d, 0)$

can be predicted as long as we know the conversion for the individual particles of original size d for any time t. The latter quantity, $\alpha(d, C_B, t)$ depends in general on the concentration of fluid-phase reactant B at time t. This value can be obtained from the mass balance, equation (5), written in the form

$$C_B(t)V = C_{B_0}V - (b/a)[\alpha^{(t)}_{\text{overall}} M_{A_0}] \tag{8}$$

where M_{A_0} is the total number of moles of A initially present in the system. Alternatively, equation (8) can be expressed in terms of the molar ratio defined by equation (2.1-4), i.e.,

$$\eta = \frac{a}{b}\frac{VC_{B_0}}{m_{A_0}}$$

as

$$C_B(t)/C_{B_0} = 1 - \alpha_{\text{overall}/\eta} \tag{8a}$$

If η is small, $\alpha(d, C_B, t)$ must in general be evaluated by (a) numerical integration of an expression for $(d\alpha/dt)(d, C_B, t)$ for each size for a small time interval; (b) α_{overall} computed from equation (7); and (c) the reduction in concentration of B calculated from equation (8a), and the procedure must be repeated. If η is large, the mass balance need not be considered, and $C_B(t) = C_{B_0}$ and $\alpha(d, C_{B_0}, t)$, which can frequently be written as simple algebraic expression, can be inserted directly into equation (7) and the indicated integration performed. If a density function $f_3(d, 0)$ is not available, but size distribution information is available as weight fractions in N size intervals, $f_i = \int_{d_{i+1}}^{d_i} f_3(d, 0)\, d(d)$, $i = 1, \ldots, N$, then equation (7) can be approximated by

$$\alpha_{\text{overall}} = \sum_{i=1}^{N} \alpha(d_i, C_B, t) f_i \tag{7a}$$

where d_i^* is the average size for feed particles in the ith size interval.

Experimental Verification. Numerous examples of the application of this method to predicting the batch leaching behavior of a size distribution using monosize kinetics have been reported in the literature. One example is depicted in Figures 2.1-18a–c. In this case, chalcopyrite concentrate is being leached in an acid ferric sulfate solution. As shown in Figure 2.1-18a, the kinetics for monosize feeds are product layer diffusion controlled (through the sulfur product layer). This material was ground extremely fine in an attritor (50% finer than 0.5 μm), as shown in Figure 2.1-18b and the kinetics of ground product leaching were predicted using the kinetic parameters obtained from the monosize data[32] in conjunction with equation (7a). In this case $\alpha(d_i, C_B, t)$ can be expressed as

$$\alpha = 1 - \left[\frac{1}{2} + \cos\left(\frac{3\theta + 4\pi}{3}\right)\right]^3$$

where

$$\theta = \frac{1}{3}\cos^{-1}\left(\frac{2t}{\tau} - 1\right) \quad \text{and} \quad \tau = \frac{d^2 \rho_A}{24 D_{B,e} C_B}$$

The number of size intervals, N, used for the prediction was 20. Experimental and predicted values are compared in Figure 2.1-18c.

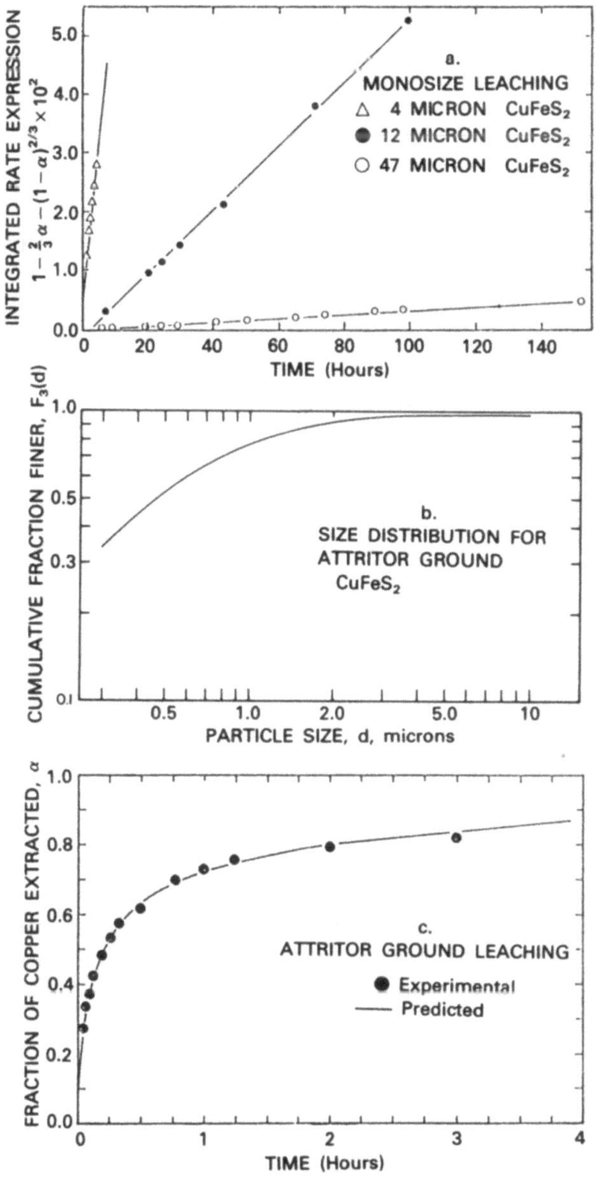

Figure 2.1-18. Experimental verification of model predictions of copper extraction from attritor ground chalcopyrite. Diffusion coefficient determined from single size fraction data (a) was used in conjunction with size distribution of attrition milled material (b) to predict the batch leaching behavior of attritor ground product according to equation (7a). Comparison of values (c).

Continuous Processes at Steady State

For continuous systems operating at steady state the accumulation of particles in the vessel is zero. In this case equation (2.1-26) reduces to

$$Q_{in}\psi_{in} - Q_{out}\psi_{out} + V\left[(\bar{B} - \bar{D}) - \sum_{j=1}^{J} \frac{\partial}{\partial \zeta_j}(v_j \bar{\psi})\right] = 0 \qquad (2.1\text{-}32)$$

Once again appropriate expressions for v_j and $\bar{B}$ and $\bar{D}$ must be substituted into equation (2.1-32) and the inlet property distribution ψ_{in} must be specified. In addition, it is necessary to specify from physical considerations the way in which ψ_{out} relates to the average distribution in the vessel $\bar{\psi}$. As in the batch case, if the kinetics v_j, $\bar{B}$, and $\bar{D}$ depend on the concentration of a consumable "reactant," a mass balance for the consumable reactant of the form

$$Q_{in}C_{i_{in}} - Q_{out}C_{i_{out}} + Vr_i = 0 \qquad (2.1\text{-}33)$$

must be solved simultaneously with the population-balance model.

In order to use equations (2.1-32) and (2.1-33) to predict the outlet conditions for a vessel at steady state, one of the two following conditions must prevail:

1. The disperse phase and continuous phase must be well mixed with a discharge composition identical to that in the vessel; in which case $\psi_{out} = \bar{\psi}$ and $C_{i_{out}} = C_i$ and the number of unknowns in the pair of equations is reduced to two. This is the classical "perfectly mixing" assumption applied to a two-phase system.

2. If perfect mixing cannot be assumed the changes which occur in the disperse phase must obey *linear kinetics* with the disperse phase completely entrained in the continuous phase. In this case equations (2.1-32) and (2.1-33) can be used to show that the outlet disperse-phase property distribution and continuous phase concentration can be obtained as weighted averages of the batch responses weighted with respect to the distribution of residence times for material in the vessel,[33,34] i.e.,

$$\psi_{out} = \int_0^{\infty} \psi_{batch}(t)E(t)\,dt \qquad (2.1\text{-}34)$$

and

$$C_{i_{out}} = \int_0^{\infty} C_{i_{batch}}(t)E(t)\,dt \qquad (2.1\text{-}35)$$

In equations (2.1-34) and (2.1-35), $\psi_{batch}(t)$ represents the distribution of particle properties which would be obtained in the process vessel operated in a batch fashion for a time t and $C_{i_{batch}}(t)$ represents the corresponding batch concentration for species i in the continuous phase. (See preceding section). The quantity $E(t)$ is the experimentally determinable residence-time distribution function, i.e., $E(t)\,dt$ gives the fraction of material in the exit stream of the vessel (disperse and continuous phases) which spends a time t to $t+dt$ in the vessel. In this case, equations (2.1-34) and (2.1-35) can be used to obtain descriptions of continuous steady-state behavior.

Two examples of the application of equations (2.1-32) through (2.1-35) to metallurgical systems are presented below. In the first example, an equation is derived to predict the size distribution produced in a continuous open-circuit ball mill. In the second example, equations are derived to predict the continuous leaching behavior of material whose intrinsic reaction kinetics are governed by a generalized shrinking-core model. Furthermore, methods of estimating intrinsic kinetic parameters from batch data are illustrated and the usefulness of this model is demonstrated through a series of computer simulations for specific systems.

Example 5. *Description of Continuous Open-Circuit Ball Milling*

Problem. Derive an equation which can be used to predict the steady-state product-size distribution from a well mixed, open-circuit ball mill (see Figure 2.1-19). The mill is to be fed an arbitrary feed-size distribution, $f_3(d)_{in}$, with a feed rate M_F. The kinetics of breakage are to be assumed to be linear. Solve this problem by beginning (a) with the general macroscopic population balance model and (b) with the residence time distribution form of the model, equation (2.1-34).

Solution

1. Write the continuous, steady-state macroscopic population-balance and mass-balance models:

$$\text{PBM:}\quad Q_{in}\psi_{in} - Q_{out}\psi_{out} + V\left[(\bar{B} - \bar{D}) - \sum_{j=1}^{J} \frac{\partial}{\partial \zeta_j}(v_j\bar{\psi})\right] = 0 \qquad (1)$$

MBM: not required

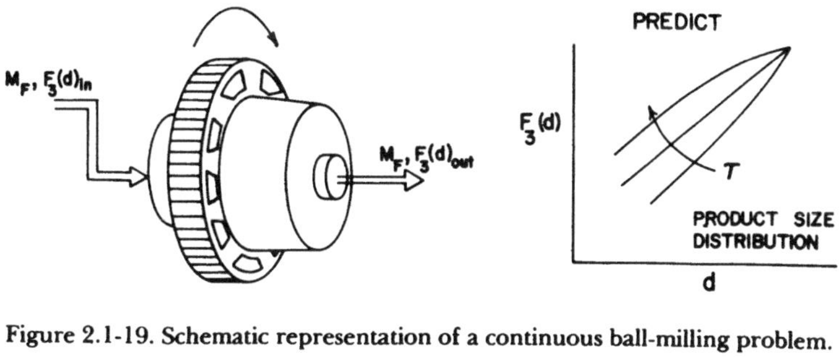

Figure 2.1-19. Schematic representation of a continuous ball-milling problem.

2. Simplify equation (1) to suit problem.
   a. Only one property of interest, i.e., particle size, $\zeta_1 = d$.
   b. No continuous changes in size, $v_1 = d(d)/dt = 0$.
   c. Flow rates of pulp into and out of the mill are identical, $Q_{\text{in}} = Q_{\text{out}} = Q$.
   d. Contents of mill well mixed and product removal from the mill is nonselective (with respect to d) $\psi_{\text{out}}(d) = \bar{\psi}(d)$. Considering a–d, equation (1) becomes

$$Q\psi_{\text{in}}(d) - Q\bar{\psi}(d) + V[\bar{B}(d) - \bar{D}(d)] = 0 \tag{1a}$$

3. Specify the form of property velocities and birth and death rates.
For linear grinding kinetics, the birth and death rates are as specified in Example 2, i.e.,

$$V\bar{D}(d) \quad S(d)\bar{\psi}(d) \tag{2}$$

$$V\bar{B}(d) = \int_d^{d_{\max}} b^0(d, d')S(d')\bar{\psi}(d')\, d(d') \tag{3}$$

4. Solve equations subject to side conditions. Substitution of equations (2) and (3) into equation (1a) gives the integral equation

$$Q\psi_{\text{in}}(d) - Q\bar{\psi}(d) + V\int_d^{d_{\max}} b^0(d, d')S(d')\bar{\psi}(d')\, d(d') - VS(d)\bar{\psi}(d) = 0 \tag{4}$$

As in Example 2, this number balance can be converted to a mass balance by multiplying equation (4) by $\rho C_3 d^3$ using equation (2.1-14) and recognizing that $\rho C_3 d^3 Q\bar{\psi}(d) = M_F f_3(d)$ and $\rho C_3 d^3 V\bar{\psi}(d) = Hf_3(d)$, i.e.,

$$M_F f_{3_{\text{in}}}(d) - M_F f_3(d) + H\left[\int_d^{d_{\max}} b(d, d')S(d')f_3(d')\, d(d') - S(d)f_3(d)\right] = 0 \tag{5}$$

where $b(d, d') = (d/d')^3 b^0(d, d')$.

A simple form of solution results if equation (5) is expressed in terms of $F_3(d)$ [by integrating equation (5) from 0 to d] and the kernel has the special form $S(d')b(d, d') = \alpha k_0 d^{\alpha-1}$. In this case,

$$F_3(d) = 1 - \frac{1 - F_{3_{\text{in}}}(d)}{1 + k_0 d^\alpha \tau} \tag{6}$$

where τ is the mean retention time for solids, H/M_F.

A more general form of solution arises from discretization of the size variable as was done in Example 2. Integration of equation (5) from $d - d_{i+1}$ to d_i with the recognition that $f_i = \int_{d_{i+1}}^{d_i} f_3(d)\, d(d)$; $i = 1, \ldots, n$, yields a set of simultaneous algebraic equations, one for each size fraction in the particulate assembly:

$$M_F f_{i_{\text{in}}} - M_F f_i + H\sum_{j=1}^{i-1} b_{ij}S_j f_j - HS_i f_i = 0 \tag{7}$$

When S_i and b_{ij} are independent of the size distribution in the mill, equation (7) represents a set of n linear algebraic equations in n unknowns, i.e., the steady-state mass fractions in the mill discharge, f_i, $i = 1, \ldots, n$. This set of equations can be written in matrix form as

$$M_F[\mathbf{I} + [\mathbf{I} - \mathbf{B}]\mathbf{S}H]\mathbf{f} = M_F\mathbf{f}_{\text{in}} \tag{8}$$

where **S** is an $n \times n$ diagonal matrix of selection functions, S_i, **B** is an $n \times n$ lower triangular matrix of breakage function b_{ij}, **I** is an $n \times n$ identity matrix, and **f** and $\mathbf{f}_{in}$ are $\mathbf{n} \times 1$ vectors of the product mass fractions, f_i, and inlet mass fractions, $f_{i_{in}}$. Equation (8) can be solved to yield the size-discretized solution:

$$\mathbf{f} = [\mathbf{I} + [\mathbf{I} - \mathbf{B}]S\tau]^{-1}\mathbf{f}_{in} \tag{9}$$

Experimental Verification. In this case, the data used to confirm the model were obtained for limestone grinding in a 30-in. × 16-in. Denver pilot scale mill, using a 760-lb equilibrium ball charge with a top size of 1.5 in. Data obtained at a feed rate of 600 lb/hr ($\tau = 8.0$ min) were used to estimate S_i and b_{ij} values by nonlinear regression.[35] In turn, these estimated values were used in conjunction with equation (9) to *predict* the size distributions produced for $M_F =$ 300 lb/hr ($\tau = 16.0$ min), 400 lb/hr ($\tau = 12$ min), and 1040 lb/hr ($\tau = 4.6$ min). As shown in Figure 2.1-20, the predicted size distributions are in good agreement with those measured experimentally.

Solution based on residence time distribution information. As observed at the beginning of this section, if the kinetics of change in a particulate system are linear and all particle sizes share a common residence time distribution (RTD), then ψ_{out} can be obtained directly from equation (2.1-34). For this example the kinetics of breakage have been assumed to be linear and the mill has been assumed well mixed, i.e., all particle sizes possess a single RTD given by[33]

$$E(t) = (1/\tau)e^{-t/\tau} \tag{10}$$

In this case, equation (2.1-34) can be transformed to its mass distribution form by multiplying both sides of the equation by $\rho C_3 d^3$ [see equation (2.1-14)] to yield

$$f_3(d) = \int_0^\infty f_3(d, t)\frac{1}{\tau}e^{-t/\tau}\,dt \tag{11}$$

For special forms of the selection and breakage function, the batch solution $f_3(d, t)$ can be substituted into equation (11) and the indicated integration performed to find the steady-state distribution. In the more general case, equation (11) can be discretized with respect to particle size by integrating both sides from d_{i+1} to d_i to yield

$$f_i = \int_0^\infty f_i(t)\frac{1}{\tau}e^{-t/\tau}\,dt \tag{12}$$

Substitution of equation (8) from Example 2 for the batch response of the ith size fraction, $f_i(t)$, and performing the indicated integration yields[37,38]

$$f_i = \sum_{j=1}^{i} A_{ij}\left(\frac{1}{1 + S_j\tau}\right) \tag{13}$$

Algebraic manipulation of equation (13) demonstrates that it is equivalent to equation (9) which was obtained directly from the population-balance equation.

Example 6. *Description of Continuous Leaching in an Agitated Vessel*

Problem. Derive a set of equations that can be used to predict the steady-state extraction of A from solid particles according to the reaction

$$a\mathrm{A}_{solid} + b\mathrm{B}_{fluid} \rightarrow c\mathrm{C}_{solid} + \text{fluid products}$$

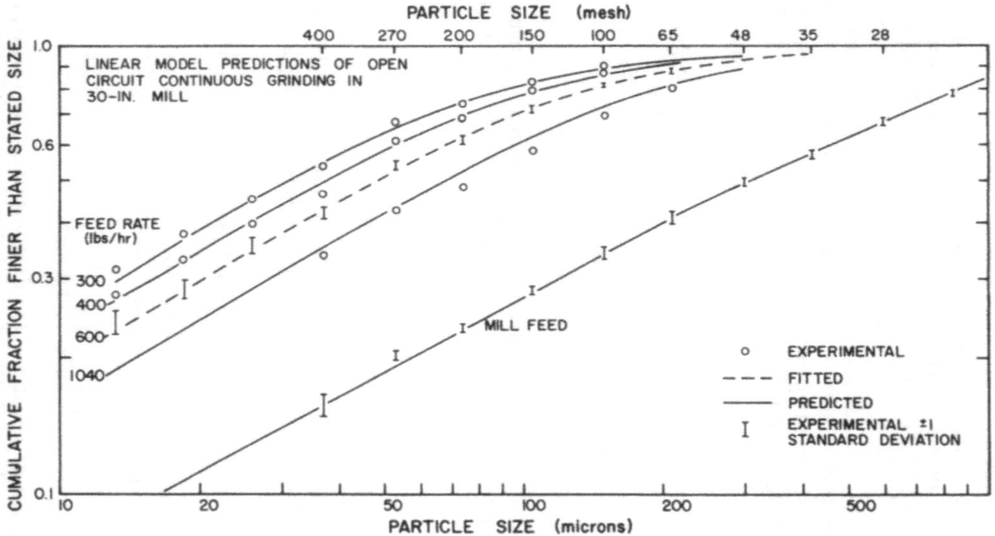

Figure 2.1-20. Experimental verification of linear grinding-model predictions of open circuit product size distributions for the ball milling of limestone in a 30-in. × 16-in. Denver mill.[36]

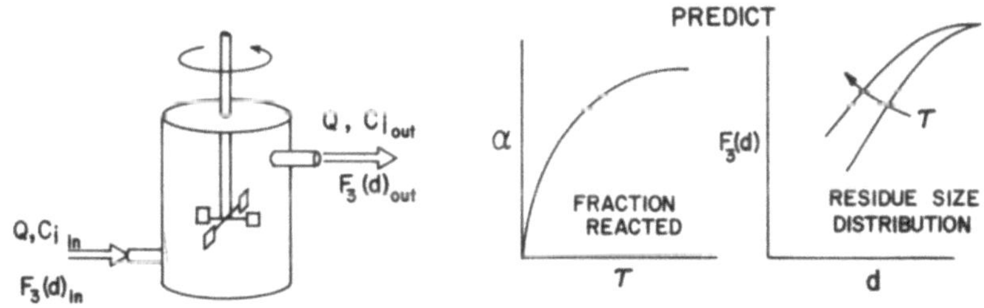

Figure 2.1-21. Schematic representation of continuously agitated leaching problem.

as it occurs in a continuously agitated leaching vessel of volume V,[12] shown in Figure 2.1-21. Assume that the reaction occurs topochemically, as illustrated in Figure 2.1-22. Assume further that the particles obey a generalized first-order shrinking-core model with any combination of film diffusion, product layer diffusion or surface reaction controlling the advancement of the reaction boundary. In this case the reaction boundary velocity for a particle of size d_p is given by

$$\frac{d(d_c)}{dt}=\frac{-2a/bC_B}{\rho_A\{1/k_g(d_c/d_p)^2+\frac{1}{2}D_{B,e}[d_c-(d_c^2/d_p)]-1/k_s\}}$$

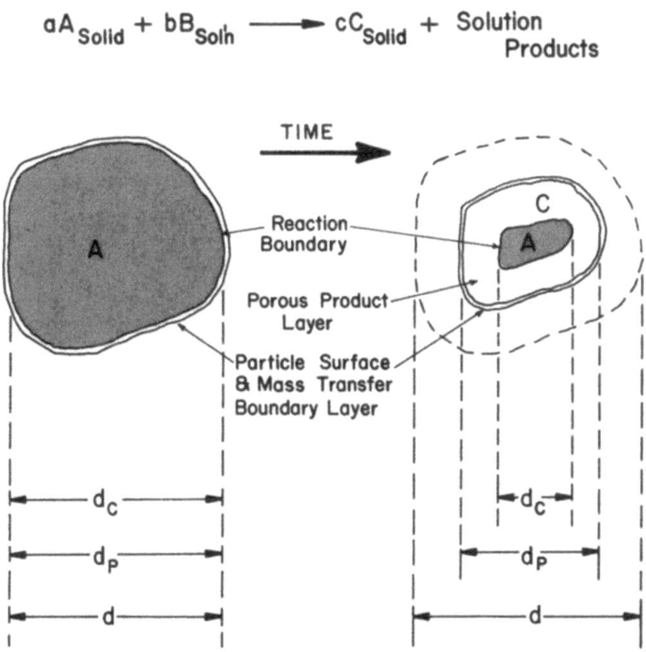

Figure 2.1-22. Schematic representation of topochemical leaching reaction kinetics.

where k_g, $D_{B,e}$ and k_s are the film mass-transfer coefficient, effective diffusivity for B in the porous product layer C, and k_s is the surface reaction-rate constant, and C_B and ρ_A are the bulk-phase concentrations of B (in solution) and the molar density of A in the original particle. The particle size, d_p, may increase, decrease, or remain the same during reaction according to the equation

$$d_p = [Zd^3 + (1-Z)d_c^3]^{1/3}$$

where Z is the Pilling–Bedworth ratio (volume of C produced per unit volume of A reacted $= (c/a)(\rho_A/\rho_C)$. For isometric particles the extent of reaction of a particle of initial size d is given by

$$\alpha = \frac{\rho_A C_3 d^3 - \rho_A C_3 d_c^3}{\rho_A C_3 d^3} = 1 - \left(\frac{d_c}{d}\right)^3$$

The particles in the feed are to have an arbitrary size distribution.

Solution

1. Write the continuous, steady-state macroscopic population-balance and mass-balance models:

$$\text{PBM:} \quad Q_{\text{in}}\psi_{\text{in}} = Q_{\text{out}}\psi_{\text{out}} + V\left[(\bar{B} - \bar{D}) - \sum_{j=1}^{J} \frac{\partial}{\partial \zeta_j}(v_j\bar{\psi})\right] = 0 \tag{1}$$

$$\text{MBM:} \quad Q_{\text{in}}C_{i_{\text{in}}} - Q_{\text{out}}C_{i_{\text{out}}} + Vr_i = 0 \tag{2}$$

2. Simplify equations (1) and (2) to suit the problem.
   a. Two properties of a particle are required to compute conversions, i.e., d_c and d. In this case, $\bar{\psi} dZ = \bar{\psi}(d_c, d) d(d_c)\, d(d) =$ the number of particles per unit volume with core size d_c which had initial size d.
   b. No catastrophic changes in d_c, i.e., no breakage or coalescence, therefore, $\bar{B} = \bar{D} = 0$.
   c. Rate of consumption of B per unit volume is equal to the rate of consumption of A times the stoichiometry factor b/a,

$$r_B = \frac{b}{a}\left[\frac{Q_{\text{in}}}{V}\iint \rho_A C_3 d_c^3 \psi_{\text{in}}\, d(d_c)\, d(d) - \frac{Q_{\text{out}}}{V}\iint \rho_A C_3 d_c^3 \psi_{\text{out}}\, d(d_c)\, d(d)\right]$$

   d. Well mixed vessel and product removal,

$$\psi_{\text{out}} = \bar{\psi}, \qquad C_{B_{\text{out}}} = C_B$$

   e. The volume flow rate in to the vessel, Q_{in}, equals the volume flow rate, Q_{out}, i.e.,

$$Q_{\text{in}} = Q_{\text{out}} = Q$$

With these simplifications, equations (1) and (2) become

$$\frac{\partial}{\partial d_c}\left[\frac{d(d_c)}{dt}\bar{\psi}(d_c, d)\right] + \frac{\partial}{\partial d}\left[\frac{d(d)}{dt}\bar{\psi}(d_c, d)\right] = \frac{Q}{V}[\psi_{\text{in}}(d_c, d) - \bar{\psi}(d_c, d)] \tag{1a}$$

and

$$C_{B_{\text{in}}} - C_B = \frac{b}{a}\left[\int_{d_{\min}}^{d_{\max}}\int_0^d \rho_A C_3 d_c^3 \psi_{\text{in}}(d_c, d)\, d(d_c)\, d(d) - \int_{d_{\min}}^{d_{\max}}\int_0^d \rho_A C_3 d_c^3 \bar{\psi}(d_c, d)\, d(d_c)\, d(d)\right] \tag{2a}$$

3. Specify the form of property velocities and birth and death rates:

Property velocities. If leaching kinetics follow a shrinking-core model with a first-order irreversible dependence on lixiviant concentration, then

$$\frac{d(d_c)}{dt} = v(d_c, d) \tag{3}$$

as given in the problem statement. The size of a particle in the inlet is a fixed quantity so that

$$\frac{d(d)}{dt} = 0 \tag{4}$$

Birth and death rates: No coalescence or breakage occurs in vessel, therefore $\bar{B}$ and $\bar{D}$ have been taken to be zero.

3. Solve equations, subject to side conditions.

Substitution of equations (3) and (4) into equation (1a) gives

$$\frac{\partial}{\partial d_c}[v(d_c, d) \cdot \bar{\psi}(d_c, d)] = \frac{1}{\tau}[\psi_{\text{in}}(d_c, d) - \bar{\psi}(d_c, d)] \tag{5}$$

Expansion of the left-hand side of this equation results in the first-order variable-coefficient, inhomogeneous differential equation:

$$\frac{\partial}{\partial d_c}\bar{\psi}(d_c, d) + \left\{\frac{\partial[\ln v(d_c, d)]}{\partial d_c} + \frac{1}{\tau v(d_c, d)}\right\}\bar{\psi}(d_c, d) = \frac{1}{\tau v(d_c, d)}\psi_{\text{in}}(d_c, d) \tag{6}$$

This equation can be solved with the integrating factor IF,

$$\text{IF}(d_c) = \frac{v(d_c, d)}{v(a, d)} \exp \int_a^{d_c} \frac{d(d_c')}{\tau v(d_c', d)} \tag{7}$$

to yield the joint density function

$$\bar{\psi}(d_c, d) = -\frac{1}{\text{IF}(d_c)} \int_{d_c}^{d} \frac{\text{IF}(d_c')}{\tau v\,(d_c', d)}\, \psi_{\text{in}}(d_c', d)\, d(d_c') \tag{8}$$

This equation can be simplified by recognizing that particles in the inlet are completely unreacted. In terms of d_c and d this requires that the core size of an entering particle d_c is equal to its particle size d. Mathematically this can be expressed as the side condition

$$\psi_{\text{in}}(d_c, d) = \psi_{\text{in}}(d) \cdot \delta(d_c - d) \tag{9}$$

where $\delta(d_c - d)$ is the Dirac delta function, i.e.,

$$\delta(d_c - d) = 0 \qquad \text{for } d_c \neq d$$

$$\int_a^b \psi_{\text{in}}(d) \cdot \delta(d_c - d)\, d(d_c) = \psi_{\text{in}}(d), \qquad d \in [a, b]$$

Substituting equation (9) into equation (8) and performing the indicated integration yields

$$\bar{\psi}(d_c, d) = -\frac{\psi_{\text{in}}(d)}{\tau v(d_c, d)} \exp \int_{d_c}^{d} \frac{d(d_c')}{\tau v(d_c', d)} \tag{10}$$

To find the marginal distribution of core sizes, d_c, for particles in the outlet stream of the vessel integrate equation (10) over d to yield

$$\bar{\psi}(d_c) = -\int_{d_c}^{d_{max}} \frac{\psi_{in}(d)}{\tau v\,(d_c, d)}\left[\exp \int_{d_c}^{d} \frac{d(d_c)}{\tau v\,(d_c', d)}\right] d(d) \tag{11}$$

In turn, the overall extent of conversion from A to products in the leaching vessel is calculated from the amount of A in the inlet (per unit volume) minus the amount of A in the outlet divided by the amount in the inlet, i.e.,

$$\bar{\alpha} = \frac{\int_{d_{min}}^{d_{max}} \rho_A C_3 d^3 \psi_{in}(d)\, d(d) - \int_0^{d_{max}} \rho_A C_3 d_c^3 \bar{\psi}(d_c)\, d(d_c)}{\int_{d_{min}}^{d_{max}} \rho_A C_3 d^3 \psi_{in}(d)\, d(d)} \tag{12}$$

By performing the indicated integration over d_c in equation (2a) and substituting equation (12) in to the resulting expression, one obtains the alternative form of the mass balance

$$C_{B_{in}} - C_B = (b/a)\bar{\alpha} M_I \tag{13}$$

where M_I is the denominator of equation (12) that gives the number of moles of A per unit volume in the inlet stream of the vessel.

It should be noted that since the kinetics of leaching have been assumed to be linear and the disperse phase completely entrained in the continuous phase, equations equivalent to equations (11–13) can be obtained from equations (2.1-34) and (2.1-35) with $E(t) = (1/\tau)e^{-t/\tau}$ and the batch responses $\psi_{batch}(t)$ and $C(t)$ as determined in the last section. Bartlett[40] has presented dimensionless reactor design correlations for idealized systems using this approach (see also Chapter 2A).

Equations (11–13) constitute a solution to the continuous leaching problem. Given an initial feed size distribution, initial concentrations, reaction information including intrinsic rate parameters (k_g, $D_{B,e}$, and k_s) and a retention time, equations (11–13) can be solved simultaneously to determine the steady-state conversion of A and the outlet lixiviant concentration. Equations (8) and (11) can be used to determine information concerning the size distribution of the leach residue if required.

Analytical solutions to equations (11–13) are possible for especially simple systems,[39] e.g., when a single mechanism is rate controlling and the feed size distribution, $\psi_{in}(d)$, allows the integration of equation (11). In most instances of practical importance, however, these equations must be solved using numerical integration techniques.

Simulations. The remainder of this example is devoted to illustrating how equations (11–13) can be used to simulate continuous leaching behavior. The nature of predictions obtained with this model is shown for two copper leaching systems.

1. Oxygen–acid leaching of chalcopyrite at elevated temperature according to the reaction

$$2\,CuFeS_2 + \tfrac{17}{2}O_2 + 2\,H^+ \rightarrow 2\,Cu^{2+} + 2\,Fe^{3+} + 4\,SO_4^{2-} + H_2O$$

2. Acid leaching of chrysocolla according to the reaction

$$CuO{\cdot}SiO_2{\cdot}2\,H_2O + 2\,H^+ \rightarrow Cu^{2+} + SiO_2(2-n)H_2O + n\,H_2O$$

In the first case, all reaction products are soluble, hence the chalcopyrite particles shrink in size during leaching with $Z = 0$ and $d_p = d_c$. In the second case, the size of the chrysocolla particles does not change during leaching because the reaction products include a hydrated silica layer with $Z = 1$. Intrinsic kinetic parameters for each system are determined from batch data for narrow size fraction feeds by minimizing the sums of squares of deviations between experimental and batch model predictions using a Flexiplex algorithm.[43] In both cases continuous leaching behavior is being assessed for a 30-in. × 30-in. cylindrical vessel agitated with a turbine-type impellar at a speed of 100 rpm. The mass transfer coefficient, k_g, for the two systems has been estimated using the method of Harriot.[41] In each of the systems the size distribution of material feeding the vessel was assumed to conform to a Gaudin–Schuhmann distribution, $F_3(d) = (d/d_{max})^m$.

Simulation A—Chalcopyrite Leaching. The kinetics of oxidation leaching of chalcopyrite at elevated temperatures and oxygen overpressures have been studied in detail by Yu.[3] His analysis suggests that for moderate agitation, the kinetics of leaching are controlled by the reaction rate of oxygen at the sulfide surface with a rate which is proportional to the P_{O_2} for pressure less than 400 psi. In order to determine an appropriate value of the surface reaction-rate constant for simulation, some of the batch data obtained by Yu for different feed sizes and oxygen pressures (see Figure 2.1-23) was used to determine a value of k_s for 125° C by nonlinear regression. In particular, for each condition depicted in the figure, the objective function

$$\phi = \sum_{j=1}^{J} [\underset{\text{experimental}}{\alpha(t_j)} - \underset{\text{model}}{\alpha(t_j)}]^2$$

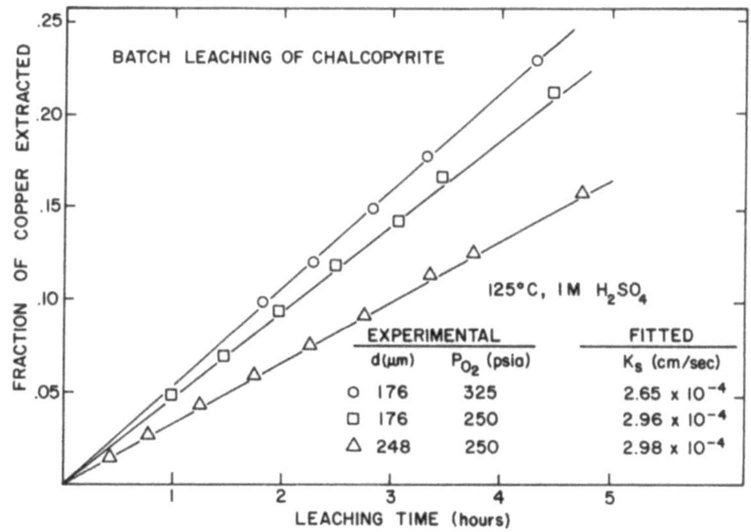

Figure 2.1-23. Estimation of k_s from batching leaching data[3] for Simulation A.

where for surface-reaction controlled kinetics

$$\underset{\text{model}}{\alpha(t_j)} = 1 - \left(1 - \frac{2k_s C_B}{d\rho_A} t_j\right)^2$$

was minimized with respect to k_s for each of the three sets of data. As shown in Figure 2.1-23, the fitted values of k_s for different conditions are closely grouped around a mean of 2.85×10^{-4} cm/sec.

The objective of the simulations which follow was to determine the relationship of basic conversion vs. retention time for chalcopyrite leaching with fixed values of feed size distribution and oxygen partial pressure. Secondly, the effect of changes in feed size distribution and oxygen partial pressure was to be evaluated for a fixed retention time. In addition, it was desired to determine the size distribution of the unleached solids (residue) in the reactor effluent. It was decided at the outset that the oxygen was to be sparged into the 30-in. reactor on demand in order to control the oxygen concentration in solution at any desired level in the vessel. This decision eliminates the need for an oxygen mass balance.

In order to apply equations (11)–(13) for continuous simulation one must simplify the reaction boundary velocity expression, substitute into equation (11), and integrate. Since the Pilling–Bedworth ratio is zero ($Z = 0$) for this case, and $a = 2$, $b = 2/17$, and $\rho_A = 0.0245$ moles/cm^3,

$$v(d_c, d) = \frac{-2(4/17)C_{O_2}}{0.0245(1/k_g + 1/k_s)}$$

For oxygen dissolved in acid solution, $C_{O_2} = KP_{O_2}$ where K has been determined for various acid concentration and temperatures.[3] Substitution of this expression into equation (11) yields

$$\bar{\psi}(d_c) = \bar{\psi}(d_p) = \frac{0.0245(1/k_g + 1/k_s)}{\tau \cdot (4/17)C_{O_2}} \int_{d_c}^{d_{max}} \psi_{in}(d)$$
$$\times \exp\left[-\frac{(d - d_c)}{\tau(4/17)C_{O_2}[0.0245(1/k_g + 1/k_s)]^{-1}}\right] d(d)$$

This integral was evaluated by Simpson's integration[44] with 500 intervals. In turn, the values for $\bar{\psi}$ were introduced into equation (12) and a second numerical integration was performed to determine the overall conversion $\bar{\alpha}$.

Figure 2.1-24 shows model predictions of the overall conversion (fraction on copper extracted), $\bar{\alpha}$, vs. retention time, τ, for an oxygen partial pressure of 250 psi at 125° C, in a one-molar acid solution when the feed consists of particles characterized by a Gaudin–Schuhman distribution with a maximum size, d_{max}, of 200 μm and a distribution modulus, m, of unity. Notice that a retention time in excess of 50 hr, is required to achieve 90% copper release under these conditions. Figure 2.1-25 shows the size distribution of the unleached chalcopyrite in the effluent for selected retention times depicted in Figure 2.1-24. Interestingly, these simulations show that the residue size distribution approaches a constant density function for high conversions.

Figure 2.1-26 shows the simulated effect of changes in the chalcopyrite feed size distribution at a fixed retention time of 50 hr. As expected, the finer the feed size distribution (d_{max} and m small), the higher the conversion. Figure 2.1-27 shows the effect of oxygen partial pressure for this leaching system. Notice that there is a significant increase in copper extraction with oxygen pressure for

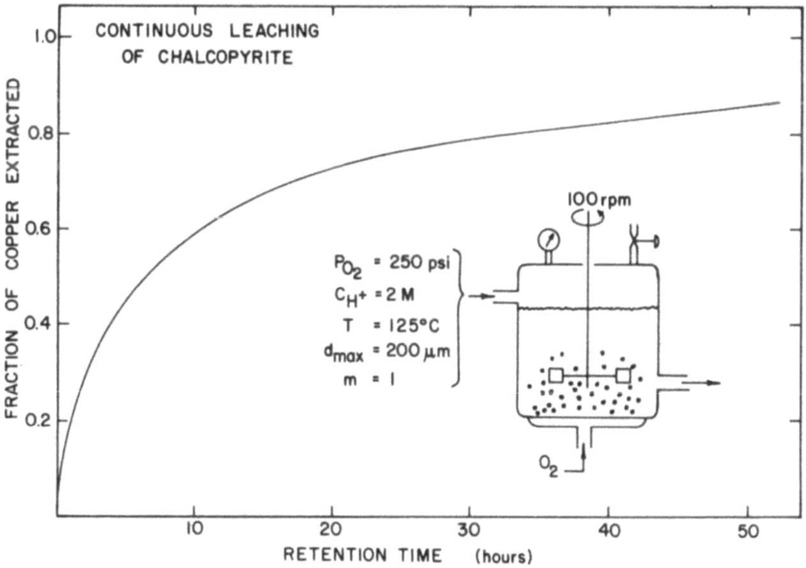

Figure 2.1-24. Predicted performance of 30×30-in. reactor effect of retention time on copper extraction for Simulation A.

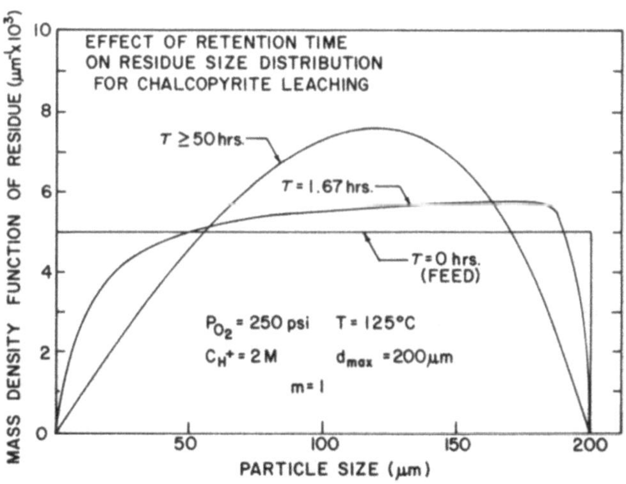

Figure 2.1-25. Predicted effect of retention time on residue size distribution for Simulation A.

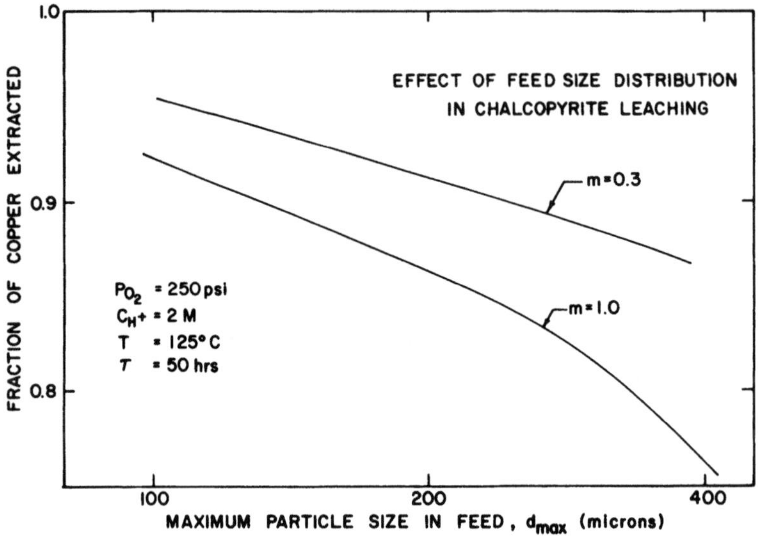

Figure 2.1-26. Predicted effect of feed size distribution parameters, d_{max} and m, on copper extraction for Simulation A.

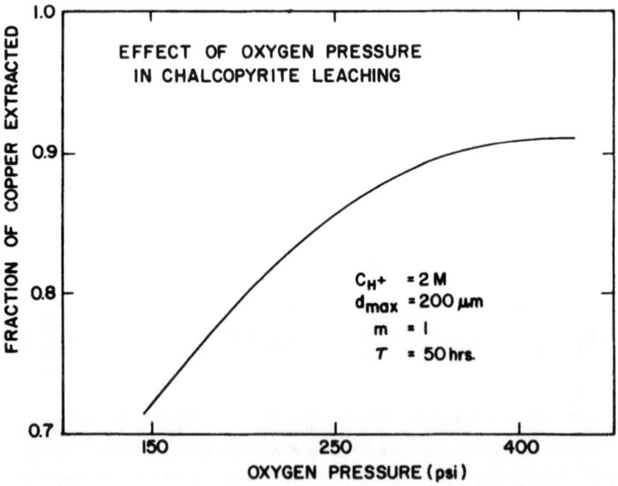

Figure 2.1-27. Predicted effect of oxygen pressure on copper extraction for Simulation A.

values below about 350 psi but that the kinetic advantage is diminished substantially for P_{O_2} values above this level.

Simulation B—Chrysocolla Leaching. The kinetics of leaching of chrysocolla with acid has been studied by Pohlman in a flow-cell apparatus.[42] His data show that individual particle leaching kinetics follow a shrinking-core model with similar resistances because of hydrogen ion diffusion through the hydrated silica product layer and chemical reaction at the chrysocolla–silica reaction boundary. In order to determine appropriate values for the intrinsic kinetic parameters the batch data for two particle sizes and three acid concentrations, shown in Figure 2.1-28, were used in conjunction with the shrinking-core model to estimate k_s and D_e by nonlinear regression. As these data were obtained in a flow cell at low Reynolds number ($N_{Re} \leq 0.15$), film diffusion was important and the nuisance parameter, k_g, had to be determined as well. In this case the objective function

$$\phi = \sum_{j=1}^{J} [\underset{\text{experimental}}{\alpha(t_j)} - \underset{\text{model}}{\alpha(t_j)}]^2$$

was minimized with respect to k_s, D_e, and k_g, using the Flexiplex algorithm for each of the three sets of data. Since the flow cell was operated with a low percentage of solids, the constant lixiviant concentration form of the batch shrinking-core model was appropriate for estimation, i.e.,

$$\alpha = 1 - \frac{1}{27}\left\{\gamma + 2(3\beta + \gamma^2)^{1/2} \cos\left[\frac{3\theta(t) + 4\pi}{3}\right]\right\}^3$$

where the quantities γ, β, and $\theta(t)$ can be defined in terms of the Sherwood number, $N_{Sh} = k_g d/D_{B,e}$, and Damköhler number, $N_{DaII} = k_s d/D_{B,e}$, as follows:

$$\gamma = \tfrac{3}{2}(1 - 2/N_{Sh})^{-1}$$

$$\beta = (6/N_{DaII})(1 - 2/N_{Sh})^{-1}$$

$$\theta(t) = \frac{1}{3}\cos^{-1}\left[\frac{2\gamma^3 + 9\gamma\beta + 27\delta(t/\tau - 1)}{2(3\beta + \gamma^2)^{3/2}}\right]$$

$$\delta = 3\left(\frac{1}{1 + 2/3N_{Sh} + 2/N_{DaII}} - \frac{1}{1/3 + 1/12N_{Sh} + N_{Sh}/N_{DaII}}\right)$$

$$\tau = \frac{1}{24}\frac{\rho_A(b/a)d^2}{C_B D_{B,e}}\left(1 + \frac{4}{N_{Sh}} + \frac{12}{N_{DaII}}\right)$$

As shown in Figure 2.1-28, the estimated values of the effective diffusivity, D_e, cluster around 2.1×10^{-7} cm^2/sec, and values of the surface reaction-rate constant, k_s, around 5×10^{-3} cm/sec.

The objective of this series of continuous simulations is to determine the relationship of basic copper extraction vs. retention time for a fixed inlet acid concentration, inlet % solids, and feed size distribution. In addition, the effects of changes in inlet acid and % solid on conversion for a fixed retention time will be examined.

In order to apply equations (11–13) to the continuous chrysocolla leaching problem one must simplify the reaction boundary velocity expression for $Z = 1$,

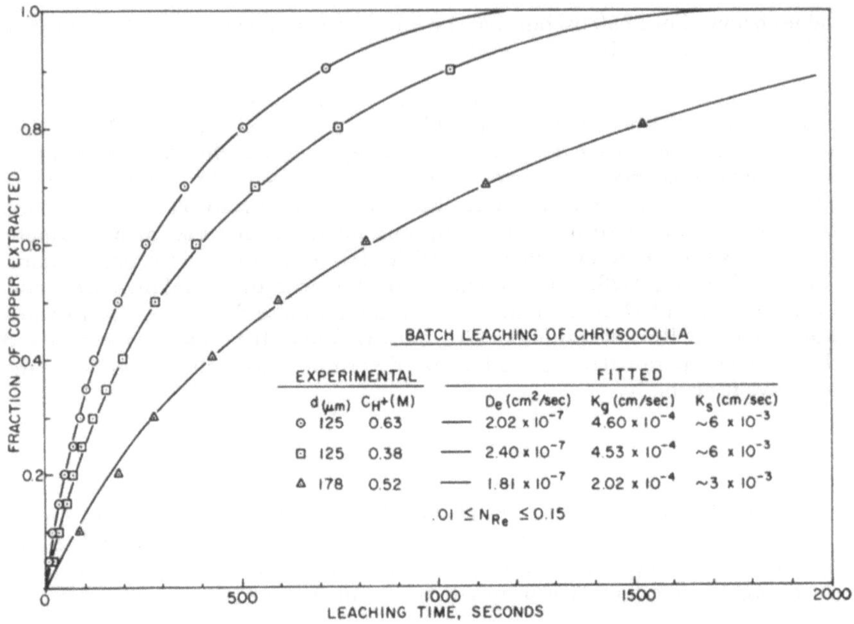

Figure 2.1-28. Estimation of k_s, $D_{B,e}$, and k_g from batch leaching data[42] for Simulation B.

$a = 1$, $b = 2$, and $\rho_A = 0.0137$ moles/cm^3 to yield

$$v(d_c, d) = \frac{-2(1/2)C_{H^+}}{0.0137[1/k_g(d_c/d)^2 + 1/2D_{H^+,e}(d_c - d_c^2/d) + 1/k_s]}$$

In this case the problem is somewhat more complicated than that of the previous set of simulations. Since the depletion of lixiviant, H^+, can not be ignored, the mass-balance equation (13) must be solved simultaneously with equations (11) and (12) in an iterative fashion. This procedure involved substituting the above expression for reactor boundary velocity into equation (11) which was in turn integrated with an initial guess for C_{H^+} in the reactor by Simpson's integration with 500 equisize intervals to obtain a set of trial values for $\bar{\psi}(d_c)$. The values obtained in this fashion were substituted into equation (12) which was similarly integrated numerically to determine the trial conversion, $\bar{\alpha}$. This trial value of $\bar{\alpha}$ was substituted into the mass-balance expression, equation (13), and the computed value of C_{H^+} was compared to the trial value used in evaluating equation (11). A new trial value for C_{H^+} was determined by a Newton–Raphson[45] procedure, and all the steps involving equations (11)–(13) repeated until C_{H^+} computed from equation (13) came within 0.01% of the corresponding trial value used in equation (11). Convergence was normally achieved in five or six iterations.

Figures 2.1-29–31 show the results of the chrysocolla simulations. Figure 2.1-29 shows the increase in extraction and decrease in hydrogen ion concentration which occurs with increasing retention time for a 3.2% solid feeds, an inlet lixiviant concentration of 0.316 $M\,H^+$, and a Gaudin–Schuhmann feed size distribution with $d_{\max} = 200\ \mu$m and $m = 1$. Note that in this system 90% copper

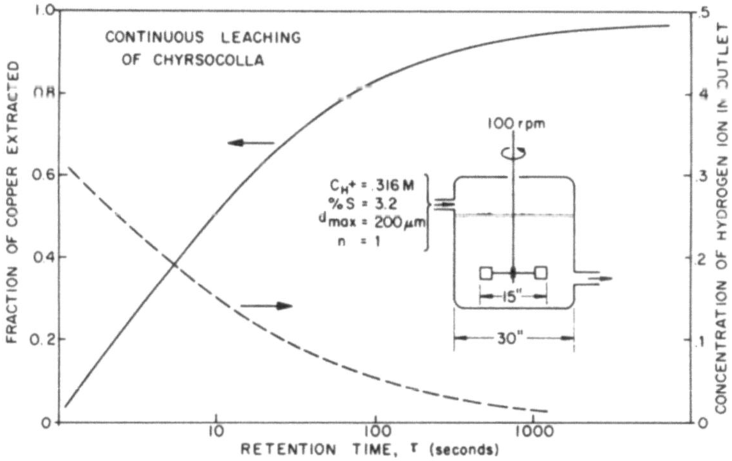

Figure 2.1-29. Predicted performance of 30×30-in. reactor effect of retention time on copper extraction and hydrogen ion depletion for Simulation B.

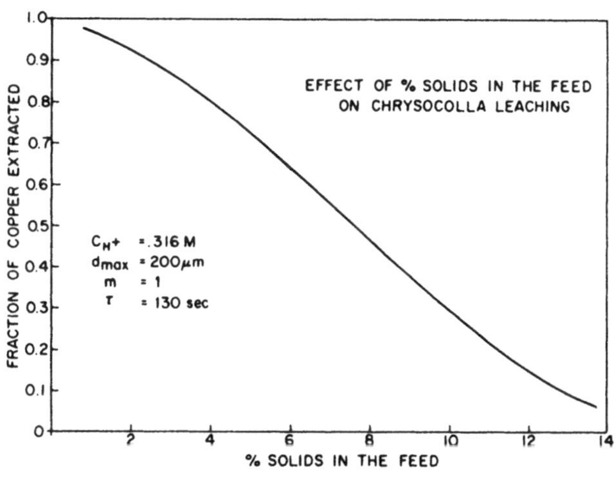

Figure 2.1-30. Predicted effect of feed percent solids on copper extraction for Simulation B.

extraction is obtained with a mean residence time of about 300 sec. Note also that for a retention time which exceeds 1000 sec, less than 5% of the H^+ introduced in the inlet remains in the outlet. This latter result shows that the frequently used simplification of "constant lixiviant concentration" is not appropriate here and can result in a substantial error in the calculation of conversions. In general, it is essential to account for fluid–solid interactions and simultaneous solution of equations (11) and (13) is required to obtain accurate predictions.

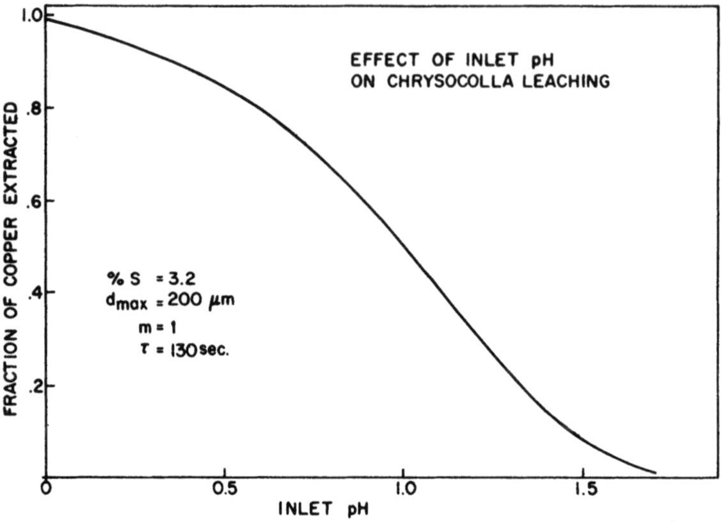

Figure 2.1-31. Predicted effect of inlet pH on copper extraction for Simulation B.

Figure 2.1-30 further illustrates the effect of fluid–solid interactions. Notice that, as the inlet concentration of chrysocolla (% solids) is increased, the fraction of copper extracted for a 130-sec retention time is predicted to decrease dramatically. Finally, Figure 2.1-31 shows the effect of inlet pH on copper extraction for a 130-sec retention time. Note that for this system an increase of 1 pH unit in the feed slurry from pH 0.5 to pH 1.5, results in a reduction in copper extraction from about 85% to less than 10%.

2.1.7. Summary and Conclusions

The purpose of Section 2.1 has been to provide a framework for describing the kinetic behavior of multiparticle systems. This framework consists of (1) a population balance to describe changes in the distribution of selected particle properties in the disperse phase, and (2) a mass balance to describe changes the concentration of any species in the continuous phase that influence the disperse phase behavior. In this type of description individual particle kinetics provide the basic building blocks for predicting the behavior of an entire assembly of particles.

The specific forms of population balance and mass balance described here are deterministic (ignore statistical fluctuations) and macroscopic (consider only the spatially averaged contents of a process vessel) in nature. These descriptions are adequate for many engineering applications in extractive metallurgy. Extended treatments involving stochastic representations[45,46] and microscopic (spatially distributed) forms[1,47] are also available.

A survey of the literature related to the mathematical modeling of extractive metallurgical systems over the last 20 years, reveals the great variety of approaches used by the investigators. Because of this variety comparisons between models developed by different investigators may be difficult, if not impossible, to mak. The population-balance framework described here provides a unified approach to the development of models of multiparticle system behavior. The final equations obtained in Examples 2 and 5 for tumbling-mill grinding and in Example 4 for heterogeneous reaction have appeared repeatedly in the literature, but their common origin in population balance has not been widely recognized. Similar statements can be made for other systems treated as examples here and for various systems of importance in extractive metallurgy, e.g., froth flotation,[48] wet pelletization,[49] and fluidized-bed roasting.[50]

The development and use of detailed mathematical models for particle processing systems is in its infancy in extractive metallurgy. Because of the tremendous potential these models have for applications in process design, optimization, and automatic control, they will undoubtedly become the focus of considerable research activity in the future.

ACKNOWLEDGMENT

The author gratefully acknowledges the assistance of several graduate students in the preparation of illustrative examples. Special thanks are owed to Messrs. R. Fernandez, D. Kinneberg, R. Mackelprang, E. Oblad, K. Rajamani, J. L. Sepulveda, and J. E. Sepulveda.

References

1. H. M. Hulburt and S. Katz, *Chem. Eng. Sci.* **19**, 555 (1964).
2. A. D. Randolph, *Can. J. Chem. Eng.* **42**, 280 (1964).
3. P. Yu, *A kinetic study of the leaching of chalcopyrite at elevated temperatures and pressures*, Ph.D. Dissertation, University of Utah, 1972.
4. C. Orr and J. M. Dallavalle, *Fine Particle Measurement*, Macmillan, New York (1959).
5. R. D. Cadle, *Particle Size*, Reinhold Publishing, Stamford, Conn. (1965).
6. R. Irani and C. Callis, *Particle Size: Measurement, Interpretation and Application*, Wiley, New York (1963).
7. T. Allen, *Particle Size Measurement*, Chapman and Hall, London (1968).
8. G. Herdan, *Small Particle Statistics*, Academic Press, New York (1960).
9. E. Parzen, *Modern Probability Theory and Its Applications*, Wiley, New York (1963).
10. M. Fisz, *Probability Theory and Mathematical Statistics*, Wiley, New York (1960).
11. H. D. Lewis and A. Goldman, *Theoretical small particle statistics*, Los Alamos Scientific Laboratory Report (1967).
12. D. M. Himmelblau, *Process Analysis by Statistical Methods*, Wiley, New York (1969).

13. A. D. Randolph and M. Larson, *Theory of Particulate Processes*, Academic Press, New York (1971).
14. R. B. Bird, W. E. Stewart, and E. N. Lightfoot, *Transport Phenomena*, Wiley, New York (1960).
15. V. G. Jenson and G. V. Jeffereys, *Mathematical Methods in Chemical Engineering*, Academic Press, New York (1963).
16. D. C. Tim and M. A. Larson, *AIChE J.* **14**, 452 (1968).
17. L. G. Austin, *Powder Technol.* **5**, 1 (1972–72).
18. T. Meloy and A. Gaudin, *Trans. AIME* **223**, 43 (1962).
19. A. Filippov, *Theory Probab. Its Appl.* (*USSR*) **6**, No. 3 (1961).
20. L. G. Austin and P. T. Luckie, *Trans. AIME* **252**, 82 (1972).
21. J. A. Herbst and T. Mika, *Proceedings IX International Mineral Processing Congress*, Prague (1970).
22. V. K. Gupta and P. C. Kapur, *Fourth European Symposium on Comminution*, H. Rumpf and K. Schonert, eds., Dechema-Monographien, Verlag Chemie (1976), p. 447.
23. K. Rajamani and J. A. Herbst, Computer evaluation of errors involved in the use of size discretized grinding models, manuscript in preparation.
24. K. J. Reid, *Chem. Eng. Sci.* **20**, 953 (1965).
25. J. A. Herbst et al., *Fourth European Symposium on Comminution*, H. Rumpf and K. Schonert, eds., Dechema-Monographien, Verlag Chemie (1972), p. 475.
26. J. A. Herbst and D. W. Fuerstenau, *Trans. AIME* **241**, 538 (1968).
27. S. K. Freidlander and C. S. Wang, *J. Colloid Interface Sci.* **22** (2), 126 (1966).
28. K. V. S. Sastry, *The agglomeration of particulate materials by green pelletization*, Ph.D. Thesis, University of California (1970).
29. J. Valentas and N. Amundsen, *Ind. Eng. Chem. Fundam.* **4**, 533 (1966).
30. R. K. Bjpai and D. Ramkrishna, *Chem. Eng. Sci.* **31**, 913 (1976).
31. D. E. Brown and K. Pitt, *Chem. Eng. Sci.* **27**, 577 (1972).
32. L. W. Beckstead *et al.*, *Trans. TMS-AIME*, **2** 611 (1976).
33. D. M. Himmelblau, *Process Analysis of Simulation*, Wiley, New York (1968).
34. J. A. Herbst and K. V. S. Sastry, unpublished results (1977).
35. J. A. Herbst, K. Rajamani, and D. Kinneberg, ESTIMILL, University of Utah, Dept. of Metallurgy (1977).
36. M. Siddique, *A kinetic approach to ball mill scale-up*, M.S. Thesis, University of Utah (1977).
37. J. A. Herbst *et al.*, *Trans. IMM* **80**, C193 (1971).
38. R. P. Gardner and K. Verghese, *Powder Technol.* **11**, 87 (1975).
39. J. A. Herbst, An approach to the modeling of continuous leaching systems, Annual AIME Meeting, New York (1975).
40. R. W. Bartlett, *Met. Trans.* **2**, 2999 (1971).
41. P. Harriot, *AIChE J.* **8**, 93 (1962).
42. S. Pohlman, *The dissolution kinetics of chrysocolla using a weight loss technique*, Ph.D. Thesis, University of Utah (1974).
43. D. M. Himmelblau and D. A. Paviani, *Operations Res.* **17**, 872 (1969).
44. L. Lapidus, *Numerical Methods for Chemical Engineers*, Wiley, New York (1962).
45. D. Ramkrishna, *Chem. Eng. Sci.* **28**, 1423 (1973).
46. D. Ramkrishna, *Chem. Eng. Sci.* **29**, 1711 (1974).
47. H. Imai and T. Miyauchi, *J. Chem. Eng. Japan* **1**(1), 77 (1968).
48. R. P. King, *S. African IMM* 341 (1973).
49. P. C. Kapur and D. W. Fuerstenau, *Ind. Eng. Chem. Process Design Develop.* **8** (1) (1969).
50. D. Kunii and O. Levenspiel, *Fluidization Engineering*, Wiley, New York (1971).

2A

Chemical Process Analysis and Design for Multiparticle Systems

Robert W. Bartlett

2A.1. Introduction

Many important metallurgical unit operations involve chemical reactions between solid particles and one or more components of a fluid phase. Leaching, roasting, gaseous reduction, and other gas–solid reactions are examples. Most commonly, the particles were produced by previous grinding and particle-size classification operations, usually on minerals.

Although kinetics are obviously often important in these process steps, there has been little application of the results of numerous laboratory kinetic studies of particle reactions in designing metallurgical process plants. We shall examine this situation by starting at the beginning of metallurgical process design. The first process design problem is about *what* is to be done. The solution is usually a synthesis of linked functional process steps, i.e., a flowsheet. After mass and heat balances are established, performance specifications must be established for each step in terms of throughput, extraction yield, etc. The metallurgical process designer must next determine *how* to conduct each step to meet the performance criteria. In the case of metallurgical reactions, this is a matter of determining the *type* of reactor and its mode of operation which may be dictated by a variety of constraints. The next question involves the *size* of the reactor to obtain the desired throughput and

Robert W. Bartlett • Stanford Research Institute, Menlo Park, California; formerly Kennecott Research Center, Salt Lake City, Utah

yield. Metallurgical reaction kinetics, including particle reaction kinetics, are only means to answering this last question and from a process-design engineering standpoint usually have no other function. Unfortunately, the study of kinetics has often been treated as an end rather than a means.

A process-design engineer faced with scaling the size of process reactors (reaction vessels) for particle systems must have more information than the particle reaction kinetics. If the particle size distribution in the feed and the residence time of particles in the process reactor are also known then, in principle, a detailed summation can be made with a computer that would calculate the appropriate reactor size for the feed rate and desired conversion of the feed. Many situations are so complex that they can be approached only in this manner, but this approach is time consuming, relatively expensive, and requires a level of expertise that may not be available to every project.

For convenience we give here the notation used in this chapter:

D_{AB}	ordinary diffusion coefficient for molecular diffusion in a fluid phase
D_{eff}	effective diffusion coefficient for diffusion in a porous solid
F	fraction of particle feed that has reacted
$1-F$	fraction of particle feed that is unreacted
j	number of stages in a series of equal-sized backmix reactors
m	Gates–Gaudin–Schuhmann grinding parameter
R	radius of particle
R_*	radius of largest particle
t	residence time
t_b, t_p	residence time for a batch or plug-flow vessel
$\bar{t}$	average residence time for a backmix reactor, or series of backmix reactors
V	reactor working volume
$\dot{v}$	volumetric flow rate
$Y(R_i)$	cumulative weight fraction
$\Delta Y(R_i)$	fraction of particles in size range between R_i and $R_i+\Delta R_i$
ε	porosity of the particle—volume fraction that is pore space
θ	dimensionless residence time, $D_{eff}t/R_*^2$
$\bar{\theta}$	dimensionless average backmix reactor residence time, $D_{eff}\bar{t}/R_*^2$
τ_D	tortuosity factor for diffusion in pores (usually $\tau_D \simeq 2$)
$\tau(R)$	time for complete reaction of particle R
τ_m	time for complete reaction of the largest particle in the aggregation
Re_p	particle Reynolds number

2A.2. The Concept of General Reactor Scale-up Criteria

Often the problem is sufficiently simple that general solutions published as dimensionless or normalized design curves can be quickly applied by the process engineer to *his* particular problem. This approach is similar to the use of dimensionless relationships in heat transfer and flow problems. For example, minerals engineers are familiar with the use of the dimensionless Reynold's number correlations for calculating settling velocities of particles.

A reactor design curve should express yield, or fraction reacted, F, as a function of a dimensionless residence time, which is related to the average residence time in the reactor. The average residence time, $\bar{t}$, is simply the reactor working volume, V, divided by the volumetric rate of throughput, $\dot{v}$:

$$\bar{t} = V/\dot{v} \tag{2A.2-1}$$

This approach has been taken for homogeneous reactions and the results appear in chemical engineering text books on reaction engineering, e.g., Levenspiel.[1]

The purpose of this chapter is to present in summary form a number of such design curves for various conditions involving heterogeneous chemical reactions with particles, define their limitations, and describe their applications.

There are four kinds of information required to describe a particulate system reacting with a component of a fluid phase and determine the reaction yield as a function of average reactor residence time: (1) the particle reactor kinetic mechanism, needed only in terms of a phenomenological description; (2) the rate-controlling step and specific rate of the particle reaction; (3) the particle-size distribution; and (4) the residence-time distribution of the particles in the reactor (fluid bed roaster, leach tank, etc.).

Fortunately, there are special types in each of these four categories that can be described by mathematical equations and occur frequently in many mineral systems, at least as adequate engineering approximations.

2A.2.1. Particle Phenomenological Kinetics

Only two types are of interest, both based on equidimensional particles (quasi-spherical). Obviously, asbestos and mica do not fit but most other minerals do.

When a valuable mineral is substantially liberated and is the major constituent of the particle, the subsequent metallurgical or chemical reaction is likely to initiate at the external surface of the particle and involve a shrinking, unreacted particle *core*, which may be exposed to the fluid or surrounded with a solid *shell* of reaction product. For example, during the roasting of metal sulfide flotation concentrates, an oxide shell usually surrounds the unreacted sulfide mineral core. In other cases, such as oxidative leaching of sulfide concentrates, the chemical reaction products are dissolved by the fluid phase and a reaction product shell does not form.

Often, a relatively small amount of rock breakage without liberating the ore mineral is sufficient to permit liquid or gaseous chemical extraction of metals from rocks in which the extractable minerals are distributed in a gangue mineral matrix. The valuable mineral phase undergoing reaction is usually a minor phase constituent of the heterogeneous rock particle. The reacting fluid must penetrate pores of the

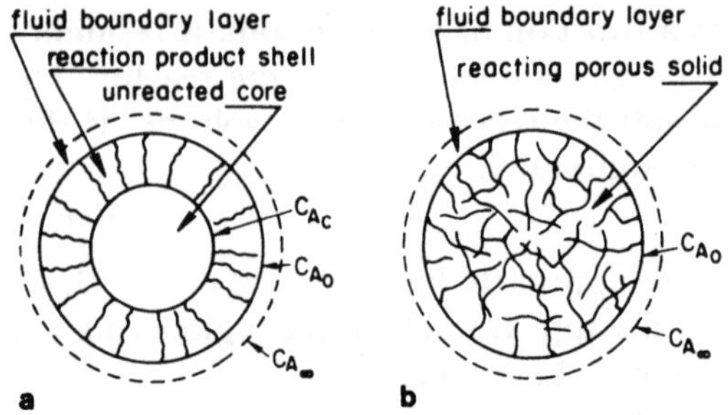

Figure 2A.2-1. Quasi-spherical models of solid particle reaction. (a) Conversion (core–shell) model; (b) extraction (porous solid) model.

rock particle and the chemical reaction occurs primarily on internal rather than external surfaces of the particle. This *porous solid* model constitutes the other major category of solid particle reactions.

For convenience and brevity, the term *conversion* will be used to designate reactions with particles following the core–shell model in the design curves to be presented, whereas the term *extraction* will be used to designate extractions following the porous solid model. Both models are illustrated in Figure 2A.2-1.

2A.2.2. Rate-Controlling Steps in Particle Reactions

In the case of particle conversion reactions there are three possible rate-controlling steps: (1) mass transport of reactant or product species in the fluid boundary layer adjacent to the particle; (2) the heterogeneous chemical reaction on the surface of the shrinking unreacted core; and (3) diffusion through the reaction product shell. For particle extraction reactions there are also three possible rate-controlling steps: (1) mass transport in the fluid boundary layer; (2) heterogeneous reaction on the internal surfaces of the particle; and (3) internal pore diffusion.

All of these six possible rate-controlling steps are included in considering the appropriate transfer equations for a sphere under four conditions: (1) a constant flux from the surface of a sphere of constant radius; (2) a constant flux from the surface of a shrinking sphere with the change in radius determined by the consumption of material (mass balance); (3) quasi-steady-state diffusion through a thickening shell with constant chemical potential at the boundaries and with the rate of shell thickening governed by the rate of mass transfer through the shell; and (4) non-steady-state diffusion (extraction) from a sphere of constant radius. These four transfer conditions and the corresponding rate-controlling steps are summarized in Table 2A.2-1.

The use of the mathematical relations for these rate-controlling steps in generating the design curves to be presented is limited to cases where only one rate-controlling mechanism is encountered throughout the particle reaction. When mixed kinetics are encountered, a specific

Table 2A.2-1. Summary of Particle Conversion and Extraction Processes and Design Curves

Transfer condition	Rate-controlling step: Conversion	Rate-controlling step: Extraction	Reaction time dependence on particle size	Design curves: Batch and plug flow reactor	Design curves: Backmix reactor	Design curves: Backmix reactor, multistage
Constant flux from a sphere of constant radius	Boundary-layer mass transport (only when a reaction product shell forms)	Boundary-layer mass transport Internal surface reaction	$\tau(R) \propto R$	Figure 2A.3-1 $F = f(t_p/\tau_m)$	Figure 2A.3-5 $F = f(\bar{t}/\tau_m)$	Figure 2A.3-9 $F = f(\bar{t}/\tau_m)$
Constant flux from a sphere of shrinking radius	Surface chemical reaction		$\tau(R) \propto R$	Figure 2A.3-2	Figure 2A.3-6	Figure 2A.3-10
	Boundary-layer mass transport (only when no reaction product shell forms for fixed bed and fluidized bed where the constant mass transfer coefficient is valid)			$F = f(t_p/\tau_m)$	$F = f(\bar{t}/\tau_m)$	$F = f(\bar{t}/\tau_m)$
Quasi-steady-state diffusion through a shell of constant external radius surrounding a shrinking core	Diffusion through the reaction-product shell		$\tau(R) \propto R^2$	Figure 2A.3-3 $F = f(t_p/\tau_m)$	Figure 2A.3-7 $F = f(\bar{t}/\tau_m)$	Figure 2A.3-11 2A.3-12 $F = f(\bar{t}/\tau_m)$
Non-steady-state diffusion from a sphere		Internal pore diffusion	$t_{\%}(R) \propto R^2$	Figure 2A.3-4 $F = f(D_{eff}t/R_*^2)$	Fgure 2A.3-8 $F = f(D_{eff}\bar{t}/R_*^2)$	Figure 2A.3-13 $F = f(D_{eff}\bar{t}/R_*^2)$

computer model for the system must be written. In addition, these systems are constrained to a constant surface-chemical reaction rate or to a constant chemical diffusing potential. This generally requires that the reaction be carried out at constant temperature or at a range of temperatures over which there is little change in reaction rate (or diffusivity). Although this constraint is severe, it is often met as an adequate engineering approximation in industrial leaching, roasting, and gaseous reduction reactions, which account for most of the metallurgical processes involving mineral particles. This is because operations are often conducted within a narrow temperature range that is dependent on either a control system or the throughput and energetics of the system. Furthermore, there is often an average concentration of oxygen (roasting) or other primary reactant that is constant over time for the continuous reactors of industrial interest. This latter condition is always met in an ideal backmix reactor because of the mixing effect.

2A.2.3. Particle Size Distribution

Only two feed size distributions are of general interest, the single (uniform) particle size and the Gates–Gaudin–Schuhmann distribution of feed particle sizes. The GGS distribution adequately describes the output of a grinding operation. Log-normal distributions of particle sizes have also been considered[2] but are usually not encountered in mineral systems and are not reviewed here.

There have been a large number of studies of the particle-size distribution resulting from grinding several ores, minerals, and other brittle solids. It has been observed that a logarithmic plot of the cumulative fraction of material finer than a given size vs. the corresponding size is a straight line with a slope, m, of 1 or slightly less.[3] The observed lower limit of the slope, m, is 0.7. This relationship has been expressed as the Gates–Gaudin–Schuhmann (GGS) equation[4]:

$$Y(R_i) = (R_i/R_*)^m \qquad (0.7 < m < 1.0) \qquad \text{(2A.2-2)}$$

where $Y(R_i)$ is the cumulative weight fraction finer than size R_i, and R_* is the maximum size after grinding. It should be noted that for a given material either continued grinding or grinding of a broad size range rather than a narrow size range simply results in a shift of the logarithmic plot without a change in slope.

Deviations from the straight-line GGS logarithmic relation are often observed among the largest particles comprising approximately the upper 15% of the weight fraction. Modifications of the GGS equation proposed to cover the largest sizes are reviewed by Harris[3] and Herbst and Fuerstenau.[5] In the case of closed-circuit grinding with size classification a considerable fraction, usually much larger than 15 wt %, of the particles entering the classifier are not allowed to pass but are returned to the grinding mill. In such instances the GGS equation is an adequate description of the size distribution eventually entering the reactor with R_* the largest size in the classifier overflow (fines). In the case of open-circuit grinding, the error resulting from use of the GGS equation is slight and because it predicts more large particles than are actually present, the GGS equation is a conservative size distribution for reactor design. The GGS equation has been used in computing the design curves that will follow. Values of the slope, m, between 0.7 and 1.0 have been used to test the sensitivity of reactor conversion efficiency on variations in the grinding parameter, m.

Differentiation of the GGS equation gives the fraction of feed, $\Delta Y(R_i)$, in the size range between R_i and $R_i + \Delta R_i$:

$$\Delta Y(R_i) = [m(R_i)^{m-1}\,\Delta R_i]/(R_*)^m \tag{2A.2-3}$$

and the sum of the discrete fractions is

$$\sum_{\text{all } R_i} \Delta Y(R_i) = 1$$

2A.2.4. Residence Time Distribution in the Reactor

The residence time distribution of the particles in the reactor depends on the flow pattern in the reactor vessel and whether the reaction is a batch process or carried out continuously. For batch and continuous, cocurrent, plug flow reactors, all particles in the particle aggregation experience the same residence time except for those particles that are completely reacted prior to experiencing the full reactor residence time, i.e., there is no backmixing or premature escape of particles from the reactor. An ideal backmixed (CSTR) reactor, e.g., an agitated leach tank or fluidized bed, can also be described by the residence time distribution equation of its contents including the particles. There are also residence-time equations describing a series of equal-sized backmix reactors, which are appropriate to a series of equal-sized agitated leach tanks.

2A.3. Computed Dimensionless Design Curves

The common types discussed in the previous section permit several combinations which have been computed and made available as dimensionless conversion–extraction yield curves. Details of the mathematical relations and computing methodology are found in the literature.[1,2,6–9] The results are applicable to ores and minerals and to a wide variety of other comminuted materials. The design curves relate conversion–extraction efficiency to a normalized reactor residence parameter consisting of the reactor residence time (average residence time for backmix reactors) divided by the time, τ_m, required to completely react the characteristic particle in the feed. Because the design curves are normalized they can be used for a variety of process design problems within the general limits that are outlined in this chapter.

To determine the reactor size that will provide a selected conversion–extraction efficiency or to ascertain the conversion–extraction efficiency yielded by a given reactor size, the designer must know only (1) *the feed rate*; (2) *the radius of the characteristic particle in the feed*; (3) *which type of reaction mechanism is rate controlling*; *and* (4) *the time*, τ_m, *for complete reaction of the characteristic particle.* The last two items usually must be determined from an appropriate laboratory kinetic study or its equivalent. The characteristic size is the largest particle size in the GGS distribution.

The appropriate design curves for a single particle size and for a GGS particle size distribution in the process feed are presented in Figures 2A.3-1–2A.3-13. Table 2A.2-1 is an index for selecting the appropriate design curve for the rate-controlling step and reactor-flow condition of any particular application. The relationship between the time required for complete particle reaction and particle radius, R, is also given in Table 2A.2-1.

The concept of a time for complete reaction is meaningless for nonsteady-state diffusion in a porous particle, which is the fourth and last transfer condition of Table 2A.2-1. The time required for a given percent extraction is proportional to R^2. For this case, the fraction extracted is expressed as a function of the dimensionless time parameter, θ or $\bar{\theta}$, where

$$\theta = D_{\text{eff}} t / R_*^2 \tag{2A.3-1}$$

$$D_{\text{eff}} = D\varepsilon / \tau_D \tag{2A.3-2}$$

The effective diffusivity, D_{eff}, can be derived from a theoretical estimate of the diffusivity using semiempirical correlations or from an experimental fit of the design curve using the experimentally derived relationships of (1) fraction extracted, (2) particle size, and (3) reaction time.

Figures 2A.3-1 to 2A.3-4 are fraction reacted or F curves for batch and continuous plug-flow reactors. In each case the upper curve corresponds to a single-particle size feed. The lower set of curves are for GGS particle-size distributions in the feed at various values of the GGS grinding parameter, m.

Figures 2A.3-5 to 2A.3-8 are F curves for a continuous flow, single-stage backmix reactor. In each figure the lower set of curves represents the GCS size distribution.

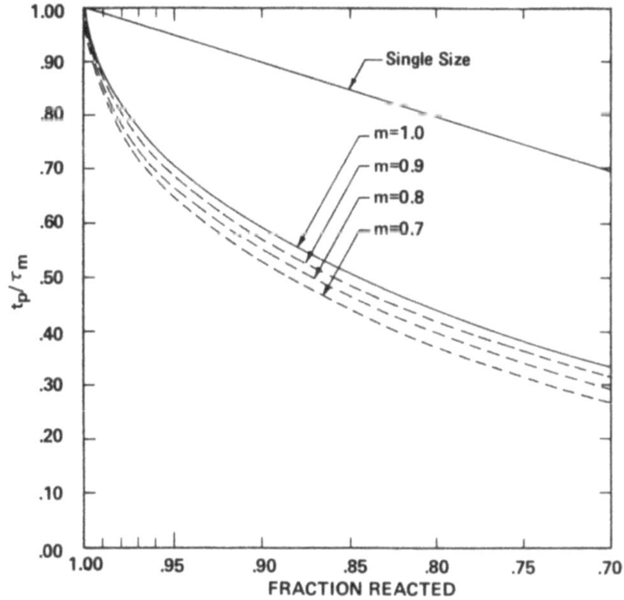

Figure 2A.3-1. Plug-flow and batch-reactor efficiency curves: extraction with boundary-layer transport rate controlling; extraction with internal surface chemical reaction-rate controlling; conversion with boundary-layer transport-rate controlling for constant particle radii.

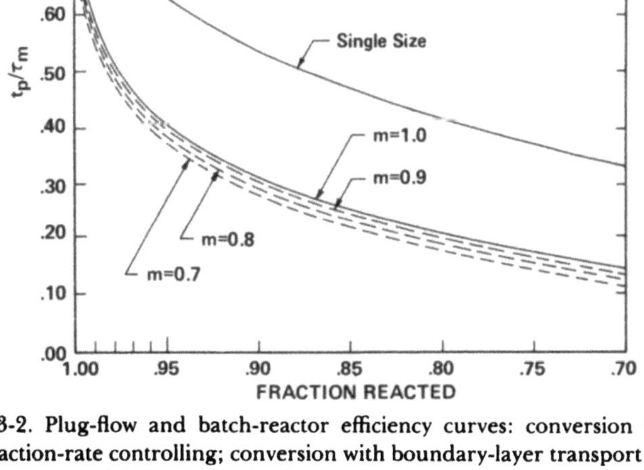

Figure 2A.3-2. Plug-flow and batch-reactor efficiency curves: conversion with surface chemical-reaction-rate controlling; conversion with boundary-layer transport rate controlling for shrinking particle radii (provided $Re_p < 50$).

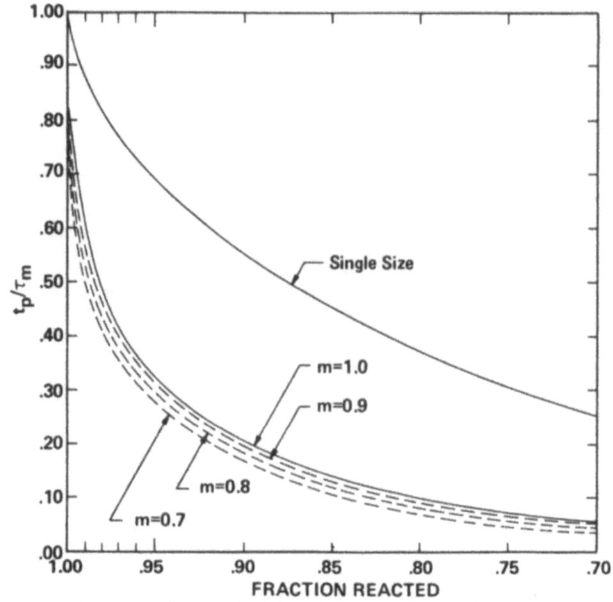

Figure 2A.3-3. Plug-flow and batch-reactor efficiency curves: conversion with diffusion through the reaction product shell rate controlling.

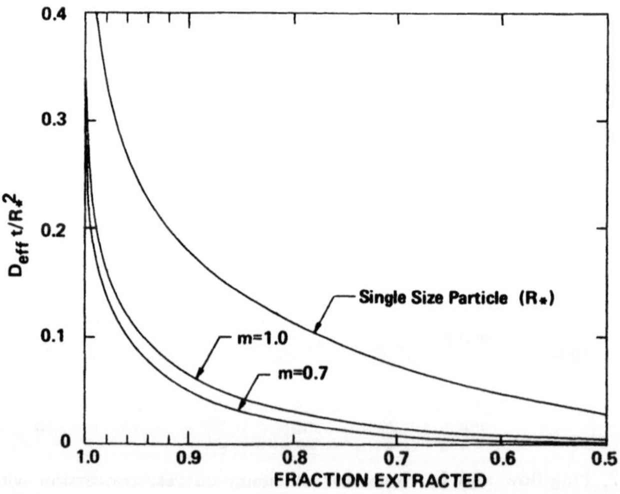

Figure 2A.3-4. Batch and continuous plug flow reactor extraction curves for particles when pore diffusion is rate controlling.

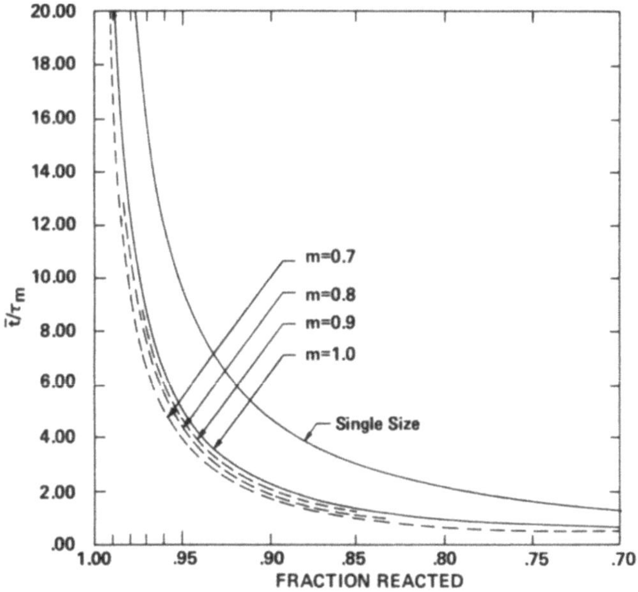

Figure 2A.3-5. Single backmix flow reactor efficiency curves: extraction with boundary-layer transport rate controlling; extraction with internal surface chemical reaction rate controlling; conversion with boundary-layer transport rate controlling for constant particle radii.

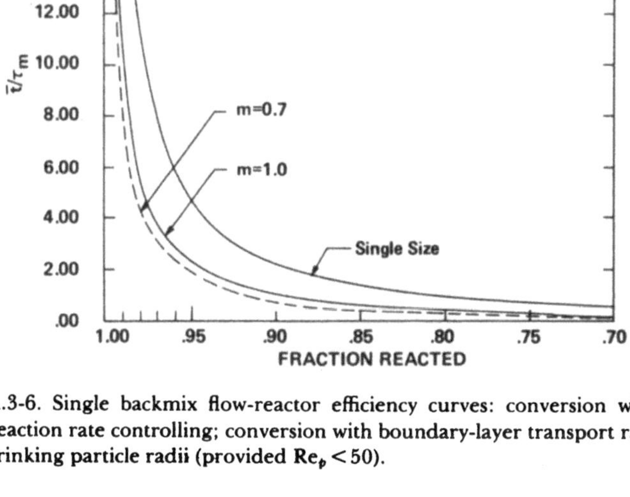

Figure 2A.3-6. Single backmix flow-reactor efficiency curves: conversion with surface chemical reaction rate controlling; conversion with boundary-layer transport rate controlling for shrinking particle radii (provided $Re_p < 50$).

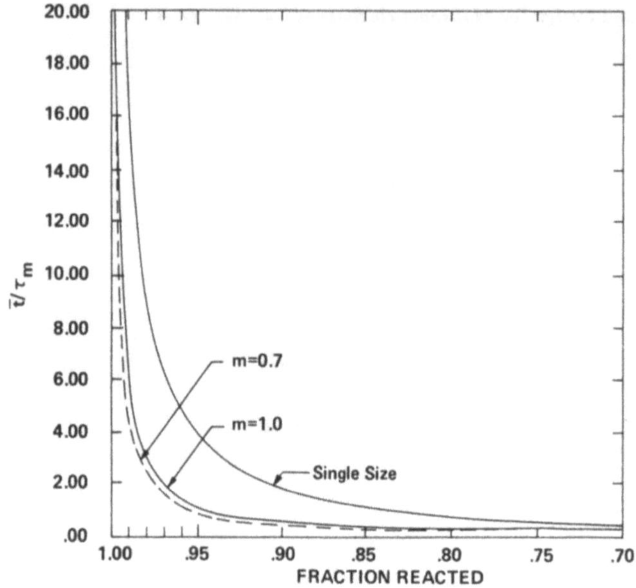

Figure 2A.3-7. Single backmix flow-reactor efficiency curves: conversion with diffusion through the reaction-product-shell rate controlling.

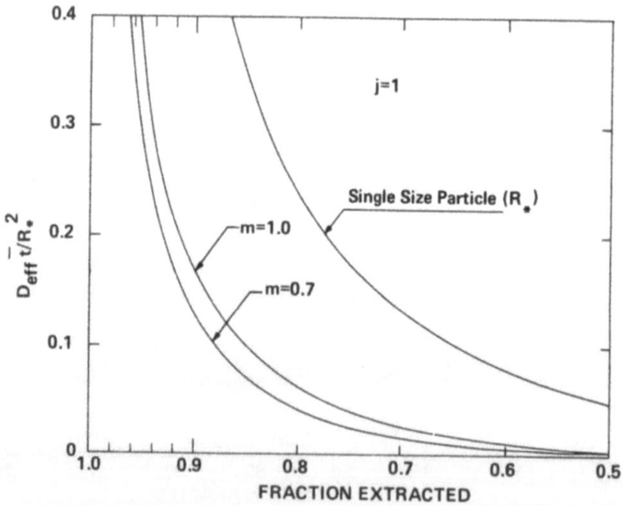

Figure 2A.3-8. Backmix continuous flow-reactor extraction curves for particles when pore diffusion is rate controlling.

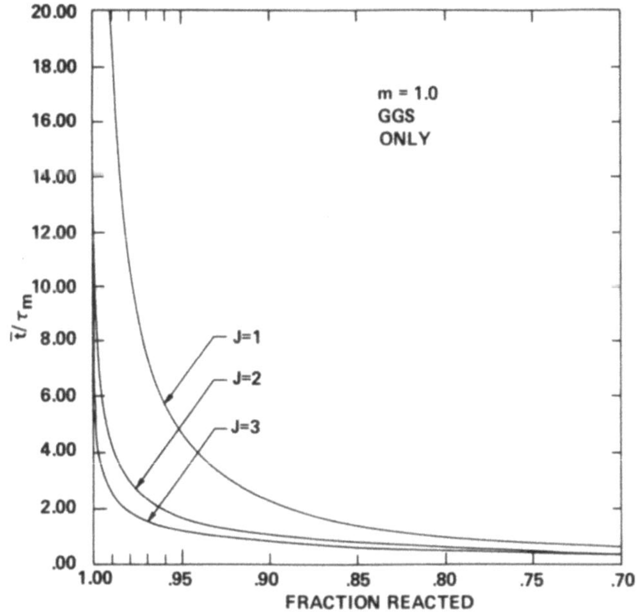

Figure 2A.3-9. Multistage backmix flow-reactor efficiency curves (J = number of stages): extraction with boundary-layer transport rate controlling; extraction with internal surface chemical reaction rate controlling; conversion with boundary-layer transport rate controlling for constant particle radii.

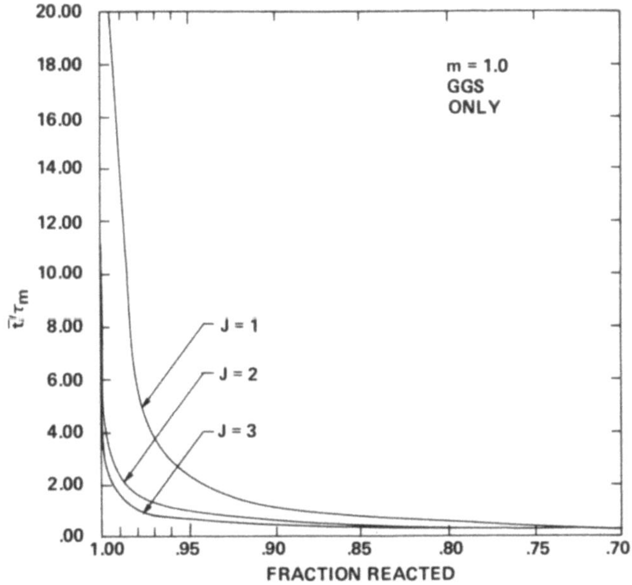

Figure 2A.3-10. Multistage backmix flow-reactor efficiency curves (J = number of stages): conversion with surface chemical reaction rate controlling; conversion with boundary-layer transport rate controlling for shrinking particle radii (provided $Re_p < 50$).

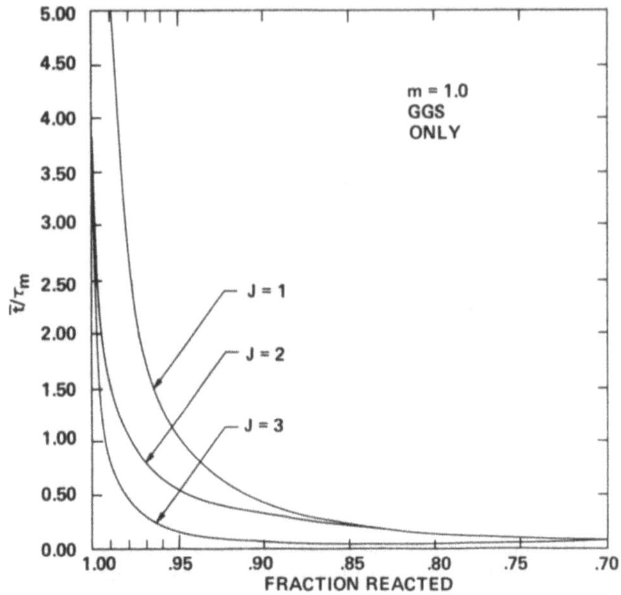

Figure 2A.3-11. Multistage backmix flow-reactor efficiency curves (J = number of stages): conversion with diffusion through the reaction product shell rate controlling.

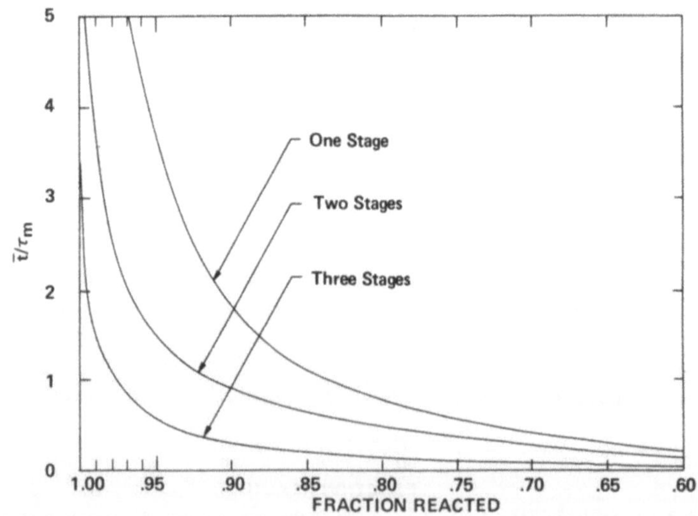

Figure 2A.3-12. Chemical conversion for uniform diameter particles in equal-sized ideal backmix reactors in series with particle shell-diffusion rate controlling.

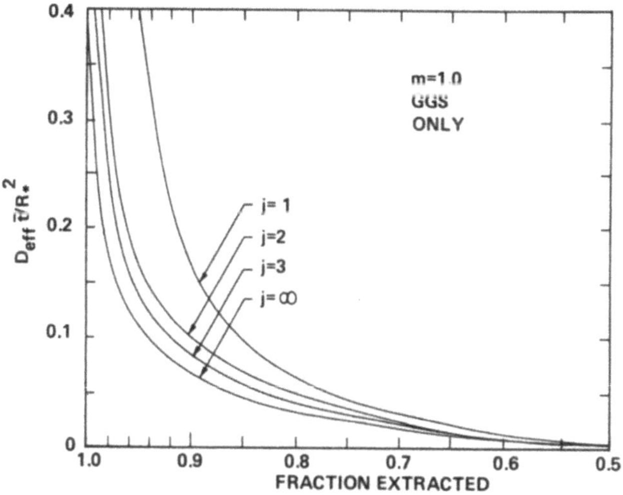

Figure 2A.3-13. Multistage backmix reactor extraction for particles when pore diffusion is rate controlling.

Figures 2A.3-9–2A.3-13 contain *F* curves for continuous flow, multistage backmix reactors. The curves in Figure 2A.3-12 are for a single-particle size feed. The remaining figures in this group pertain to GGS particle-size distributions in the feed.

2A.4. Application of F Curves in Process Design

2A.4.1. Determining the Kinetic Data

Application of the *F* curves should begin in the laboratory stage when the rate mechanism and kinetic data are being developed. For example, single-size particles in the form of a narrow sieve cut may be investigated with respect to the process step in a batch reactor, usually at constant temperature and reacting chemical potential. Extractions at various times may be determined and compared with the design curves to help determine the rate mechanism. The effect of particle size on yield may also be determined. Suppose that in a series of experiments with a particular process it is found that the time required for complete particle reaction doubles with a doubling of the particle size. This indicates either boundary-layer mass transport or a surface chemical reaction. For single-size particles a comparison of the fraction reacted, *F*, versus reaction time in the region of more than 80% reacted with Figures 2A.3-1 and 2A.3-2 would indicate which of these particle reaction mechanisms is operative.

The design curve fit indicates the reaction mechanism, but these data must be consistent with other valid kinetic data such as microscopic, structural, and chemical information before it can be accepted.

2A.4.2. Scale-Up

The laboratory study must provide τ_m as a function of the experimental value of R_* which can be used to extrapolate to τ_m for the characteristic radius of the industrial particle-size distribution. Alternatively, D_{eff} is determined experimentally when diffusion in a porous solid particle is rate controlling.

Dimensionless reactor times are determined from the desired yield, F, and the design curve (Figures 2A.3-1–2A.3-13). Then the average reactor residence time, t_p or $\bar{t}$, is determined from the dimensionless time and τ_m or D_{eff}. From the average residence time and specified throughput the reactor volume can be determined by equation (2A.2-1). Also, by looking at a series of these relationships, economic optimization of reactor size (cost) vs. extraction efficiency (revenue) can be obtained.

2A.4.3. Discussion

It is not expected that these design curves would usually be used by themselves to design a large industrial plant, but they should be used in a pilot plant design and in planning and evaluating the pilot plant test program. They can also be helpful in the intelligent design, evaluation, and scale-up of pilot plant reactors that deviate from ideal plug-flow or backmix-flow behavior. For example, a preliminary plant design on the assumption of ideal flow behavior in the process reactor and use of these design curves may show that a large capital outlay will be required and that the process is uneconomic. In this instance the large expenditure of building and operating a pilot plant would not be justified. In other instances, a preliminary design may indicate an economic process and favor further pilot plant evaluation.

These design curves allow the transposing of conversion–extraction efficiencies among reactors of different types or from particles of a single size to the size distributions resulting from grinding. This makes it relatively easy to make cost comparisons of different competing reactor types for implementation of a given process. For example, simple batch agitator leaching tests of an ore could be used to obtain the type of particle reaction morphology and $\tau(R)$. Then, this information could be used to calculate the required size of a continuous tube digestor or a series of agitators for continuous leaching. In most instances, such as those where the continuous flow reactor is approximately ideal, further tests would not be required.

The design results are as reliable as the input data, most particularly the reliability of τ_m, the rate mechanism constancy, and the degree to which flow in the reactor approaches either of the ideal types, which may be thought of as limiting extremes in flow behavior. The generation of the F curves is computationally exact and should be viewed as merely a summation of the reaction behavior of all particles over their time-distribution elements; see references 7 and 9 for computational details.

2A.5. Application Examples

Example 1. In batch leaching tests a residence time of 2 hr is required to obtain 95% dissolution of metal in a pulp of 25 wt % solids. The solid density is 2.5 g/cm^3. I wish to process 10 tons per 8-hr shift in a continuous tube digester using the same pulp density. (a) What size (in ft^3) should I use to obtain the same yield, 95%? (b) What can be said about the size of a continuous agitation tank as an alternative leach vessel? (c) What can be said about the size of a series of six equal-sized agitators to accomplish this continuous leaching task?

Answer: (a) A tube digester with a high aspect ratio is identical with a plug flow reactor; therefore,

$$t_p = t_b = 2\text{ hr}$$

$$\text{pulp} = \begin{cases} 25\text{ wt \% solids} & (10\text{ tons/8 hr} \times 2\text{ hr}) \\ 75\text{ wt \% water} & (30\text{ tons/8 hr} \times 2\text{ hr}) \end{cases}$$

$$\text{vessel holds:} \begin{cases} 2.5\text{ tons solids} = 5000\text{ lb} \\ 7.5\text{ tons water} = 15{,}000\text{ lb} \end{cases}$$

$$\text{volume of pulp} = \frac{15{,}000\text{ lb}}{62.5\text{ lb/ft}^3} + \frac{5000\text{ lb}}{2.5(62.5\text{ lb/ft}^3)} = 272\text{ ft}^3$$

(b) It is impossible to determine the equivalent agitator vessel size because we do not know the rate-controlling leaching mechanism.

(c) The case of 6 agitated tanks in series is different because regardless of the rate mechanism *all* particles experience at least 98% of the average residence time in the six-tank system. Hence, the active volume of this system needs to be about the same, 272 ft^3(45.3 ft^3 per tank).

Example 2. A new copper concentrate leaching process yields the following data at 90 °C for 100 mesh (closely sized) chalcopyrite concentrate in batch tests:

Leach time, hr:	114	144	192	240
% Leached:	70	79	91	97

Because of the slow leaching rate and discrete minerals, only two rate processes for a core–shell model are suspected, i.e., surface chemical reaction or shell diffusion. (a) Which rate mechanism best fits the data? (b) What average residence time is required for a three-stage, equal-sized series of leach vessels with a feed of −100-mesh ground particles to obtain 99% conversion? (c) If the pulp volumetric flow rate is 150 gal/min, what should be the working volume of each tank? (d) If you ground concentrate to −325 mesh before leaching, how much smaller should the vessels be?

Table 2A.5-1. Possible Rate Mechanism Curve Data

Given data			Figure 2A.3-2		Figure 2A.3-3	
F	$1-F$	t_p	t_p/τ_m	τ_m	t_p/τ_m	τ_m
0.70	0.30	114	0.325	350	0.26	438
0.79	0.21	144	0.41	351	0.35	411
0.91	0.09	192	0.56	343	0.575	334
0.97	0.03	240	0.695	345	0.77	312
			Average	347 hr		

Note: (a) t_p/τ_m is obtained from $1-F$ and the F curves. Then τ_m is computed from the measured t_p.
(b) Since the values of τ_m from Figure 2A.3-2 are more consistent, whereas the values of τ_m from Figure 2A.3-3 trend, we conclude that in Figure 2A.3-2, surface chemical reaction is rate controlling.

Answer: (a) The possible rate mechanism curves for a batch reactor are shown in Figures 2A.3-2 and 2A.3-3 for single-sized particles (see Table 2A.5-1).

(b) τ_m is the same, since the maximum size is 100 mesh. We use Figure 2A.3-10 for $F=0.99$ (99%), $\bar{t}/\tau_m = 1.5$; hence, $\bar{t} = 1.5\times 347$ hr $= 520$ hr.

(c) System volume = 150 gal/min × 60 × (520 hr) = 4,680,000 gal; one tank = 4,680,000/3 = 1,560,000 gal working volume. This would have to be divided into several parallel trains and may be uneconomic.

(d) $\tau_m \propto R_*$

$$100 \text{ mesh} = 0.147 \text{ mm diam}$$

$$325 \text{ mesh} = 0.044 \text{ mm diam}$$

$$\tau_{m325} = \frac{0.044}{0.147}\tau_{m100}$$

$$\tau_{m325} = \frac{0.044}{0.147} 347 \text{ hr} = 103.9 \text{ hr}$$

Therefore, vessel size = (0.044/0.147)1,560,000 = 466,940 gal (29.9%).

Example 3. The lime concentrate pellet roast (LCPR) process involves pelletizing a mixture of lime and copper concentrate. The pellet is roasted with simultaneous fixation of the sulfur and subsequently leached to recover copper.[10] The pellet roasting follows the core–shell model with diffusion of oxygen through the roasted shell controlling the rate. Closely screened 4-mesh pellets are made and the kinetic data show that it takes 30 min to completely roast this size pellet. (a) What leach efficiency would you expect if a fluid bed roaster is used with a 4-hr residence time? (b) Actual pilot-plant tests at a 4-hr residence time yield somewhat less, 96.5% leaching extraction. What pilot-plant extraction would you expect on doubling the feed rate? (c) Commercial plans are to provide 2-hr average retention time in the fluid-bed roaster and a hot "well" where pellets discharged from the fluid bed cool slowly in air but remain above the minimum reaction temperature for 20 min. The hot well acts like a shaft furnace and there is no fluidization. What extractions do you expect to obtain commercially?

Answer. (a) $\bar{t}/\tau = 240\ \text{min}/30\ \text{min} = 8$. Using Figure 2A.3-7, $F = 97.5\%$ reacted.

(b) Using Figure 2A.3-7, $\bar{t}/\tau_m = 4.0$; $F = 95\%$; extraction $= (95/97.5)(96.5) = 94.0\%$.

(c) Based on Figure 2.2-4, $t_p/\tau_m = 0.67$, $F = 0.94$ for the hot well. However, because of the variable reaction time distribution on exiting the fluid bed roaster and entering the hot well, this problem cannot be solved exactly without a computer, but

$$0.95 < F < 0.95 + 0.05(0.94) = 0.997$$

References

1. O. Levenspiel, *Chemical Reaction Engineering*, Chap. 12, Wiley, New York (1962).
2. R. W. Bartlett, N. G. Krishnan, and M. C. Van Hecke, *Chem. Eng. Sci.* **28**, 2179–2186 (1973).
3. C. C. Harris, *Trans. SME–AIME* **241**, 343–358 (1968).
4. R. Schuhmann, *Trans. SME-AIME* **217**, 22–25 (1960).
5. J. A. Herbst and D. W. Fuerstenau, *Trans. SME-AIME* **241**, 538–548 (1968).
6. D. Kunii and O. Levenspiel, *Fluidization Engineering*, Chap. 7, Wiley, New York (1969).
7. R. W. Bartlett, *Met. Trans.* **2**, 2999–3006 (1971).
8. J. Crank, *Mathematics of Diffusion*, Oxford University Press, London (1964), pp. 85–86.
9. R. W. Bartlett, *Met. Trans.* **3**, 913–917 (1972).
10. *J. Metals*, December (1973).

3

Hydrometallurgical Processes

Sec. 3.1. Milton E. Wadsworth
Sec. 3.2. Milton E. Wadsworth
Sec. 3.3. Jan D. Miller

3.1. Principles of Leaching (MEW)

3.1.1. Introduction

Although the classification of hydrometallurgical processes is arbitrary, it is convenient to consider two general areas: (1) the leaching of ores, and (2) the leaching of concentrates. In some cases these ores or concentrates may be subjected to some pretreatment such as roasting or reduction to improve the extraction. By definition, an ore deposit is a naturally occurring mineral deposit which can be treated economically. Under this definition the leaching of low-grade materials, normally considered to be waste products, would fall into the first category and would, if leached at a profit, be termed an ore. Ores within this definition may be subdivided into low-grade materials and moderate-to-high-grade ores. The first would refer to materials of sufficiently low grade that it is not economic to subject them to additional treatment such as fine grinding and concentration, although sizing may be carried out. The diagram in Figure 3.1-1 illustrates the classification of hydrometallurgical treatment according to the above definition.

From the point of view of rate processes, a variety of conditions can alter the response of an individual mineral particle. For example, when leaching concentrates, high pulp densities are desirable to minimize the

Milton E. Wadsworth and ***Jan D. Miller*** • Department of Metallurgy and Metallurgical Engineering, University of Utah, Salt Lake City, Utah

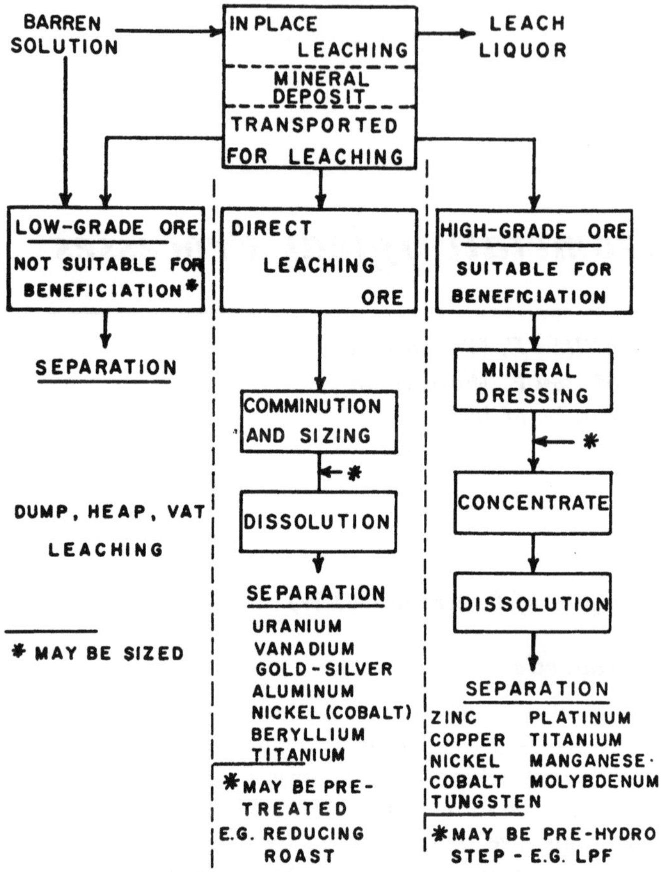

Figure 3.1-1. General classification of ores.

size of the reactor used; however, the solubility of lixiviant or product limit the effective pulp density. Agitation of pulps is usually necessary to maximize the kinetics and short reaction times are desirable for economic reasons. The same mineral particle embedded in an ore fragment may release its values in a time period extending to months or even years (for massive low-grade deposits) and still be economically feasible. Thus, conditions under which a given mineral is treated hydrometallurgically may vary broadly.

In general, a reaction-rate study can provide one of two kinds of information. From a practical point of view, studies carried out on a given system approximating conditions to be encountered commercially provide valuable engineering data leading from bench-scale to pilot tests and finally to full-scale operation. A second and more fundamental outcome of a detailed rate study arises from the successful interpretation of rates in terms of mechanism. If sufficient information is available, the fundamental approach inherently has the ability to predict results under a variety of conditions as well as to indicate fundamental concepts.

The majority of hydrometallurgical and pyrometallurgical systems are heterogeneous with the consequence that the slow step in a series of reactions may also be heterogeneous, i.e., involve solid–liquid, liquid–liquid, or solid–gas interfaces. In some cases, the slow step in a heterogeneous system may be homogeneous if diffusion through one of the phases is rate controlling. A detailed treatment of chemical kinetics is not possible here and only those concepts pertinent to hydrometallurgical systems will be covered specifically.

The first recorded reaction-rate experiments were conducted by Wenzel[1] in which he studied the dissolution of copper and zinc in acid solutions. Wilhelmy[2] developed equations describing the rate of inversion of sucrose. A formalized concept of the reaction-rate constant was provided by van't Hoff[3] and Arrhenius[4] some 40 years after the work of Wilhelmy.

In its exponential form the Arrhenius equation is

$$k = \nu e^{-\Delta E/RT} \qquad (3.1\text{-}1)$$

It is apparent from the equation (3.1-1) that the logarithm of the specific rate, k, should be linear with reciprocal temperature. The term ν is the *frequency factor* and ΔE is the *energy of activation.*

From a practical viewpoint, reaction rates may be expressed by simple differential equations. Engineering applications of these equations permit interpolation and extrapolation of measured rate data to a variety of conditions of temperature, geometric configuration, and pressures or concentrations.

Reactions in solutions must follow a sequence of steps, of which one or more may be rate controlling. These are: (1) diffusion of reactants towards each other, (2) reaction of reactants with one another, and (3) diffusion of products away from each other (Glasstone *et al.*[5]). Temperature coefficients can readily separate Steps 1 and 3 from Step 2 since the activation energy for solution diffusion is usually of the order of 5 kcal or less. Solution reactions in which Step 2 is operative may be identified from temperature data since the activation energies involved usually occur between 10 and 25 kcal. In many cases, diffusion may be eliminated as a slow step by increased agitation of the solution or by lowering the temperature. The relatively small variation of rate with temperature in diffusion reactions may result in diffusional control at higher temperatures. The change of slope in an Arrhenius plot (log k vs. $1/T$) to a smaller value as temperature increases is consistent with a change in the slow step from Step 2 to Step 1 or 3.

It should also be pointed out that several Step-2-type reactions might prevail in a given system because of interdependent equilibrium or steady-state reactions with reactants, intermediates, or products of reaction. In aqueous systems, pH effects related to hydrolysis of salts of weak acids or weak bases may be important. A complete evaluation of the kinetic system requires knowledge of these side reactions.

3.1.2. Heterogeneous Kinetics of Importance in Hydrometallurgy

3.1.2.1. Diffusion at Surfaces

Almost without exception rate-controlling processes in hydrometallurgical systems are heterogeneous, i.e., involve reactions at interfaces. The term *homogeneous* implies that a given reaction occurs completely within one phase of a system. Very often, however, an apparent homogeneous reaction may in reality be heterogeneous since the walls of the container or some solid surface which must always be present may act as a catalyst. One of the first considerations in the interpretation of so-called *homogeneous reactions* is to determine whether or not external effects are involved. In a heterogeneous system, the slow step in a series of reactions may occur in one phase and of itself be homogeneous. For example, the corrosion of a metal, dissolution of a mineral, or hydrogen precipitation of a metal may involve solution diffusion.

Basically, diffusion is a process which tends to equalize concentration within a single phase. The driving force is the concentration gradient within the phase, and the concentration at any time is related to the diffusion coordinate, the concentration gradient, and the influence of concentration and temperature upon the coefficient of diffusion. It was recognized by Fick[6] that diffusion closely parallels heat flow. Fick's first law of diffusion is

$$J = -D\frac{\partial C}{\partial x} \qquad (3.1\text{-}2)$$

where J is the amount of material diffusing per unit of time in a direction perpendicular to a reference plane having unit cross-sectional area, C the concentration, x the position coordinate measured perpendicular to the reference plane, and D the coefficient of diffusion. In cgs units, D has the dimensions $\text{cm}^2\,\text{sec}^{-1}$. The quantity J is molecules $\text{sec}^{-1}\,\text{cm}^{-2}$ or moles $\text{sec}^{-1}\,\text{cm}^{-2}$, depending upon whether or not C is measured on molecules per cm^3 or moles per cm^3. The quantity J may be expressed in the form

$$J = \frac{1}{A}\frac{dn}{dt} \qquad (3.1\text{-}3)$$

in which n is the amount (e.g., molecules or moles), t is time, and A is the area of the reference plane. Combining equations (3.1-2) and (3.1-3) results in the general expression for Fick's first law:

$$\frac{dn}{dt} = -DA\frac{\partial C}{\partial X} \tag{3.1-4}$$

The negative sign indicates a decrease in n with time. Integration of equation (3.1-4) under steady-state conditions, $J = \text{constant}$, gives

$$\frac{1}{A}\frac{dn}{dt} = J = -D\frac{\Delta C}{\Delta x} \tag{3.1-5}$$

In most hydrometallurgical systems little error is encountered in considering the coefficient of diffusion as constant and independent of concentration. In systems involving dissolution or precipitation of a solid, diffusion through a zone adjacent to the solid–liquid interface may be rate controlling. Under constant conditions of agitation, the thickness of this zone remains constant and a steady-state condition is soon attained in which the amount of material entering the zone balances that leaving the zone.

Figure 3.1-2 represents the diffusion boundary adjacent to the solid surface with concentration (dashed line) across the boundary varying linearly according to equation (3.1-5). The true concentration is indicated by the solid line which illustrates the errors associated with the simplifying assumptions. If x is measured perpendicular to and positive

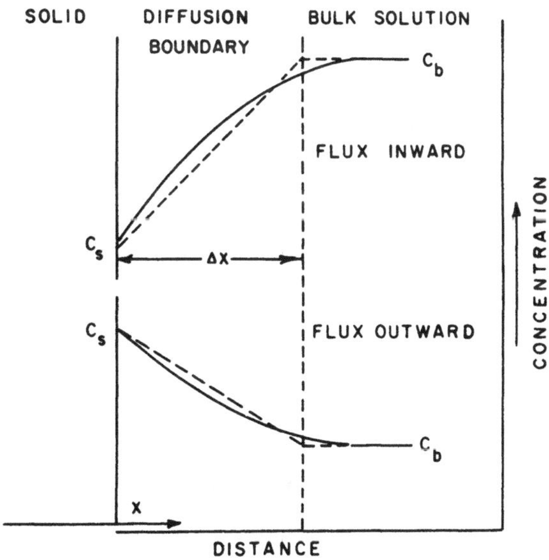

Figure 3.1-2. Steady-state diffusion boundary at solid–liquid interface.

in the direction away from the solid surface, equation (3.1-5) becomes

$$\frac{C_b - C_s}{\Delta x} = -\frac{J}{D} \quad \text{(diffusion toward surface)} \tag{3.1-6}$$

where C_b is the bulk concentration and C_s the surface concentration. Also

$$\frac{C_b - C_s}{\Delta x} = +\frac{J}{D} \quad \text{(diffusion away from surface)} \tag{3.1-7}$$

Normally, Δx will assume some average value under conditions of constant agitation. For constant flux, C_b must be held essentially constant. Essentially constant concentration results for systems having relatively large concentrations and large solution-to-solid ratios so that solution depletion or back reactions because of the buildup of reactants are negligible. Flow systems with good agitation, as an approximation, operate also with a constant steady-state concentration equivalent to that leaving the reactor.

It is often possible to eliminate stirring as a variable in a rate study if agitation is great enough. Figure 3.1-3 illustrates the variation in rate with stirring speed often encountered in the dissolution of solids. It should be emphasized, however, that the elimination of agitation as a variable is no assurance that diffusion as the rate controlling step has also been eliminated. As agitation is increased, $\Delta x \to \delta$, where δ is the effective minimum thickness of a limiting boundary layer of solution adjacent to the solid surface. Diffusion across this film may still be rate controlling. Under these conditions calculated mass transfer rates or measurement

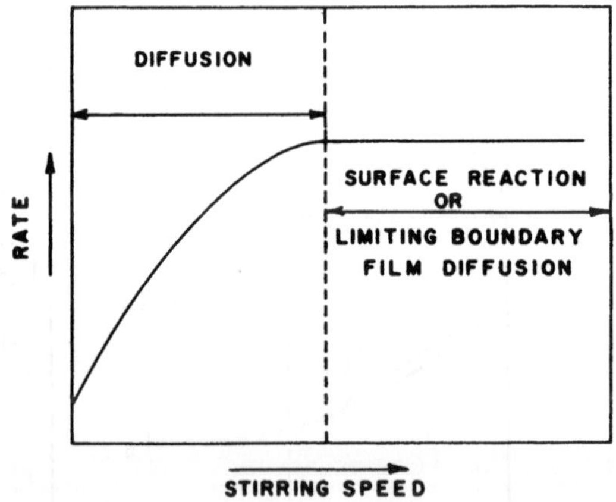

Figure 3.1-3. Rate of reaction as related to stirring speed illustrating elimination of stirring as a variable at high stirring speeds.

of the activation energy is necessary to determine whether diffusion through the limiting boundary or a surface reaction is rate controlling. Wagner[7] and Lebrun[8] studied the kinetics of crystal growth by increasing agitation until no further increase in rate was observed. They concluded that the minimum boundary thickness obtained was between 0.002 and 0.012 cm. Calculated thicknesses from basic laws of diffusion, using the Stokes–Einstein diffusion equation (Kortum and Bockris[9]) suggest that the diffusion boundary in unstirred solutions is approximately 0.05 cm, diminishing to approximately 0.001 cm under conditions of violent agitation. Using a thickness of 10^{-3} cm for the limiting boundary layer, it has been shown by Halpern[10] that the maximum rate attainable for a diffusion-controlled reaction at the solid–solution interface is approximately 10^{-1} moles $cm^{-2}\,h^{-1}$ when the bulk solution activity is unity. Similarly, at medium temperatures and pressures, this rate would diminish by a factor of 10^2–10^3 in the case of gases dissolved in aqueous solutions because of the limitations of solubility.

The effect of stirring on metal electrode kinetics under conditions of laminar and turbulent flow has been reviewed by Vetter.[11] For laminar flow, parallel to an electrode surface, the thickness of the limiting diffusion layer, δ, increases with distance, l, from the leading edge of the electrode measured parallel to the electrode surface. The limiting boundary thickness according to Levich[12] is

$$\delta = 3l^{1/2}\, v_\infty^{-1/2}\, \nu^{1/6}\, D^{1/3} \tag{3.1-8}$$

where v_∞ is the solution velocity of the bulk solution parallel to the surface and at large distances from the surface, ν the kinematic viscosity, and D the diffusion coefficient. The kinematic viscosity is the ratio of liquid viscosity divided by liquid density. Consequently δ is not constant over the surface. Also the limiting-boundary thickness does not correspond to the total interval over which the velocity of solution diminishes because of the electrode-wall effect. Prandtl[13] has shown the flow-boundary thickness, δ_{Pr}, to be given by

$$\delta_{\mathrm{Pr}} = 3l^{1/2}\, v_\infty^{-1/2}\, \nu^{1/2} \tag{3.1-9}$$

Figure 3.1-4 is a normalized plot of C/C_b and v/v_∞ according to equations (3.1-8) and (3.1-9) for x values perpendicular to the electrode surface. Clearly the diffusion boundary thickness is much less than the flow boundary thickness. The boundary thickness for flat–flat plates must be considered as some average $\bar{\delta}$ value over the surface dependent upon the electrode length, L. Levich[12] has shown that for rotating disks of radius r, for $\delta \ll r$, the δ is constant over the surface for laminar

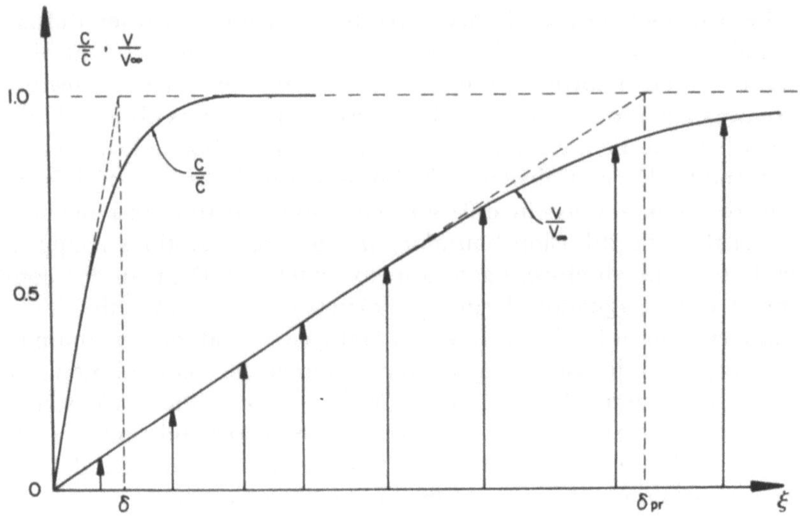

Figure 3.1.4. Distribution of concentration (C/C_b) and flow rate (V/V_∞) with laminar flow as a function of distance x from a plane surface.

flow. The value of δ is given by

$$\delta = 1.61\,\omega^{-1/2}\,\nu^{1/6}\,D^{1/3} \tag{3.1-10}$$

where ω is the angular velocity of the rotating disk.

Mass transport under conditions of turbulent flow has not been solved exactly. In the case of turbulent flow, a continuous variation in concentration over the interval δ does not occur. The concentration fluctuates at each coordinate position so that the concentration at this position is a time-average concentration. Vetter[11] discusses the Vielstich equation for turbulent flow adjacent to a flat surface, which is

$$\delta \sim l^{0.1}\,v_\infty^{-0.9}\,\nu^{17/30}\,D^{1/3} = l\,\mathrm{Re}^{-0.9}\,\mathrm{Pr}^{-1/3} \tag{3.1-11}$$

where Re is the Reynolds number, vl/ν, and Pr is the Prandtl number, ν/D. The limiting thickness is far less dependent on l so that an average δ value may be applied over the reacting surface. The average flux over the electrode surface of length L is given by

$$\bar{J} = \frac{1}{L}\int_0^L J(l)\,dl \tag{3.1-12}$$

Several investigators[11] have confirmed that $\bar{\delta}$ is approximately 10^{-3} cm.

The value of D varies but slightly for a large number of liquids.[14] In aqueous solutions D has values ranging from approximately 0.3×10^{-5} to 3.5×10^{-5} cm^2/sec at room temperature. In molten salts, D

ranges between 2×10^{-5} to 11×10^{-5} cm^2/sec. The value of D changes greatly for very complex molecules, decreasing by a factor of 10 to 10^2 for high polymers in water.

3.1.2.2. *Sample Geometry*

In heterogeneous systems the importance of sample geometry cannot be overemphasized. This is particularly true for oxidation, reduction, dissolution, and precipitation reactions in which the surface of a solid is itself receding or advancing during the course of the reaction. If the surface does not change position, as is true for a solid catalyst, and some electrochemical reactions, sample geometry is of less importance inasmuch as the overall area does not change during the course of the reaction. In heterogeneous systems, rates of reaction are often related to both the total surface and the number of reactive sites per unit area available for reaction. The surface concentration may be measured in moles or molecules per cm^2. More specifically, the rate itself may be determined by the number of potentially reactive sites which are occupied by reactants if surface equilibrium or steady-state adsorption contributes to the reaction. The general rate expression in which surface area is involved is, written for a first-order reaction,

$$dn/dt = -ACk_0k' \quad (3.1\text{-}13)$$

where A is the surface area of a solid, C the solution concentration, k' the rate constant, and k_0 the concentration of potentially reactive surface sites. Equation (3.1-13) represents conditions for which products of reaction do not form protective coatings. The true surface area A may not be known and is difficult to measure for nonmetals having a low surface area. If the area is large enough, standard gas-adsorption techniques may be used for area determination.(15) The surface areas of metals may be measured by capacitance measurements in solution.(16) The geometric area is that area which may be measured directly with a micrometer or some suitable scale. The true surface area is related to the geometric area by a surface-roughness factor f. Rate data may be normalized in terms of the apparent or geometric surface area if f does not change during the course of the reaction. If A in equation (3.1-13) is the geometric area, k_0 contains the surface-roughness factor. The negative sign in equation (3.1-13) implies that n, the total amount of reactant, is decreasing with time.

Sample geometry greatly influences the variation in surface area during the course of a reaction. Samples small in one dimension such as flat plates or disks have a minimum variation in area. Those having the three geometric coordinates equal will have the greatest variations in area, assuming that the rate of reaction is the same in all directions. Geometric variations associated with isometric shapes have been studied by Spencer and Topley.[17] Eyring *et al.*[18] have treated this problem for the burning of grains of explosives in which the radius of the burning grain varies linearly with time. The examples given here are for spheres but the final results are equally applicable to cubes, octahedrons, or any isometric shape. The rate of reaction at the surface of a sphere of radius r may be expressed by the equation

$$dn/dt = -4\pi r^2 C k_0 k' \tag{3.1-14}$$

where n is the number of moles (or molecules) remaining in the unreacted core. The rate constant, k', has the dimensions liters $\sec^{-1}$ if the concentration is in moles per liter, or the units $\text{cm}^3 \sec^{-1}$ if the concentration is in moles per cm^3. The total number of moles, n, in the unreacted sphere is

$$n = \frac{4\pi r^3}{3V} \tag{3.1-15}$$

where V is the molar volume and is equal to M/ρ, where M is the molecular weight and ρ the density. Equation (3.1-15) may be differentiated with respect to time and equated to equation (3.1-14) resulting in the expression

$$dr/dt = -VCk_0k' = -\mathcal{R}_l \tag{3.1-16}$$

where $\mathcal{R}_l$ is the linear rate, and Vk_0k' is the linear rate constant, k_l ($\text{cm}^4\,\text{mole}^{-1}\,\sec^{-1}$). If C is in moles cm^{-3}, $\mathcal{R}_l$ has the units cm $\sec^{-1}$. If C is constant, equation (3.1-16) represents the constant velocity of movement of the reaction interface, which is the basic definition of linear kinetics. If r_0 is the initial radius of the sphere and α is the fraction reacted, it may be shown that

$$\alpha = 1 - \frac{r^3}{r_0^3} \tag{3.1-17}$$

which upon differentiation with respect to time becomes

$$\frac{d\alpha}{dt} = -\frac{3r^2}{r_0^3}\frac{dr}{dt} \tag{3.1-18}$$

Combining equations (3.1-16), (3.1-17), and (3.1-18) gives equation

$$\frac{d\alpha}{dt} = \frac{3Ck_l}{r_0}(1-\alpha)^{2/3} \tag{3.1-19}$$

For the conditions $\alpha = 0$ when $t = 0$, equation (3.1-19) may be integrated for constant concentration giving

$$1-(1-\alpha)^{1/3}=kt \tag{3.1-20}$$

where $k = Ck_l/r_0$ (time^{-1}). A plot of the left side of equation (3.1-20) against t should result in a straight line having a slope k with units of reciprocal time. If the rate of reaction of a single grain follows equation (3.1-20), a composite rate for many grains will also follow it, providing all of the grains have the same initial diameter.

In applying equation (3.1-20) to a pulp containing a distribution of particle sizes, the weight fraction, w_i, of each particle-size distribution must be known. If $\bar{r}_{0i}$ represents the average initial radius of the particle in weight fraction w_i, equation (3.1-20) is

$$1-(1-\alpha_i)^{1/3}=\frac{Ck_l}{\bar{r}_{0i}}t \tag{3.1-21}$$

where α_i now refers to the fraction reacted in weight fraction w_i. The total reacted is given by the summation

$$\alpha=\sum_i w_i\alpha_i \tag{3.1-22}$$

For shattered, crushed, or ground products it is often possible to express the relationship between weight fraction finer than a given size d (diameter) according to the Schuhmann[(19)] relationship

$$y=(d/K)^m \tag{3.1-23}$$

where y is the fraction finer than size d, m is a constant (the distribution modulus), and K is a constant (the size modulus). If the particle size distribution follows equation (3.1-23), a plot of $\log y$ versus $\log d$ is a straight line and the values of m and K may be determined. It is thus possible to selected weight fractions, w_i, for which the average particle radius, $\bar{r}_{0i}$, may be calculated according to equation (3.1-23). Figure 3.1-5 illustrates Schuhmann plots for chalcopyrite ($CuFeS_2$) concentrate as received and after 6 hr of ball milling.

More complicated geometric shapes may be treated graphically by calculating the change in area with the fraction reacted. In the case of cylinders, two extreme cases provide a similar correlation between α and t. If h_0 is the initial height of a cylinder and r_0 is the initial radius, the extreme cases represent thin disks in which r_0 may be considered essentially constant and long cylinders in which h_0 may be considered essentially constant. For very thin disks or plates, where $r_0 \gg h_0$, it may be shown that for constant concentration,

$$\alpha=\frac{2Ck_l}{h_0}t \tag{3.1-24}$$

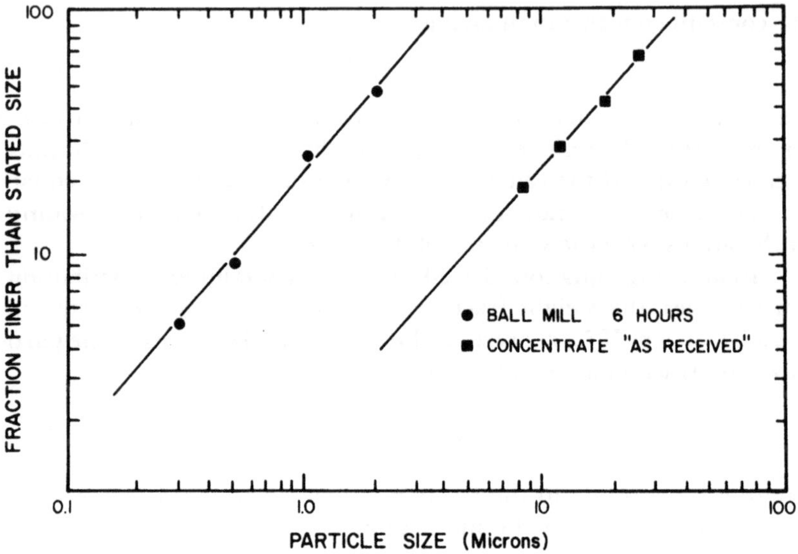

Figure 3.1-5. A Schuhmann plot showing the particle size distribution of Pima chalcopyrite concentrate as received and after 6 h of ball milling.[20]

and for long cylindrical or needlelike particles, where $r_0 \ll h_0$,

$$1-(1-\alpha)^{1/2} = \frac{Ck_l}{r_0}t \tag{3.1-25}$$

Equations (3.1-24) and (3.1-25) apply to the conditions in which the radius and thickness vary linearly with time. Geometric factors for diffusion through products of reaction will be treated separately.

Figure 3.1-6 is a plot of results obtained[21] for the leaching of chalcopyrite ($CuFeS_2$) in an autoclave at 160°C according to equation (3.1-20). The overall reaction under the conditions of this experiment is

$$CuFeS_2 + \tfrac{17}{4}O_2 + \tfrac{1}{2}H_2SO_4 = CuSO_4 + \tfrac{1}{2}Fe_2(SO_4)_3 + \tfrac{1}{2}H_2O \tag{3.1-26}$$

Pressure was maintained constant. The variation of α with t is also shown in Figure 3.1-6. These results indicate that the reaction boundary is moving at constant velocity (linear kinetics) with a rate constant k (min^{-1}). The linear rate $\mathcal{R}_l$ is equal to kr_0, and has, in this case, the units cm min^{-1} if r_0 is measured in cm. The rate $\mathcal{R}_l$ is the linear velocity of movement of the reaction interface. Figure 3.1-7 shows several similar plots for chalcopyrite at various oxygen partial pressures illustrating the fact that k contains concentration units, the contribution of which must be determined experimentally. Clearly k is not the specific rate constant

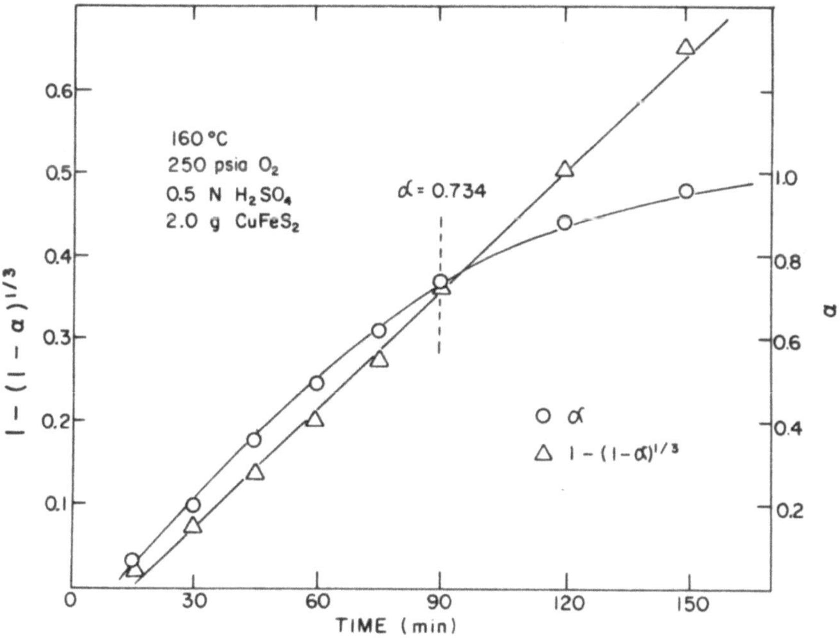

Figure 3.1-6. Comparison of α with $1-(1-\alpha)^{1/3}$ as a function of time.

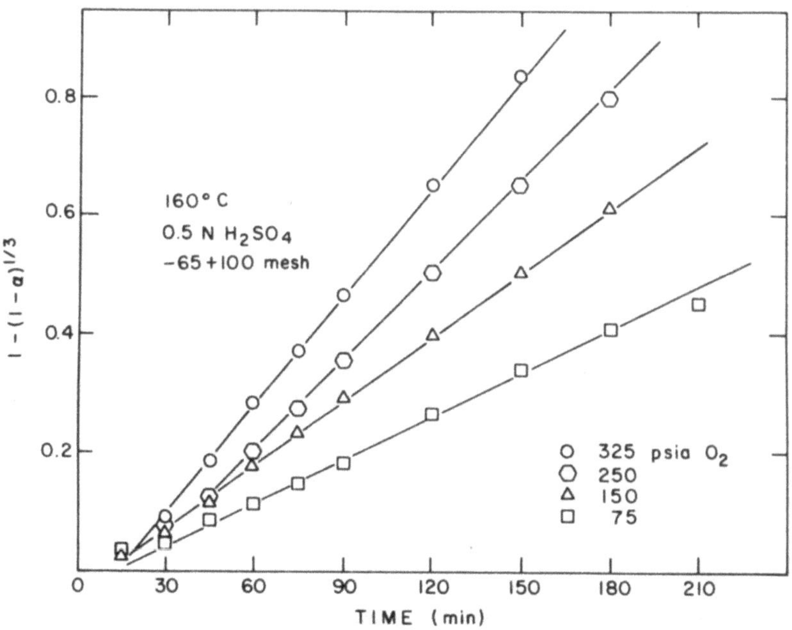

Figure 3.1.-7. Plot of $1-(1-\alpha)^{1/3}$ vs. time for different oxygen partial pressures at 160° C.

since it is constant only if the concentration (or partial pressure) remains constant.

The development of equations (3.1-5), (3.1-21), (3.1-24), and (3.1-25) presumes the rate is a surface reaction at a receding interface. Similar kinetics would be observed for diffusion through a limiting boundary layer of constant thickness, δ, at the receding interface. In this case the kinetic expression must include the stoichiometry factor, σ, which represents the number of moles of the diffusing species required for each mole of metal value released by the reaction. In any complex kinetic system the progress of the reaction may be monitored by measuring the rate of disappearance of reactants or the rate of appearance of products of reaction. The amount of each depends upon the overall stoichiometry of the reaction. This is best expressed in terms of the rational rate, $\imath$, which is the rate of appearance or disappearance of a given component divided by its linear stoichiometric coefficient. For example in regard to equation 3.1-26,

$$\imath = \frac{dn\,(CuFeS_2)}{dt} = \frac{4}{17}\frac{dn(O_2)}{dt} = -\frac{dn\,(Cu)}{dt} = -\frac{1}{V_s}\frac{d(Cu^{2+})}{dt} \tag{3.1-27}$$

where V_s is the solution volume. Therefore, if diffusion through a limiting boundary film is rate controlling, equation (3.1-14) would be expressed as

$$\frac{dn}{dt} = -\frac{4\pi r^2 D}{\sigma\delta}C \tag{3.1-28}$$

where C is the bulk solution concentration assumed here to be much greater than C_s. In comparing equation (3.1-14) with equation (3.1-28) it may be seen that

$$\frac{D}{\sigma\delta} = k_0 k' \tag{3.1-29}$$

and for spheres (C = constant)

$$1-(1-\alpha)^{1/3} = \frac{CVD}{\sigma\delta r_0}t = \frac{\mathcal{R}_l}{r_0}t \tag{3.1-30}$$

The rate of movement of the reaction interface is

$$\mathcal{R}_l = Ck_l = \frac{CVD}{\sigma\delta} \tag{3.1-31}$$

Therefore it is not possible from the kinetics alone to determine if k_l contains the rate constant for a surface reaction or includes diffusion

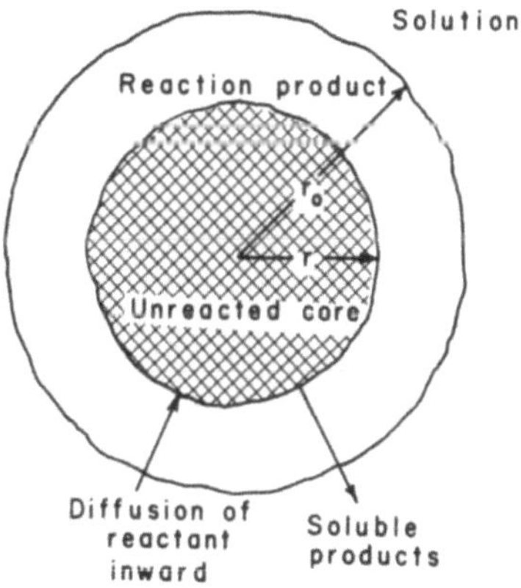

Figure 3.1-8. Particle reacting topochemically with diffusion through remanant structure or products of reaction as the rate-controlling step.

through a limiting boundary layer of solution of constant thickness adjacent to the mineral surface. The effect of temperature on the observed kinetics or the rate of mass transfer must be evaluated to distinguish between the two.

The rate of loss of mass per unit area, $\mathcal{R}_m$, is related to the linear rate by

$$\mathcal{R}_m = \mathcal{R}_l \rho \qquad (\text{mass length}^{-2}\ \text{time}^{-1}) \tag{3.1-32}$$

If products of reaction form on a reacting surface, diffusion through these products may be rate limiting. A special case is one in which the volume of products formed fills the same volume as the reacted mineral portion. In hydrometallurgical processes this normally would mean that porosity in the product compensates for the net mass removal during leaching. This case is applicable in many instances to the leaching of oxide and sulfide ores, as well as to the reaction of some pure mineral particles. The net result is an unreacted shrinking core with a reacted diffusion layer surrounding it, as illustrated in Figure 3.1-8. The original radius is r_0 and the radius of the shrinking core r.

Considering the mineral particle essentially as a sphere, the rate of reaction may be represented by the equation

$$\frac{dn}{dt} = -\frac{4\pi r^2}{\sigma} D \frac{dC}{dr} \tag{3.1-33}$$

where n is the number of moles of unreacted mineral in the core and σ is the stoichiometry factor. Equation (3.1-33) can be integrated, assuming steady-state conditions for all values of r between r and r_0, resulting in the expression

$$\frac{dn}{dt} = -\frac{4\pi D C r r_0}{\sigma(r_0 - r)} \tag{3.1-34}$$

where the concentration at the interface is small compared with C. From equations (3.1-15) and (3.1-33) the general expression for the velocity of movement of the reaction boundary in terms of the radius of the unreacted core is

$$\frac{dr}{dt} = -\frac{VDCr_0}{\sigma r(r_0 - r)} \tag{3.1-35}$$

Equation (3.1-35) combined with equations (3.1-17) and (3.1-18) gives a general expression for the rate in terms of fraction reacted,

$$\frac{d\alpha}{dt} = \frac{3VDC}{\sigma r_0^2} \frac{(1-\alpha)^{1/3}}{1-(1-\alpha)^{1/3}} \tag{3.1-36}$$

which may be integrated for the boundary conditions $\alpha = 0$ when $t = 0$, resulting in the expression

$$1 - \frac{2}{3}\alpha - (1-\alpha)^{2/3} = \frac{2VDC}{\sigma r_0^2} t \tag{3.1-37}$$

Equation (3.1-37) has been found to apply to the leaching of finely ground chalcopyrite ($CuFeS_2$) by ferric sulfate.[20] Essentially stoichiometric quantities of elemental sulfur form according to the reaction

$$CuFeS_2 + 4\,Fe^{3+} = Cu^{2+} + 5\,Fe^{2+} + 2\,S^0 \tag{3.1-38}$$

The sulfur which forms adheres to the mineral surface forming a tightly bonded diffusion layer, explaining the exceptionally low reaction rate for the ferric sulfate leaching of chalcopyrite.

Figure 3.1-9 illustrates the fraction reacted, α, versus time for monosize $CuFeS_2$ for 12- and 47-μ particle sizes when reacted in 0.5 molar ferric sulfate at 90°C. The same data are plotted in Figure (3.1-10) according to equation (3.1-37). The excellent correlation is apparent. In addition the slopes vary with r_0^{-2} as required by the diffusion model.

In the examples cited previously, it has been assumed that the concentration remains constant. Such a condition is realized if the concentration and solution volumes are great enough that solution depletion is negligible during the reaction. Similarly concentration may be maintained constant if dissolved gases are involved in the reaction and if constant overpressures are maintained as in autoclaving. In addition, in multiple-stage-agitation reactors the concentration may be considered constant under conditions of good agitation. The average concentration then is that of the effluent from the reactor. If the concentration varies as the reaction proceeds, it must be included in the integral of the basic rate expression. For surface controlled reactions or diffusion through a limiting boundary film (assuming spherical

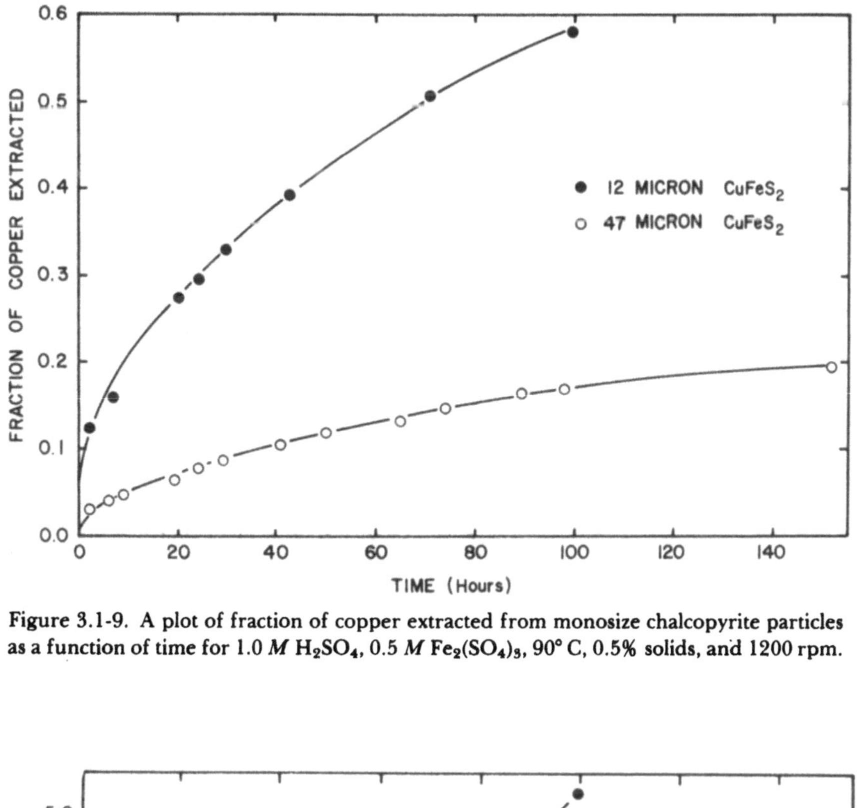

Figure 3.1-9. A plot of fraction of copper extracted from monosize chalcopyrite particles as a function of time for 1.0 M H_2SO_4, 0.5 M $Fe_2(SO_4)_3$, 90° C, 0.5% solids, and 1200 rpm.

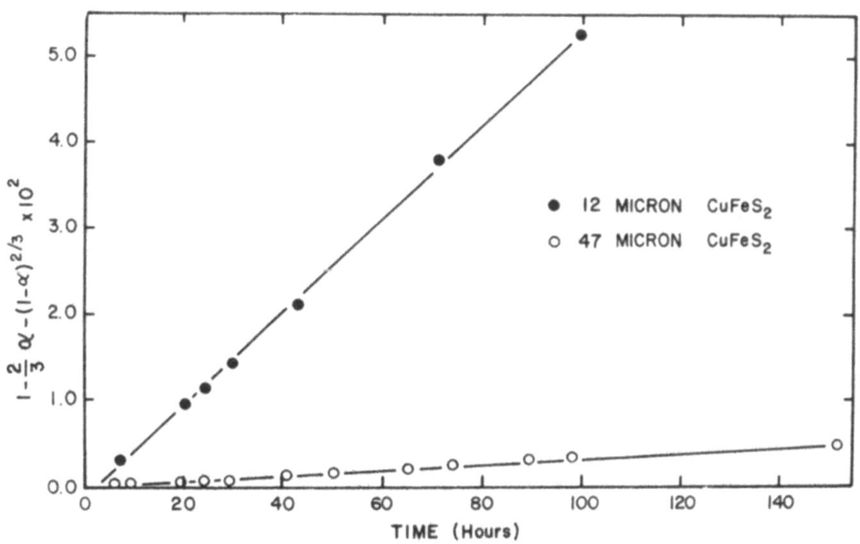

Figure 3.1-10. A plot of $1-(2/3)\alpha-(1-\alpha)^{2/3}$ for monosize chalcopyrite particles as a function of time for the data shown in Figure 3.1-9.

geometry) equation (3.1-19) must now include the concentration of reactant such that

$$\frac{d\alpha}{dt}=\frac{3k_l}{r_0}(1-\alpha)^{2/3}C_0(1-\sigma b\alpha) \tag{3.1-39}$$

where C_0 is the initial concentration, σ is the stoichiometry factor, $b=n_0/V_sC_0$, and n_0 is the total number of moles of mineral in the system. Equation (3.1-39) has an exact integral if stoichiometric amounts of mineral and lixiviant are present in the system. For this special case $\sigma b=1$, and equation (3.1-39) becomes on integration

$$1-(1-\alpha)^{-2/3}=-\frac{2k_l}{r_0}C_0t \tag{3.1-40}$$

Similarly for diffusion through product or remnant layers equation (3.1-36) becomes

$$\frac{d\alpha}{dt}=\frac{2VDC_0}{\sigma r_0^2}\frac{(1-\alpha)^{1/3}(1-\sigma b\alpha)}{1-(1-\alpha)^{1/3}} \tag{3.1-41}$$

For the special case where $\sigma b=1$, equation (3.1-41) may be integrated, resulting in the relationship

$$\frac{1}{3}\ln(1-\alpha)-[1-(1-\alpha)^{-1/3}]=\frac{VDC_0}{\sigma r_0^2}t \tag{3.1-42}$$

For the general case where $\sigma b\neq 1$, equations (3.1-39) and (3.1-41) must be integrated using numerical methods.

A special case worth noting is that in which the amount of mineral reacted in a given stage or cycle is small. Examples are multiple-stage reaction of particles flowing from one reactor to another or dump leaching in which dissolution and oxidation cycles follow each other over long periods of time. In such cases, for an appropriate time interval, the amount of dissolution may be sufficiently small that the surface area may be considered essentially constant. There are few systems in which the area is constant for large values of α; however, essentially constant area is observed when the solid reactant is flat or plateletlike. Furthermore, preferred reaction in one crystallographic direction would result in an essentially constant area, providing the particles are made up of single crystals. An important example of constant A is in cementation reactions, such as reduction of Cu^{2+} by iron where detinned sheet iron scrap is used. If area is considered constant

$$dn/dt=-A_0Ck_0' \tag{3.1-43}$$

where $k_0k'=k_0'$; A_0 is the total area of all particles, considered constant, and n is the total number of moles of solid reactant.

The rate of change of concentration is related to equations (3.1-43), using the rational rate relationship, resulting in the equation

$$\frac{dC}{dt}=\frac{\sigma}{V_s}\frac{dn}{dt}=-\frac{\sigma A_0}{V_s}Ck'_0 \qquad (3.1\text{-}44)$$

which when integrated becomes

$$\log\frac{C}{C_0}=-\frac{A_0k'}{2.303\ V_s}t \qquad (3.1\text{-}45)$$

where k' now contains the stoichiometry factor σ, or $k'=\sigma k'_0$.

As before, if the reaction is controlled by diffusion through a limiting boundary film, equation (3.1-45) becomes, for $C_s \ll C$,

$$\log\frac{C}{C_0}=-\frac{A_0D}{2.303\ \delta V_s}t \qquad (3.1\text{-}46)$$

By comparison, $k'=D/\delta V_s$ (length^{-2} time^{-1}). For most cementation reactions the rate controlling step is solution diffusion for which k' may be approximated using the appropriate D and δ values.[22,23]

An interesting example of solution depletion with constant area for a system in which the kinetics are controlled by transport through products of reaction was given by Bauer *et al.*[24] for the initial stage dissolution of chalcopyrite using ferric ion as the oxidant. Chalcopyrite reacts according to equation (3.1-38), consuming 4 moles of ferric ion for each mole of copper produced. The kinetics are controlled by the transport of ferric ion through the residual sulfur layer. If the extent of the reaction is small enough that the area may be considered constant the rate expression is

$$\frac{dn_{\text{Fe}}}{dt}=-A_0D_{\text{Fe}}\frac{[\text{Fe}^{3+}]}{\Delta x}=\phi\frac{d\Delta n}{dt} \qquad (3.1\text{-}47)$$

where Δn is micrograms of copper dissolved per gram of chalcopyrite, and ϕ contains the stoichiometry factor and appropriate conversion units. If Δx is proportional to Δn, the amount of chalcopyrite reacted, then $\Delta x = b\ \Delta n$. Also $[\text{Fe}^{3+}]=[\text{Fe}^{3+}]_0[1-a\ \Delta n]$, where a is determined from the stoichiometry of the reaction. Using these substitutions, equation (3.1-47) becomes

$$\frac{d\Delta n}{dt}=\frac{A_0D_{\text{Fe}}}{\phi}\frac{[\text{Fe}^{3+}]_0(1-a\Delta n)}{b\Delta n} \qquad (3.1\text{-}48)$$

which may be integrated for the boundary conditions $\Delta n=0$ when $t=0$

resulting in the expression

$$\frac{\Delta n^2}{1-a\Delta n}-\frac{1}{a^2}\left[\frac{1}{1-a\Delta n}+4.606\log(1-a\Delta n)-(1-a\Delta n)\right]=k_p t \qquad (3.1\text{-}49)$$

where all constant terms are included in k_p. Figure 3.1-11 shows the plot of the left-hand side of equation (3.1-49) versus time for the data of Baur *et al.*[24]

At the beginning of the reaction or for very large initial concentrations of ferric ions, $a\Delta$ is very much less than unity, whereupon equation (3.1-49) becomes

$$\Delta n^2 = k_p t \qquad (3.1\text{-}50)$$

which is the commonly observed parabolic rate law.

Figure 3.1-12 illustrates the results obtained by Mishra[25] for the oxidation of chalcopyrite in which wetted surfaces of chalcopyrite were exposed to oxygen. The amount of reaction was sufficient that the surface area did not decrease measurably. The resultant kinetics are parabolic in which the (amount of oxygen)^2 versus time provides a linear correlation of the data in agreement with equation (3.1-50). The products of the reaction, constituting the diffusion barrier, were elemental sulfur and basic iron sulfates ($Fe_2O_3 \cdot y\, SO_3 \cdot z\, H_2O$) formed by the

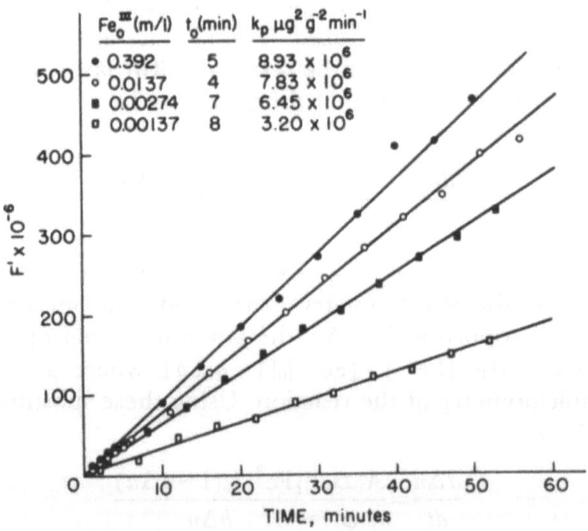

Figure 3.1-11. Plot of ferric sulfate data according to equation (3.1-49) for determination of parabolic rate constants.

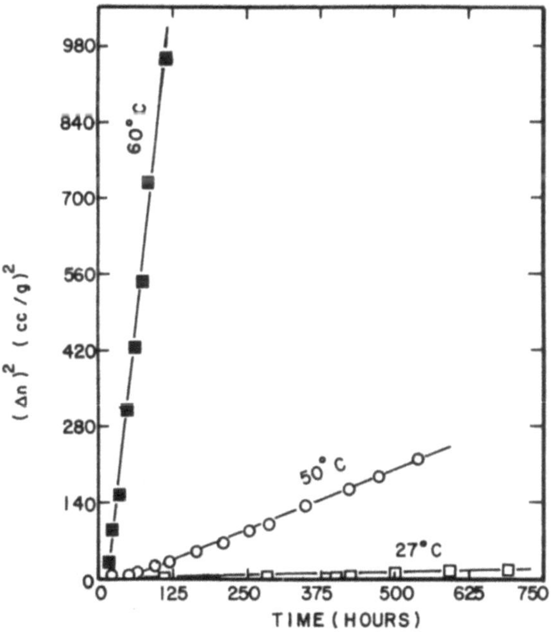

Figure 3.1-12. Parabolic plot of oxygen uptake (cm^3) per gram of chalcopyrite at 27, 50, and 60° C.

reactions

$$CuFeS_2 + O_2 + 4\,H^+ = Cu^{2+} + Fe^{2+} + 2S^0 + 2H_2O \qquad (3.1\text{-}51)$$

$$CuFeS_2 + 4\,O_2 = Cu^{2+} + Fe^{2+} + 2\,SO_4^{2-} \qquad (3.1\text{-}52)$$

$$2\,Fe^{2+} + y\,SO_4^{2-} + \tfrac{1}{2}\,O_2 + (2 - y + z)\,H_2O = Fe_2O_3 \cdot y\,SO_3 \cdot z\,H_2O + (4 - 2\,y)\,H^+ \qquad (3.1\text{-}53)$$

The value of y is variable, representing a series of compounds having y values between zero and four, depending upon temperature, pH, and concentration.[26] In dump leaching, sulfate concentration will be very high during oxidation cycles resulting in the deposition of a series of iron oxides and sulfates.

3.1.2.3. Mixed Kinetics

Any reaction system represents a series of steps of which one or more may be rate controlling. There are several examples in hydrometallurgical systems in which more than one rate process is involved in the observed kinetics. Examples may be cited in cementation, leaching of sulfide mineral fragments, and in the leaching of low-grade ore fragments. Mixed surface reaction plus mass transport through diffusion layers have been observed to contribute simultaneously to the overall kinetics in several systems. If n represents the total number of moles of solid reactant in all particles and A is the total area, the surface reaction may be represented by the equation

$$dn/dt = -AC_sk_s \tag{3.1-54}$$

in which C_s is the surface concentration and is different from that in the bulk because of the diffusion boundary surrounding the particle, and k_s is the reaction-rate constant for the surface reaction (Figure 3.1-2). Mass transport through the diffusion boundary is represented by equation (3.1-4) where C is the concentration of reactant in the bulk solution. The value for C_s may be solved under steady-state conditions using equations (3.1-6) and (3.1-54) with the appropriate stoichiometry factor. The value for C_s then may be substituted in equation (3.1-54) resulting in the combined expression

$$\frac{dn}{dt} = -\frac{CA}{\sigma\Delta x/D + 1/k_s} = -CAk_0' = V\frac{dC}{dt} \tag{3.1-55}$$

where Δx is the thickness of the diffusion boundary. The constant k_0' in equation (3.1-55) is represented by the sum of two resistances (reciprocal rate constants) for series reactions, namely, mass transport through the diffusion layer and the surface reaction as indicated by the equation

$$k_0' = \left(\frac{\alpha\Delta x}{D} + \frac{1}{k_s}\right)^{-1} \tag{3.1-56}$$

If area and Δx remain constant during the course of the reaction, equation (3.1-55) becomes a simple first-order rate expression. The first-order rate constant, k_0', will have a complex temperature dependence since it contains both D and k_s. Normally D (solution-diffusion coefficient) is less sensitive to temperature change than k_s which has a larger activation energy. Consequently as temperature increases, k_s will become much larger than D. It is apparent from equation (3.1-55) that diffusion will be expected to be rate controlling at high temperatures and surface reaction controlling at low temperatures, with mixed control for intermediate temperatures.

If the reaction results in the formation of products on the surface through which diffusion occurs and if the area is considered to be constant, equation (3.1-55) becomes

$$\frac{dn}{dt} = -\frac{CA}{b(n_0 - n)/D + 1/k_s} \tag{3.1-57}$$

where $\sigma\Delta x = b(n_0 - n)$, n_0 is the number of moles of reacting material in all mineral fragment at the beginning of the reaction, and n is the number of moles remaining at any time t. The quantity $n_0 - n$ represents the amount reacted, Δn. Equation (3.1-57) may be integrated for constant concentration and area resulting in the expression

$$\frac{\Delta n^2}{A_0Ck_p}+\frac{\Delta n}{A_0Ck_s}t \tag{3.1-58}$$

which represents the sum of a parabolic and linear rate and was developed earlier by Wagner and Grunewald[27] for the oxidation of copper. The rate constant k_p is equal to $2D/b$. Equation (3.1-58) may be written in the form

$$\frac{t}{\Delta n}=\frac{\Delta n}{A_0Ck_p}+\frac{1}{A_0Ck_s} \tag{3.1-59}$$

which would predict that $t/\Delta n$ plotted versus Δn should result in a straight line, the slope of which contains the reciprocal of the parabolic rate constant k_p and the intercept contains the surface rate constant k_s. The Wagner–Grunewald equation has been found to explain initial oxidation of chalcopyrite with Fe^{3+} (see Figure 3.1-13[24]).

In considering the leaching of mineral particles, it is often possible to treat the kinetics using the steady-state approximation. Any number of first-order steady-state processes may be treated in sequence with the result that each kinetic expression is separately additive. The general

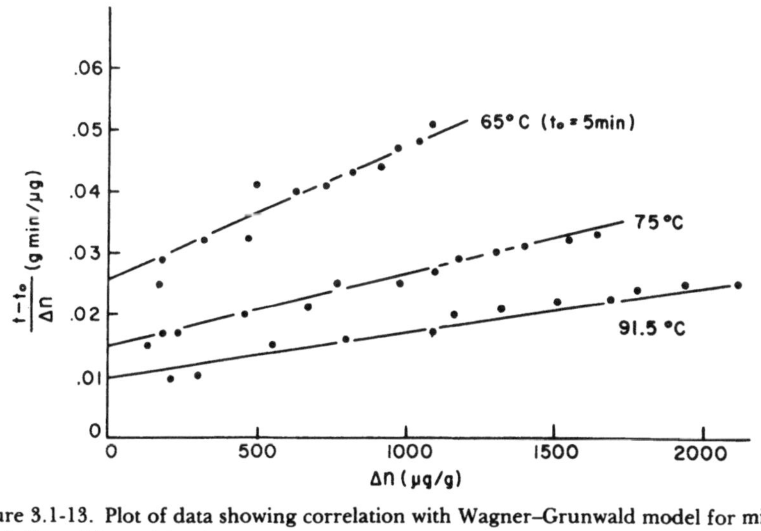

Figure 3.1-13. Plot of data showing correlation with Wagner–Grunwald model for mixed parabolic and linear kinetics.

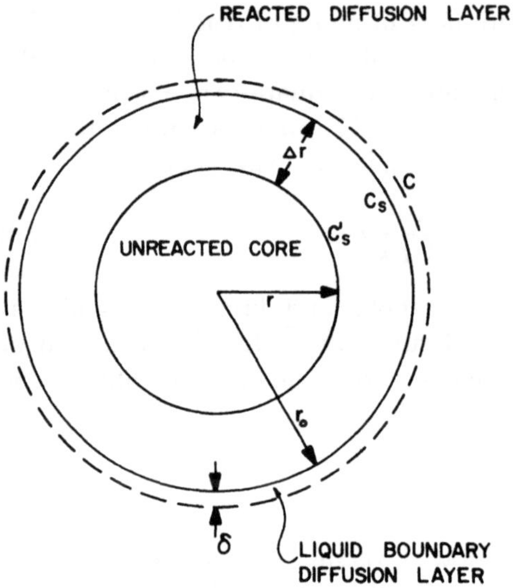

Figure 3.1-14. Reaction of a mineral particle with shrinking core and liquid boundary diffusion.

case in hydrometallurgical systems treating individual mineral particles, as in agitation leaching, would have to include solution diffusion of reactant through a liquid boundary film, diffusion through solid reaction products, surface reaction, and diffusion of products away. Since back-reaction kinetics are rarely important in hydrometallurgical leaching reactions only the first three steps will be included in a general solution. For spherical geometry the three kinetic processes in series are (see Figure 3.1-14)

Boundary Diffusion
$$\frac{dn}{dt} = -\frac{4\pi r_0^2 D_s(C - C_s)}{\sigma\delta} \tag{3.1-60}$$

Diffusion through Products
$$\frac{dn}{dt} = -\frac{4\pi D' r r_0 (C_s - C'_s)}{\sigma(r_0 - r)} \tag{3.1-61}$$

Surface Reaction
$$\frac{dn}{dt} = -4\pi r^2 C'_s k'_0 \tag{3.1-62}$$

where D_s is the coefficient of diffusion for bulk solution diffusion, D' is the effective diffusion coefficient for diffusion through products of reaction, k'_0 is the surface reaction rate constant, σ is the stoichiometry factor, and n is the number of moles of unreacted metal value remaining in the core at any time t. Under the steady-state approximation each of the above rates in series are equal, resulting in the combined expression

(neglecting back-reaction kinetics)

$$\frac{dn}{dt} = -\frac{4\pi r_0^2 D_s C}{\sigma\delta\left[1 + \frac{r_0(r_0 - r)D_s}{\delta r D'} + \frac{D_s r_0^2}{\sigma\delta k_0' r^2}\right]} \tag{3.1-63}$$

Using equation (3.1-15), (3.1-17), and (3.1-63) the rate may be expressed in terms of fraction reacted, or

$$\frac{dn}{dt} = \frac{3CV}{\frac{\sigma\delta r_0}{D_s} + \frac{r_0^2\sigma(1-(1-\alpha)^{1/3}]}{D'(1-\alpha)^{1/3}} + \frac{r_0}{k_0'(1-\alpha)^{2/3}}} \tag{3.1-64}$$

Equation (3.1-64) may be applied to a slurry of monosized particles of average radius r_0. For a particle size distribution, $\alpha = \alpha_i$ for the ith size r_{0i}. The total fraction reacted, w_i, then is summed as shown in equation (3.1-22). For constant concentration, equation (3.1-64) may be integrated giving

$$\frac{\sigma\delta}{D_s}\alpha + \frac{3\sigma r_0}{2D'}\left[1 - \frac{2}{3}\alpha - (1-\alpha)^{2/3}\right] + \frac{1}{k_0'}[1-(1-\alpha)^{1/3}] = \frac{3CVt}{r_0} \tag{3.1-65}$$

It is evident from equation (3.1-65) that the integrated form simply includes the sum of the expression for boundary diffusion, diffusion through solid products, and the surface reaction.

In a batch reactor, concentration varies according to the stoichiometry of the system, in which case the concentration in equation (3.1-63) is $C = C_0(1 - \sigma b\alpha)$, as before. The equation may then be integrated numerically.

3.1.2.4. *Temperature Dependence*

In logarithmic form the Arrhenius equation, equation (3.1-1), is

$$\ln k = \frac{\Delta E}{R}\left(\frac{1}{T}\right) + \ln A \tag{3.1-66}$$

It is apparent that $\ln k$ values plotted vs. $(1/T)$ should result in a straight line of negative slope, since ΔE must be positive. Since the measured slope equals $-\Delta E/R$, the value ΔE is obtained by multiplying the measured slope by R. If common logarithms are used, the slope must be multiplied by $2.303R$. In differential form equation (3.1-66) becomes

$$d \ln k = \frac{\Delta E}{RT^2}\, dT \tag{3.1-67}$$

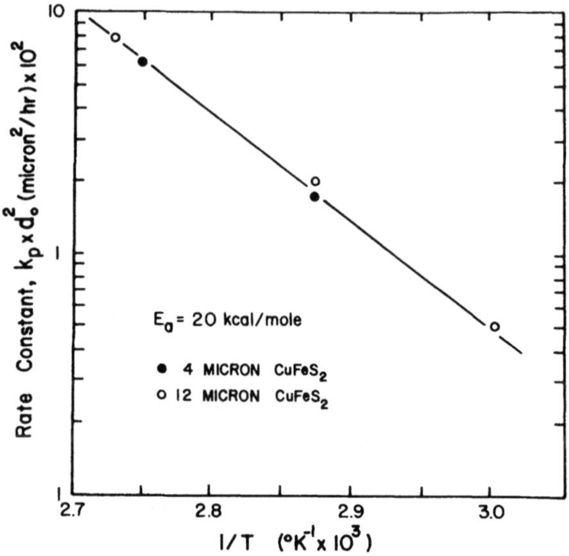

Figure 3.1-15. An Arrhenius plot for the leaching of 12-μ monosize chalcopyrite particles showing the effect of temperature on k'. Leaching conditions are the same as those reported in Figure 3.1-9.

and upon integration

$$\ln\frac{k_2}{k_1}=\frac{\Delta E}{R}\left(\frac{T_2-T_1}{T_1-T_2}\right) \tag{3.1-68}$$

If the specific rate constant is known for two temperatures, ΔE may be calculated directly by equation (3.1-68).

Figure 3.1-15 is an Arrhenius plot showing $\log k$ vs. T^{-1} according to equation (3.1-66) for the reaction of ferric sulfate with chalcopyrite. The experimental activation energy is 20 kcal/mole.

3.1.2.5. Absolute Reaction Rate Theory

According to absolute reaction-rate theory,[28] reactants are considered to be in equilibrium with an activated complex which represents an intermediate of definite configuration and which forms before product formation. There is an activation energy necessary to form the transition intermediate before the product can form. The rate, therefore, is related to the concentration of the intermediate and, in turn, is controlled by a pseudoequilibrium condition with the reactants. Such a reaction is represented by the equation

$$a\mathrm{A}+b\mathrm{B}\leftrightarrows \mathrm{C}^{\ddagger}\rightarrow \text{products} \tag{3.1-69}$$

The mass action constant for this equilibrium is represented by

$$K^{\ddagger} = \frac{C^{\ddagger}}{C_A^a C_B^b} \tag{3.1-70}$$

The basis for absolute reaction rate theory rests on the fact that the ratio of concentrations, as indicated in equation (3.1-70), be related to the ratio of partition functions corrected for zero-point energy. The contribution of absolute-reaction-rate theory in terms of associated thermodynamic functions adds much detail in the evaluation of heterogeneous systems. If the kinetics are monitored by the rate of appearance of products in solution, according to absolute-rate theory, the rate may be expressed by

$$\text{rate} = dn/dt = \nu C^{\ddagger} \tag{3.1-71}$$

in which n represents the number of moles of product, ν is the frequency of reaction, and $C^{\ddagger}$ is the concentration of the activation complex. According to absolute-rate theory, the frequency of reaction of all activated complexes is the same for a given temperature and may be represented in terms of the Boltzmann constant, k, and the Planck constant, h, by the equation

$$\nu = \frac{kT}{h} \tag{3.1-72}$$

whereupon the rate expression may be obtained by combining equations (3.1-70) and (3.1-72) resulting in the expression

$$\text{rate} = \frac{kT}{h} C^{\ddagger} = \frac{kT}{h} C_A^a C_B^b K^{\ddagger} \tag{3.1-73}$$

where the specific rate constant, independent of concentration, is given by

$$k' = \frac{kT}{h} K^{\ddagger} \tag{3.1-74}$$

The thermodynamic equilibrium constant, $K_0^{\ddagger}$, is given by the equation

$$K_0^{\ddagger} = \frac{C^{\ddagger}}{C_A^a C_B^b} \frac{\gamma^{\ddagger}}{\gamma_A^a \gamma_B^b} \tag{3.1-75}$$

which can be related to the mass-action constant by the equation

$$K_0^{\ddagger} = K^{\ddagger} \frac{\gamma^{\ddagger}}{\gamma_A^a \gamma_B^b} = \exp\left[-\frac{\Delta G^{\ddagger}}{RT}\right] = \exp\left[-\frac{\Delta H^{\ddagger}}{RT}\right] \exp\left[\frac{\Delta S^{\ddagger}}{R}\right] \tag{3.1-76}$$

where $\Delta G^{\ddagger}$, $\Delta H^{\ddagger}$, and $\Delta S^{\ddagger}$ are the free energy, enthalpy, and entropy of activation, respectively. The rate expression then becomes

$$\text{rate} = \frac{kT}{h} \frac{C_A^a C_B^b \gamma_A^a \gamma_B^b}{\gamma^{\ddagger}} \exp\left[-\frac{\Delta H^{\ddagger}}{RT}\right] \exp\left[\frac{\Delta S^{\ddagger}}{R}\right] \tag{3.1-77}$$

or in general terms

$$\text{rate} = \frac{kT}{h\gamma^{\ddagger}} \prod_i a_i^{\nu_i} \exp\left[-\frac{\Delta H^{\ddagger}}{RT}\right] \exp\left[\frac{\Delta S^{\ddagger}}{R}\right] \tag{3.1-78}$$

The specific rate constant, k' (sec^{-1}), is

$$k' = \frac{kT}{h} \exp\left[-\frac{\Delta H^{\ddagger}}{RT}\right] \exp\left[\frac{\Delta S^{\ddagger}}{R}\right] \tag{3.1-79}$$

In logarithmic form the specific rate constant becomes

$$\log \frac{k'}{T} = -\frac{\Delta H^{\ddagger}}{2.303R}\left(\frac{1}{T}\right) + \log\left(\frac{k}{h} \exp\left[\frac{\Delta S^{\ddagger}}{R}\right]\right) \tag{3.1-80}$$

If $\log(k'/T)$ is plotted versus reciprocal temperature, the enthalpy of activation, $\Delta H^{\ddagger}$, may be determined from the slope. In hydrometallurgical systems where temperatures are low the difference between the experimental activation energy, ΔE, and the enthalpy of activation, $\Delta H^{\ddagger}$, will usually be less than 1 kcal. The advantage of absolute-reaction-rate theory is that the evaluation of the frequency factor separate from the entropy term can provide valuable information regarding the number of reaction sites in heterogeneous systems and help identify the entropy contribution of various steps in the system.

It should be pointed out that ΔE or $\Delta H^{\ddagger}$ may in reality be only apparent values. In homogeneous reactions the reactant involved in the true rate step may be some intermediate equilibrium species. In this case the enthalpy of the equilibrium constant would be contained in the apparent $\Delta H^{\ddagger}$. Similarly in heterogeneous systems equilibrium adsorption may occur prior to a surface reaction whereupon the apparent enthalpy of activation may contain the enthalpy of adsorption. Because of the exponential contribution of enthalpy, associate equilibria or steady-state reactions provide separate enthalpy terms which are additive. Similarly the entropy of activation includes a summation of entropy terms for associated equilibria and steady-state processes.

In heterogeneous systems it may not be possible to determine $\Delta S^{\ddagger}$ unambiguously. For example, a solid dissolution reaction must include the true surface area of the solid. If the geometric area is used the surface roughness factor cannot be separated from the entropy unless it is measured separately. Not all surface reaction sites may be potentially reactive at the same time. As a consequence the entropy may be in error by very large factors. For example, there are approximately 10^{15} sites/cm^2 on a solid surface. If only 10^{10} of these sites are reactive (dislocations, impurity centers, etc.), an error of some 23 e.u. results if not properly accounted for. These factors must be separated indirectly. It is often possible to estimate the entropy for a given mechanism, whereupon remaining factors such as the number of potentially reactive sites may be approximated.

3.1.2.6. Reversible Reactions

It is interesting to consider forward and back reactions in a system in which the build-up of products may affect the overall rate. Consider an equation of the form

$$aA + aB \rightleftharpoons C^{\ddagger} \rightleftharpoons cC + dD \qquad (3.1\text{-}81)$$

The rate is the difference between the forward and the back reactions and, assuming the activity coefficients to be unity, the net rate may be expressed by

$$\text{rate} = \frac{kT}{h}(\vec{C}^{\ddagger} - \overleftarrow{C}^{\ddagger}) \qquad (3.1\text{-}82)$$

where $\vec{C}^{\ddagger}$ represents the concentration of activated complexes which are capable of proceeding with product formation and $\overleftarrow{C}^{\ddagger}$ represents the concentration of activated complexes which are capable of reacting in the direction from products to initial reactants. It is apparent from equation (3.1-82) that at equilibrium the concentration of activated complexes capable of going on to products is just equal to the concentration of those capable of returning to reactants. The partition functions needed to describe the activated complex is independent of the direction of the reaction but the pseudoequilibrium is related (in the case of reactants) only to the activated complex capable of continuing on to products. The same condition is true regarding the relationship between products and the activated complex capable of returning to the reactants.

Figure 3.1-16 represents the standard free-energy configuration along the reaction coordinate showing the relationship between reactants, the activated complex, and the product. The r–r base line is an arbitrary line of zero energy to which all standard free-energy measurements must be made. For example, the value $G_r^{\ddagger}$ represents the free energy of the activated complex in reference to the base line, $\Delta G_f^{\ddagger}$ represents the free energy of activation for the forward reaction, and $\Delta G_b^{\ddagger}$ represents the free energy of activation for the back reaction; ΔG^0

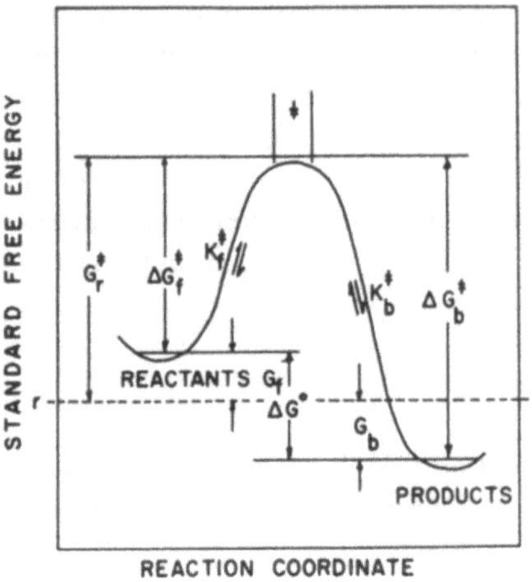

Figure 3.1-16. Energy relationships showing activation energies for forward and back reactions according to absolute reaction-rate theory.

represents the standard molar-free-energy difference between reactants and products, and G_f and G_b are the values of free energy of reactants and products, respectively, compared to the reference base line; $K_f^\ddagger$ is the equilibrium constant for the forward reaction, and $K_b^\ddagger$ represents the equilibrium constant for the back reaction. From Figure 3.1-16 it is apparent that the following relationships exist:

$$\Delta G_f^\ddagger = G_r^\ddagger - G_f \tag{3.1-83}$$

$$\Delta G_b^\ddagger = G_r^\ddagger - G_b \tag{3.1-84}$$

$$\Delta G^0 = G_b - G_f \tag{3.1-85}$$

Using the relationships indicated in Figure 3.1-16, equation (3.1-82) becomes

$$\text{rate} = \frac{kT}{h}\left(C_A^a C_B^b \exp\left[-\frac{\Delta G_f^\ddagger}{RT}\right] - C_C^c C_D^d \exp\left[-\frac{\Delta G_b^\ddagger}{RT}\right]\right) \tag{3.1-86}$$

By factoring out terms related to the forward reaction, equation (3.1-86) becomes

$$\text{rate} = \frac{kT}{h} C_A^a C_B^b \exp\left[-\frac{\Delta G_f^\ddagger}{RT}\right]\left(1 - \frac{C_C^c C_B^b \exp\left[-(G_r^\ddagger - G_b)/RT\right]}{C_A^a C_B^b \exp\left[-(G_r^\ddagger - G_f)/RT\right]}\right) \tag{3.1-87}$$

It is apparent from equation (3.1-87) that the ratio of exponential terms within the parentheses becomes $\exp\left[\Delta G^0/RT\right]$. Using the relationship $x = \exp\left[\ln x\right]$, the right-hand term within the parentheses of equation (3.1-82) becomes

$$\left(G_b - G_f + RT \ln \frac{C_C^c C_D^d}{C_A^a C_B^b}\right)\frac{1}{RT} \tag{3.1-88}$$

Since the free energy for the overall reaction is given by

$$\Delta G = \Delta G^0 + RT \ln \frac{C_C^c C_D^d}{C_A^a C_B^b} \tag{3.1-89}$$

equation (3.1-87) becomes

$$\text{rate} = \frac{kT}{h} C_A^a C_B^b \exp\left[-\frac{\Delta G_f^\ddagger}{RT}\right]\left(1 - \exp\left[\frac{\Delta G}{RT}\right]\right) \tag{3.1-90}$$

In equation (3.1-90) ΔG for the overall reaction may be determined from the thermodynamics for the system and provide insight as to whether or not back reactions may be expected to be important. Using the approximation that, if exp $[\Delta G/RT]$ is less than 0.01, back reactions are negligible, it is apparent from this limit that back reactions are negligible if $\Delta G < 10\,\text{T}$. In hydrometallurgical systems, which may be considered to extend from 0° C to approximately 200° C or higher in autoclave systems, it is apparent that back reactions usually are not important. For instance, at room temperature ΔG must be more negative than approximately −3000 kcal/mole for back reactions to be important in the system. Thus, as a generalization in hydrometallurgical systems, back reactions are rarely of importance. Exceptions to this are ion-exchange and solvent extraction where the partition coefficients involve free-energy changes which are quite small and therefore operate near equilibrium. In very high temperature processes such as slag metal equilibria, at 1500° C for example, ΔG must be more negative than −15,000 kcal mole^{-1} before the back reaction is negligible. Consequently, high-temperature processes may operate fairly close to equilibrium, whereas it is expected that hydrometallurgical processes operate under conditions well displaced from equilibrium.

3.1.2.7. Electrochemical Reactions

Electrochemical reactions, as distinct from chemical reactions, involve electrons which react at the interface of a solid phase, capable of electron conduction, and a solution of electrolytes. This particular element of the overall electrode process is called a *transfer reaction*. More specifically electrical charge is transferred between the electrode and the electrolyte by electrons in redox reactions and by positively charged metal ions in the case of metal–metal ion electrodes.[29] Associated equilibrium and sequential chemical reactions may be involved in the total process and nonelectrochemical reactions may, in fact, be rate determining. Transport of ions by diffusion may also be rate determining even though influenced by potentials associated with the electrical double layer at the electrode–electrolyte interface. The influence of potential in the electrical double layer is a fundamental feature of electrode reactions since it will increase or diminish activation free energies of individual reactions, depending upon the sign and magnitude of the charge of the reacting species and its position within the electrical double layer.

The transport of a metal ion in solution to a lattice position in the cathode during reduction involves a series of reactions of which one or more may be rate controlling. If there are n such steps and steps n_i to n_j are rate controlling, the free-energy difference is given by the chemical potentials such that

$$\Delta G = \mu_j - \mu_i \tag{3.1-91}$$

Furthermore, if all steps prior to n_i and following n_j are essentially at equilibrium, the free-energy difference is the same as that for the overall process, or

$$\Delta G = \mu_j - \mu_i \cong \mu_c - \mu_a \tag{3.1-92}$$

where μ_c and μ_a are the cathodic and anodic chemical potentials for the total process. At equilibrium the chemical potentials are equal and $\Delta G = 0$, resulting in a zero net current density.

Figure 3.1-17 depicts the flow of electrons and ions between the electrode and the electrolyte. The net current density I is the sum of the partial current densities I_+ and I_-.[29] If $|I_-| > I_+$, I is negative, and the net process is cathodic. If $|I_-| < I_+$, I is positive, and the net process is anodic. It follows that for a net cathodic process the potential of the electrode E is less than E_0 and for a net anodic process $E > E_0$. The overvoltage η is by definition the difference between E, the voltage when there is a net current density, and E_0 the voltage when the net current density is zero; i.e., $\eta = E - E_0$ and is positive for anodic currents and

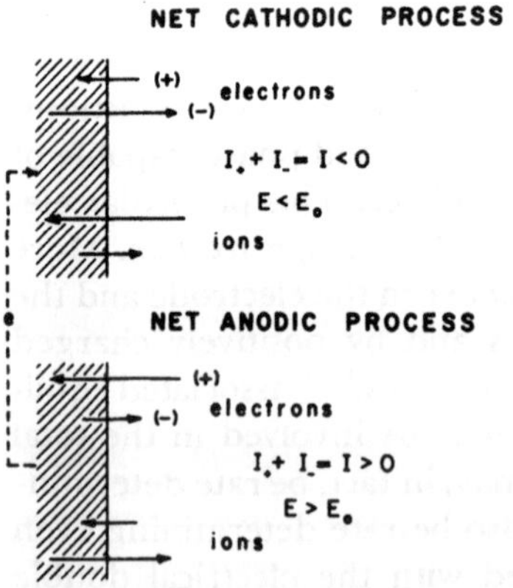

Figure 3.1-17. Flow of electrons and ions between electrode and electrolyte for net cathodic and net anodic processes.

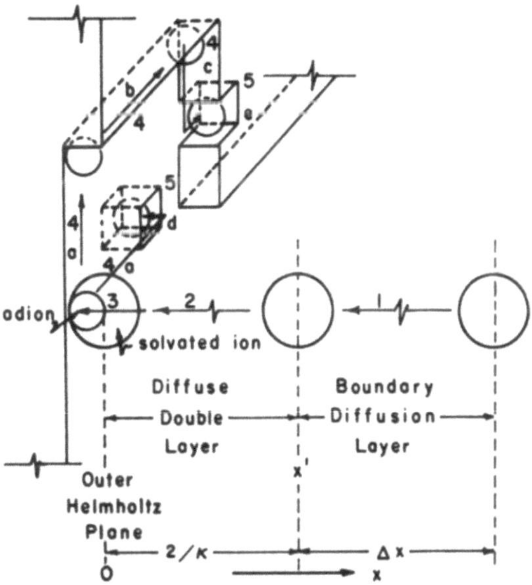

Figure 3.1-18. Illustration of transport processes from bulk solution to crystallization in lattice sites: (1) transport through boundary diffusion layer; (2) transport through diffuse double layer; (3) transfer reactions; (4) diffusion on surfaces (a), edges (b) and kinks (c); and (5) crystallization at surface vacancies (d) and edge vacancies (e).

negative for cathodic currents. A net current density of zero does not imply zero current density but represents the equilibrium condition where the partial current densities are equal, $E = E_0$ and $\eta = 0$. The partial current density at equilibrium is called the *exchange-current density*, I_0. The overvoltage may result from any one of the n steps in the overall process, including, in addition to the transfer reaction, crystallization, surface diffusion, chemical reactions, and solution diffusion. Accordingly, the overvoltage is termed *transfer overvoltage, crystallization overvoltage, diffusion overvoltage*, or *reaction overvoltage*.[28]

Figure 3.1-18 illustrates the diffusion of an ion to be reduced at a metal electrode, showing boundary diffusion, diffusion through the electrical double layer (where κ is the half-thickness of the electrical double layer), adsorption, and surface diffusion to various surface sites and ultimately to a final lattice position. The outer Helmholtz plane represents the position of the center of closest approach of solvated ions.

The transport of an ion to or away from the surface of an electrode is influenced by the potential gradient at the interface. Figure 3.1-19 illustrates the potential gradient at the electrode surface. If $\Delta\phi$ is the increase in potential across the electrical double layer, the free energy is increased by the amount of $zF\Delta\phi$, where F is the Faraday constant and z

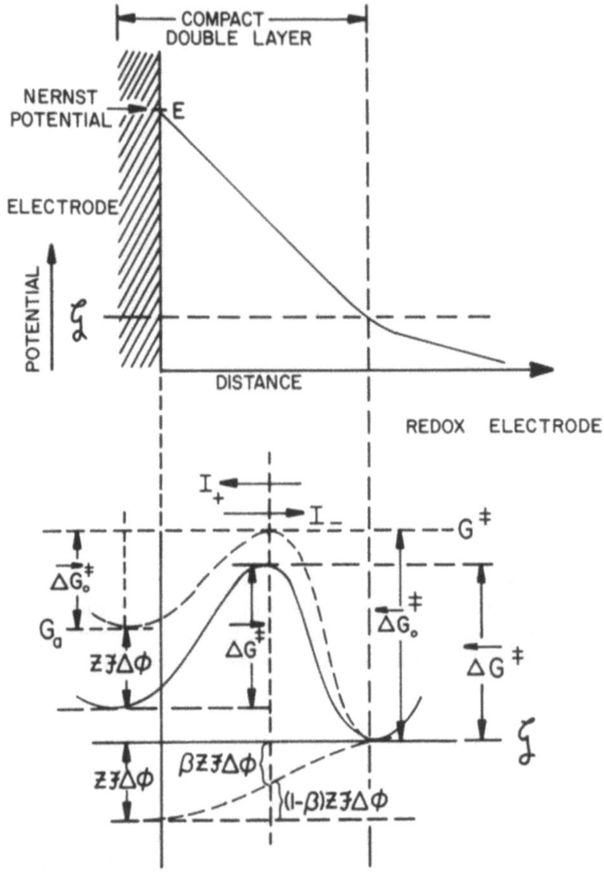

Figure 3.1-19. Dependence of the anodic ($\Delta\vec{G}_0$‡) and cathodic ($\Delta\overleftarrow{G}_0$‡) energies of activation on the potential difference $\Delta\phi$ across the compact double layer.[30]

is the charge, to the value G_a. It is apparent that the potential decreases the activation energy for the anodic direction ($\Delta\vec{G}^{\ddagger}$) by $(1-\beta)zF\Delta\phi$ and increases $\Delta\overleftarrow{G}^{\ddagger}$ by the amount $\beta zF\Delta\phi$, where β is the transfer coefficient. The potential difference $\Delta\phi$ is given by

$$\Delta\phi = E - \zeta \tag{3.1-93}$$

where ζ is the zeta potential. If we consider $\zeta = 0$, $\Delta\phi = E$. The net current density, $I = I_{-} + I_{+}$, and may be expressed for a single-electron transfer process, $z = -1$, by the rate expression

$$I = k_a C_a \exp\,[\beta FE/RT] - k_c C_c \exp\,[-(1-\beta)FE/RT] \tag{3.1-94}$$

where k_a and k_b are the anodic and cathodic rate constants in the absence of potential. Equation (3.1-94) is the Butler–Volmer equation[31,32] commonly applied to electrochemical rate processes.

When the electrode is in equilibrium with the solution, $I = 0$, $E = E^0$, and $I_+ = |I_-| = I_0$. Equation (3.1-94) expressed in terms of the exchange-current density, I_0, is

$$I = I_0[\exp(\beta F\eta/RT) - \exp[-(1-\beta)F\eta/RT)] \qquad (3.1\text{-}95)$$

Equation (3.1-95) reduces to the Tafel equations[33] for large anodic currents ($\eta \gg 0$)

$$\eta = \frac{RT}{\beta F}\ln I_0 + \frac{RT}{\beta F}\ln I \qquad (3.1\text{-}96)$$

and for large cathodic currents ($\eta \ll 0$)

$$\eta = \frac{RT}{(1-\beta)F}\ln I_0 + \frac{RT}{(1-\beta)F}\ln |I| \qquad (3.1\text{-}97)$$

The concept of mixed potentials is important in explaining many hydrometallurgical reactions involving metal sulfides and oxides which are good electronic conductors (semiconductors). The mixed potential results from the fact that anodic and cathodic regions or domains are in electrical contact (short-circuited) so that the mixed potential, E_m, falls between the equilibrium potential values of the two electrode processes. Consequently the region for which $\eta < 0$ reacts cathodically, and the region for which $\eta > 0$ reacts anodically. The mixed potential does not obey the Nernst equation since it is a nonequilibrium value representing conditions where the total anodic current just equals the total cathodic current. This is illustrated in Figure 3.1-20 which shows the cathodic and anodic I-potential curves for chalcopyrite with Nernst potential, E_{0a}. Also shown is the curve for a metal-electrode reaction, such as the Fe^{3+}–Fe^{2+} couple, having a Nernst potential, E_{0c}. Although E_{0a} and E_{0c} are Nernst potentials, E_m is not. The mixed potential corresponds to the voltage where the anodic current density balances the cathodic current density or

$$I_a = -I_c \qquad (3.1\text{-}98)$$

The current densities I_a and I_c are normalized in terms of the total area, A. Actually the current passing through the cathode of area A_c must be equal to the current passing through the anode of area A_a. The net cathodic and anodic regions are depicted in Figure 3.1-17. If the anodic and cathodic areas are separate and distinguishable but electrically in contact, when, for example, two conducting minerals are in contact, the system represents a galvanic couple. If the anodic and cathodic areas are part of the same surface, so that $A = A_a + A_c$, the system represents a

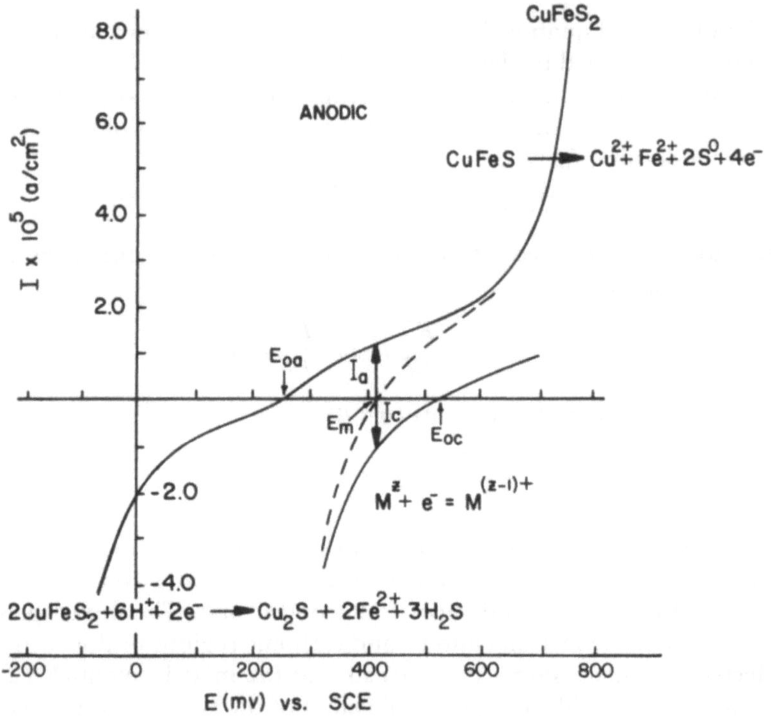

Figure 3.1-20. Cathodic and anodic I-potential curves for $CuFeS_2$ and metal ion oxidant showing mixed potential, E_m.

corrosion couple. This is illustrated by the two stages of reaction of chalcocite in which the reactions follow in sequence:

$$\textit{Stage I} \qquad Cu_2S \rightarrow CuS + Cu^{2+} + 2\,e^- \qquad (3.1\text{-}99)$$

$$\textit{Stage II} \qquad CuS \rightarrow Cu^{2+} + S^0 + 2\,e^- \qquad (3.1\text{-}100)$$

Figure 3.1-21 illustrates the two stages. Actually equations (3.1-99) and (3.1-100) represent an oversimplification since several defect structures form; however, considering the net overall equation as indicated, it is characteristic of an electrochemical coupling that reaction stages follow in sequence. The first stage represents a galvanic couple. The anodic surface of area A_a is represented by Boundary I, Figure 3.1-21. The cathodic surface of area A_c is separated by the CuS layer (capable of electron conduction) and is indicated by Boundary II. In Stage II anodic and cathodic sites make up Boundary III and represent a corrosion couple. At low temperatures, the two stages are distinctly in sequence. At high temperatures, there is some overlap of the reaction stages. This may be explained by an electrochemical model in terms of expected

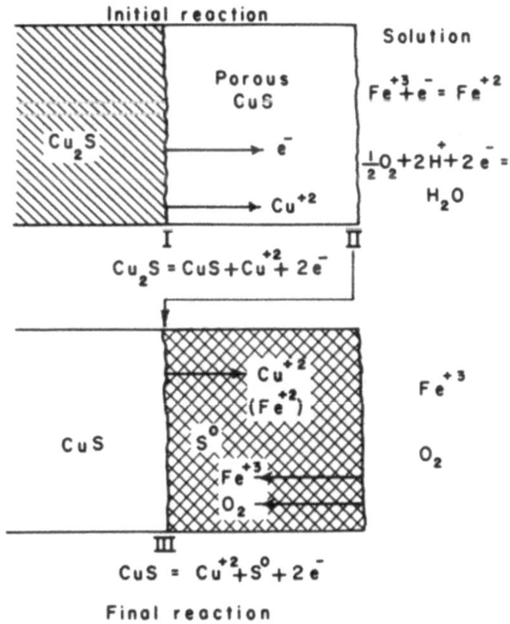

Figure 3.1-21. Surface layers formed during the anodic dissolution of chalcocite (Cu_2S) and covellite (CuS).

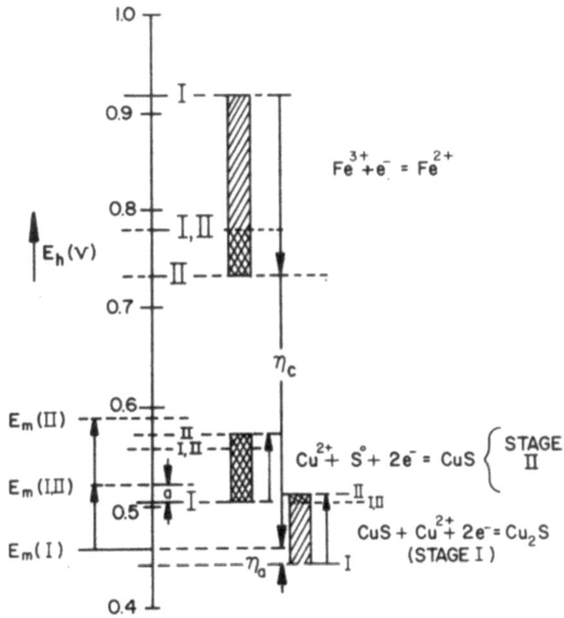

Figure 3.1-22. Illustrations of half-cell and mixed potential variations during leaching of chalcocite with ferric ion. Initial conditions: 0.1 moles Cu_2S per liter; $[Cu^{2+}] = 0.001$; $[Fe^{3+}] = 0.5$; $[Fe^{2+}] = 0.001$.

mixed potentials. Figure 3.1-22 illustrates the range of variation of the half-cell potentials for the two stages of leaching of Cu_2S. The system depicted contains initially 0.1 moles Cu_2S per liter and 0.5 moles Fe^{3+}, 0.001 moles Fe^{2+}, and Cu^{2+}. Considering solutions as ideal, the E_h for the Fe^{3+}–Fe^{2+} half-cell would vary from 0.917 V at the beginning of the first stage (I) to 0.781 V at the end of the first stage or beginning of the second stage (I, II) and then to 0.735 V at the end of the second stage II. The voltage for Stage I, equation (3.1-99), is more negative than that of Stage II, equation (3.1-100), and is found thermodynamically. Figure 3.1-23 is the Pourbaix diagram for the system showing the instability of elemental sulfur in the presence of Cu_2S. Consequently, if the mixed potential initially falls below the range of potentials for Stage II, S^0 will be unstable in the presence of cuprous ion. This is to be expected, as shown by E_m(I) in Figure 3.1-22, since slow discharge of ferric ion is the main kinetic process in the system. Therefore the cathodic overpotential, η_c, should be large and the anodic overpotential, η_a, should be small. The shaded portions of Figure 3.1-22 illustrate the variation in half-cell potentials as a result of the change of chemistry of the solutions during

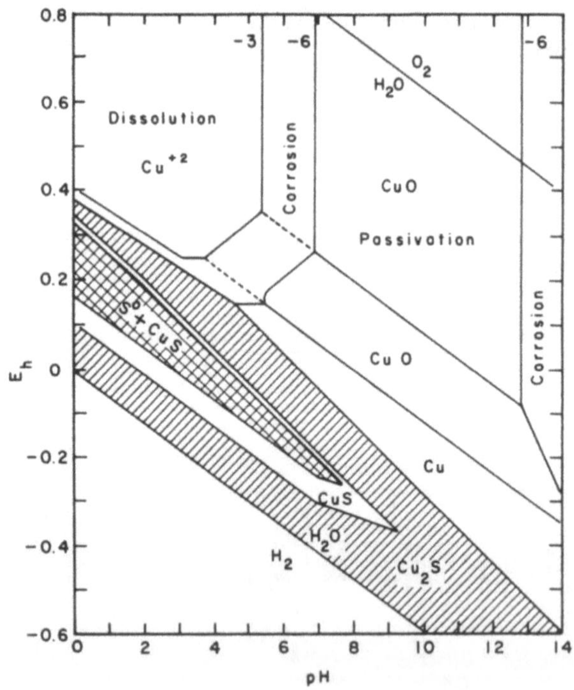

Figure 3.1-23. Pourbaix diagram of Cu-O-S-H_2O system showing stable copper–sulfur phases at $\sum S = 10^{-1}$.

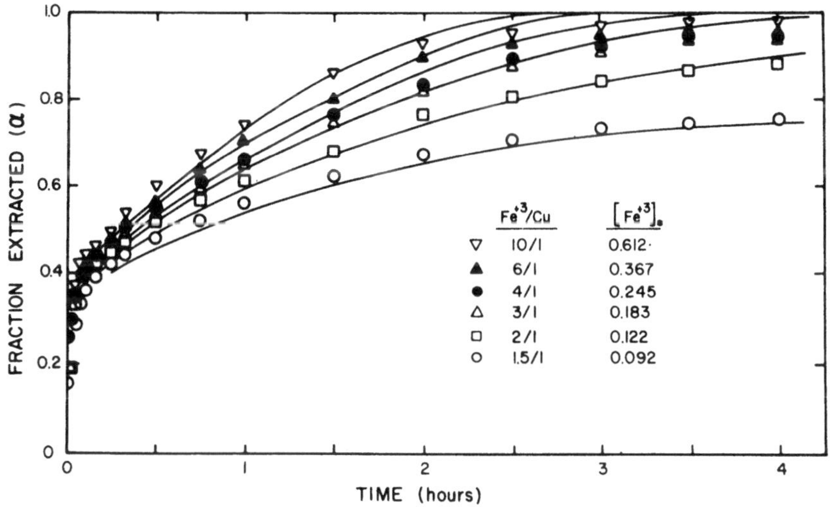

Figure 3.1-24. Results of leaching Cu_2S in ferric sulfate solutions at 90° C for various Fe^{3+}/Cu ratios.

the course of the reaction. As the half-cell voltage increases, $E_m(I)$ moves to position $E_m(I, II)$ at the end of the first-stage reaction. Actually $E_m(I, II)$ may be above the initial Stage II voltage before Stage I is finished, thus permitting S^0 to form. At low temperatures, the two stages are clearly distinct. At high temperatures, E_m is higher causing an overlap of the two stages as illustrated by (a) in Figure 3.1-22. Once Stage I is complete, the mixed potential becomes more positive and the reaction of CuS to form S^0, equation (3.1-100), is favored.

If rate equations for the individual cathodic and anodic reactions can be formalized in terms of a mixed potential, it is possible to evaluate the contribution of the mixed potential to the observed kinetics. Consider, for example, the second stage of leaching of Cu_2S according to equation (3.1-100). The cathodic reaction is

$$Fe^{3+} + e^- = Fe^{2+} \tag{3.1-101}$$

Figure 3.1-24 illustrates second-stage leaching kinetics obtained by Marcantonio[34] using ferric sulfate solutions. The kinetics are controlled by electrochemical reactions at the CuS–S^0 interface. The cathodic current for ferric ion discharge is given by the Butler–Volmer equation (assuming single-electron transfer)

$$\begin{aligned} i_c = {} & A_c z_c F[Fe^{2+}]\vec{k}_c \exp[\beta_c FE/RT] \\ & - A_c z_c F[Fe^{3+}]\overleftarrow{k}_c \exp[-(1-\beta_c)FE/RT] \end{aligned} \tag{3.1-102}$$

where A_c is the cathodic surface and E is the mixed potential. For large cathodic currents

$$i_c = -A_c z_c F[\mathrm{Fe}^{3+}]\overleftarrow{k}_c \exp[-(1-\beta_c)FE/RT] \qquad (3.1\text{-}103)$$

Similarly for large anodic currents

$$i_a = A_a z_a F\overrightarrow{k}_a \exp[\beta_a FE/RT] \qquad (3.1\text{-}104)$$

In addition, $A = A_a + A_c$ (corrosion couple) and $i_0 = i_c$. For these reactions, $z_c = 1$ and $z_a = 2$. The mixed-potential term may be evaluated giving

$$\exp\left[-\frac{(1-\beta_c+\beta_a)FE}{RT}\right] = \frac{2A_a\overrightarrow{k}_a}{[\mathrm{Fe}^{3+}]A_c\overleftarrow{k}_c} \qquad (3.1\text{-}105)$$

From the rational-rate expressions

$$\frac{dn(\mathrm{Cu}^{2+})}{dt} = \frac{dn(\mathrm{Fe}^{3+})}{2dt}$$

and letting $A_a/A_c = b$ constant, the general rate expression becomes

$$\frac{dn(\mathrm{Cu}^{2+})}{dt} = -k'A[\mathrm{Fe}^{3+}]^n \qquad (3.1\text{-}106)$$

where $n = \beta_a/(1-\beta_c+\beta_a)$ and

$$k' = \frac{\overleftarrow{k}_c b}{2(b+1)}\left(\frac{2\overrightarrow{k}_a}{b\overleftarrow{k}_c}\right)^{[1-\beta_c/(1-\beta_c+\beta_a)]} \qquad (3.1\text{-}107)$$

Marcantonio[34] has applied equation (3.1-106) successfully to the kinetics of Figure 3.1-24 using cylindrical geometry and has found $n \cong 0.5$. For the Fe^{3+}–Fe^{2+} system β_c is reported to be 0.58,[30] giving a value of $\beta_a = 0.42$ for the anodic transfer coefficient.

3.1.3. Leaching of Metals

The electrochemical nature of the dissolution of gold and silver in cyanide solutions was recognized very early by Christy[35,36] who measured the potentials of several metals in cyanide solutions. Boonstra[37] and Thompson[38] proposed electrochemical models in which gold dissolves totally at anodic sites, whereas oxygen is reduced at cathodic sites. Flow of current between the net cathodic and net anodic sites occurs as shown in Figure 3.1-17 with reactions proceeding as

Anodic $$\mathrm{Au} = \mathrm{Au}^+ + e^- \qquad (3.1\text{-}108)$$

$$\mathrm{Au}^+ + 2\,\mathrm{CN}^- = \mathrm{Au(CN)}_2^- \qquad (3.1\text{-}109)$$

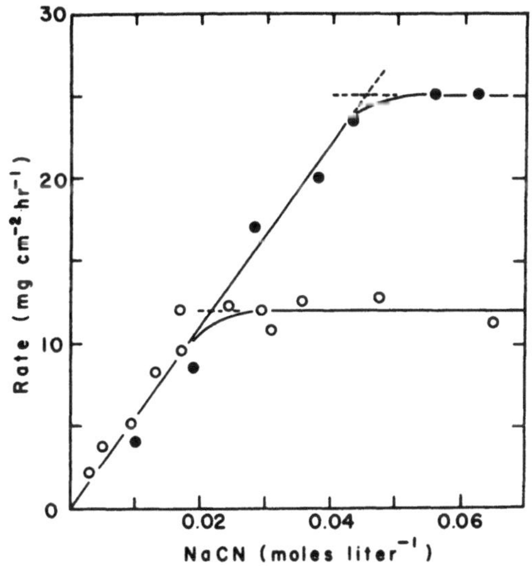

Figure 3.1-25. Effect of cyanide concentration and oxygen pressure (○, 3.40; ●, 7.40 atm) on the rate of dissolution of silver at 24° C.

Cathodic $$O_2 + 2\,H^+ + 2\,e^- = H_2O_2 \qquad (3.1\text{-}110)$$

$$H_2O_2 + 2\,H^+ + 2\,e^- = H_2O \qquad (3.1\text{-}111)$$

The discharge of H_2O_2, equation (3.1-111), is slow resulting in the buildup of H_2O_2 in solution.[39] Several previous investigators have shown that the kinetics are diffusion controlled for both gold and silver cyanidation.[40,41] The rate of reaction involves liquid-boundary diffusion of cyanide ion to anodic sites for high oxygen concentration, and boundary diffusion of oxygen to cathodic sites for high cyanide concentrations. For a fixed oxygen partial pressure the rate increases with increasing cyanide concentration, finally approaching a plateau value for which the rate is proportional to the oxygen partial pressure (Figure 3.1-25), indicating a shift from predominantly anodic to predominantly cathodic overpotential.

According to the stoichiometry, the rational rate, $\imath$, can be expressed by the relations

$$\imath = \frac{dn_{Au}}{dt} = \frac{1}{2}\frac{dn_{CN^-}}{dt} = 2\frac{dm_{O_2}}{dt} \qquad (3.1\text{-}112)$$

If A_a and A_c refer to the area of anodic and cathodic surface sites at any given time t,

$$\frac{A_a}{2}\frac{D_{CN^-}}{\delta}([CN^-] - [CN^-]_s) = 2A_c\frac{D_{O_2}}{\delta}([O_2] - [O_2]_s) \qquad (3.1\text{-}113)$$

where D_{CN^-} and D_{O_2} are the coefficients of diffusion, δ is the diffusion boûndary thickness, and $[CN^-]_s$ and $[O_2]_s$ refer to surface concentrations. Gold assumes a mixed potential dependent upon the polarization potentials related to the surface concentrations. If the surface concentrations are small compared to the bulk concentrations

$$\imath = \frac{2AD_{O_2}[CN^-]K_hD_{CN} - D_{O_2}}{\delta(D_{CN} - [CN^-] + 4\,P_{O_2}K_hD_{O_2})} \tag{3.1-114}$$

where $A = A_a + A_c$ is the total area, P_{O_2} is the oxygen partial pressure, and K_h is the Henry constant. A limiting rate can be defined for the condition[39]

$$\frac{[CN^-]}{K_hP_{O_2}} = \frac{[CN^-]}{[O_2]} = 4\frac{D_{O_2}}{D_{CN^-}} \tag{3.1-115}$$

The ratio $D_{O_2}/D_{CN} \cong 1.5$ ($D_{O_2} = 2.76 \times 10^{-5}$; $D_{CN^-} = 1.83 \times 10^{-5}\,cm^2\,sec^{-1}$), giving a value of 6 for the $[CN^-]/[O_2]$ ratio. Experimental values range between 4.5 and 7.5.

Similar results were observed by Hultquist[42] for the dissolution of copper in cyanide solutions. Copper forms complex cyanide ions, $Cu(CN)_2^-$, $Cu(CN)_3^{2-}$, and $Cu(CN)_4^{3-}$, with $Cu(CN)_3^{2-}$ being the most stable. The reactions proceed electrochemically,

Anodic $$Cu = Cu^+ + e^- \tag{3.1-116}$$

$$Cu^+ + 3\,CN^- = Cu(CN)_3^{2-} \tag{3.1-117}$$

Cathodic $$\tfrac{1}{2}O_2 + H_2O + e^- = \tfrac{1}{2}H_2O_2 + OH^- \tag{3.1-118}$$

The rational rate thus becomes

$$\imath = \frac{1}{3}\frac{dn_{CN}}{dt} = 2\frac{dn_{O_2}}{dt} \tag{3.1-119}$$

and

$$\imath = -\frac{2AP_{O_2}[CN^-]K_hD_{CN}\text{-}D_{O_2}}{\delta[D_{CN}\text{-}[CN^-] + 6\,P_{O_2}K_hD_{O^2}]} \tag{3.1-120}$$

Now the limiting rate should correspond to the ratio $[CN^-]/[O_2] \cong 9$. Figure 3.1-26 illustrates results obtained for copper at 40° C for various oxygen partial pressures and 0.07 and 0.05 *M* KCN. Extrapolation to points (a) and (b) gives $[CN^-]/[O_2]$ ratios of 8.94 and 9.41, respectively. This supports the reaction in which three cyanide ions are involved in

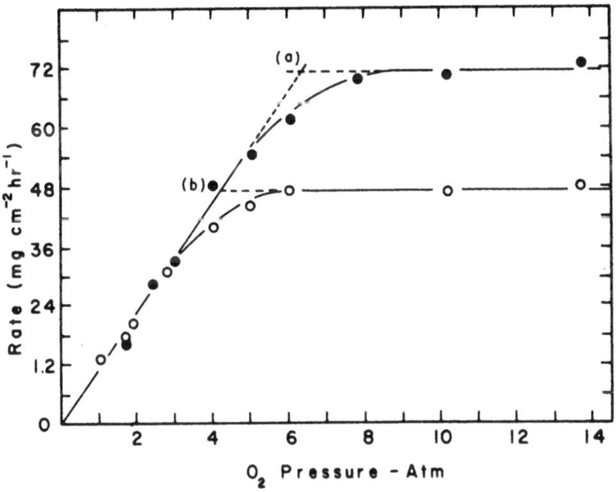

Figure 3.1-26. Effect of cyanide concentration (●, 0.07; ○, 0.05 mole liter^{-1}) and oxygen pressure on the rate of dissolution of copper at 40° C. Key: (a) 6.2 atm, $O_2 = 7.83 \times 10^{-3}$ *M*; (b) 4.2 atm, $O_2 = 5.31 \times 10^{-3}$ *M*.

the copper complex and O_2 is discharged to H_2O_2. These results parallel those found for the cyanidation of gold. In the copper system, however, hydrogen peroxide has not been reported to build up in solution.

The effect of temperature on the rate of cyanidation supports the diffusion model with shift from anodic to cathodic diffusion. Table 3.1-1 summarizes the results of several investigators for the gold, silver, and copper systems.

Table 3.1-1. Measured Activation Energies Associated with the Electrochemical Dissolution of Metals in Cyanide and Ammoniacal Solutions[a]

Metal	Oxygen diffusion	Cyanide species	Ammonia species	Reference
Gold	3.5–4.0	3.5–4.0		Kudryk and Kellogg,[40] 1954
Silver	2.4	—		Deitz and Halpern,[41] 1953
Copper	2.1	4.65		Hultquist,[42] 1957
Copper	1.33		5.54	Halpern,[43] 1953
Nickel	1.92		—	Shimakage and Morioka, 1971

[a] Values are in kcal/mole.

The dissolution of copper and nickel in ammoniacal solutions is similar to cyanidation, metal complexes being formed which are stable in basic solutions. Copper and nickel form $Me(NH_3)_n^{2+}$, where $n = 1$–4 for copper and $n = 1$–6 for nickel. Oxygen is required for the reaction to proceed.

Figure 3.1-27 illustrates results obtained by Halpern[43] for the dissolution of copper, showing the effect of oxygen pressure at various NH_3 concentrations. The curves are similar to those observed for cyanidation, but the $[NH_3]/[O_2]$ limiting rate values indicated by the break in the curves are of an order of magnitude larger than would be expected for diffusional processes alone. The oxygen dependence is similar to that observed in cyanide systems and can be attributed to diffusional control. The plateau-rate values at different NH_3 concentrations indicate very high concentrations of NH_3 for the reaction to proceed. This clearly indicates a surface reaction. Halpern interpreted the kinetics in terms of an adsorption model involving oxygen and both NH_3 and NH_4^+, since the rate was found to increase with the total ammonia concentration. More probably, the reaction is electrochemical[39] involving kinetic control by oxygen diffusion to cathodic sites at low oxygen pressure which then changes to anodic rate control at higher oxygen pressures. According to the electrochemical model, the anodic process, consistent with the kinetics observed, would involve adsorption

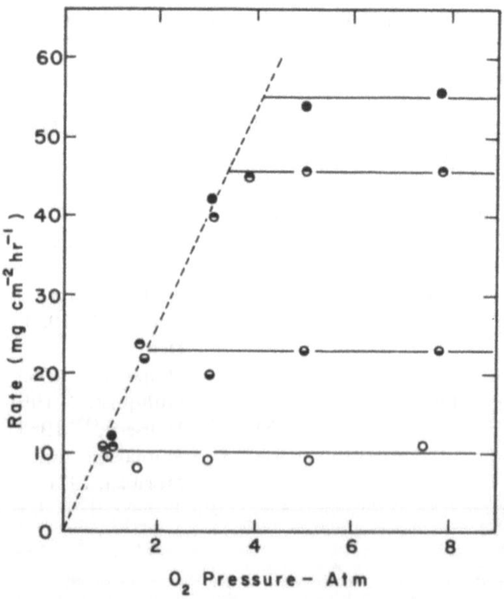

Figure 3.1-27. Effect of oxygen pressure on the rate of dissolution of copper at various NH_3 concentrations (○, 0.26; ◒, 0.52; ◓, 0.74; ●, 1.00 M liter^{-1}); $T = 26°$ C.

of neutral ammonia or ammonia ions as indicated by Halpern but not on the same surface sites as the oxygen. The overall process can be viewed as follows:

$$\textit{Anodic:}\quad Cu + NH_3 \rightarrow CuNH_3^{2+} + 2\,e^- \qquad \text{(slow)} \qquad (3.1\text{-}121)$$

or

$$Cu + NH_4^+ \rightarrow CuNH_3^{2+} + H^+ + 2\,e^- \qquad \text{(slow)} \qquad (3.1\text{-}122)$$

$$CuNH_3^{2+} + 3\,NH_3 \rightarrow Cu(NH_3)_4^{2+} \qquad \text{(fast)} \qquad (3.1\text{-}123)$$

$$\textit{Cathodic:}\quad \tfrac{1}{2}\,O_2 + H_2O + e^- = \tfrac{1}{2}\,H_2O_2 + OH^- \qquad (3.1\text{-}124)$$

The rational rate would be

$$\imath = \frac{1}{4}\frac{dn_{NH_3}}{dt} = 2\frac{dn_{O_2}}{dt} \qquad (3.1\text{-}125)$$

resulting in the general rate expression

$$\imath = -\frac{2AD_{O_2}[O_2](k_{NH_4^+}[NH_4^+] + k_{NH_3}[NH_3])}{8D_{O_2}[O_2] + \delta k_{NH_4^+}[NH_4^+] + \delta k_{NH_3}[NH_3]} \qquad (3.1\text{-}126)$$

where k_{NH_3} and $k_{NH_4^+}$ are the rate constants and δ is the thickness of the diffusion layer.

At low oxygen pressures, equation (3.1-126) becomes

$$\imath = \frac{2AD_{O_2}}{\delta 10^3}[O_2] \qquad (3.1\text{-}127)$$

where $[O_2]$ is in moles liter^{-1}. At high oxygen pressures, equation (3.1-126) becomes

$$\imath = -\tfrac{1}{4}A(k_{NH_3}[NH_3] + k_{NH_4^+}[NH_4^+]) \qquad (3.1\text{-}128)$$

Equations (3.1-127) and (3.1-128) are consistent with the findings of Halpern. The limiting rate, corresponding to the break in the curves of Figure 3.1-27, represents the case for $[NH_4^+] \cong 0$, where equations (3.1-127) and (3.1-128) are equal, so that

$$\frac{[NH_3]}{[O_2]} = \frac{8D_{O_2}}{10^3\,\delta k_{NH_3}} \qquad (3.1\text{-}129)$$

From the curves of Figure 3.1-27, $[NH_3]/[O_2] \cong 240$, and k_{NH_3} from the data of Halpern, is 1.46×10^{-6} mole cm^{-2} sec^{-1}. From this it is found that $\delta \cong 0.6 \times 10^{-3}$ cm, which is consistent with the model for oxygen diffusion through a limiting-boundary film to cathodic sites for low oxygen pressure and surface reaction for high oxygen pressure. The slow process evidently does not involve charge transfer since the rate is independent of the $Cu(NH_3)_4^{2+}$ concentration in solution and becomes totally independent of oxygen pressure at high pressure. This analysis does not lend support to H_2O_2 formation. If O_2 reduction to H_2O occurred in the electrochemical release of copper, the calculated value of δ would be 1.2×10^{-3} which is likewise consistent.

In the case of copper, the rate involving NH_4^+ was approximately 18 times that for NH_3. Halpern attributed this to the favored protonation of the surface because of adsorbed oxygen. Just why the rate for NH_4^+ is faster is not clear, and, as will be shown, is just opposite that for nickel. Similar results were observed by Halpern *et al.*(44) for copper using complexing agents other than ammonia. Ethylenediamine, glycinate, α-alaninate, and β-alaninate all adsorbed similar to ammonia and the rate of dissolution was of the same order as the stability constants of the complexes formed.

Shimakage and Morioka(45) investigated the dissolution of nickel in ammoniacal solutions at elevated pressures and temperatures and proposed the following electrochemical model:

Anodic $$Ni \rightarrow Ni^{2+} + 2\,e^- \qquad (3.1\text{-}130)$$

$$Ni^{2+} + nNH_3 \rightarrow Ni(NH_3)_n^{2+} \qquad (3.1\text{-}131)$$

with cathodic discharge of oxygen; with large excess of NH_3, $n \rightarrow 6$. Equilibrium conditions are readily established at the nickel surface, resulting in a rapid initial dissolution and a decrease in rate as OH^- builds up. For the reaction to continue, OH^- must be removed, which is readily accomplished by adding NH_4^+ to the solution and providing buffering action according to the reaction

$$OH^- + NH_4^+ \rightarrow NH_4OH \rightarrow NH_3 + H_2O \qquad (3.1\text{-}132)$$

Although NH_4^+ permits the reaction to continue, it was noted that it lowered the rate of reaction, having an opposite effect to that for copper. The separate contributions of NH_3 and NH_4^+ to the rate were not evaluated. The decrease in rate when adding $(NH_4)_2SO_4$ was attributed to the decrease in solubility because of the increased ionic strength. This explanation is not consistent with the results observed for copper, which suggest that NH_4^+ has a specific retarding or blocking effect in the case of nickel.

3.1.4. Leaching of Sulfides

The most abundant iron sulfide mineral is pyrite, FeS_2. McKay and Halpern[46] investigated the dissolution of FeS_2 in acid solutions for temperatures greater than 100° C and found that it reacts heterogeneously with oxygen producing both ferrous and ferric ions in solution plus elemental sulfur in varying amounts dependent upon the temperature, oxygen potential, and pH of the solution. The formation of ferric iron in solution occurred from the homogeneous oxidation of ferrous iron by oxygen.

The reactions of importance are

$$FeS_2 + O_2 \rightarrow FeSO_4 + S^0 \qquad (3.1\text{-}133)$$

and

$$2\,FeS_2 + 7\,O_2 + 2\,H_2O \rightarrow 2\,FeSO_4 + 2\,H_2SO_4 \qquad (3.1\text{-}134)$$

The rate of reaction follows first-order kinetics with regard to oxygen

$$d[FeS_2]/dt = -kAP_{O_2} \qquad (3.1\text{-}135)$$

where A is the surface area of the pyrite, and

$$k = 0.125\exp[-13{,}300/RT] \qquad (3.1\text{-}136)$$

having the units moles $cm^{-2}\ atm^{-1}\ min^{-1}$.

McKay and Halpern explain the reaction according to a model involving oxygen adsorption on the reacting surface site

$$FeS_2 + O_2(aq) \rightleftharpoons FeS_2{\cdot}O_2 \qquad (3.1\text{-}137)$$

$$FeS_2{\cdot}O_2 + O_2(aq) \xrightarrow{\text{slow}} [FeS_2{\cdot}2\,O_2]^{\ddagger} \rightarrow FeSO_4 + S^0 \qquad (3.1\text{-}138)$$

Equation (3.1-138) accounts for equal formation of SO_4^{2-} and S^0. Sulfur is subsequently oxidized according to the reaction

$$S^0 + 4\,H_2O + \tfrac{3}{2}\,O_2 = SO_4^{2-} + 3\,H_2O + 2\,H^+ \qquad (3.1\text{-}139)$$

Low pH retards the reaction, accounting for higher S^0 production; and high pH favors the reaction, accounting for lower S^0 production and increased formation of H_2SO_4.

Peters and Majima[47] investigated the electrochemical reaction of pyrite. The rest potential was 0.62 V, which makes pyrite more noble than any other sulfide mineral and suggests that it should enhance the anodic dissolution of other minerals in electrical contact. For the anodic dissolution at potentials above 0.62 V but below the potential necessary to discharge oxygen, the major reaction at room temperature was

$$FeS_2 + 8\,H_2O \rightarrow Fe^{3+} + 2\,SO_4^{2-} + 16\,H^+ + 15\,e^- \qquad (3.1\text{-}140)$$

and no elemental sulfur was formed. It was concluded therefore that elemental sulfur does not form by an electrochemical process during the pressure dissolution of FeS_2. This is in contrast to results observed for Ni_3S_2, CuS, PbS, and $Fe_{1-x}S$ (pyrrhotite), which do form elemental sulfur during anodic dissolution.

Mathews and Robins[48] investigated the oxidation of pyrite with ferric sulfate. The overall stoichiometry corresponds to the reaction

$$FeS_2 + 8\,H_2O + 14\,Fe^{3+} \rightarrow 15\,Fe^{2+} + 2\,HSO_4^- + 14\,H^+ \qquad (3.1\text{-}141)$$

which has important implications in dump leaching. Ferric ion produced in regions of high oxygen potential, but which migrate to regions low in oxygen, can react with pyrite according to equation (3.1-141) and produce acid *in situ*. The rate of reaction follows first-order kinetics[48] and is influenced by the total amount of iron in solution. The reaction is retarded by hydrogen ion and has an activation energy of 20.5 kcal/mole. The rate is given by

$$\frac{d[Fe^{3+}]}{dt} = -k\frac{WS[Fe^{3+}]}{V[Fe^T][H^+]^{0.44}}\exp\left[-\frac{20{,}500}{RT}\right] \quad \text{(moles/min)} \qquad (3.1\text{-}142)$$

where W, S, and V are the weight, surface area, and solution volume (WS/V has dimensions m^2/liter), and $k = 1.47 \times 10^8$. The fact that the rate is proportional to the ratio of ferric to total iron, and the -0.44 order on hydrogen ion is strong evidence of an electrochemical reaction, in which the pyrite dissolves anodically according to the reaction

$$FeS_2 + 8\,H_2O = Fe^{2+} + 2\,SO_4^{2-} + 16\,H^+ + 14\,e^- \qquad (3.1\text{-}143)$$

Preferential galvanic attack which can occur when one mineral is in contact with another mineral or metal is well illustrated by Peters *et al.*,[49] who found pyrite to accelerate the anodic dissolution of galena (PbS). Figure 3.1-28 illustrates potentiostatic polarization curves obtained for pyrite and galena. Galena decomposes anodically according to the reaction

$$PbS \rightarrow Pb^{2+} + S^0 + 2\,e^- \qquad (3.1\text{-}144)$$

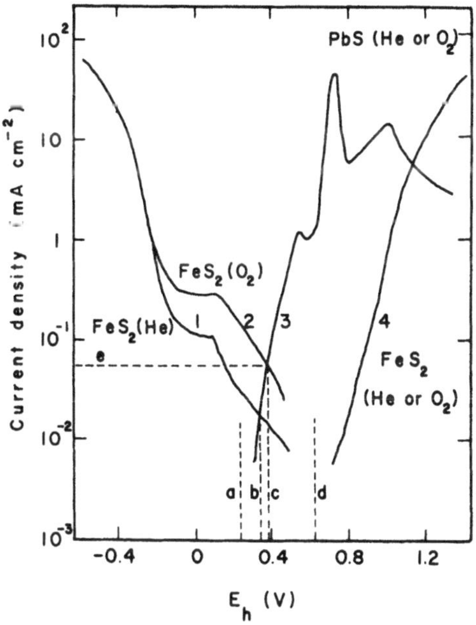

Figure 3.1-28. Potentiostatic polarization curves of pyrite and galena, 1.0 *M* $HClO_4$, 25° C: (a) rest potential of galena; (b) pyrite–galena mixed potential in helium; (c) pyrite–galena mixed potential in oxygen; (d) rest potential of pyrite; and (e) current density in oxygen.

In the presence of oxygen the anodic branch of the PbS curve (Curve 3, Figure 3.1-28) crosses the cathodic branch of FeS_2 (Curve 2), establishing mixed potential *c* with a galvanic current *e*. Cathodic discharge occurs at the pyrite surface according to the reaction

$$\tfrac{1}{2}\,O_2 + 2\,H^+ + 2\,e^- = H_2O \qquad (3.1\text{-}145)$$

with PbS becoming the sacrificial anode. Conversely, pyrite is protected by the galena. Similar results are expected for covellite (CuS) and sphalerite (ZnS). Figure 3.1-29a illustrates results obtained in the pyrite–galena system.

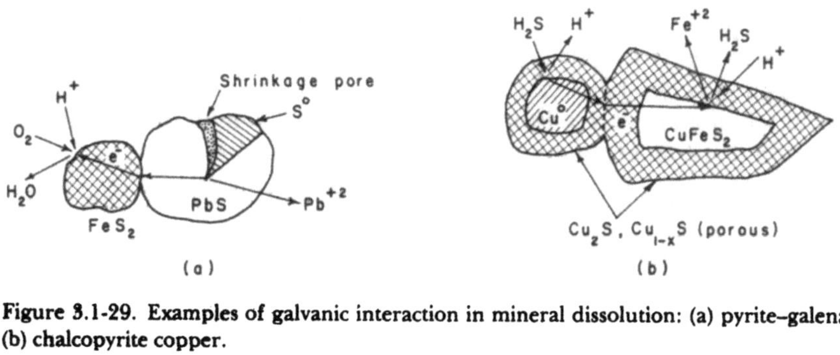

Figure 3.1-29. Examples of galvanic interaction in mineral dissolution: (a) pyrite–galena, (b) chalcopyrite copper.

Another example of the galvanic effect in mineral dissolution was observed by Shirts.[50] Chalcopyrite ($CuFeS_2$) reacts cathodically if it comes into contact with a metal having a more negative potential. The rest potential of $CuFeS_2$ is approximately 0.46–0.56 V. Therefore if contacted with metals such as Cu, Fe, Pb, Zn, etc., having more negative potentials, the chalcopyrite reacts cathodically,

$$2\,CuFeS_2 + 6\,H^+ + 2\,e^- = Cu_2S + 2\,Fe^{2+} + 3\,H_2S \qquad (3.1\text{-}146)$$

The anode in this case is not provided cathodic protection but is induced to react anodically. For copper the reaction is

$$2\,Cu + H_2S = Cu_2S + 2\,H^+ + 2\,e^- \qquad (3.1\text{-}147)$$

The copper sulfide products are actually chalcocite (Cu_2S) plus $Cu_{1-x}S$, with at least two identifiable defect structures. Figure 3.1-29b illustrates the copper–chalcopyrite galvanic interaction.

Hiskey and Wadsworth[51] explained the galvanic conversion of $CuFeS_2$ according to equations (3.1-146) and (3.1-147). The rate was proportional to the anodic surface area (Cu) initially, but as the $CuFeS_2$ reacted cathodically its surface area diminished and discharge at the cathodic surface became rate limiting as the reaction proceeded. Furthermore, the rate was proportional to hydrogen ion activity to the one-half power. The kinetics follow the general rate expression

$$\frac{dn}{dt} = \frac{a_{H^+}{}^{1/2} A_c A_a k_1}{(A_c + k_2 A_a)^{1/2} (k_3 A_c + k_4 A_a)^{1/2}} \qquad (3.1\text{-}148)$$

The apparent half-power dependence on hydrogen ion results from a systematic shift in the mixed potential during the course of the reaction. The details of several additional studies on the leaching of sulfides were recently reviewed by Wadsworth[52,53] and are not repeated here.

3.1.5. Leaching of Oxides

In general the kinetics of leaching of oxide minerals is dependent upon the activity of hydrogen ions in the system. The surface area and more complex geometric factors are also involved. In the case of chrysocolla, Pohlman and Olson[54] found that a shrinking core model, equation (3.1-65), explained their results. The reaction is

$$CuO\cdot SiO_2\cdot 2\,H_2O + 2\,H^+ \rightarrow Cu^{2+} + SiO_2\cdot nH_2O + (3-n)H_2O \qquad (3.1\text{-}149)$$

and the surface reaction plus diffusion terms of equation (3.1-65) so that

$$1 - \frac{2}{3}\alpha - (1-\alpha)^{2/3} + \frac{\beta}{r_0}[1 - (1-\alpha)^{1/3}] = \frac{\gamma[H^+]}{r_0^2} t \qquad (3.1\text{-}150)$$

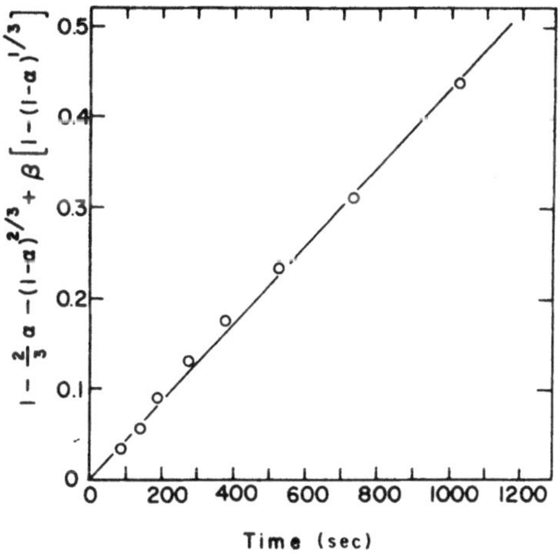

Figure 3.1-30. Plot of data according to reaction zone model [equation (3.1-138)] for −100 + 200-mesh chrysocolla at 28.5° C, pH = 0.22 ($\beta = 0.459$).

The remaining silica lattice provided a diffusion barrier with a moving reaction boundary. Figure 3.1-30 illustrates the data plotted according to equation (3.1-150). Pure oxide minerals which leach without forming products of reaction should follow surface-reaction-rate control or diffusion through a limiting boundary film for which equation (3.1-150) becomes

$$1-(1-\alpha)^{1/3}=\frac{\gamma[H^+]}{\beta r_0} \quad (3.1\text{-}151)$$

The dissolution of metal oxides normally does not involve oxidation–reduction couples. An exception is the dissolution of cuprous oxide or cuprite (Cu_2O) in acid solutions.[55] In the absence of oxygen the reaction is

$$Cu_2O+2\,H^+=Cu^{2+}+Cu^0+H_2O \quad (3.1\text{-}152)$$

in which the oxidation–reduction couple is completed by the disproportionation of the cuprous copper. Two parallel rates were observed and explained in terms of the adsorption of sulfuric acid followed by surface decomposition of the surface species or reaction with a proton to account for the observed pH dependence. In the presence of oxygen, cuprite reacts to form cupric ions in solution according to the reaction

$$Cu_2O+4\,H^+ +\tfrac{1}{2}\,O_2=2\,Cu^{2+}+2\,H_2O \quad (3.1\text{-}153)$$

If the same rate-controlling process accounts for both processes, the rational rate is given by

$$\imath=-\frac{d[Cu_2O]}{dt}=\frac{d[Cu^{2+}]}{dt}=\frac{1}{2}\,\frac{d[Cu^{2+}]O_2}{dt} \quad (3.1\text{-}154)$$

where the subscript O_2 refers to the rate of appearance of cupric ions in the presence of oxygen. The fact that Cu^{2+} appears at twice the rate in the presence as in the absence of oxygen indicates the same rate-controlling process is involved. Since Cu_2O is a semiconductor the dissolution process probably is electrochemical in nature, although such has not been demonstrated.

Oxidation is required for the dissolution of urania (UO_2) in both acid and basic circuits. It has been proposed by Mackay and Wadsworth,[56] that in acid circuits reacted surface sites rapidly hydroxylate according to the reaction

$$_s|O{-}U{-}O + H_2O \rightleftharpoons |O{-}U\begin{matrix}\diagup OH \\ \diagdown OH\end{matrix} \qquad \text{(rapid hydroxylation)} \tag{3.1-155}$$

$$s|O{-}U\begin{matrix}\diagup OH \\ \diagdown OH\end{matrix} + O_2(aq) \xrightarrow{\text{slow}} UO_2^{2+} + HO_2^- + OH^- \tag{3.1-156}$$

and that oxidation occurs at the mineral surface. This mechanism is similar to one previously proposed by Forward and Halpern[57] for the aqueous oxidation of U_3O_8 in acid solutions. Habashi[39,58] proposed that the dissolution of O_2 occurs by an electrochemical reaction

$$\textit{Anodic} \qquad UO_2 \rightarrow UO_2^{2+} + 2\,e^- \tag{3.1-157}$$

$$\textit{Cathodic} \qquad O_2 + 2\,H_2O + 4\,e^- \rightarrow 4\,OH^- \tag{3.1-158}$$

However, no proof of an electrochemical mechanism was offered. Nicol, Needes, and Finkelstein[59] have carried out an excellent study of the leaching of UO_2 in both acid and basic media, demonstrating clearly the electrochemical character of the reactions involved. Ferric ion was the oxidant so that in addition to equation (3.1-157) the Fe(III)/Fe(II) half-cell reaction must be considered

$$Fe(III) + e^- = Fe(II) \tag{3.1-159}$$

When Fe(II) is very low the rate is proportional to an Fe(III) reaction order of 0.73. At higher Fe(II) concentrations the rate is proportional to $\{[Fe(III)]/[Fe(II)]\}^{1/2}$, and at still higher Fe(II) concentration the rate is proportional to $\{[Fe(III)]/[Fe(II)]\}^{1.1}$. All of these results follow directly from the electrochemical nature of the dissolution processes for which $I_c = I_a$, thus establishing a mixed potential which varies systematically as the solution concentrations vary. An important feature of the study was that the complex iron monosulfate ion, $FeSO_4^+$, is the most electroactive iron species in solution. Formation of $Fe(SO_4)_2^-$ at higher sulfate values diminished the rate.

In basic carbonate solutions the electrochemical reactions are [59]

$$UO_2 + 3\,CO_3^{2-} = UO_2(CO_3)_3^{4-} + 2\,e^- \qquad (3.1\text{-}160)$$

$$M + 2\,e^- = M^{2-} \qquad (3.1\text{-}161)$$

where M refers to various oxidants used. The order of leaching rates for several oxidants studied was

$$Fe(CN)_6^{3-} > H_2O_2 > Cu(NH_3)_4^{2+} > O_2$$

and was proportional to the oxidant concentration to the one-half power.

The dissolution of bauxite, a naturally occurring ore containing trihydrated and monohydrated aluminum oxide, is a necessary purification step leading to the production of aluminum. The kinetics of dissolution of the trihydrated oxide gibbsite ($Al_2O_3{\cdot}3\,H_2O$) in basic (NaOH) solution was measured by Glastonbury.[60] The dissolution of gibbsite in basic solution occurs by the reaction

$$Al(OH)_3 + NaOH = NaAlO_2 + 2\,H_2O \qquad (3.1\text{-}162)$$

The rate is chemically controlled; for a heterogeneous process occurring at the gibbsite surface with an activation energy of 23.9 kcal mole^{-1} it was found to follow the rate equation

$$\text{rate} = 4.60 \times 10^5\, A[OH^-]^{1.78} \exp\left[-\frac{23{,}850}{RT}\right] (g\text{-atoms Al})\ \text{sec}^{-1} \qquad (3.1\text{-}163)$$

The fractional order of 1.78 results from an analysis of the data and does not correct for activity coefficients or the decrease in area which occurs most likely during the course of the reaction. It seems reasonable to assume therefore that the reaction is second order with respect to (OH^-) which probably breaks hydroxyl bridges splitting out water. The second order can be explained in terms of the amphoteric nature of $Al(OH)_3$, which in basic solutions may be in equilibrium with hydroxyl ions according to the reaction

$$_s|\underset{\phi}{Al_2(OH)_6} + OH^- \xrightarrow{K_1} {}_s|\underset{\theta}{Al_2O_2(OH)_3^-} + 2\,H_2O \qquad (3.1\text{-}164)$$

If the slow process involves the reaction of OH^- with the partially dehydroxylated surface, then we have

$$_s|\,\underset{\theta}{Al_2O_2(OH)_3^-} + OH^- \xrightarrow{k} 2\,AlO_2^- + 2\,H_2O \qquad (3.1\text{-}165)$$

If ϕ and θ represent the fractions of the surface that are $_s|\,Al_2(OH)_6$ and $_s|\,Al_2O_2(OH)_3^-$, respectively, then

$$K_1 = \theta/\phi[OH^-] \qquad (3.1\text{-}166)$$

The rate is

$$\text{rate} = \theta[OH^-]k_0k \qquad (3.1\text{-}167)$$

where k_0 is the number of reactive sites per cm^2 and k is the rate constant. Solving for ϕ ($\phi + \theta = 1$), equation (3.1-167) becomes

$$\text{rate} = \frac{[OH^-]^2K_1}{1 + K_1[OH^-]}k_0k \qquad (3.1\text{-}168)$$

Under the condition $K_1[OH^-] \ll 1$, equation (3.1-168) becomes second order. The partial contribution of $[OH^-]$ in the denominator may also account for the apparent 1.78 order observed.

3.2. *Dump and In Situ Leaching Practices (MEW)*

3.2.1. *Introduction*

In recent years the general field of hydrometallurgy has received renewed attention, initiated by two major developments. First, consideration of the environment has focused strongly on the minerals producers because of localized emission and earth disturbance. Though small in the total of environmental damage, the minerals and metals industry is making a concerted effort to find and develop new technology. Second, scarcity of materials, particularly raw material sources within the United States, has resulted in renewed exploration activity.

The direct dissolution of metal values from an ore deposit is an intriguing concept considering the vast mineral wealth of the earth's crust which cannot be mined economically by normal techniques. Detailed rate studies of such systems are few, but significant progress is being made. Scale-up from laboratory to field is a monumental extrapolation requiring knowledge of mineral and ore fragment kinetics, hydrology, gas-flow characteristics, and complex solution chemistry.

Bartlett[61] has developed the basis for a pore-diffusion, mineral kinetics model under non-steady-state conditions. In its general application, this model has greatest potential in providing a fundamental first approximation of extraction kinetics from known porosity data, individual mineral kinetics, and solution chemistry. Useful steady-state models have been employed by Braun *et al.*[62] for deep solution mining of copper sulfides, by Roman *et al.*[63] for leaching of oxide ores, and by Madsen *et al.*[64] for leaching of Cu_2S under dump-leaching conditions. The steady-state solution is based upon a shrinking core model and is valid for the case in which mineral kinetics are relatively rapid compared to the effective diffusion rates in an ore fragment. Cathles and Apps[65] have extended the shrinking-core model to include thermal effects as predicted from the chemical kinetics. The thermal gradients in turn influence air convection within the deposit.

3.2.2. Leaching Systems

3.2.2.1. Conventional Leaching practice

The leaching of low-grade ores, as in copper dump leaching, represents an area in which significant advancement has been made in recent years. The leaching of low-grade materials is one of the oldest metallurgical operations known, but its physics and chemistry have been little understood. Heap leaching and vat (percolation) leaching are usually carried out on oxide ores somewhat higher in value than ores considered for dump leaching. Figure 3.2-1 illustrates each type of leaching operation.

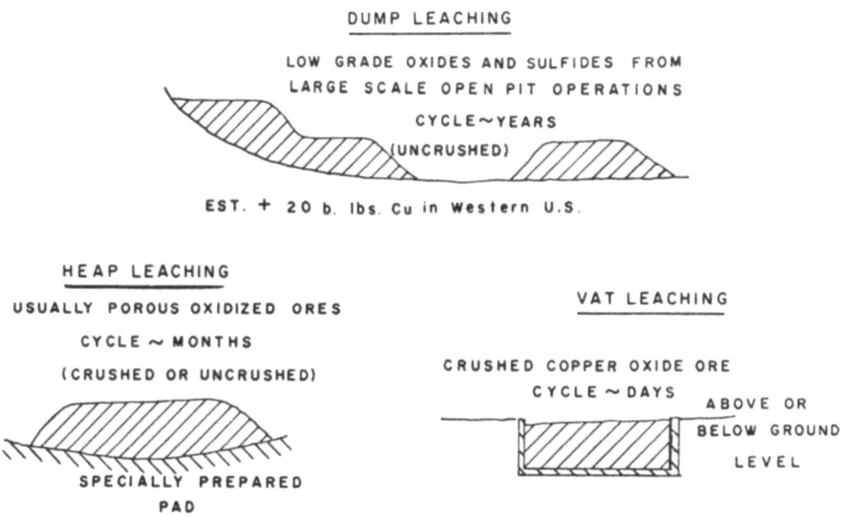

Figure 3.2-1. Various types of leaching.

3.2.2.2. Solution Mining Systems

Recently, in-place (*in situ*) leaching of ore deposits has received increased emphasis and appears to be an area in which significant advances will be made in the near future. This type of hydrometallurgical operation is often referred to as *solution mining*. It is useful to consider dump leaching as part of solution mining since the physical and chemical features are similar to the leaching of fragmented or rubblized deposits in place. Much of what is to be expected from *in situ* extraction can be derived from current experience and practice in dump leaching.

Deposits amenable to *in situ* leaching may be classified into the three general groupings shown in Figure 3.2-2.

I. Surface dumps or deposits having one or more sides exposed, and deposits within the earth's crust but above the natural water table.

II. Deposits located below the natural water table but accessible by conventional mining or well-flooding techniques.

III. Deposits below the natural water table and too deep for economic mining by conventional methods.

Dump leaching is placed in the first classification. Type II is characteristic of what is to be expected in the near future. Type III is expected to develop more slowly.

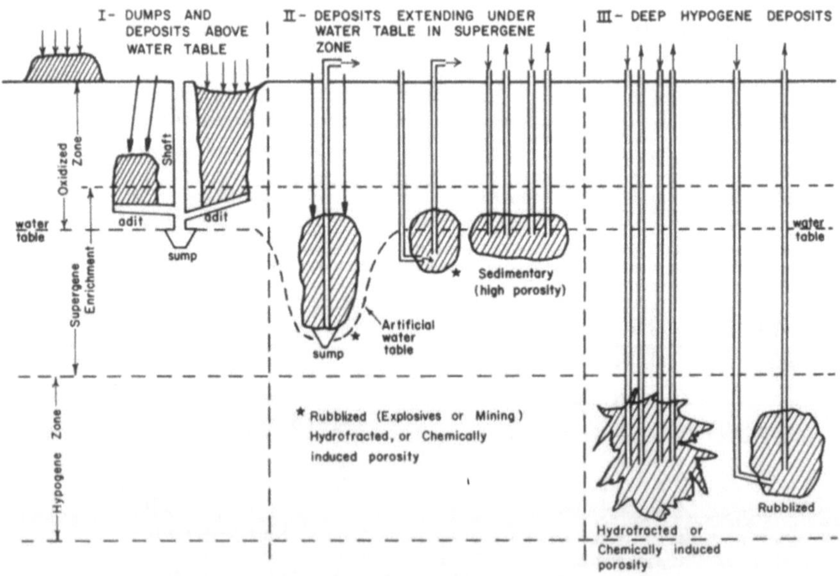

Figure 3.2-2. Three sets of conditions for solution mining.

Type I would be the leaching of a fractured ore body near the surface above the natural water table in the surrounding area. This would apply to mined out regions of old mines such as a block-caved portion of a copper mine, or regions which have been fractured by hydrofracturing or by the use of explosives. The chemistry and physical requirements would be essentially the same as in dump leaching.

Type II refers to the leaching of deposits which exist at relatively shallow depths, less than approximately 500 ft, and which are under the water table. Such deposits will have to be fractured in place and dewatered so they may be subjected to alternate oxidation and leach cycles or percolation leaching, although the use of special oxidants may eliminate the drainage cycle. This is a special problem requiring a complete knowledge of the hydrology of the region. Water in the deposit, if removed during the oxidation cycle, must be processed, stored, and returned under carefully controlled conditions. An alternate method of leaching would be by flooding as described for Type III below. An important, rapidly developing example of Type II is the application of flooding using wells distributed on a grid such as is currently under field test for uranium extraction.[66]

The third general type (Type III) of solution mining is represented by deep deposits below the water table and below approximately 500 ft in depth. The ore body is shattered by conventional or nuclear devices. Again, the hydrology of the region must be well known for proper containment of solutions. This represents a unique situation in that the hydrostatic head will increase the oxygen solubility to the point that the direct oxygen oxidation of sulfide minerals becomes possible.

3.2.3. Rate Processes

3.2.3.1. Leaching of Sulfide Ores

General Leaching Model. Virtually all researchers agree that the leaching of ore fragments involves penetration of solution into the rock pore structure. The kinetics thus involve diffusion of lixiviant into the rock where reaction with individual mineral particles occurs. The kinetics are thus complicated by changing porosity, pH, and solution concentration. Bartlett[61] applied the continuity equation to a system as described by Type III. The system is one in which a copper porphyry ore containing chalcopyrite as the major copper sulfide mineral, is subjected to oxidation under conditions proposed by researchers at the Lawrence Livermore Laboratory (LLL) for deep solution mining.[62] Diffusion of oxygen and subsequent reaction with mineral particles combine to give a non-steady-state concentration gradient within the ore

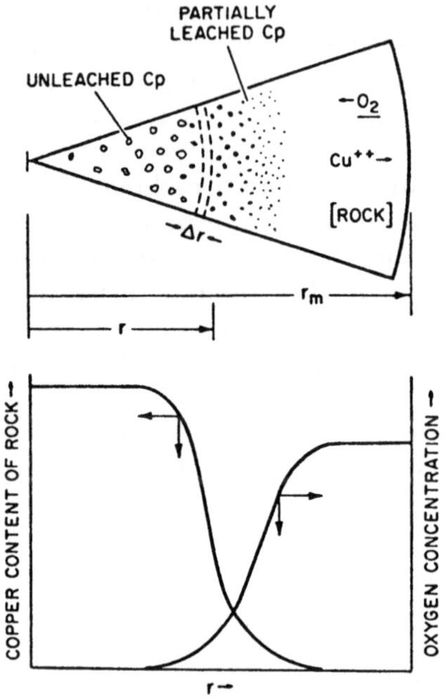

Figure 3.2.3. Schematic drawing illustrating the leaching of a rock containing disseminated chalcopyrite (Cp).

fragment. This treatment provides a basis for modeling of all similar leaching systems. The shrinking core concept used by several researchers is a special case of this general treatment. Figure 3.2-3 illustrates the mineral fragment showing different regions at some time, t. Oxygen and copper gradients vary markedly over a broad zone in which chalcopyrite (Cp) particles extend from partly to totally reacted. The fragment, considered as spherical, may be viewed in cross section, as divided into annular rings of thickness Δr. If J is the flux of dissolved reactant diffusing into and out of this incremental volume the continuity equation for an individual ore fragment is

$$\varepsilon \frac{\partial C}{\partial t} = -\mathcal{R} + \frac{1}{A} \frac{\partial J}{\partial r} \tag{3.2-1}$$

where A is the area at radius r, and $\mathcal{R}$ is the rate of loss of reactant per unit volume because of reaction with mineral particles. Actually, this is a summation of rate processes for all mineral types and sizes, within the incremental volume. The porosity ε must be included since only the solution volume is involved. Combining equation (3.2-1) with Fick's laws gives for spheres

$$-\varepsilon \frac{\partial C}{\partial t} = \mathcal{R} + D'\left(\frac{\partial^2 C}{\partial r^2} + \frac{2}{r} \frac{\partial C}{\partial r}\right) \tag{3.2-2}$$

The evaluation of $\mathcal{R}$ requires a knowledge of the kinetics of dissolution of individual mineral processes. Bartlett[61] assumes these particles follow a log-normal size-distribution law. The evaluation of diffusion terms is accomplished by finite difference procedures for each rock fragment size. Using this method a matrix is developed from which the oxygen concentration profile may be determined for any given time, t. Knowing the oxidant concentration for each incremental volume makes it possible to calculate the rate of reaction within each volume for finite time Δt. The summation of metal release values for each incremental volume in turn summed over each finite time interval results in an evaluation of metal recovery versus time. Figure 3.2-4 illustrates percent extraction versus time[61] for a variety of porosities using this method.

Reaction Zone Model. If the rate of reaction of individual mineral particles is fast enough, the region of partially leached minerals, Figure 3.2-3, is fairly narrow. Such a condition is illustrated in Figure 3.2-5 where a reaction zone of thickness δ moves topochemically inward during the course of the reaction. According to the reaction zone model steady-state diffusion occurs through the reacted outer region and is equal to the rate of reaction within the reaction zone itself. This model was developed by Braun *et al.*[62] for the same ore as considered by Bartlett.[61] The effective area of mineral particles within the moving reaction zone is assumed to be essentially constant and independent of the mineral-particle-size distribution since new particles in each size fraction will begin to leach at the leading edge of the reaction zone just as similar particles are completely leached at the tail of the reaction zone. The rate of reaction within the reaction zone may be expressed by the

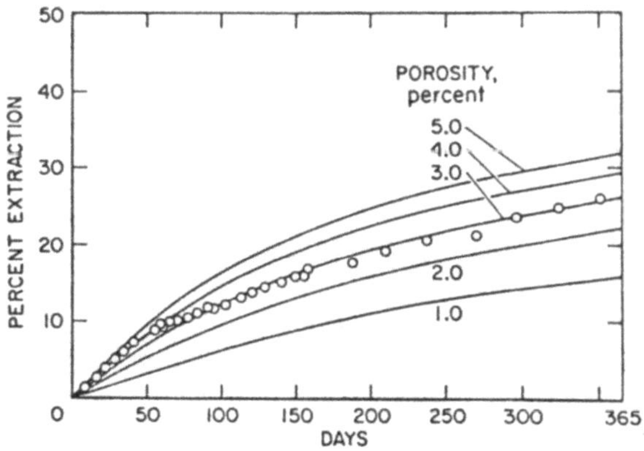

Figure 3.2.4. Computed and experimental extractions for the LLL-San Manuel test for 1 year.[62]

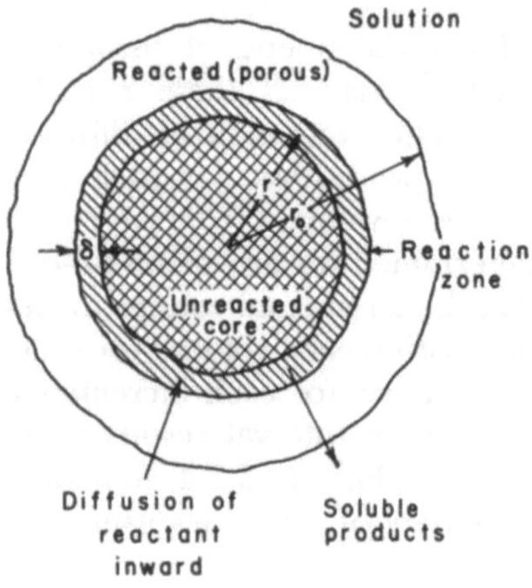

Figure 3.2.5. Reaction zone model with reaction zone thickness δ at concentration C_s.

equation

$$\frac{dn}{dt} = -\frac{4\pi r^2 \delta n_p A_p}{\phi} C_s k_s \tag{3.2-3}$$

where

n = moles of leachable mineral,
t = time,
n_p = number of mineral particles per unit volume of rock,
A_p = average area per particle in the reaction zone,
k_s = rate constant of mineral particle,
C_s = average concentration in the reaction zone, and
N = Avogadro's number.

Diffusion through pores to the reaction zone may be expressed by the equation

$$\frac{dn}{dt} = \frac{4\pi r^2 D' K_h}{\phi\sigma} \frac{dC}{dr} \tag{3.2-4}$$

where

D' = effective coefficient of diffusion,
K_h = Henry constant,
C = bulk solution concentration,
ϕ = geometric factor to account for deviation in sphericity, and
σ = stoichiometry factor (number of moles of reactant required per mole of metal released).

Equation (3.2-4) may be integrated for steady-state transport where for a given geometrical configuration the reaction dn/dt is constant for all values of r between r and r_0. The result is

$$\frac{dn}{dt}=-\frac{4\pi D r r_0 K_h}{\phi(r_0-r)\sigma}(C-C_s) \tag{3.2-5}$$

Equations (3.2-3) and (3.2-5) are similar to equations (3.1-61) and (3.1-62) for mixed diffusion plus surface reaction when a sharp reaction interface results. As before, these equations may be combined under the steady-state approximation giving for ore fragment of radius, r_i,

$$\frac{dn_i}{dt}=-\frac{4\pi r_i^2}{\phi_{0i}}\left[\frac{1}{G\beta}+\left(\frac{\sigma}{D'}\right)\left(\frac{r_i}{r_{i0}}\right)(r_{i0}-r_i)\right]^{-1} \tag{3.2-6}$$

where G is the grade (weight fraction of copper sulfide mineral),

$$G=\frac{\delta A_p r_p \rho_p}{3\rho_r}; \qquad \beta=\frac{3\rho_r \delta k_s}{r_p \rho_p} \tag{3.2-7}$$

where

r_p = average mineral particle radius,
ρ_p = mineral particle density, and
ρ_r = bulk rock density.

Equation (3.2-6) may be integrated numerically and summed over all ore fragment sizes, equation (3.1-22). Weathering of large ore fragments roughened the surface. This was compensated by a correction term giving a variation of ϕ_{0i} with time. Figure 3.2-6 illustrates the copper extractive curve calculated according to equation (3.2-6) using numerical integration. The sample weighed 5.8×10^6 g and varied in particle size from 0.01 to 16 cm.

Madsen *et al.*[64] applied the reaction zone model to the leaching of Butte ore for particles up to 6 in. in diameter. Columns, 5 ft in diameter,

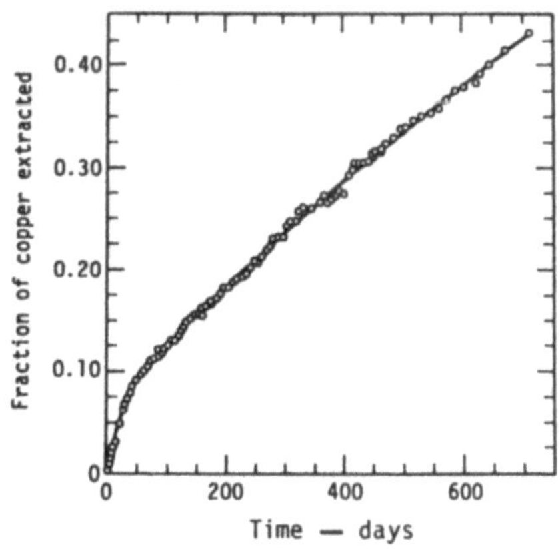

Figure 3.2-6. Extraction of copper from coarse-sized primary sulfide ore at 90° C and 400 psia.

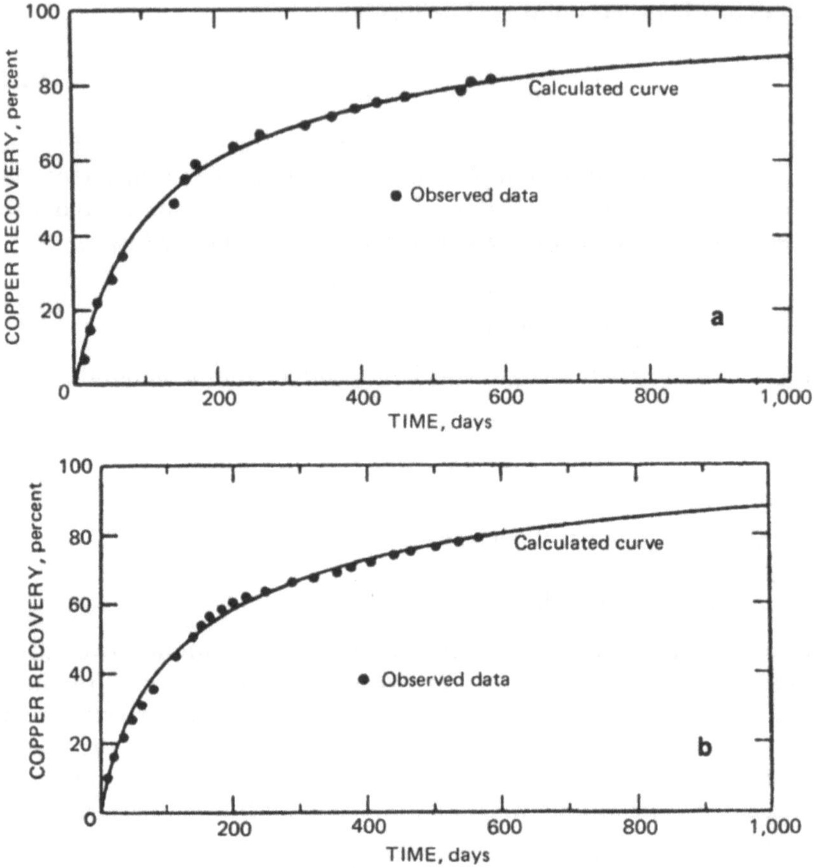

Figure 3.2-7. Modeling of dump leaching kinetics. (a) Copper recovery from sized monzonite ore (−6 in., +½ in.). (b) Copper recovery from sized quartz monzonite ore (−6 in., +½ in.).

containing 5 tons of ore were used. The principal copper mineral was Cu_2S. Equation (3.2-6) in integrated form was used. Integration leads to the equation

$$1-\frac{2}{3}\alpha_i-(1-\alpha_i)^{2/3}+\frac{\beta'}{Gr_{0i}}[1-(1-\alpha_i)^{1/3}]=\frac{\gamma' t}{Gr_{i0}^2} \quad (3.2\text{-}8)$$

where α_i is the fraction reacted for the ith particle size. Also

$$\beta'=\frac{2D'}{\sigma\beta} \quad (3.2\text{-}9)$$

$$\gamma'=\frac{2MD'C}{\rho_r\sigma\phi_{i0}} \quad (3.2\text{-}10)$$

where M is the molecular weight of the copper sulfide mineral. Figures 3.2-7a and 3.2-7b illustrate the calculated curves according to equation (3.2-8) for two types of ore.

3.2.3.2. Leaching of Oxide Ores

Roman *et al.*[63] have provided a model for the leaching of copper oxide ores using the shrinking-core model. They found that the surface reaction is fast so that only the diffusion term need by considered. Figure 3.2-8 illustrates a leach heap with a separate column of unit cross-sectional area, "unit heap," removed. Figure 3.2-9 illustrates the variation in acid concentration, copper concentration, and recovery for various times, t. Since acid is continually consumed the situation is very different from the case of dump leaching where ferric ion is the principal lixiviant and is generated continuously by bacterial oxidation of ferrous iron in solution. The "unit heap" may be divided into reaction volume sections, shown in Figure 3.2-10. For $j=n$, the concentration entering the reaction volume is C_{n-1}^0 and that leaving is C_n^0. The consumption of acid is related to the copper recovery by the expression

$$-\frac{dA}{dt}=a\frac{dCu}{dt} \tag{3.2-11}$$

where $a=(\text{wt acid})/(\text{wt Cu})$ consumed in the reaction. The value of a may be determined experimentally. The diffusion equation, equation

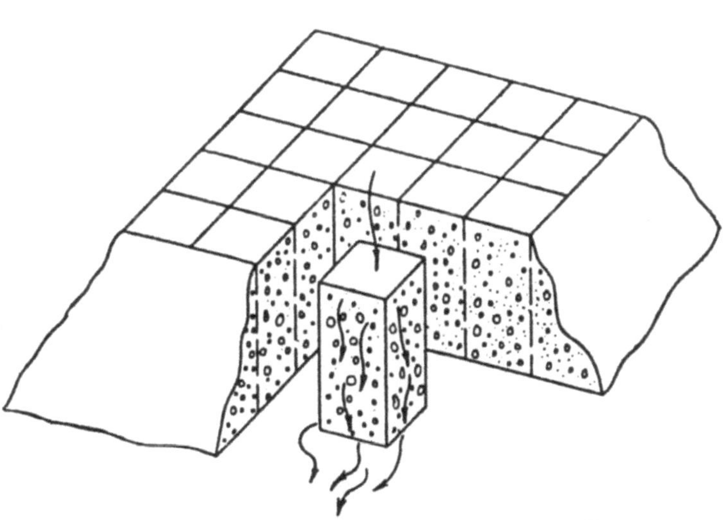

Figure 3.2-8. Schematic diagram of a leach heap.

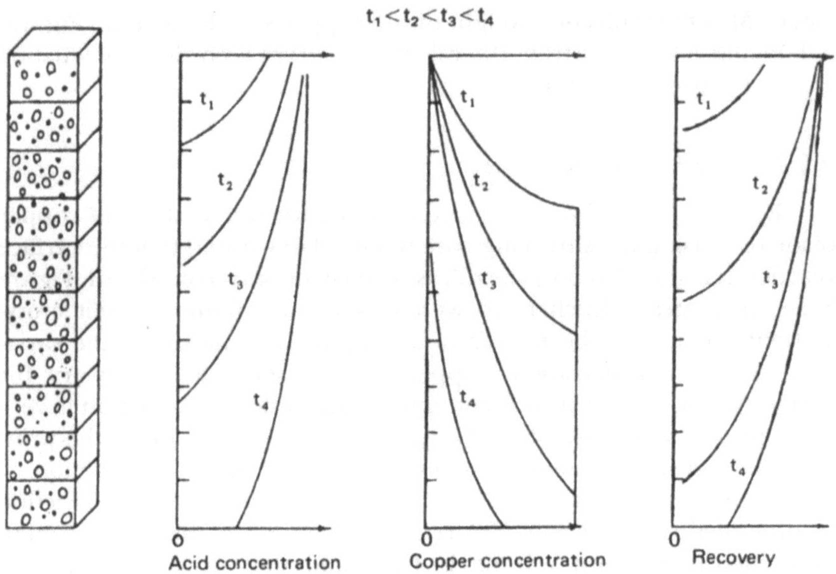

Figure 3.2-9. Schematic diagram of a *unit heap* and acid concentration, copper concentration, and recovery as a function of time and position in the heap.

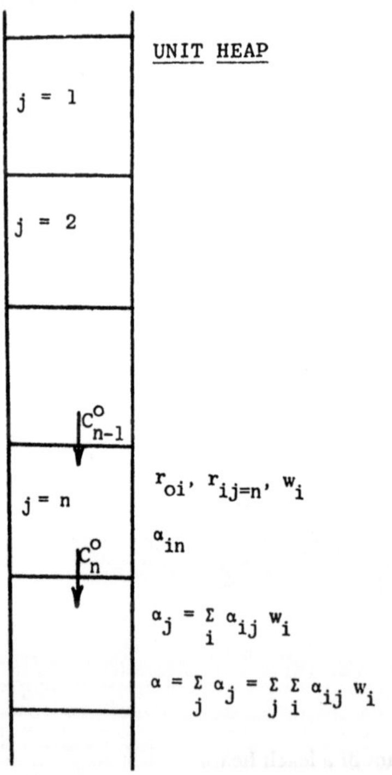

Figure 3.2-10. Volume increments in vertical section of oxide ore. Each increment is assumed to have the same initial size distribution.

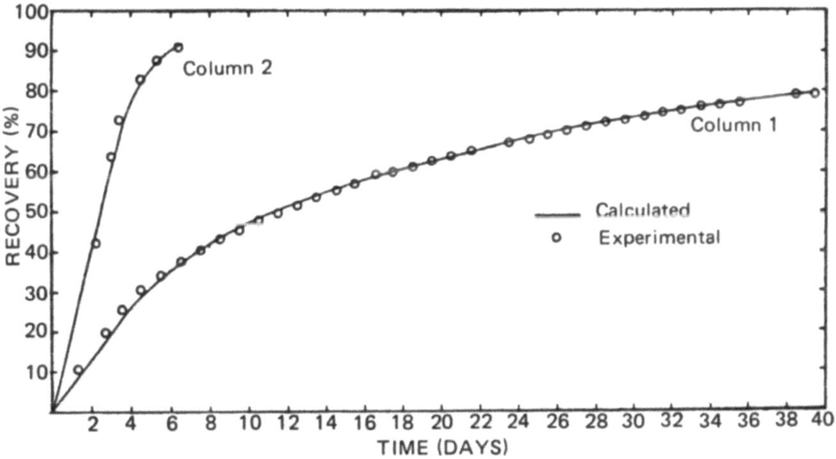

Figure 3.2-11. Recovery vs. time for two experimental columns.

(3.1-37), according to the Roman model becomes

$$\frac{d\alpha_{ij}}{dt}=\frac{3D'C_j^0(1-\alpha_{ij})^{1/3}}{r_{0i}^2\rho_r a\phi[1-(1-\alpha_{ij})^{1/3}]} \tag{3.2-12}$$

where α_{ij} refers to the fraction recovered from the *i*th particle size in volume increment *j*, *G* is the grade, ρ_r is the ore density, ϕ is a geometry factor, and D' is the effective diffusivity. Equation (3.2-12) may be evaluated numerically and summed over all unit volumes of the "unit heap." Within one incremental volume

$$\alpha_j=\sum_i \alpha_{ij}\omega_i \tag{3.2-13}$$

where ω_i is the weight fraction of particle size *i*. Summing over all values *j* gives the total fraction recovered

$$\alpha=\sum_i a_j=\sum_j\sum_i \alpha_{ij}\omega_i \tag{3.2-14}$$

Figure 3.2-11 illustrates calculated and experimental data for two column leach tests using a computer evaluation of equations (3.2-12) and (3.2-14).

3.3. Cementation (JDM)

An important part of all hydrometallurgical processes is the treatment of the leach liquor for solution concentration and/or purification prior to metal recovery. The techniques used vary considerably, depending on the composition of the solution and other factors, not the least of which is the subsequent recovery process. Concentration and purification processes include crystallization, precipitation, cementation, ion exchange, and solvent extraction. These reactions are, for the most part, heterogeneous reactions involving particulate systems which vary from the relatively simple solubility phenomenon in the case of crystallization to the complex electrochemical phenomenon operative in cementation systems.

Similarly, the reaction kinetics are varied and the rate of reaction can be limited by diffusion through the mass transfer boundary layer (MTBL), pore diffusion, and/or surface reaction. As in other heterogeneous reaction systems, study of agitated particulate assemblages is important from a pragmatic standpoint for it can provide empirical correlations for scale-up procedures. However, the determination of intrinsic kinetic parameters by such experimentation is, at best, difficult and the results may not be of great accuracy. Consequently, many experimental techniques have been devised to provide more accurate information regarding intrinsic reaction-rate constants and to provide a convenient way of analyzing the relative importance of mass transfer in these systems. Even for these experimental techniques which generally have well-defined geometries and fluid flow patterns, the interpretation of kinetic data still can be misleading as will be shown by detailed consideration of cementation reactions.

Cementation is used extensively as a means of primary metal recovery in the case of cadmium, copper, gold, and silver, and as an electrolyte purification technique in electrolytic recovery systems. Cementation can be classified as an electrochemical reaction and the analysis of cementation systems is possible from electrochemical theory.

3.3.1. Electrochemical Reactions

An electrochemical reaction is distinct from a chemical redox reaction in that charge transfer does not occur at the same physical position but rather the half-cell reactions are separated by some finite distance; consequently the solid phase must be an electrical conductor. The kinetics of electrochemical reactions, such as cementation, not only involve the activation energy associated with the chemical process but also an activation energy effect associated with the charge-transfer process. This latter effect arises because a charge carrier (metal ion or electron) is transported through an interfacial electrical potential gradient developed by the system.

In its most elementary form, the overall cementation reaction, a net cathodic reaction with respect to the noble metal M_1,

$$M_1^{+Z_1} + \frac{Z_1}{Z_2} M_2^0 \rightarrow M_1^0 + \frac{Z_1}{Z_2} M_2^{+Z_2} \tag{3.3-1}$$

can be represented by the two half-cell reactions:

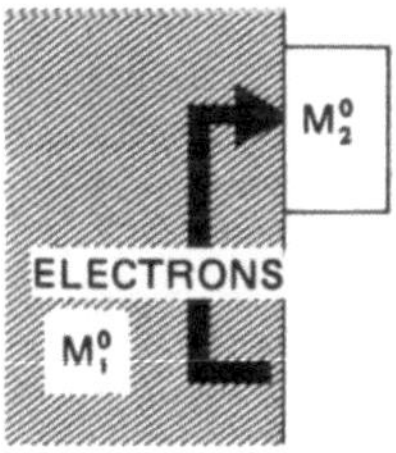

$$M_1^0 \rightleftharpoons M_1^{+Z_1} + Z_1 e \tag{3.3-2}$$

$$\frac{Z_1}{Z_2} M_2^0 \rightleftharpoons \frac{Z_1}{Z_2} M_2^{+Z_2} + Z_1 e \tag{3.3-3}$$

Now each of these half-cell reactions can be considered as a separate reaction involving both a forward and reverse direction. However, remember that the half-cells are short-circuited so that the amount reacted and the rate of consumption of M_2 for this ideal case must be related to the amount reacted and rate of production of M_1 by the stoichiometry of the reaction.

$$\frac{dn_{M_1}}{dt} = \frac{Z_1}{Z_2} \frac{dn_{M_2}}{dt} \tag{3.3-4}$$

where

n_{M_1}, n_{M_2} = number of moles of metal ion,
t = time, and
Z_1, Z_2 = valence.

Detailed analysis of electrochemical half-cells has been given by Vetter,[67] and Bockris and Reddy.[68] Application of these concepts to electrochemical cementation reactions has been presented by Wadsworth.[69]

For *either one* of the separate half-cells the reaction kinetics can be considered in terms of absolute reaction-rate theory. Absolute reaction-rate theory considers the rate process, in which the reactant species is transformed to the product species, to involve an intermediate high-energy state. This high-energy state is referred to as the activation state and any chemical species which acquires this energy of activation is referred to as the activated complex. Only a small number of reactant or product species acquire this energy at any instance and the activated complex cannot be isolated or observed directly by experimentation.

The reactants are assumed to be in equilibrium with the activated complex and the rate at which the reaction proceeds in a given direction

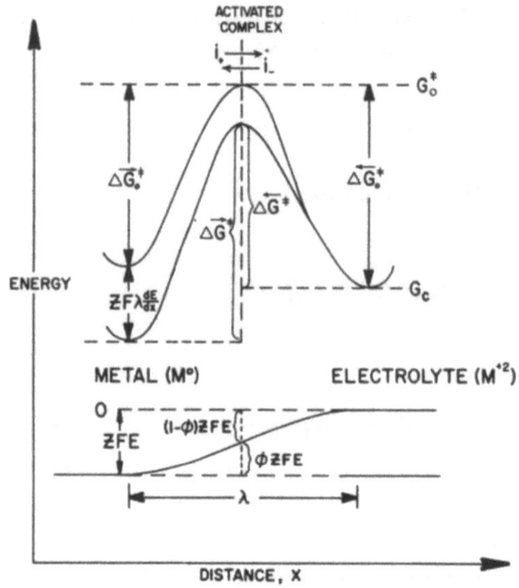

Figure 3.3-1. Shift in activation energy barrier due to applied potential field for net cathodic half-cell.[67,68]

is determined by the concentration of the activated complex, which in turn is determined by the magnitude of the energy barrier.

$$\text{reactants} \underset{\text{equilibrium}}{\rightleftharpoons} \text{activated complex} \underset{\substack{\text{rate}\\\text{limiting}}}{\longrightarrow} \text{products}$$

The activation energy barrier for the net cathodic half-cell [equation (3.3-2)] of a cementation reaction is represented schematically in Figure 3.3-1. The diagram has been constructed so that the reduction of the cation is thermodynamically favored, that is, the lower free-energy state is associated with the metal rather than the metal ion. The unique feature of this diagram is the contribution of the potential field to the activation energy which accompanies any electrochemical reaction. In this case [equation (3.3-2)], the potential is seen to enhance the reduction, or reverse, reaction of the half-cell by lowering the activation energy in the reverse direction from $\Delta\overleftarrow{G}_0^{\ddagger}$ to $\Delta\overleftarrow{G}^{\ddagger}$ and raising the activation energy for the forward direction from $\Delta\overrightarrow{G}_0^{\ddagger}$ to $\Delta\overrightarrow{G}^{\ddagger}$. The following terms are used in the diagram and in subsequent equations:

i_-, i_+ = cathodic and anodic current densities

$\Delta\overleftarrow{G}_0^{\ddagger}$, $\Delta\overrightarrow{G}_0^{\ddagger}$ = activation energies for the cathodic and anodic directions in the absence of a potential gradient

$\Delta\overleftarrow{G}^{\ddagger}$, $\Delta\overrightarrow{G}^{\ddagger}$ = activation energies for the cathodic and anodic directions in the presence of a potential gradient

ϕ = transfer coefficient

λ = distance between minima on either side of the activated state

F = Faraday constant
E_0 = equilibrium reversible electrode potential in the absence of current
E = potential in presence of a net current
η = overvoltage, $E_0 - E$
z = charge transfer valence.

Usually other interfacial potentials such as the ξ potential, are much less than E, the potential associated with the charge transfer reaction and consequently can be neglected. If the potential gradient is constant, then $dE/dx = E/\lambda$. Furthermore, the transfer coefficient is determined by the shape of the energy barrier, which is usually symmetrical. In essence, this means that the decrease in activation energy in the cathodic direction due to the potential gradient is equivalent to the increase in activation energy in the anodic direction and $\phi = 0.5$.

For the general case, it can be seen from the diagram in Figure 3.3-1 that the activation energy for the cathodic direction of the half-cell [equation (3.3-2)] decreases by $(1-\phi)zFE$, whereas the activation energy for the anodic direction increases by ϕzFE. For the electrochemical discharge of the metal ion, M^+, the activation energy $\Delta G^\ddagger$, can be expressed in terms of the chemical contribution, $\Delta G_0^\ddagger$, and the electrical contribution, $zF\lambda(dE/dx) \simeq zFE$:

$$\Delta \vec{G}_0^\ddagger + zFE = \Delta \vec{G}^\ddagger + (1-\phi)zFE$$

$$\Delta \vec{G}^\ddagger = \Delta \vec{G}_0^\ddagger + \phi zFE \tag{3.3-5}$$

$$\Delta \overleftarrow{G}^\ddagger = \Delta \overleftarrow{G}_0^\ddagger - (1-\phi)zFE \tag{3.3-6}$$

Considering the equilibrium established with the activated complex for both the cathodic and anodic directions of the half cell, the equilibrium constants can be written in terms of the chemical and electrical energy barriers.

$$\overleftarrow{K}^\ddagger = \frac{\overleftarrow{C}^\ddagger}{C_c} = \exp\left(-\frac{\Delta \overleftarrow{G}^\ddagger}{RT}\right) = \exp\left(-\frac{\Delta \overleftarrow{G}_0^\ddagger}{RT}\right) \exp\left[\frac{(1-\phi)zFE}{RT}\right] \tag{3.3-7}$$

$$\vec{K}^\ddagger = \frac{\vec{C}^\ddagger}{C_a} = \exp\left(-\frac{\Delta \vec{G}^\ddagger}{RT}\right) = \exp\left(-\frac{\Delta \vec{G}_0^\ddagger}{RT}\right) \exp\left(-\frac{\phi zFE}{RT}\right) \tag{3.3-8}$$

Other useful relationships that come from absolute reaction-rate theory are the expressions for the specific rate constants in the absence of a potential field.

$$\overleftarrow{k}' = \frac{kT}{h} \exp\left(-\frac{\Delta \overleftarrow{G}_0^\ddagger}{RT}\right) \tag{3.3-9}$$

$$\vec{k}' = \frac{kT}{h} \exp\left(-\frac{\Delta \vec{G}_0^\ddagger}{RT}\right) \tag{3.3-10}$$

Using these equations, the concentrations of the activated complex can be expressed as

$$\overleftarrow{C}^{\ddagger} = C_c \frac{h}{kT} \overleftarrow{k}' \exp\left[\frac{(1-\phi)zFE}{RT}\right] \tag{3.3-11}$$

$$\overrightarrow{C}^{\ddagger} = C_a \frac{h}{kT} \overrightarrow{k}' \exp\left(-\frac{\phi zFE}{RT}\right) \tag{3.3-12}$$

3.3.1.1. Butler–Volmer Equation

The Butler–Volmer equation applicable to electrolytic as well as electrochemical reactions, can be derived from these basic postulates of absolute reaction-rate theory. The rate process for a half-cell written as an oxidation reaction can be formulated in terms of the cathodic (reverse) and anodic (forward) direction of the half-cell reaction. It can be shown for the unit area that

$$\left(\frac{dn}{dt}\right)_{\text{cathodic direction}} = \frac{i_-}{ZF} = \overleftarrow{C}^{\ddagger} \lambda \frac{kT}{h} \kappa \tag{3.3-13}$$

$$\left(\frac{dn}{dt}\right)_{\text{anodic direction}} = \frac{i_+}{ZF} = \overrightarrow{C}^{\ddagger} \lambda \frac{kT}{h} \kappa \tag{3.3-14}$$

where

$\overleftarrow{C}^{\ddagger}$, $\overrightarrow{C}^{\ddagger}$ = the concentrations of the activated complex for the cathodic and anodic directions, respectively,
k = the Boltzmann constant, 1.38×10^{-16} erg/°K,
T = absolute temperature,
h = the Planck constant, 6.6×10^{-27} erg-sec, and
κ = the transmission coefficient.

The net current density, i, for this half-cell reaction

$$\mathrm{M} \underset{i_-}{\overset{i_+}{\rightleftharpoons}} M^+ + \mathrm{e}^- \tag{3.3-15}$$

is the difference between the partial current density in the anodic direction and the partial current density in the cathodic direction,

$$i = i_+ - i_- \tag{3.3-16}$$

provided that the anodic and the cathodic areas are equivalent. After substitution for the partial current density expressions, equations (3.3-13) and (3.3-14), the net current density for the half-cell can be written as

$$i = (\overrightarrow{C}^{\ddagger} - \overleftarrow{C}^{\ddagger}) zF\lambda (kT/h) \tag{3.3-17}$$

Next, substitution of the expressions for the concentrations of the activated complexes, given by equations (3.3-11) and (3.3-12) results in what is one form of the Butler–Volmer equation

$$i = \vec{k}C_a \exp\left(-\frac{\phi zFE}{Rt}\right) - \overleftarrow{k}C_c \exp\left[\frac{(1-\phi)}{RT} zFE\right] \qquad (3.3\text{-}18)$$

$$\vec{k} = zF\lambda\vec{k}' \qquad (3.3\text{-}19)$$

$$\overleftarrow{k} = zF\lambda\overleftarrow{k}' \qquad (3.3\text{-}20)$$

In this expression C_c is the concentration of the metal ion and C_a is a constant characteristic of the solid metal and representing, in essence, the surface concentration of discharge sites.

At some half-cell potential, E_0, the partial current densities are equal in magnitude and opposite in direction. The value of either partial current density is designated the equilibrium exchange-current density, i_0,

$$i_0 = i_- = \overleftarrow{k}C_c \exp\left[\frac{(1-\phi)zFE_0}{RT}\right] = i_+ = \vec{k}C_a \exp\left(-\frac{\phi zFE_0}{RT}\right) \qquad (3.3\text{-}21)$$

The Butler–Volmer equation can then be written in terms of the exchange-current density, i_0, and the overpotential, $\eta = E_0 - E$. The net current for the half-cell reaction is

$$i = i_0\left\{\exp\left(-\frac{\phi zF\eta}{RT}\right) - \exp\left[\frac{(1-\phi)zF\eta}{RT}\right]\right\} \qquad (3.3\text{-}22)$$

Now in a system such as cementation the hypothetical half-cell which we have been discussing is short-circuited to another more reactive half-cell which undergoes a net anodic, or oxidation reaction [equation (3.3-3)]. The theoretical discussion given for equation (3.3-2) applies equally well to the more reactive half-cell, equation (3.3-3) and its activation energy barrier is similar to Figure 3.3-1. Of course, it is apparent under these conditions that the net current for the cathodic half-cell must be equal to the net current for the anodic half-cell. One of the most important aspects of this type of electrochemical reaction is the fact that, unlike electrolytic reactions in which the potential can be fixed externally, the short-circuited cementation reaction and other electrochemical reactions acquire a mixed potential, similar to what is known as a corrosion potential. This is not an equilibrium potential such as is observed at the exchange current density for a single half-cell reaction, but rather a steady-state mixed potential the magnitude of which arises from the relative activation energy barriers for each of the half-cells. The steady-state mixed potential is a complex function of the forward and reverse rate constants for each half-cell, as well as reactant and product concentrations and the cathodic/anodic area ratio. In the absence of significant internal and ohmic resistance, each half-cell sees the same mixed potential.

Cementation reaction kinetics can be analyzed in terms of electrochemical theory, using the concept of a steady-state mixed potential, E_m, which the system develops. The half-cells of the hypothetical cementation reaction under consideration, equation (3.3-1) are presented in the current density–potential diagram in Figure 3.3-2. The position of these curves may be dependent on the composition of the electrolyte, the concentrations of $M_1^{+Z_1}$ and $M_2^{+Z_2}$. Consequently, the current–potential diagram for the short-circuited system, shown by the dashed line, also varies as the sum of the respective half-cell current–potential curves. Under these circumstances the mixed potential can vary between the two reversible potentials, E_{0,M_2} and E_{0,M_1}. Its exact position is determined by the fact that $I_{M_1} = -I_{M_2}$ for a system at steady state. Furthermore, as the reaction proceeds, the half-cell current–potential plots change and the mixed potential drifts.

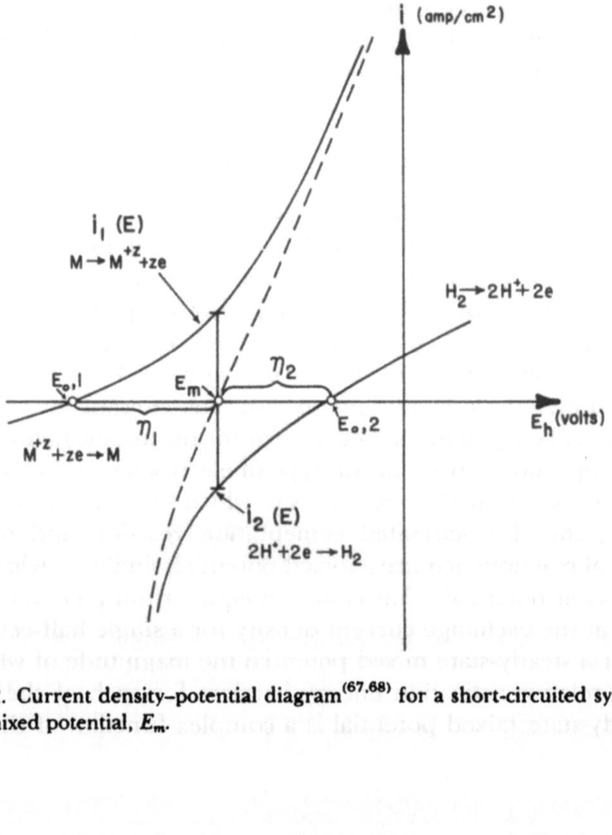

Figure 3.3-2. Current density–potential diagram[67,68] for a short-circuited system which achieves a mixed potential, E_m.

The mixed potential can be treated quantitatively by using the Butler–Volmer equation for each half-cell

$$I_{M_1} = A_c \overleftarrow{k}_c \exp\left(-\frac{\phi z F E_m}{RT}\right) - A_c \overleftarrow{k}_c (M_1^{+Z_1}) \exp\left[\frac{(1-\phi) z F E_m}{RT}\right] \quad (3.3\text{-}23)$$

$$I_{M_2} = A_a \overrightarrow{k}_a \exp\left(-\frac{\phi z F E_m}{RT}\right) - A_a \overleftarrow{k}_a (M_2^{+Z_2}) \exp\left[\frac{(1-\phi) z F E_m}{RT}\right] \quad (3.3\text{-}24)$$

For a symmetrical energy barrier, the transfer coefficient equals 0.5. Setting, $I_{M_1} = -I_{M_2}$, equations (3.3-23) and (3.3-24) can be solved to give

$$\exp\left(\frac{FE_m}{RT}\right) = \frac{A_c \overrightarrow{k}_c + A_a \overrightarrow{k}_a}{A_a \overleftarrow{k}_a (M_2^{+Z_2}) + A_c \overleftarrow{k}_c (M_1^{+Z_1})} \quad (3.3\text{-}25)$$

and the complex nature of the mixed potential is revealed. The expression can be greatly simplified if the back reaction is considered negligible, that is large overvoltages exist. In this case the expression becomes

$$\exp\left(\frac{FE_m}{RT}\right) = \frac{A_a \overrightarrow{k}_a}{A_c \overleftarrow{k}_c (M_1^{+Z_1})} \quad (3.3\text{-}26)$$

In an actual system it is conceivable that not only the concentrations of the metallic ions could vary, but also the anodic and cathodic areas.

Finally it is worthwhile to consider what the rate expression might be for a cementation reaction controlled by the cathodic discharge of the noble metal ion $M_1^{+Z_1}$. In terms of the rate of change in the number of moles of M_1, the rate expression is

$$\left(\frac{dn}{dt}\right)_{M_1} = \frac{I_{M_1}}{Z_1 F} = A_c\left[\overrightarrow{k}_c \exp\left(\frac{FE_m}{2RT}\right) - \overleftarrow{k}_c (M_1^{+Z_1}) \exp\left(\frac{FE_m}{2RT}\right)\right] \quad (3.3\text{-}27)$$

3.3.1.2. Evans Diagrams

A convenient way to think about cementation reaction kinetics is to plot the current–potential curves for the respective half-cells involved in the cementation reaction, as shown by Power.[70] These diagrams are referred to as *Evans diagrams* and have been used in the analysis of corrosion reactions. In essence, the diagrams are simply the superposition of two current–potential (polarization) curves. A typical polarization curve for an arbitrary metal is shown in Figure 3.3-3. At very low currents the potential, E_0, is equivalent to the reversible electrode potential. As E is made more negative (cathodic) than E_0, the metal ions begin to be reduced. At some point the potential begins to vary linearly with log i. This region of the polarization curve is known as the *Tafel region* and the slope of the linear portion of the polarization

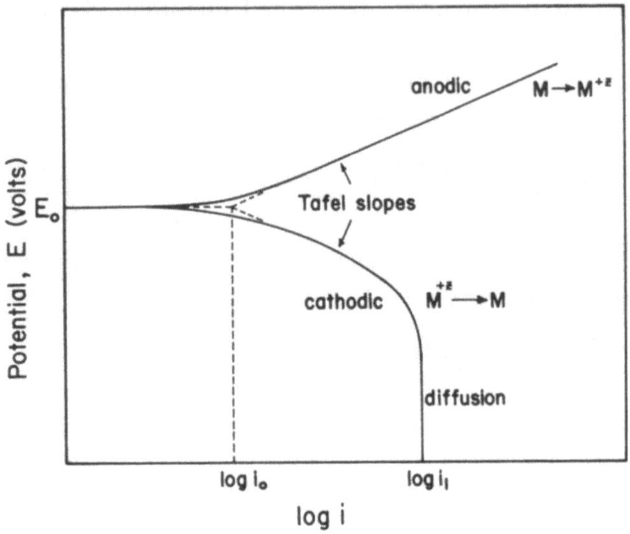

Figure 3.3-3. Polarization curves for an arbitrary metal showing cathodic and anodic reaction.[70]

curve is known as the *Tafel slope*. For large negative values of potential, the current density reaches a maximum value, that is, the *limiting current density*. Under these conditions the current is determined by the rate of transport of metal ions to the surface. Similar regions can be identified for the anodic behavior of the system when the potential is made more positive than E_0. In this case, however, the limiting current density is reached only when the metal surface becomes saturated with respect to a salt of the metal. From the linear region of polarization curves, the Tafel equation is defined,

$$\eta = a \pm b \log i \tag{3.3-28}$$

where a and b are constants, and η is the overpotential (the difference between the potential at some current i and the reversible electrode potential, $E_0 - E$). Alternatively, the Tafel equation can be written from the Butler–Volmer equation (3.3-22) in terms of the exchange-current density,

$$\eta = b \log (i/i_0) \tag{3.3-29}$$

The polarization curves can be determined experimentally or calculated, provided the exchange-current density and transfer coefficients are known.

The use of Evans diagrams in cementation systems is best illustrated by some examples taken from Power.[70] In these systems it is assumed that the polarization curves are independent, the cathodic and anodic areas are equal, and there is no ohmic resistance between cathode and anode. Furthermore, the diagrams are constructed for mass transfer rates calculated for a rotating disk at 100 rad/sec. Details of the rotating disk system are discussed in Section 3.3.2. The most important feature of these diagrams is that they demonstrate quite clearly whether the reaction will be controlled by an electrochemical surface reaction.

For the first example, consider the (Cu^{2+}/Fe) Evans diagram presented in Figure 3.3-4. The point of intersection of the polarization curves determines the mixed potential for the reaction and the rate-controlling step. Because the polarization curves intersect the cathodic copper curve in the diffusion-limited region, the reaction would be expected to be controlled by mass transfer of cupric ion to the reaction surface. Many investigators found this to be true from experimental cementation studies.

As the other example, consider the (Fe^{2+}/Zn) system shown in Figure 3.3-5, for which a different conclusion would be reached. In this case the point of intersection may occur in the Tafel region of the polarization curve and under these conditions the reaction would be electrochemically controlled.

The criterion for chemical control is intersection of the anodic and the cathodic curves in the Tafel region. An idealized Evans diagram can be constructed as shown in Figure 3.3-6 and the point at which rate control switches from mass transfer to chemical reaction can be evaluated. Solution of the respective Tafel equations for cathodic and anodic

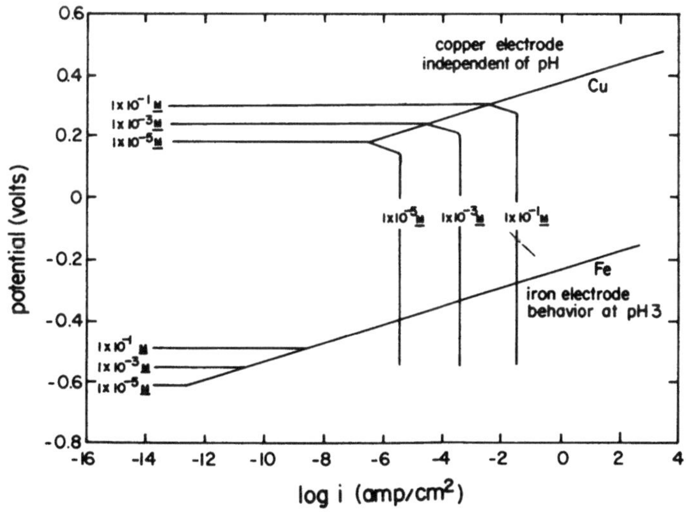

Figure 3.3-4. Evans diagram for the Cu^{2+}/Fe system, mass transfer controlled. Polarization curves constructed with data from Hurlen.[71]

half-cells at the transition mixed potential results in

$$\Delta E_0 = b\left[\log\left(\frac{i_l}{i_0}\right)_c + \log\left(\frac{i_l}{i_0}\right)_a\right] \tag{3.3-30}$$

The variables in equation (3.3-30) take on the following values:

$b \cong 0.03$ V/decade current density for many metals (Cu, Fe, Zn)
$i_l \cong 10^2$ amp/m^2 (0.1M solution)
$i_0 = 10^1$ amp/m^2 (Cu)$\rightarrow 10^{-4}$ amp/m^2 (Fe)

Consequently, for cementation reactions a good rule-of-thumb would be:

$\Delta E_0 < 0.06$ V for electrochemical control

$\Delta E_0 > 0.36$ V for mass transfer control

Between these values, the exchange-current densities must be known in order to predict the rate-limiting step. As a further approximation, the standard electrode potentials ΔE_0^0 of the respective half-cells can be used to approximate ΔE_0. When these values are calculated it is found that most cementation systems have a difference in standard-electrode potential of greater than 0.36 V. Furthermore, experimental investigations indicate that most cementation reactions are controlled by mass transfer processes as indicated by the results presented in Table 3.3-1.

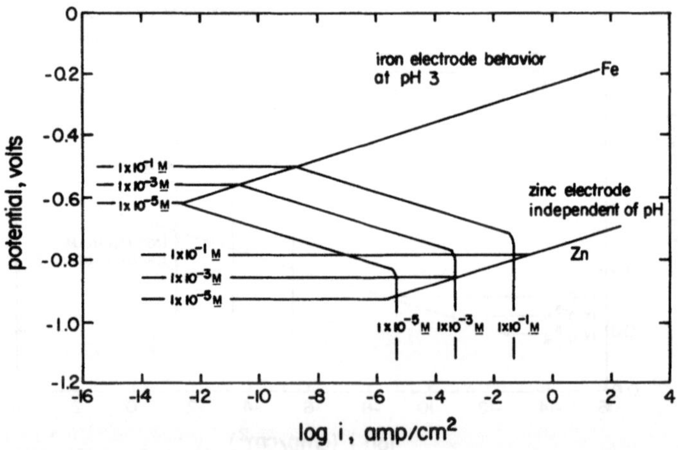

Figure 3.3-5. Evans diagram for the Fe^{2+}/Zn system, electrochemical reaction controlled. Polarization curves constructed with data from Hurlen.[71]

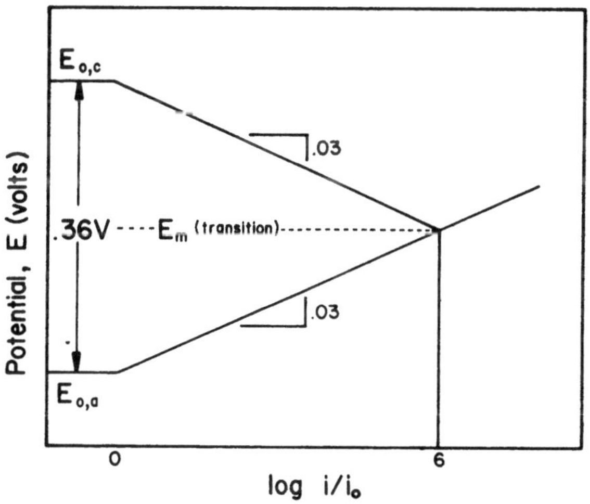

Figure 3.3-6. Idealized Evans diagram which represents the transition from mass transfer control to electrochemical reaction control.[70]

3.3.1.3. Rate Control in Cementation Reactions

From the analysis of cementation reactions in terms of Evans diagrams it appears that the reaction kinetics of only a few systems should be controlled by an electrochemical reaction mechanism. Furthermore, the general irreversibility of these electrochemical reactions was previously established by Wadsworth,[69] in which it was shown

Table 3.3-1. Data for Selected Cementation Systems at 25° C[69]

Systems	ΔE^0_0, V	Activation energy, kcal/mole	Rate constant, cm/sec
Ag^+/Cu	0.46	2.0–5.0 (12)	$2.5–6.0\times10^{-2}$
Ag^+/Cu(CN^-)	1.83	3.7–5.8	1.5×10^{-2}
Ag^+/Fe(Cl^-)	1.29	3.0	2.2×10^{-2}
Ag^+/Zn(CN^-)	0.95	5.5	5.5×10^{-2}
Ag^+/Zn	1.56	2.0–6.0 (12)	$2.6–5.2\times10^{-2}$
Bi^{3+}/Fe	0.76	4.5–7.6	2.9×10^{-2}
Cd^{2+}/Zn	0.36	4.0–4.7	$0.54–1.1\times10^{-2}$
Cu^{2+}/Fe	0.75	3.1–5.1	$0.6–0.9\times10^{-2}$
Cu^{2+}/In	0.83	2.3	5.9×10^{-2}
Cu^{2+}/Ni	0.57	2.7–3.7 (14.2–19.0)	$0.25\times1.0\times10^{-2}$
Cu^{2+}/Zn	1.10	3.1	$1.6–2.1\times10^{-2}$
Ni^{2+}/Fe	0.21	7.0	$\sim1\times10^{-4}$
Pb^{2+}/Fe	0.31	12.0	—
Pb^{2+}/Zn	0.64	—	0.64×10^{-2}
Pd^{2+}/Cu	0.49	9.5–7.4	$0.36–2.3\times10^{-2}$

that if $Z_1 \Delta E_0^0$ is greater than 0.3 V, the back reaction kinetics should not be important.

Experimental results confirm these conclusions, although, as will be seen later, surface deposit effects can lead to inaccurate interpretation of kinetic data. Notice in Table 3.3-1 that both systems (Ni^{2+}/Fe and Pb^{2+}/Fe), which exhibit a difference in standard-electrode potentials of less than 0.36 V, have activation energies which support a surface-reaction control mechanism. Further evidence in the case of the Ni^{2+}/Fe system includes a rather small rate constant of less than 1×10^{-4} cm/sec. All other systems have a difference in standard-electrode potentials, ΔE_0^0, in excess of 0.36 V and the reaction kinetics appear, at least under most circumstances, to be limited by aqueous-phase mass transfer, as supported by measured activation energies and rate constants.

Virtually all kinetic data for cementation reactions obey a first-order-rate law of the following form, as would be expected for mass transfer control:

$$dn_1/dt = -kAC_1 \tag{3.3-31}$$

The disappearance of the noble metal ion can be studied in a variety of experimental geometries, an important consideration due to the importance of mass transfer in these systems. The experimental geometries will be discussed in the next section. The integrated rate expression then assumes the following common form in terms of the noble metal ion concentration,

$$\log\left[\frac{C_1(t)}{C_1(0)}\right] = \frac{-kA}{2.303\ V}t \tag{3.3-32}$$

where

$C_1(0)$ and $C_1(t)$ = initial concentration and concentration at time t of noble metal ion
k = rate constant (cm/sec)
A = reaction surface area (cm^2)
V = solution volume (cm^3)
t = reaction time (sec)

3.3.2. *Mass Transfer—Hydrodynamics*

For reactions limited by diffusion through the mass transfer boundary layer, the hydrodynamic flow regime at the reacting surface is of considerable importance in understanding the reaction kinetics. Mass transfer in an aqueous phase has been analyzed in detail by Newman[72] and Bockris and Reddy.[68] Basically in these systems mass is transferred by

1. diffusion, i.e., motion of both charged and uncharged species due to concentration gradients,
2. migration, i.e., motion of charged species in an electric field, and
3. convection, i.e., transfer of mass due to the bulk motion of the liquid phase.

The flux of some species, i, is given then by

$$\underset{\text{}}{\mathbf{J}_i} = \underset{\text{diffusion}}{-D_i \nabla C_i} \quad \underset{\text{migration}}{-z_i u_i F C \nabla \phi} \quad \underset{\text{convection}}{+C_i \mathbf{v}} \tag{3.3-33}$$

where

$\mathbf{J}_i$ = the flux of species i (moles/cm^{2}sec),
D_i = diffusion coefficient of species i (cm^2/sec),
C_i = concentration of species i (moles/cm^3),
x = direction in which mass is being transferred,
z_i = valence, or charge of species, including sign,
U_i = mobility of species i (cm^2-mole/J sec),
F = Faraday's constant (96,500 Cs/equiv),
ϕ = electrostatic potential (V), and
$\mathbf{v}$ = liquid velocity vector (cm/sec).

In order to solve the flux equation, the liquid velocity vectors must be known as a function of the systems' coordinates. The equation of motion, a form of the Navier–Stokes equation for constant density and viscosity, spatially defines the fluid flow regime

$$\rho \frac{d\mathbf{v}}{dt} = -\nabla p + \mu \nabla^2 \mathbf{v} + \rho \mathbf{g} \tag{3.3-34}$$

where

ρ = liquid density (g/cm^3),
p = pressure (dyne/cm^2),
μ = viscosity (g/cm sec), and
$\mathbf{g}$ = gravitational acceleration (cm/sec^2).

The left-hand side of equation (3.3-34) represents the change in momentum the liquid element experiences due to gravitational, viscous, and pressure forces. The equation of motion, equation (3.3-34), together with a material balance on the liquid, the continuity equation $\nabla \mathbf{v} = 0$, may allow for the determination of the liquid velocity vector.

Because the kinetics of most cementation reactions appear to be controlled by a mass transfer process, several experimental geometries and methods which allow for the evaluation of the respective mass transfer coefficients will be considered. The principles discussed, of course, are not limited to cementation reactions but could apply to any heterogeneous reaction in which mass transfer in the liquid phase limits the reaction rate.

3.3.2.1. Rotating Disk

The rotating disk system has been used with good success in the quantitative study of electrode kinetics and more recently in the study of cementation kinetics. The rotating disk system is unique for kinetic studies, laminar flow being maintained even to Reynolds numbers ($r^2\omega/\nu$) of 2×10^5, where

r = the radial position on the disk (cm),
ω = angular velocity (rad/sec), and
ν = kinematic viscosity (cm^2/sec).

Furthermore, the hydrodynamic flow in the system is well defined and can be calculated from first principles as done by Levich in 1942.[73] As a result of this well-defined hydrodynamic flow, a theoretical expression can be developed from which the rate of mass transfer to the disk surface can be calculated. This important fact provides additional insight into distinguishing the rate-limiting step for a heterogeneous reaction such as cementation.

A theoretical treatment of the rotating-disk system is given by Riddiford.[74] Analysis of the system begins by consideration of the Navier–Stokes equation in cylindrical coordinates for liquid flow about a plane disk at $y = 0$. The coordinate system is shown superimposed on the schematic of the flow regime in Figure 3.3-7. The velocity components of the system are v_y, v_θ, and v_r which can be explicitly determined for the following simplifying assumptions:

1. $d\mathbf{v}/dt = 0$, incompressible fluid at steady state;
2. the velocity components are independent of θ because of axial symmetry; and
3. v_y is independent of r.

The boundary conditions for this system can be realized from the fact that close to the disk's surface the fluid moves in rotation with the disk, whereas at distances further away from the disk, the fluid is thrown outward radially in laminar flow parallel to the plane of the disk. Consequently, at even larger values of y this outward radial flow must be balanced by an axial flow as y approaches infinity. The flow pattern is depicted schematically in Figure 3.3-7. Thus two boundary conditions

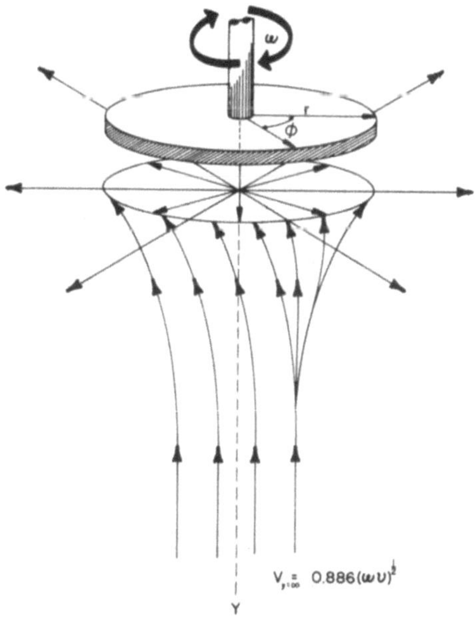

Figure 3.3-7. Schematic diagram of the fluid flow for a rotating disk system.

can be identified:

1. $y = 0$: $v_r = 0$, $v_\theta = \omega r$, $v_y = 0$
2. $y = \infty$: $v_r = 0$, $v_\theta = 0$, $v_y = \text{constant}$

Under these conditions the Navier–Stokes equation is reduced, resulting in the following component velocity equations in terms of the functions F, G, and H which are defined by the dimensionless variable $\gamma = (\omega/\nu)^{1/2}y$,

$$v_r = r\omega F(\gamma)$$

$$v_\theta = r\omega G(\gamma)$$

$$v_y = (\nu\omega)^{1/2} H(\gamma)$$

The functions F, G, and H have been evaluated by numerical integration and are presented in Figure 3.3-8. At distances very far from the disk the fluid flows toward the disk's surface at constant velocity:

$$v_y = -0.866(\omega\nu)^{1/2} \tag{3.3-35}$$

The negative sign is indicative of the established coordinate system and the velocity vector is directed toward the disk. As the fluid increment is considered closer to the disk's surface, the magnitude of v_y is reduced and velocity components v_θ and v_r increase. At a certain point, v_r reaches a maximum and begins to decrease in magnitude. Finally, at the surface $y = 0$ the only prevalent velocity component is v_θ.

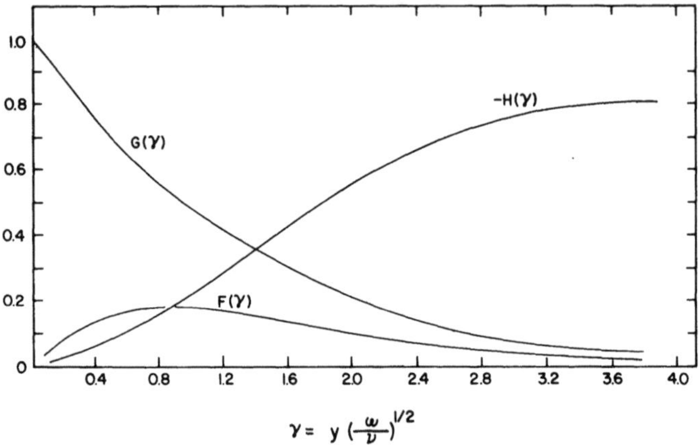

Figure 3.3-8. Plot of the functions F, G, and H which determine component velocities V_r, V_θ, and V_y as a function of distance from the disk surface.[74]

The hydrodynamic boundary layer is defined as the distance at which the liquid gains 10% of the surface velocity, i.e., $v_\theta = 0.1\ \omega r$. This corresponds to a distance of $y = 2.8(\nu/\omega)^{1/2}$. This distance should be distinguished from the mass transfer boundary layer (MTBL), i.e., the distance across which a concentration gradient may be established during a heterogeneous reaction. However, the MTBL is determined by the hydrodynamic boundary layer as well as the diffusion coefficient of the reacting species and the viscosity of the liquid phase.

$$\mathrm{MTBL} = 3.6\left(\frac{D}{\nu}\right)^{1/3}\left(\frac{\nu}{\omega}\right)^{1/2}$$

Normally, in cementation systems the hydrodynamic boundary layer is approximately one order of magnitude larger than the mass transfer boundary layer.

With the velocity vectors, the flux expression for mass transfer given by equation (3.3-33) can be solved to predict the rate of mass transfer. Consider the transfer of a molecular species at steady state. In this case the potential gradient contribution to mass transfer vanishes because the diffusing species is uncharged.

Consideration of an ionic species in a solution of a binary electrolyte and the potential gradient contribution to mass transfer results in a similar expression to that to be derived, except that the expression involves the effective diffusivity of the electrolyte.

$$D_{\mathrm{eff}} = \frac{D_+D_-(z_+ - z_-)}{z_+D_+ - z_-D_-} \qquad (3.3\text{-}36)$$

In the case of three ionic components, if the diffusing species is in dilute concentration relative to the other species, then the potential gradient will be independent of the concentration gradient of species i. That is, the potential gradient will be constant and will be determined by the indifferent electrolyte concentration. The treatment of solutions which contain more than three components becomes even more complex.

It should be clear from the system's symmetry that the concentration is independent of θ and r. Consequently, at steady state, the flux equation (3.3-33) can be expressed as

$$D_i \frac{d^2 C_i}{dy^2} = v_y \frac{dC_i}{dy} \tag{3.3-37}$$

which is equivalent to Fick's second law as it should be for unidimensional mass transfer. Integration of equation (3.3-37) is possible using the following axial velocity expression which holds for small values of y:

$$v_y = -(\omega\nu)^{1/2}(0.51\gamma^2) = 0.33\gamma^3 + 0.103\gamma^4 + \cdots \tag{3.3-38}$$

The boundary conditions for equation (3.3-37) are

$$y = 0: \qquad C = 0 \quad \text{and} \quad v_y = 0$$

$$y = \infty: \qquad C = C_b \quad \text{and} \quad v_y = -0.886(\omega\nu)^{1/2}$$

After the introduction of dummy variables, Riddiford[74] shows the integration of equation (3.3-37) from which dC/dy can be determined, and the expression for the mass flux at the interface is found to be

$$J_{y=0} = -D\left(\frac{dC}{dy}\right)_{y=0} = 0.62\, D^{2/3}\, \nu^{-1/6}\, \omega^{1/2} C \tag{3.3-39}$$

A number of cementation-rate studies have been reported in which the reaction kinetics were studied with a rotating disk. The results of these studies invariably support the contention presented in the Section 3.3.1: for most systems the rate of cementation is limited by mass transfer. According to equation (3.3-39), the mass transfer coefficient should be

$$k = 0.62\, D^{2/3}\, \nu^{-1/6}\, \omega^{1/2} \tag{3.3-40}$$

A plot of k vs. $\omega^{1/2}$ is presented in Figure 3.3-9 for a number of cementation systems which have different noble metal ion diffusivities.[75] The lines represent a plot of equation (3.3-40) with the following values for diffusion coefficients at 25° C selected from the literature:

Cation	Cu^{2+}, Cd^{2+}	Pb^{2+}	Ag^+
D, cm^2/sec	7.3×10^{-6}	9.5×10^{-6}	16.5×10^{-6}

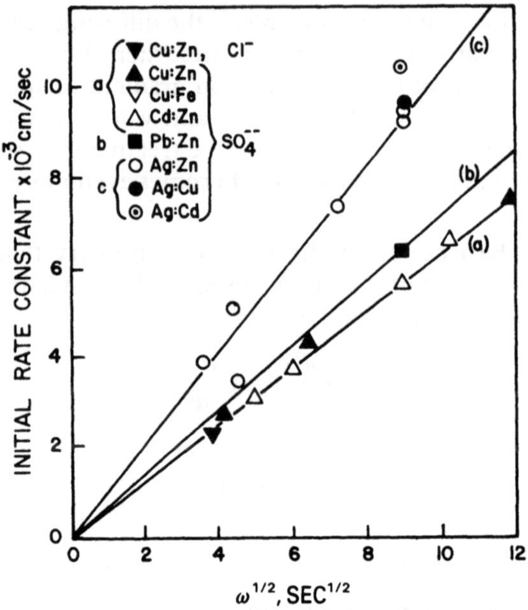

Figure 3.3-9. Demonstration of the half-order dependence of the angular velocity on the mass transfer coefficient as predicted by the Levich equation for rotating disks for a number of cementation systems.[75]

The symbols represent the experimental data. Excellent agreement between theory and experiment is observed and was expected from electrochemical theory since for each of the systems reported the difference in the standard-electrode potential of the respective half-cells exceeds 0.36 V. (See Table 3.3-1.)

Generally, the systems are more complicated than has been indicated. Specifically, competing reactions, such as equilibrium complexation, hydrolysis, and chemical redox reactions can make the analysis of kinetic data difficult. Furthermore, the formation of a surface deposit can cause an increase in the apparent reaction-rate constant. The form of this deposit can be particularly sensitive to the temperature and electrolyte composition. Details of this effect will be considered later.

3.3.2.2. Rotating Cylinder

The major difficulty encountered for this particular geometry, unlike the rotating disk system, is that the flow regime becomes turbulent at very low Reynolds numbers,

$$\mathrm{Re} = vd/\nu \qquad (3.3\text{-}41)$$

where v is the peripheral velocity of the cylinder, d is the cylinder diameter, and ν is the kinematic viscosity. On this basis the transition to turbulent flow for a rotating cylinder begins to occur at $\mathrm{Re} \simeq 10^1$, whereas for the rotating disk the transition occurs at $\mathrm{Re} \simeq 10^5$. For a rotating cylinder, turbulent flow would certainly be established at $\mathrm{Re} = 20$, which, for a cylinder of 2.0-cm diameter in an aqueous solution at room temperature ($\nu = 10^{-2}$ cm^2/sec), would correspond to a peripheral speed of only 0.1 cm/sec, or approximately 1.0 rpm.

Empirical correlations of mass transfer data have established the following dimensionless relationship for a rotating cylinder in turbulent flow:

$$\mathrm{St} = \text{constant } \mathrm{Re}^a \, \mathrm{Sc}^b \tag{3.3-42}$$

where

St = Stanton number, k/v,
Re = Reynolds number, vd/ν, and
Sc = Schmidt number, D/ν,

and a varies from -0.3 to -0.4, whereas b varies from -0.59 to -0.67. Eisenberg *et al.*[76] developed an empirical equation which has had good success in experimental electrochemical studies. The equation, written in terms of the mass flux, J, is as follows:

$$J = 0.079\, v^{0.7}\, d^{-0.3}\, \nu^{-0.344}\, D^{0.644} C \tag{3.3-43}$$

where

J = mass flux (moles/cm^2 sec),
v = peripheral velocity (cm/sec),
d = cylinder diameter (cm),
ν = kinematic viscosity (cm^2/sec),
D = diffusion coefficient (cm^2/sec), and
C = concentration (moles/cm^3).

Cornet and Kappesser[77] have independently verified this empirical equation in other electrochemical studies and it should be adequate to evaluate mass transfer coefficients for cementation reactions.

Gabe and Robinson[78] have derived a mass transfer equation quite similar to that which has been determined empirically. The derivation is based on the Prandtl–von Karman model which divides the fluid regime into three zones: I. a laminar flow layer next to the surface-Nernst diffusion layer; II. a viscous damping transition layer which damps out eddy current effects; and III. completely developed turbulence far from the surface-Prandtl's hydrodynamic layer. The theory for turbulent mass transfer at a rotating cylinder basically involves the determination

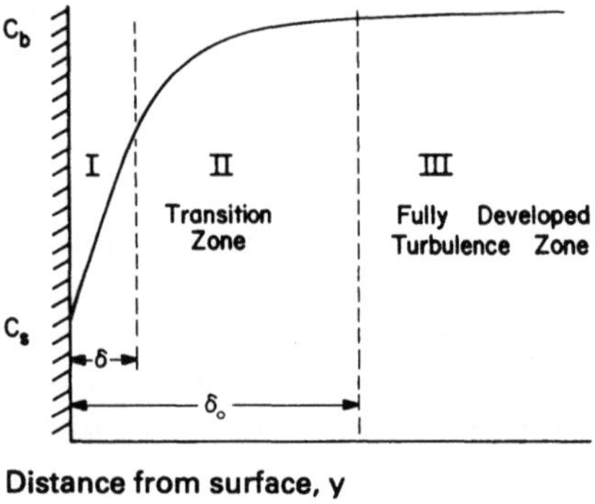

Figure 3.3-10. Flow regime zones for rotating cylinder system.[78]

of a turbulent diffusion coefficient, D_{turb}, analysis of Fick's law for the concentration gradient in each zone, and demonstration that the greatest resistance to mass transfer occurs in the laminar flow layer (Zone I) in which discrete molecular interactions prevail. See Figure 3.3-10.

In the case of Zone III, it is assumed for fully developed turbulence, according to the "mixing length" theory of Prandtl, that the turbulent diffusivity, D_{turb}, is equivalent to the apparent eddy viscosity, ε. Diessler[79] has shown that the eddy viscosity is dependent on the peripheral velocity and distance from the surface

$$\varepsilon = n^2 y v \tag{3.3-44}$$

where

n = integer coefficient,
y = radial distance, and
v = velocity.

For mass transfer to a rotating cylinder, in order to account for the influence of the radius of curvature, the factor arctan (ay/d) is introduced, where a is a constant and d is the cylinder diameter. Thus, for Zone III,

$$D^{\text{III}}_{turb} = n^2 y v \arctan(ay/d) \tag{3.3-45}$$

which is independent of concentration since complete mixing occurs in Zone III ($y \geq \delta_0$ and $C^{\text{III}} = C_b$).

In the case of Zone II, again using the results from Diessler,[79] the "mixing length" theory of Prandtl, and considering the radius of curvature, it can be shown that

$$D_{turb} = n^2 y v\left[1 - \exp\left(\frac{-n^2 y v}{\nu}\right)\right] \arctan\left(\frac{ay}{d}\right) \tag{3.3-46}$$

When d is finite and y is small, this expression can be reduced to

$$D_{\text{turb}} = \frac{an^4 y^3 v^2}{\nu d} = \frac{by^3 v^2}{\nu d} \tag{3.3-47}$$

where

$$b = an^4$$

This general expression for the turbulent diffusion coefficient can be used together with Fick's law (assuming the net flow velocity does not change, i.e., the damping only changes the eddy currents) to express the concentration in Zone II:

$$J = D\frac{dC}{dy}$$

$$\int_{C^{\text{II}}}^{C_b} dC = \int_y^{\delta_0} \frac{J}{D} dy$$

$$C^{\text{II}} = C_b + \frac{J\nu d}{2bv^2}\left(\frac{1}{\delta_0^2} - \frac{1}{y^2}\right) \tag{3.3-48}$$

Consider, finally, the laminar flow (Zone I) the Nernst layer in which $y < \delta$ and diffusion is only influenced by molecular interactions. Evaluation of Fick's law in this layer yields

$$\int_{C_s}^{C^I} dC = \frac{J}{D} \int_0^y dy$$

and

$$C^I = C_s + \frac{J}{D} y \tag{3.3-49}$$

when $y = \delta$, then $C^I = C^{\text{II}}$, and equation (3.3-48) can be set equal to equation (3.3-49), the mass flux is found to be

$$J = \frac{D(C_b - C_s)}{\delta - \dfrac{\nu d D}{2bv^2}(1/\delta_0^2 - 1/\delta^2)} \tag{3.3-50}$$

Analysis of the second term in the denominator suggests that it is significantly smaller than δ, so that as an approximation

$$J = \frac{D(C_b - C_s)}{\delta} \tag{3.3-51}$$

The Nernst-layer thickness can be defined from application of equation (3.3-47) at the point of transition from Zone I to Zone II where $D_{\text{turb}} = D$ and $y = \delta$

$$\delta = (D\nu d/bv^2)^{1/3} \qquad (3.3\text{-}52)$$

Substitution into equation (3.3-51) yields

$$J = b'v^{0.67}\, d^{-0.33}\, \nu^{-0.33}\, D^{0.67}\, (C_b - C_s) \qquad (3.3\text{-}53)$$

which is in good agreement with the empirical equation (3.3-43) discussed previously.

A number of cementation rate studies have been made using rotating cylinders but few investigators have analyzed their results in terms of predicted mass transfer rates for turbulent flow at a rotating cylinder. Power[70] was one of the first investigators to examine experimental cementation-rate data in this fashion. However, there are two major complications which must be considered. The first factor which may lead to erroneous interpretation of cementation rate data is the surface deposit and its morphology. As mentioned previously, this effect is common to all experimental techniques and will be discussed in the next section. The second factor which may cause difficulty in the analysis of cementation data for a rotating cylinder is the end effect. This becomes important only when the length-to-diameter ratio, l/d, reaches a critical value. Power[70] examined this effect for the Cu^{2+}/Zn system and found that for 25° C and a peripheral speed, $v = 138$ cm/sec, the critical l/d ratio is approximately unity, as shown in Figure 3.3-11. Notice that for l/d ratios of less than unity the experimental mass transfer coefficient increases significantly, almost linearly with decreasing l/d.

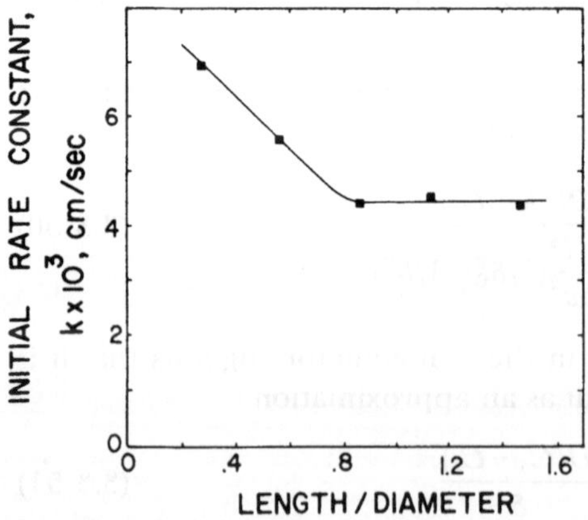

Figure 3.3-11. The effect of length/diameter ratio on the experimental mass transfer coefficient for the Cu^{2+}/Zn rotating cylinder system.[70]

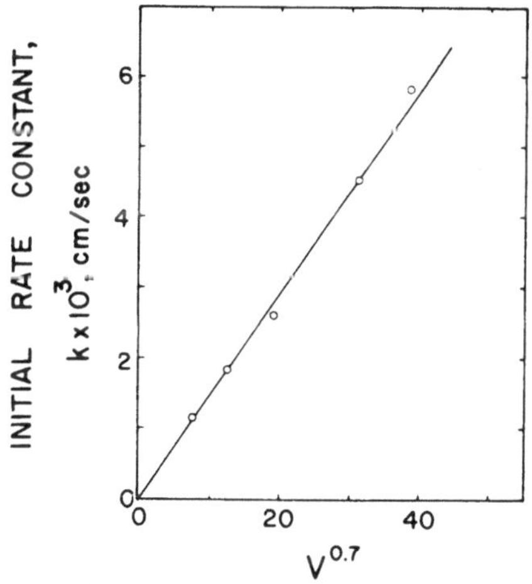

Figure 3.3-12. Demonstration of the 0.7th power dependence of the peripheral velocity on the mass transfer coefficient for the Cu^{2+}/Zn rotating cylinder system.(70)

According to equation (3.3-43), the mass transfer coefficient for rotating cylinders, k, should be,

$$k = 0.079\, v^{0.7}\, d^{-0.3}\, \nu^{-0.344}\, D^{0.644} \qquad (3.3\text{-}54)$$

provided the aforementioned complications are avoided. The dependence on the peripheral velocity to the 0.7th power has been demonstrated for the Cu^{2+}/Zn system, as shown in Figure 3.3-12. A review of the literature indicates that for studies in which surface deposit effects have been eliminated the calculated mass transfer coefficient agrees well with the experimental values as shown in Table 3.3-2.

To further illustrate the critical nature of the l/d ratio, it can be seen that for the last two systems listed in Table 3.3-2 the l/d ratio is

Table 3.3-2. Comparison of Experimental vs. Calculated Mass Transfer Coefficients for Selected Cementation Systems Using Rotating Cylinder Geometry

System	Reference	D, cm^2/sec	v, cm/sec	d, cm	l/d	Mass transfer coefficient, cm/sec: Experimental	Calculated
Cu^{2+}/Zn	70	0.67	19.5	3.5	1.14	1.1×10^{-3}	1.0×10^{-3}
Cu^{2+}/Fe	80	0.67	99.7	0.64	40	4.2×10^{-3}	5.1×10^{-3}
Ag^{+}/Zn	81	1.65	166	1.75	1.47	8.3×10^{-3}	9.7×10^{-3}
Cd^{2+}/Zn	82	0.47	497	7.6	0.13	1.23×10^{-2} $7.2\times10^{-3\,a}$	6.0×10^{-3}
Ag^{+}/Cu	83	1.65	500	7.4	0.13	2.6×10^{-2} $1.5\times10^{-2\,a}$	1.4×10^{-2}

[a] Mass transfer coefficients corrected for end effects.

significantly less than the critical value of unity determined by Power[70] and, as might be expected, the experimental mass transfer coefficient is greater by a factor of two than the calculated value. From the data in Figure 3.3-11 an end-effect correction factor of 0.58 can be calculated for an l/d ratio of 0.13. Applying this empirical correction factor to the measured mass transfer coefficients, corrected values of 7.2×10^{-3} cm/sec for Cd^{2+}/Zn and 1.5×10^{-2} cm/sec for Ag^{+}/Zn are obtained which agree with the mass transfer coefficients calculated from equation (3.3-43).

3.3.2.3. *Suspended Particles*

Cementation on suspended particles is of considerably greater interest than the rotating disk or the rotating cylinder systems in that frequently cementation reactions on an industrial scale are accomplished in stirred reactors with particulate metal powder, e.g., Cd^{2+}/Zn, Cu^{2+}/Fe, Ag^{+}/Zn, and Cu^{2+}/Ni. However, the theoretical treatment of mass transfer to particles suspended in a stirred reactor has not been fully developed. Nevertheless, the minimum mass transfer coefficient, k^*, in which the slip velocity would be equivalent to that for free-falling particles, can be calculated on a semitheoretical basis for monosize spheres from the physical properties of the aqueous and solid phase and the diffusion coefficient of the reactant. The free-fall mass transfer coefficient can then be empirically corrected for design and operating variables of the reactor.

Basically, the semitheoretical correlation[84] takes the following form for free-falling spherical particles:

$$\mathrm{Sh} = 2 + 0.6\,\mathrm{Re}^{0.5}\,\mathrm{Sc}^{0.33} \tag{3.3-55}$$

where

Sh = Sherwood number, k^*d/D,
Re = Reynolds number, vd/ν, and
Sc = Schmidt number, D/ν.

Figure 3.3-13 is a plot of the above correlation in terms of the mass transfer coefficient vs. particle size. Notice that in the range of 100 to 1000 μ the mass transfer coefficient is nearly constant because the terminal settling velocity in this particle-size region is approximately proportional to the diameter. Under these conditions the transfer coefficient is independent of particle size. For large particles, the terminal settling velocity varies with the square root of particle size shown by equation (3.3-56). Consequently the mass transfer coefficient should be dependent on the particle diameter to the −0.25 power. For

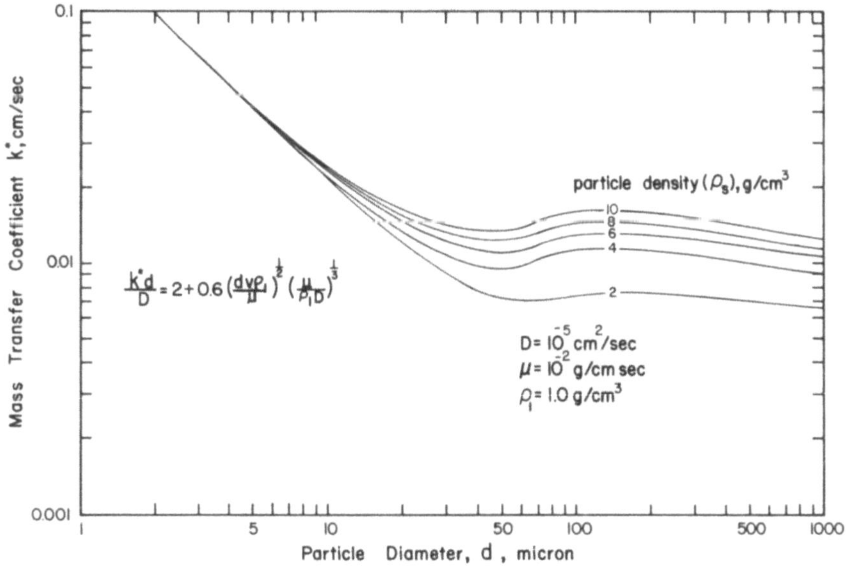

Figure 3.3-13. Plot of mass transfer coefficient vs. particle diameter for free-falling spheres; minimum mass transfer coefficient, k^*.

small particles, less than $10\,\mu$, the second term of equation (3.3-55) becomes negligible and the Sherwood number approaches a value of 2.0, the value for diffusion in an infinite stagnant medium. Under these conditions the mass transfer coefficient varies inversely with particle size to the first power.

The diagram shown in Figure 3.3-13 is constructed using the terminal settling velocity to calculate the Reynolds number and should give the minimum transfer coefficient for a suspended particle. The actual coefficient is dependent on reactor design and stirrer speed, or power input, whereas the effect of the physical properties can be predicted from equation (3.3-55) or Figure (3.3-13). Harriot[85] described a method for prediction of mass transfer coefficients in stirred reactors using the following procedure:

1. Calculate the terminal settling velocity and the particle Reynolds number

$$v = \left[\frac{4d(\rho_s - \rho_l)}{3Q\rho_l} g\right]^{1/2} \tag{3.3-56}$$

where

v = velocity (cm/sec),
d = particle diameter (cm),

Q = drag coefficient,
ρ_s = density of particle (g/cm^3),
ρ_l = density of liquid (g/cm^3), and
g = gravitational acceleration (cm/sec^2).

In the case of laminar flow, Re $\leqslant$ 1.0, the terminal settling velocity can be calculated from Stokes' law:

$$v = \frac{d^2(\rho_s - \rho_l)g}{18\mu}$$

2. Determine the minimum transfer coefficient, k^*, for free-falling spheres from equation (3.3-55) and/or Figure 3.3-13.

3. Select the proper ratio of a "standard" mass transfer coefficient to the minimum mass transfer coefficient, k_{200}/k^*, from Figure 3.3-14. The "standard" conditions of agitation intensity are

$$ND_I^{2/3}\left(\frac{D_I}{T}\right)\left(\frac{T}{Z}\right)^{1/3} = 200 \text{ in.}^{2/3} \text{ min}^{-1} \qquad (3.3\text{-}57)$$

and

$$(D_I/T) = 0.5 \qquad (3.3\text{-}58)$$

where

N = stirrer speed (rpm),
D_I = impeller diameter (in.),
T = tank diameter (in.), and
Z = suspension depth (in.).

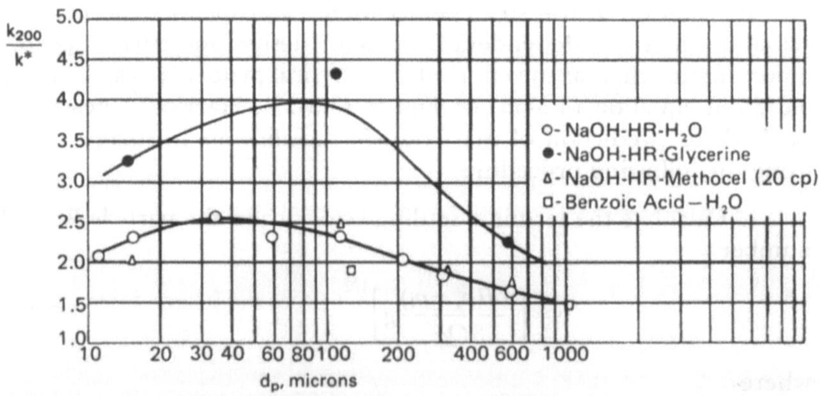

Figure 3.3-14. Relation between standard mass transfer coefficient, k_{200}, and minimum mass transfer coefficient, k^*.[85] The data are for a 4-in. tank, 2-in. turbine, and 300 rpm.

After selection of the proper ratio, k_{200}/k^*, the standard mass transfer coefficient, k_{200}, can be calculated:

$$k_{200} = k^*\left(\frac{k_{200}}{k^*}\right) \tag{3.3-59}$$

4. Extrapolate k_{200} to k for the given agitation intensity and specific particle size, using the correlation presented in Figure 3.3-15.

It appears from a review of the literature that cementation reaction kinetics have not been extensively studied using monosize particles and consequently the correlation remains untested in cementation systems. Almost all cementation studies involving particulate metal in a stirred reactor have used an unspecified particle-size distribution. However, in one study of the Cu^{2+}/Fe system, Nadkarni *et al.*[86] did report cementation kinetics with monosize particles of granulated cast iron. The mass transfer coefficient calculated from the reported data at 26° C is found to be 1.2×10^{-2} cm/sec for a particle diameter of 200 μ. From the plot in

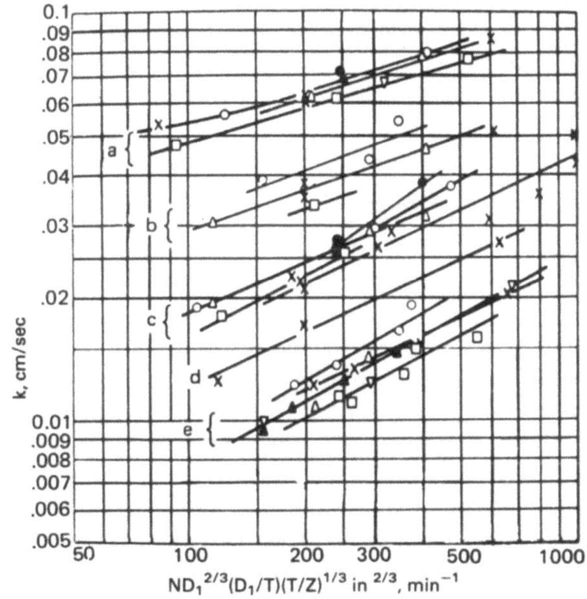

Figure 3.3-15. Relation between mass transfer coefficient and agitation intensity, $ND_I^{2/3}(D_I/T)(T/Z)^{1/3}$, for various particle sizes[85]: (a) 15.3 μm; (b) 34.5 μm; (c) 59 μm; (d) 115 μm; (e) 304 μm.

Symbol:	□	▽	△	▲	●	○	×
D:	2	1.5	3	7	2	4	2
T:	8	4	8	21	4	8	4
T/Z:	1	1	1	1	1	1	0.57

Figure 3.3-13 the minimum mass transfer coefficient k^* is found to be 1.4×10^{-2} cm/sec. The "standard" mass transfer $k_{200}=k^*(k_{200}/k^*)=1.4\times10^{-2}\times2=2.8\times10^{-2}$ cm/sec. Finally, using the correlation presented in Figure 3.3-15, the mass transfer coefficient corresponding to Nadkarni's experimental conditions can be determined. In order to make this last part of the calculation, some approximations regarding reactor size and impeller diameter must be made:

$$ND_I^{2/3}\left(\frac{D_I}{T}\right)\left(\frac{T}{Z}\right)^{1/3}\simeq 100\ \text{in.}^{2/3}\ \text{min}^{-1}$$

From Figure 3.3-15, for a 200-μ particle diameter it is then found that the predicted mass transfer coefficient is approximately 2×10^{-2} cm/sec compared to the experimental value of 1.2×10^{-2} cm/sec. The prediction is made for spherical particles and for irregularly shaped particles that the coefficients would be somewhat smaller.

Calculation of mass transfer coefficients for particles suspended in stirred reactors is not nearly as accurate as the calculation of mass transfer coefficients for rotating disks and rotating cylinders. Again, for cementation systems, as always, complications associated with surface deposit formation must be carefully considered in the analysis of rate data.

3.3.3. Surface Deposit Effects

During the past several years the importance of the surface deposit in the interpretation of cementation rate data has been clearly demonstrated in a quantitative manner. Unlike many other heterogeneous reaction systems, cementation reactions are unique in that the reaction product usually does not impede the reaction progress but rather frequently enhances the reaction kinetics. This phenomenon can be attributed to the electrochemical nature of the reaction and the structure of the surface deposit.

Similar to electrolytic deposition processes the cementation surface deposit may be influenced by hydrodynamics, concentration, temperature, and ligand and organic colloid addition. In electrolysis, of course, the current density may be an independent variable and it in itself can determine the nature of the deposit.

Surface deposits formed in cementation systems are quite similar to electrolytic deposits because of the electrochemical nature of the reaction. In cementation systems the current density is *not* an independent variable and consequently some interesting surface deposit effects are observed. Under special circumstances, i.e., conditions which lead to a tight, smooth, coherent deposit, the cementation reaction may become inhibited presumably because of product ion diffusion resistance. Generally, however, the surface deposit is either dendritic or botroiydal and the reaction kinetics are enhanced significantly due to the presence of the deposit. The significance of the surface deposit and this rate enhancement, which has been observed by several investigators, will be discussed in terms of hydrodynamics, concentration, temperature, and particle–particle interaction effects.

3.3.3.1. Hydrodynamics

Solution flow is an important variable which controls the type of deposit obtained in electrolytic cells. It has been shown[87] in the case of zinc deposition from cyanide solution that, as the flow rate of solution over the cathode increases at constant current density, the nature of the electrolytic deposit changes from dendritic, to botroiydal, to smooth as shown by the sequence of photographs presented in Figure 3.3-16. This

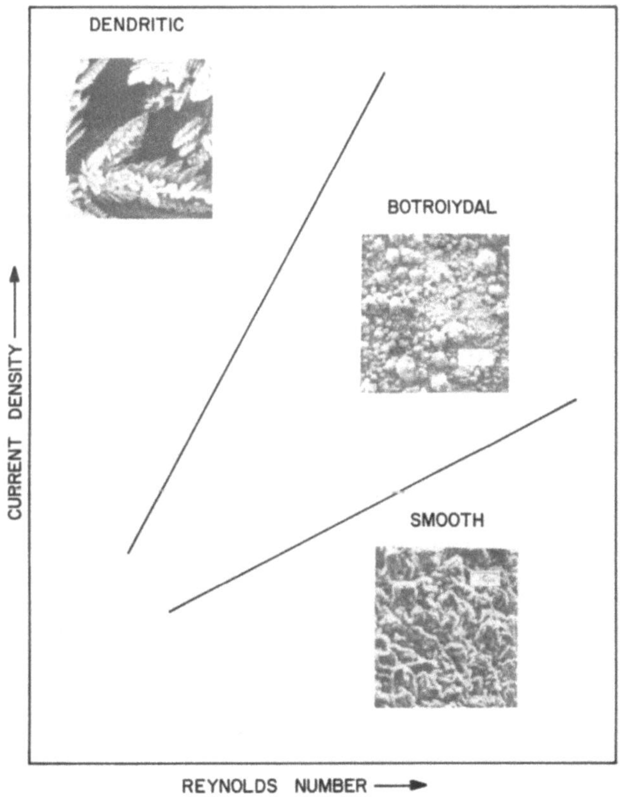

Figure 3.3-16. The influence of fluid flow and current density on the nature of the electrolytic deposit.

"phase" diagram characterizes the nature of the surface deposit with respect to current density and Reynolds number.

The effect of fluid flow on the nature of the surface deposit in cementation systems has not been studied extensively. In the Cu^{2+}–Fe rotating-disk system the surface deposit has been shown to vary with respect to radial position on the disk's surface. For laminar flow, $(r^2\omega/\nu) < 10^5$, at all positions on the disk surface there is a gradual change from a coarse deposit at the disk center (small Reynold's number approaching dendrite formation) to a finer textured deposit at the edge of the disk (large Reynolds' number) as shown in Figure 3.3-17.

For the case in which the flow changes from laminar to turbulent at some radial position of the disk, there is an abrupt rather than a gradual change in surface-deposit character, and the influence of the transition to the turbulent flow regime on the surface deposit is most pronounced. Under certain circumstance, the coarse deposit at the center of the disk does not shear off but rather grows continually to form a cone which, in one instance, is reported to reach a height of 0.5 cm.[88] The results from this study indicate in a qualitative fashion that fluid flow not only determines the extent of surface deposit removal by shear forces on a macroscopic basis, but also determines at a microscopic level the basic character and structure of the surface deposit, namely nucleation and growth phenomena.

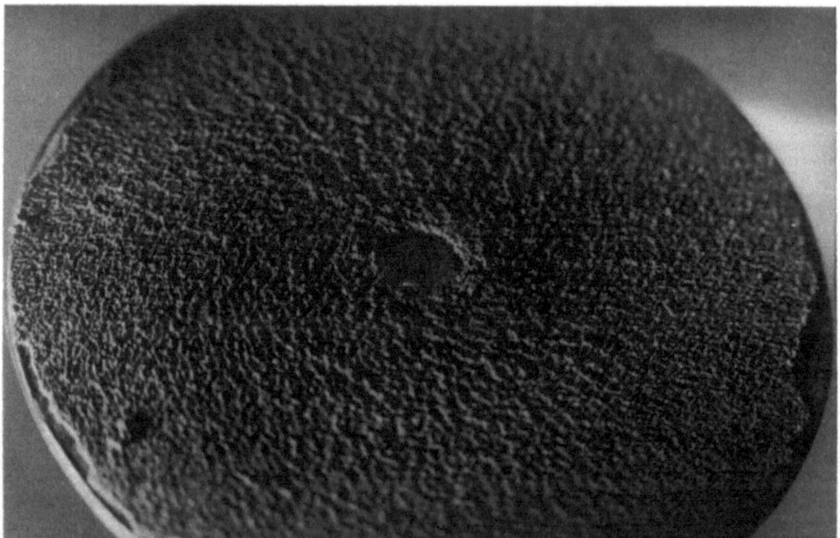

Figure 3.3-17. Surface deposit for a rotating disk in laminar flow regime, Cu^{2+}/Fe system.[88]

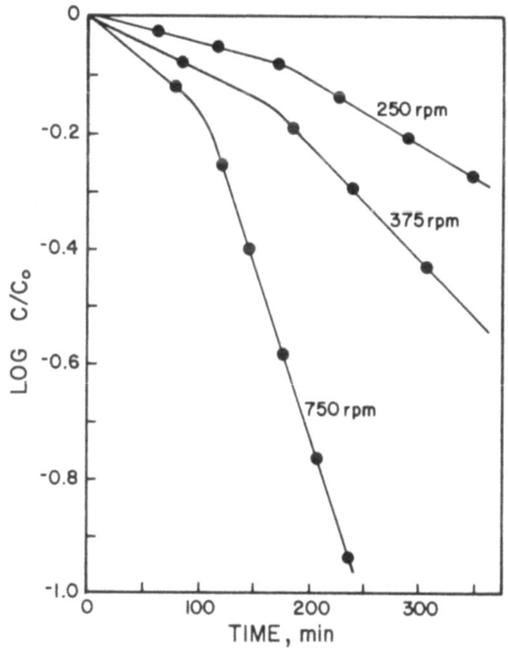

Figure 3.3-18. First-order reaction plot for rotating disk; Cd^{2+}/Zn system.[75]

3.3.3.2. Concentration

In many systems a characteristic effect of the noble metal ion concentration on cementation reaction kinetics has been observed. These systems include Cd^{2+}/Zn, Pb^{2+}/Zn, Ag^{+}/Zn, Cu^{2+}/Zn, and Cu^{2+}/Fe. Without exception, for small initial concentrations of the noble metal ion first-order plots of a type similar to that obtained for the Cd^{2+}/Zn system (rotating disk), shown in Figure 3.3-18, are obtained.[75] Two stages of reaction are apparent. The initial stage of reaction corresponds to deposition on a relatively smooth surface. During this stage of reaction the deposit structure and thickness has not significantly affected the reaction kinetics and the reaction kinetics follow Levich's equation for a rotating disk system:

$$J = 0.62\, D^{2/3}\, \nu^{-1/6}\, \omega^{1/2} C \qquad (3.3\text{-}39)$$

For example, theoretically calculated mass transfer coefficients,

$$k = 0.62\, D^{2/3}\, \nu^{-1/6}\, \omega^{1/2} \qquad (3.3\text{-}40)$$

agree quite well with experimental values from initial-stage kinetics for a

Table 3.3-3. Comparison of Mass Transfer Coefficients at 750 rpm Determined from Rotating Disk Data[75] with Theoretical Values Predicted by Equation (3.3-40)

System	Mass transfer coefficient, cm/sec		
	Theoretical	1st stage	2nd stage
$Cu^{2+}/Zn, Cu^{2+}/Fe, Cd^{2+}/Zn$	4.5×10^{-3}	5.6×10^{-3}	2.15×10^{-2}
$Ag^{+}/Zn, Ag^{+}/Cd, Ag^{+}/Cu$	7.2×10^{-3}	8.9×10^{-3}	

number of reactions studied with the rotating disk as shown by the values presented in Table 3.3-3.

At some critical point of reaction progress, which corresponds to a critical deposit mass per unit area, a second stage of enhanced first-order kinetics is observed as shown in Figure 3.3-18. The increased rate of reaction is due to the development of a structured surface deposit which

Figure 3.3-19. Comparison of the nature of the surface deposits for 1st stage (top) and 2nd stage (bottom) reaction kinetics; rotating disk Cu^{2+}/Zn system.[70]

effectively increases the surface area available for a reaction. Under these circumstances theoretical mass transfer coefficients, calculated on the basis of the geometric area of a flat disk are smaller than those determined experimentally, as shown in Table 3.3-3. The change in surface character is vividly illustrated by the photographs (400×) for the Cu^{2+}/Zn system presented in Figure 3.3-19. The photograph of the surface during the initial stages of reaction (shown after 5 min reaction from a solution initially $1.56\times10^{-4}M$ Cu^{2+}) reveals the original structure of the zinc surface with scratches present from the polishing preparation before reaction. No trace of the original surface structure can be observed during the second stage of reaction (shown after 130 min reaction from a solution initially $1.56\times10^{-4}M$ Cu^{2+}). The botroiydal nature of the deposit is similar to that observed under certain circumstances in electrolytic deposition as discussed at the beginning of this section. Strickland and Lawson[75] suggest that the critical deposit mass density, at which the transition to enhanced kinetics occurs, is approximately 0.3 mg/cm^2 for most systems. This value is consistent with the value obtained by Ingraham and Kerby[82] who report that the transition to enhanced, second-stage, first-order kinetics occurs at 0.25 mg/cm^2. The rate enhancement continues to increase with deposit mass density up to 5 mg/cm^2 in the case of Cu^{2+}/Zn, above which it appears that a steady-state surface profile is reached and the second-stage kinetics no longer increases with the extent of reaction.

This phenomenon is reflected in Figure 3.3-20 by the fact that as the initial cupric concentration increases, the second-stage reaction rate increases and is reached in a much shorter time interval. At initial concentrations above 100 ppm, rate enhancement is no longer found to be a function of initial concentration. Consequently, for high initial concentrations of noble metal ions, the first-stage reaction kinetics may not be seen, the reason being, of course, that the critical deposit mass density is reached almost immediately. Further evidence of the effect of initial noble metal ion concentration on rate enhancement is given by the kinetic response of the system when the reductant metal is given an initial surface deposit, a "strike." The case study is the Cu^{2+}/Fe system in which a first-order plot for an initial cupric concentration of 10 ppm is shown in Figure 3.3-21 (solid circles). Even after 60 min reaction time the surface deposit mass density is approximately 0.6 mg/cm^2 and the rate enhancement, or second-stage kinetics, is not observed. On the other hand, for 200 ppm (see solid squares) the critical deposit mass density is reached after 3 min and the initial stage of reaction in the absence of surface deposit effects is not observed. Notice, importantly, that when the iron surface is given an initial "strike" several times greater than the critical deposit mass density, and the reaction then

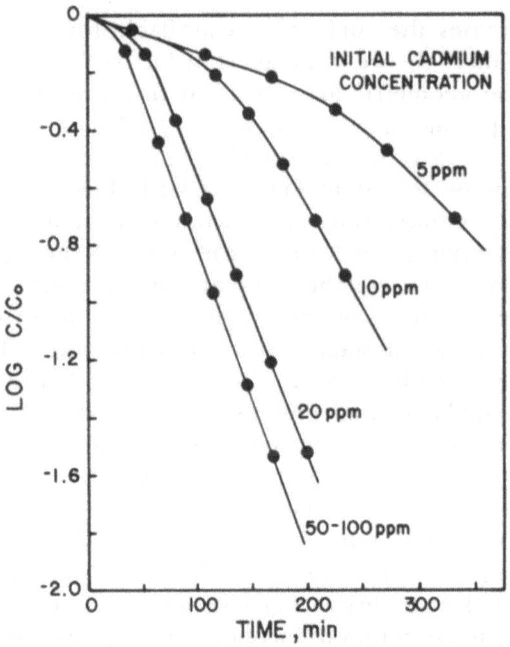

Figure 3.3-20. The effect of initial concentration on 1st-stage reaction kinetics; rotating disk Cd^{2+}/Zn system.

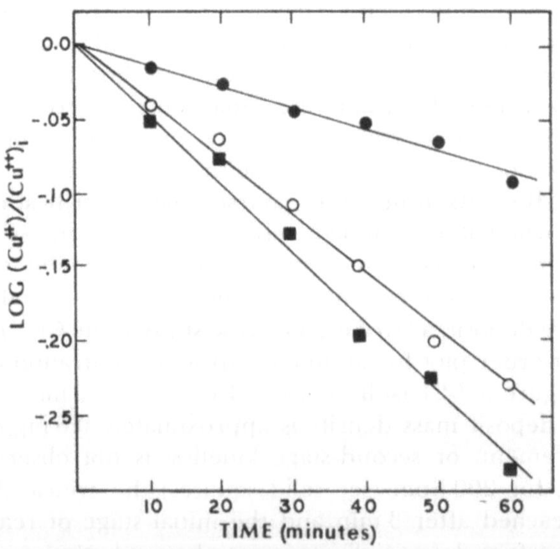

Figure 3.3-21. Influence of a surface deposit (*strike*), produced at 200 ppm Cu^{2+}, on the cementation rate at 10 ppm Cu^{2+}, a temperature of 45° C, and a pH of 1.90. *Key*: ●, 1 hr at $(Cu^{2+})_i = 10$ ppm; ○, 10-min "strike" at 200 ppm (Cu^{2+}), then 1 hr at $(Cu^{2+})_i = 10$ ppm; ■, 1 hr at $(Cu^{2+})_i = 200$ ppm.

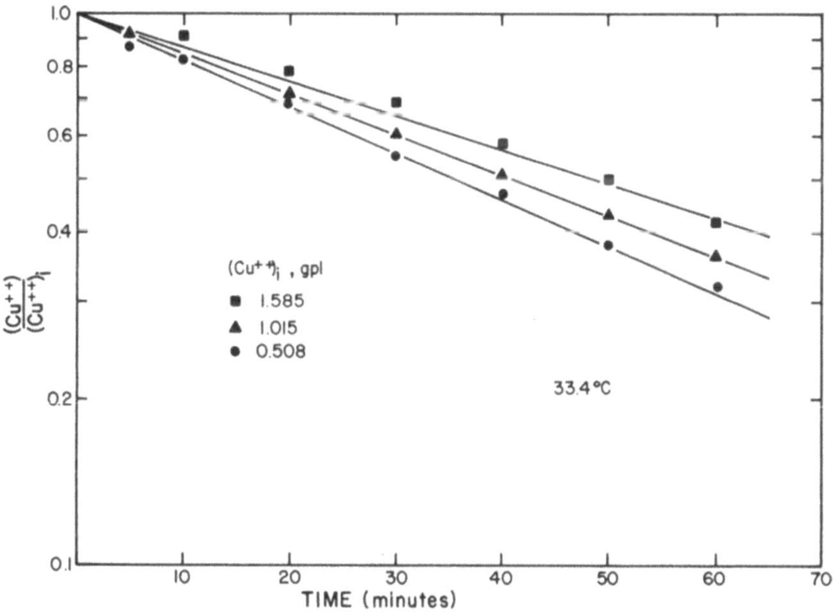

Figure 3.3-22. Initial concentration effect on 2nd-stage reaction kinetics; Cu^{2+}/Fe system.[86]

followed, the observed reaction kinetics for an initial cupric concentration of 10 ppm are almost equivalent to the observed reaction kinetics for an initial cupric concentration of 200 ppm.

The concentration of noble metal ion, and probably the ionic strength in general, can influence cementation reaction kinetics not only as they relate to the structure and mass density of the surface deposit, but also as these parameters influence the diffusion coefficient and activity of diffusing species. Ideally, it would be expected that the kinetics of diffusion through the MTBL would be independent of initial concentration of noble metal ion in the absence of a time-dependent surface deposit effect. However, it appears that such is not necessarily the case. For example, consider the Cu^{2+}/Fe system shown in Figure 3.3-22. Clearly, these data are for the second stage of reaction in which the surface deposit has been fully developed to some steady-state profile. Consequently, the minor effect of the initial cupric concentration must be accounted for in some manner such as activity- or diffusion-coefficient corrections. In the case of activity coefficients, according to absolute reaction-rate theory, the reactants are in equilibrium with the activated state and the apparent rate constant for a nonideal system can be evaluated by the following expression:

$$k_{\text{apparent}} = \frac{kT}{h} K^{\ddagger} \frac{\pi_i \gamma_i^{j_i}}{\gamma^{\ddagger}} \tag{3.3-60}$$

Table 3.3-4. Second-Stage Rate Constants for the Cu^{2+}/Fe System Taken from Figure 3.3-22 and Corrected for Nonideal Behavior

Initial copper concentration, g/liter	k_{apparent}, cm/sec	γ_{Cu}^{2+}	k_{ideal}, cm/sec
0.508	2.33×10^{-2}	0.48	4.85×10^{-2}
1.015	1.66×10^{-2}	0.44	4.25×10^{-2}
1.585	1.67×10^{-2}	0.39	4.28×10^{-2}

where (kT/h) is the ideal rate constant, $K^{\ddagger}$ is the equilibrium constant between reactants and activated species, T is the absolute temperature (°K), h is Planck's constant, k is Boltzmann's constant, $\Pi_i\gamma_i^{\nu_i}$ is the product of the activity coefficients of the reactants, $\gamma^{\ddagger}$ is the activity coefficient of the activated species, and k_{apparent} is the apparent reaction-rate constant.

For a diffusion-controlled reaction such as the Cu^{2+}/Fe cementation system the only reactant is the diffusing species, i.e., the cupric ion. Furthermore, because of the normally low concentrations of the activated species it is assumed that it behaves ideally, and the activity coefficient for the activated complex, $\gamma^{\ddagger}$, is unity. The above expression can then be reduced to

$$k_{\text{apparent}} = k_{\text{ideal}}\gamma_{Cu}^{2+} \tag{3.3-61}$$

where k_{ideal} is the rate constant predicted by ideal behavior. This expression can be used to correct for nonideal conditions and accounts successfully for the observed decrease in rate constant at higher initial copper concentrations, as shown in Table 3.3-4.

Alternatively, the concentration dependence of the mass transfer coefficient can be explained in terms of the concentration dependence of the self-diffusion coefficient.

3.3.3.3. Temperature

Surface deposit effects on cementation reaction kinetics have been shown to have a significant temperature dependence. Consider the Arrhenius plots for the Cu^{2+}/Fe rotating cylinder system presented in Figure 3.3-23. The open squares represent the standard kinetic response of a rotating cylinder experiment where the rate constant, or mass transfer coefficient, is calculated on the basis of the smooth geometric surface area of the iron cylinder (7 cm^2). At first glance it would appear from consideration of the activation energies that, as the temperature decreases, the reaction changes from MTBL diffusion

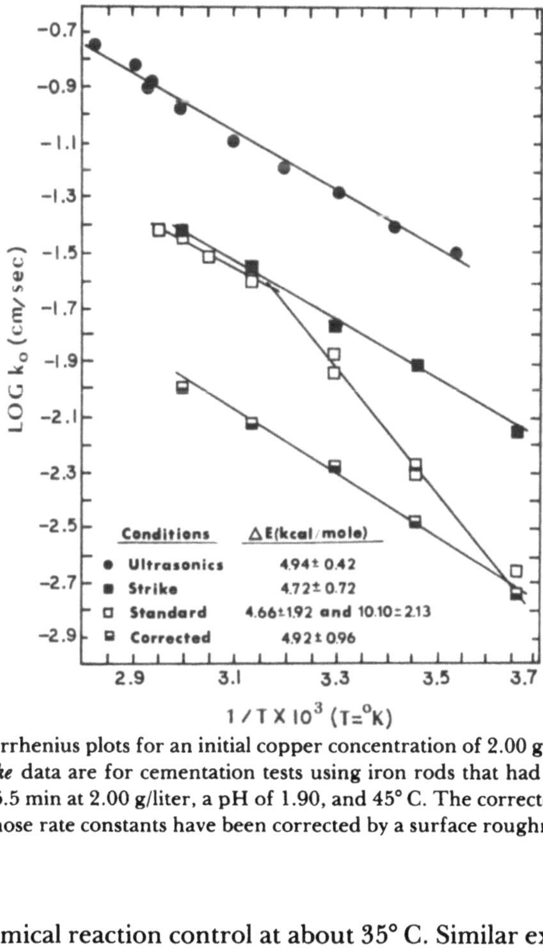

Figure 3.3-23. Arrhenius plots for an initial copper concentration of 2.00 g/liter and a pH of 1.90. The *strike* data are for cementation tests using iron rods that had been given an initial strike for 6.5 min at 2.00 g/liter, a pH of 1.90, and 45° C. The corrected data are for standard tests whose rate constants have been corrected by a surface roughness factor (see Table 3.3-5).

control to chemical reaction control at about 35° C. Similar experimental results were observed by Calara[88] for the Cu^{2+}/Fe rotating disk system. This behavior has been also observed in the Cu^{2+}–Ni[89] and Pd^{2+}–Cu[90] systems. Analysis of these systems in terms of standard electrode potentials suggest that reaction kinetics should be controlled by mass transfer.

Calculation of the theoretical mass transfer coefficients are much lower than those observed experimentally. For example, in the case of the Cu^{2+}/Fe system at 45° C the calculated mass transfer coefficient for rotating cylinders is 7.7×10^{-3} cm/sec [see equation (3.3-54)] compared to the 2.5×10^{-2} cm/sec observed experimentally, which suggests that the reaction is going three times as fast as theoretically possible. The reason for this apparent anomalous behavior is the effect of the surface deposit. These particular experiments were performed at a cupric concentration for which enhanced, second-stage reaction kinetics is observed. The temperature dependence of the surface deposit is substantiated by both direct and indirect evidence.

As for indirect evidence, it is seen from Figure 3.3-23 that when the reaction is studied in an ultrasonic field the mass transfer coefficient is increased significantly because of a reduction in the effective MTBL thickness from the microstreaming which accompanies the ultrasonic field. In addition, the dissipation of ultrasonic energy at the cylinder surface continually removes the surface deposit and the apparent region of chemical reaction control is no longer observed; an activation energy of approximately 5 kcal/mole is obtained over the entire temperature range studied as shown by the solid circles in Figure 3.3-23. Furthermore, the effect of temperature on the surface deposit and how it affects reaction kinetics is illustrated by experiments in which the iron cylinder is given a "strike," an initial surface deposit, of copper at 45° C. The established surface maintained its form at all other temperatures for which the reaction was studied, even at 0° C, and an activation energy of 5 kcal/mole is observed over the entire temperature range (solid squares).

Direct evidence of the surface deposit's temperature dependence is revealed by microscopic examination of the surface for the standard cementation experiment. A distinct change in the nature of the deposit at a critical temperature of 35° C is observed in the set of microphotographs shown in Figure 3.3-24 (approximately equivalent deposit mass densities, 10 mg/cm^2, in considerable excess of the critical deposit mass density of 0.3–0.6 mg/cm^2). Notice that below 35° C for the cross-sectional and surface photographs the botroiydal deposit becomes more dense and consolidated as the temperature is lowered, whereas above 35° C the deposit maintains a constant uniform porosity and surface area at all temperatures. Basically these observations support the concept that the area term, A, in the rate expression is a function of temperature, depending, under certain circumstances, on the relative magnitudes of the growth and nucleation rates. From the cross-sectional photographs, the actual surface area can be measured and the rate constants corrected with respect to the true surface area. The results of these calculations are presented in Table 3.3-5.

When the standard rate constants are corrected in this fashion and plotted according to the Arrhenius equation, an activation energy of 4.9 kcal/mole is obtained as shown by half-filled squares in Figure 3.3-23; and thus, the activation energy for the standard cementation data actually supports boundary-layer diffusion as the rate-limiting step in the cementation reaction. Furthermore, when the rate data are corrected for the area increased because of the surface deposit the experimental rate constants, or mass transfer coefficients, agree quite well with

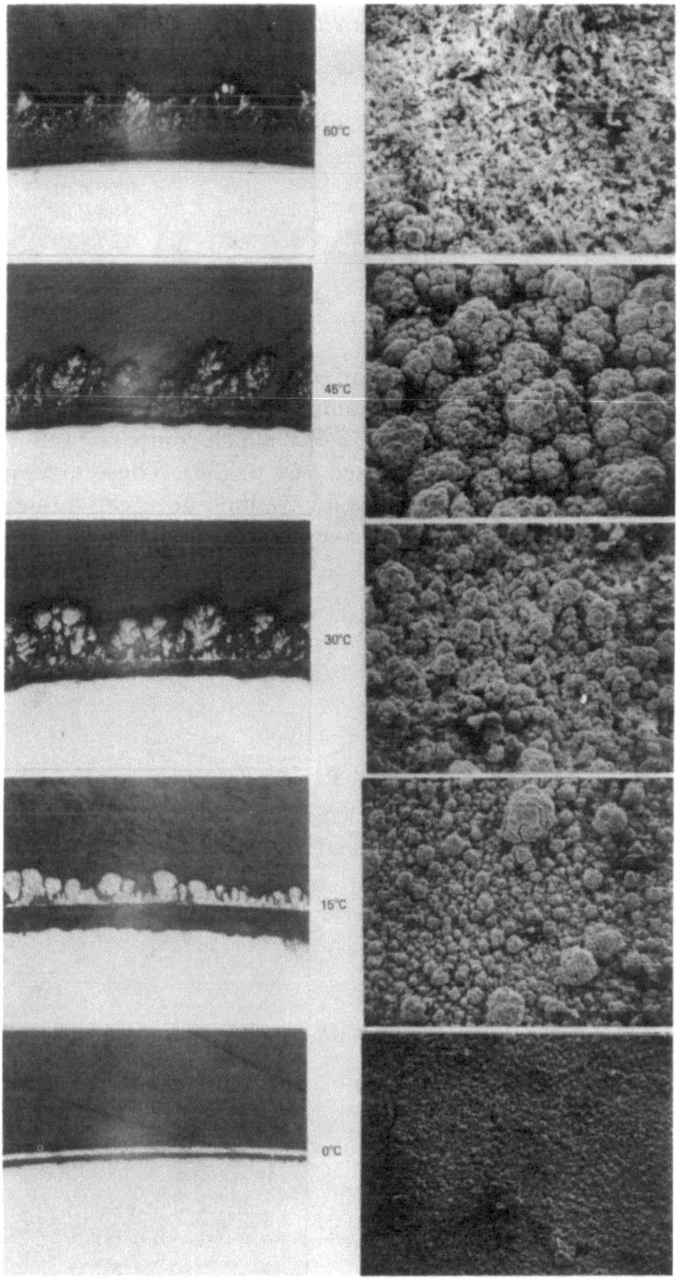

Figure 3.3-24. Photomicrographs taken using an optical microscope (left column) and a scanning electron microscope (right column) of copper deposits produced at a Cu^{2+} concentration of 2.00 g/liter and a pH of 1.90 at various temperatures. Scale: 1 cm = 70 μ.

Table 3.3-5. Rate Constants Corrected for a Surface Roughness Factor[80]

Temp., °C	$k_{apparent}$, cm/sec	Roughness factor	$k_{corrected}$, cm/sec
0	0.20×10^{-2}	1.0	0.20×10^{-2}
15	0.52×10^{-2}	1.5	0.34×10^{-2}
30	1.32×10^{-2}	2.4	0.54×10^{-2}
45	2.70×10^{-2}	3.6	0.75×10^{-2}
60	3.10×10^{-2}	3.4	0.92×10^{-2}

the mass transfer coefficients that would be predicted from semi-theoretical considerations. For example, at 45° C the corrected rate constant is 7.5×10^{-3} cm/sec (Table 3.3-5) which compares well with the theoretical value of 7.7×10^{-3} cm/sec (see p. 235). These experimental results illustrate vividly how surface deposition effects can influence the interpretation of rate data for heterogeneous reactions.

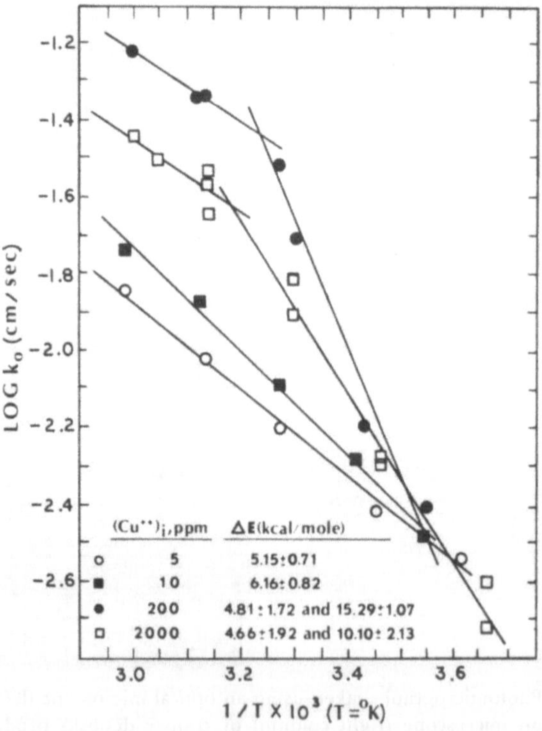

Figure 3.3-25. Arrhenius plots for various initial Cu^{2+} concentrations.

On the basis of the foregoing, it would be expected that an Arrhenius plot at low cupric concentrations, where the first-stage reaction response is observed, should be linear over the entire temperature region. The expected, simultaneous interaction of temperature and concentration is revealed by the Arrhenius plots in Figure 3.3-25. Again, the effect of the surface deposit as influenced by temperature and concentration is markedly demonstrated.

3.3.3.4. Particle–Particle Interactions

Finally, one other important aspect of cementation reaction kinetics, which relates to surface deposit effects, is particle–particle interaction. This phenomenon is of considerable importance from the standpoint of reactor design for all types of reactors, such as launders, stirred tank reactors, and cementation cones. Studies by Fisher and Groves[91] demonstrate this phenomenon quite well. In these investigations copper cementation was studied in a rotating drum using iron finishing nails as the reductant.

In essence these studies reveal, as might be expected in view of the previous discussion, that particulate conductors enhance the cementation reaction kinetics. The conducting particles act to extend the cathodic surface area and promote copper deposition even though the cupric ion may never reach the iron surface. Rate enhancement in the presence of particulate conductors occurs because of a "fluidized cathode mechanism," i.e., an effective increase in cathodic area, provided good particle–particle contact is maintained for electron conduction. See the schematic presented in Figure 3.3-26.

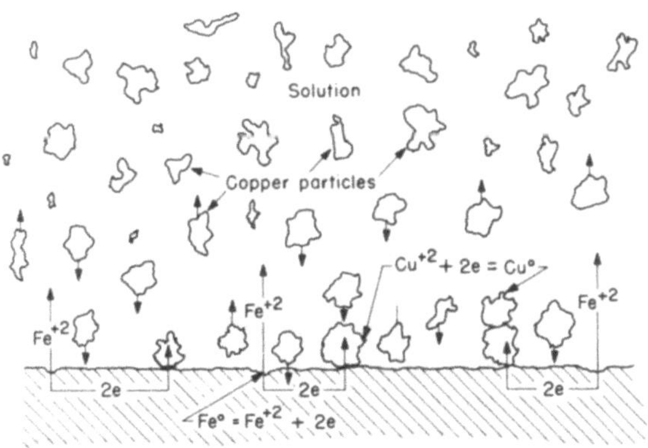

Figure 3.3-26. Schematic representation of *fluidized cathode mechanism* operative in cementation systems.[91]

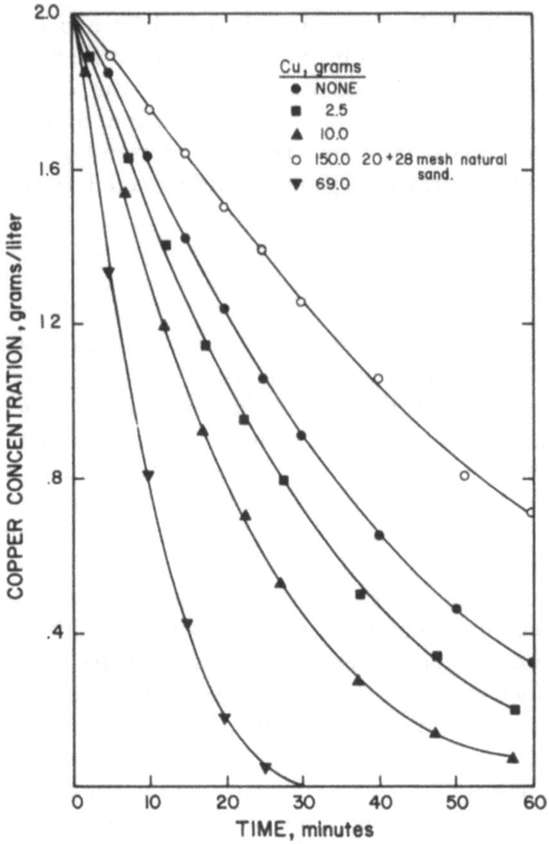

Figure 3.3-27. The effect of conducting and nonconducting particles on the reaction kinetics Cu^{2+}/Fe system.

Typical reaction kinetics for the drum reactor in the absence of conducting particles are shown by the solid circles in Figure 3.3-27. Addition of particulate conductors, such as cement copper and gold powder, significantly increase the reaction kinetics, whereas the addition of nonconductors causes a decrease in the reaction kinetics. The decrease in rate observed in the presence of nonconducting sand particles is presumably due to the fact that not only are they nonconductors and hence do not increase the cathodic area, but in addition the sand particles abrade the iron surface removing the surface deposit and tend to eliminate the second-stage rate enhancement attributed to surface deposit effects. Furthermore, in the presence of nonconducting particulates, short-circuiting of the fluid bed is prevented and the "fluidized cathode mechanism" is not operative.

As would be expected, the rate enhancement is directly related to the weight of particulate conductors up to some saturation point. More accurately, the rate enhancement should be proportional to the number of conductive particles, the collision rate, and time of contact. The result of these studies indicate that packed-bed reactors such as cementation cones or any other system which would provide for good particle–particle, electrochemical interaction are ideal. Further consideration should be given to designing with inert conductors to catalyze the electrochemical cementation reaction.

3.3.3.5. Conclusion

Surface deposit effects demand careful consideration in the analysis of cementation rate data. Several explanations for the rate enhancement because of surface deposit have been tendered:

1. decrease in diffusion path length, MTBL,
2. microturbulence in the interstices of the surface deposit, and
3. increase in effective surface area.

Of these, it appears that the first can be discounted because the shape of the effective MTBL is distorted when the scale of roughness is comparable to the thickness of the MTBL, which is the case for second-stage cementation kinetics. The second explanation is questioned because the surface pores are too small to be able to accommodate eddy currents of the size normally encountered in turbulent flow systems. Consequently, the rate enhancement is most probably due to an increase in effective area, as demonstrated by the behavior of the Cu^{2+}/Fe system.[80]

References

1. K. F. Wenzel, Lehre von der Chemischen Affinität der Körper, reported by A. Findlay, *A Hundred Years of Chemistry*, 2nd ed., Duckworth, London (1977).
2. L. F. Wilhelmy, *Ann. Physick* **81**, 413 (1850).
3. J. H. van'Hoff, *Etudes de Dynamique Chimique*, F. Muller and Company, Amsterdam (1884).
4. S. Arrhenius, *Z. Phys. Chem.* **4**, 226 (1889).
5. S. Glasstone, K. J. Laidler, and H. Eyring, *The Theory of Rate Processes*, McGraw-Hill Book Company, New York (1941), pp. 400–401.
6. A. Fick, *Pogg. Ann.* **94**, 59 (1855).
7. C. L. Wagner, *Z. Phys. Chem.* **71**, 401 (1910).
8. J. Lebrun, *Bull. Sci. Acad. Roy. Belge* 953 (1913).
9. G. F. Kortum and J. O'M. Bockris, *Textbook of Electrochemistry*, Vol. 2, Elsevier Publishing Company, Amsterdam (1951), pp. 403–405.
10. J. Halpern, *J. Metals, Trans. AIME*, **209**, 280 (1957).

11. K. J. Vetter, *Electrochemical Kinetics*, Academic Press, New York (1967), pp. 189–193.
12. V. G. Levich, *Physiochemical Hydrodynamics*, 2nd ed., Moscow (1959).
13. L. Prandtl, *Phys. Z.* **11**, 1072 (1910).
14. W. Jost, *Diffusion in Solids, Liquids and Gases*, Academic Press, Inc., New York (1952).
15. S. Brunauer, P. H. Emmett, and E. Teller, *J. Am. Chem. Soc.* **60**, 309 (1938).
16. C. Wagner, *J. Electro. Chem.* **97**, 71 (1950).
17. W. D. Spencer and B. Topley, *J. Chem. Soc.* **50**, 2633 (1929).
18. H. Eyring, R. E. Powell, G. H. Duffy, and R. R. Parlin, *Chem. Rev.* **45**, 145 (1949).
19. R. Schuhmann Jr., *Trans. AIME* **17**, 22–25 (1960).
20. L. Beckstead *et al.*, Acid ferric sulfate leaching of attritor-ground chalcopyrite concentrates, International Symposium on Copper Extraction and Refining, J. C. Yannopoulos and J. C. Agarwal, eds., 105th AIME Annual Meeting, Las Vegas, Nevada, February 22–26, 1976.
21. P. H. Yu, C. K. Hanson, and M. E. Wadsworth, *Met. Trans.* **4**, 213–44 (1973).
22. M. E. Wadsworth, *Trans. Met. Soc.* **245**, 1381–1394 (1969).
23. R. M. Nadkarni and M. E. Wadsworth, *Trans. Met. Soc.* **239**, 1066–1074 (1967).
24. J. P. Baur, H. L. Gibbs, and M. E. Wadsworth, *U.S.B.M. RI* 7823 (1974).
25. R. Mishra, Kinetics of oxidation of sulfide minerals through thin aqueous films, Ph.D. Dissertation, Department of Mining, Metallurgical and Fuels Engineering, University of Utah, Salt Lake City, Utah (1972).
26. E. Posnjak and H. E. Merwin, *J. Am. Chem. Soc.* **44**, 1965 (1922).
27. C. Wagner and K. Grunewald, *Z. Physik. Chem.* (*B*) **40**, 455 (1938).
28. S. Glasstone, K. J. Laidler, and H. Eyring, *The Theory of Rate Processes*, McGraw-Hill Book Company, New York (1941), pp. 184–191.
29. K. J. Vetter, The determination of electrode reaction mechanisms by the electrochemical reaction orders, in *Transactions of the Symposium on Electrode Processes*, E. Yeager, ed., John Wiley and Sons, Inc., New York, (1959), pp. 47–65.
30. K. J. Vetter, *Electrochemical Kinetics—Theoretical and Experimental Aspects*, Academic Press, New York (1967).
31. J. A. V. Butler, *Trans. Faraday Soc.* **19**, 729–733 (1924).
32. T. Erdey-Gruz and M. Volmer, *Z. Physick. Chem.* (A) **150**, 203–213 (1930).
33. J. Tafel, *Z. Physik. Chem.* **50**, 641–712 (1905).
34. P. Marcantonio, Kinetics of dissolution of chalcite in ferric sulfate solutions, Ph.D. Thesis, Department of Mining, Metallurgical and Fuels Engineering, University of Utah (1975), in preparation.
35. S. B. Christy, *Trans. AIME* **26**, 735 (1896).
36. S. B. Christy, *Trans. AIME* **30**, 864 (1900).
37. B. Boonstra, *Korros. Metallschutz* **19**, 146 (1943).
38. P. F. Thompson, *Trans. Electrochem. Soc.* **91**, 41 (1947).
39. F. Habashi, *Principles of Extractive Metallurgy*, Vol. 2, Gordon and Breach, New York (1970).
40. V. Kudryk and H. H. Kellogg, *J. Metals*, **6**, 541 (1954).
41. G. A. Deitz and J. Halpern, *J. Metals*, **5**, 1109 (1953).
42. A. E. Hultquist, The reaction kinetics of the corrosion of copper by cyanide solutions, M.S. Thesis, University of Utah, Salt Lake City, Utah (1957).
43. J. Halpern, *J. Electrochem. Soc.* **100**, 421 (1953).
44. J. Halpern, H. Milants and D. R. Wiles, *J. Electrochem. Soc.* **106**, 657 (1959).
45. R. Shimakage and S. Morioka, *Trans. Inst. Min. Met.* **80**, C228 (1971).
46. D. R. McKay and J. Halpern, *Trans. AIME* **212**, 301 (1958).
47. E. Peters and H. Majima, *Can. Met. Quart.* **7**, 111 (1968).
48. C. T. Mathews and R. G. Robins, *Aust. Chem. Eng.* pp. 21–25, Aug. (1972).

49. E. Peters, I. H. Warren, and H. Veltman, Extractive met-hydromet: theory and practice, Tutorial symposium, M. T. Hepworth, ed., Sect. V, University of Denver (1972).
50. M. B. Shirts, J. K. Winter, P. A. Bloom, and G. M. Potter, U.S. B. M., RI 7953 (1974).
51. J. B. Hiskey and M. E. Wadsworth, *Met. Trans. AIME* **613**, 183–190 (1975).
52. M. E. Wadsworth, *So. Afr. Min. Sci. Eng.*, **4**, 36 (1972).
53. M. E. Wadsworth, *Ann. Rev. Phys. Chem.* **23**, 355–384 (1972).
54. S. L. Pohlman and F. A. Olson, A kinetic study of acid leaching of chrysocolla using a weight loss technique, in *Solution Mining Symposium*, F. F. Aplan, W. A. McKinney, and A. D. Pernichele, eds., Soc. Min. Engr. AIME, Dallas, Texas, February 25–27, 1974, Soc. Min. Engr. AIME, New York (1974), pp. 46–60.
55. M. E. Wadsworth and D. R. Wadia, *Trans. AIME* **209**, 755 (1955).
56. T. L. Mackay and M. E. Wadsworth, *Trans. AIME* **212**, 597 (1958).
57. F. A. Forward and J. Halpern, *Trans. Can. Inst. Min. Met.* **56**, 344–50 (1953).
58. F. Habashi and G. A. Thurston, *Energ. Nucl. (Milan)* **14**, 238–244 (1967).
59. M. J. Nicol, C. R. S. Needes and N. P. Finkelstein, Electrochemical model for the leaching of uranium dioxide, Extract from *Leaching and Reduction in Hydrometallurgy*, The Institute of Mining and Metallurgy, The Chameleon Press Ltd., London (1975).
60. J. R. Glastonbury, in *Advances in Extractive Metallurgy*, Inst. Mining and Metallurgy, London (1968), pp. 908–917.
61. R. W. Bartlett, *Met. Trans.* **3**, 913 (1972).
62. R. L. Braun, A. E. Lewis, and M. E. Wadsworth, *Met. Trans.* **5**, 1717 (1974).
63. R. J. Roman, B. R. Benner, and G. W. Becker, *Trans. Soc. Min. Eng.* **256**, 247 (1974).
64. B. W. Madsen, M. E. Wadsworth, and R. D. Groves, *Trans. Soc. Min. Eng.* **258**, 69–74 (1974).
65. L. M. Cathles and J. Apps, A model of the dump leaching process that incorporates by physics and chemistry, Kennecott Copper Corporation, presented at AIChE Symposium on Modeling and Analysis of Dump and In-Situ Leaching Operations, Salt Lake City, Utah, August 19, 1974.
66. L. White, *Eng. Mining J.* 73–82 (1975).
67. K. Vetter, *Electrochemical Kinetics*, Academic Press, New York (1967).
68. J. Bockris and A. Reddy, *Modern Electrochemistry, Vols.* 1 *and* 2, Plenum, New York (1973).
69. M. Wadsworth, Extractive Metallurgy Lecture, *Trans. TMS/AIME* **245**, 1381 (1969).
70. G. Power, Cementation reactions, Ph.D. Thesis, University of Western Australia, Nedlands, Western Australia (1975).
71. T. Hurlen, *Acta Chem. Scand.* **14**, 1533 (1960); **15**, 630 (1961); and **16**, 1337 (1962).
72. J. Newman, *Electrochemical Systems*, Prentice-Hall, Englewood Cliffs, N.J. (1973).
73. V. Levich, *Physiocochemical Hydrodynamics*, Prentice-Hall, Englewood Cliffs, N.J. (1962).
74. A. Riddiford, *Ad. Electrochem. Electrochem. Eng.*, **4**, 47 (1966).
75. P. Strickland and F. Lawson, *Proc. Aust. Inst. Mining Met.* **237**, 71 (1971).
76. M. Eisenberg, C. Tobias, and C. Wilke, *J. Electrochem. Soc.* **101**, 306 (1954).
77. I. Cornet and R. Kappeser, *Trans. Inst. Chem. Eng.* **47**, 194 (1969).
78. D. Gabe and D. Robinson, *Electrochemica Acta*, **17**, 1121 (1972).
79. R. Diessler, *N.A.C.A.* Report 1210 (1955).
80. J. Miller and L. Beckstead, *Trans. TMS/AIME* **4**, 1967 (1973).
81. R. Glickman, H. Mouguin, and C. King, *J. Electrochem. Soc.* **100**, 580 (1953).
82. T. Ingraham and R. Kerby, *Trans. TMS/AIME* **245**, 17 (1969).
83. E. von Hahn and T. Ingraham, *Trans. TMS/AIME*, **239**, 1895 (1967).
84. W. Ranz and W. Marshall, *Chem. Eng. Prog.*, **48**, 141 (1952).

85. P. Harriott, *AIChE J.* **8**(1), 93 (1962).
86. R. Nadkarni, C. Jelden, K. Bowles, H. Flanders, and M. Wadsworth, *Trans. TMS/AIME*, **239**, 581 (1967).
87. R. Naybour, *J. Electrochem. Soc. Electrochem. Technol.* **116**, 520 (1969).
88. J. Calara, M.S. Thesis, University of Philippines, Manila, May (1970).
89. R. Miller, Ph.D. Thesis, University of Utah, Salt Lake City, Utah (1968).
90. E. von Hahn and T. Ingraham, *Trans. TMS/AIME*, **236**, 1098 (1966).
91. W. Fisher and R. Groves, *U.S.B.M. RI* 8098 (1975).

4

Pyrometallurgical Processes

Sec. 4.1. S. E. Khalafalla
Sec. 4.2. J. W. Evans and C.-H. Koo
Sec. 4.3. H. Y. Sohn and E. T. Turkdogan
Sec. 4.4. I. B. Cutler
Sec. 4.5. E. T. Turkdogan
Sec. 4.6. C. H. Pitt

4.1. Roasting as a Unit Process (SEK)

4.1.1. Introduction

In the journey from raw materials in ores to metals as finished products, roasting represents a significant milestone. Specifically, roasting covers operations between ore dressing or concentration on the one hand and actual metal recovery and finishing on the other. Traditional roasting involves chemical reactions with the furnace atmosphere at temperatures below the fusion point of the charge or product. Reactions in the molten state to produce metal, matte, or slag are metallurgically distinguished as smelting.

In the chemical sense, roasting can effect oxidation, reduction or magnetization (controlled reduction), sulfation, or chloridization. In

S. E. Khalafalla • Twin Cities Metallurgy Research Center, U.S. Bureau of Mines, Twin Cities, Minnesota. ***J. W. Evans and C.-H. Koo*** • Department of Materials Science and Mineral Engineering, University of California, Berkeley, California. ***H. Y. Sohn*** • Department of Metallurgy and Metallurgical Engineering, University of Utah, Salt Lake City, Utah. ***E. T. Turkdogan*** • Research Laboratory, U.S. Steel Corporation, Monroeville, Pennsylvania. ***I. B. Cutler*** • Department of Materials Science and Engineering, University of Utah, Salt Lake City, Utah. ***C. H. Pitt*** • Department of Metallurgy and Metallurgical Engineering, University of Utah, Salt Lake City, Utah

the processing sense, roasting may be autogenous, blast, flash, or volatilizing, whereas in the operational sense it may be suspension or fluidizing. Roasting involves chemical changes other than those of thermal decomposition (usually included under calcination). A roast may sometimes effect drying and calcination or thermal decomposition in passing. For example, the first stages in the roasting of covellite (CuS) is really a calcination step to Cu_2S with the elimination of sulfur which escapes as a gas to the surface of each particle and burns there to SO_2. The last stages in the oxidative roasting of covellite or chalcocite involve the thermal decomposition of $CuSO_4 \cdot CuO$ to cuprite. Hence, roasting depends on the diffusion of chemical species through the products to the reaction front in each particle. Thus, in the roasting of chalcocite (Cu_2S) or the second-stage roasting of covellite (CuS) oxygen must diffuse into the particle and SO_2 out of it. In order to assist diffusion a draft must be maintained to keep the partial pressure of the gaseous products low outside the particles.

Magnetizing and chloridizing roasts are usually reducing in character and require good control of furnace atmosphere with respect to CO/CO_2 and H_2/Cl_2 ratio, respectively. Sintering is in general an oxidizing process that is used to eliminate carbonate and sulfide as well as to produce agglomerated products.

4.1.1.1. *Types of Metallurgical Roastings*

Not only does roasting involve oxidation, but it can accomplish any of the following objectives[1]:

1. Oxidation roasting to burn out sulfur from sulfides and replace it in whole or in part with lattice oxygen.
2. Magnetization roasting accomplishes controlled reduction of hematite to magnetite, thus enabling concentration of the ore by magnetic separation.
3. Sulfation roasting to convert metal sulfides or oxides to sulfates, usually prior to leaching. This is especially valuable for manganese and uranium ores.
4. Chloridation roasting to convert certain metals to their volatile chlorides.
5. Carburization roasting to prepare calcine or refractory metals for chlorination as with titanium and zirconium ores.
6. Carbonate or soda ash roasting to extract refractory elements such as chromium from their ores by forming leachable soluble compounds as sodium chromate.
7. Segregation roasting or *chlorometallization* to isolate metallic particles from the ore matrix onto a reducing surface. This is par-

ticularly useful for beneficiating copper and copper nickel oxidized ores through volatilization of their chlorides by HCl, followed by deposition in a metallic state on a carbonaceous substrate. Hydrogen chloride gas is generated by the action of hot water vapor (700–900° C) on sodium chloride in presence of quartz, a common constituent of most minerals, according to

$$2\,NaCl + SiO_2 + H_2O \rightarrow Na_2SiO_3 + 2\,HCl \tag{4.1-1}$$

Ores of metals such as antimony, bismuth, cobalt, gold, lead, palladium, silver, and tin, which form volatile chlorides or oxychlorides,[2] may respond to the segregation roasting.

8. Volatilizing roast to eliminate other metals with volatile oxides such as As_2O_3, Sb_2O_3, or ZnO which can be condensed in the colder parts of the roaster. A volatilizing roast at carefully controlled temperature is sometimes used in the ore dressing of bismuth to remove arsenic and antimony as oxides.

9. Reducing roast to convert oxide to metal in a complex matrix prior to leaching or smelting.

10. Sintering or blast roasting to modify the physical characteristics of the ore or concentrate. This usually covers agglomeration and pelletization as in the iron ore industry.

4.1.2. Roasting Operations and Furnaces

4.1.2.1. Suspension Roasting

The early primitive roasting in "heaps" and "stalls" was first replaced by the use of shaft and simple air furnaces, and later by rotating kilns and more complicated machines like the wedge multiple-hearth roaster, shown in Figure 4.1-1. In this furnace both heat and mass are transferred countercurrently as the ore progresses down the shaft and across each of a series of hearths, being continuously turned over by rotating rakes which also advances the ore across the hearths. Despite the poor accessibility of furnace gases to the surface of the finely ground ore particles, a high proportion of the reactions takes place only as the ore falls from one hearth to the next one below.

4.1.2.2. Autogenous or Flash Roasting

A natural development of the multihearth roasting was suspension roasting in which the ore was dried and preheated on two top hearths and then allowed to fall through the central section of the furnace against the oxidizing furnace gases. The modern development of this is flash roasting[3] in which the preheated ore is injected through a "burner" with preheated air rather like pulverized fuel (see Figure 4.1-2).

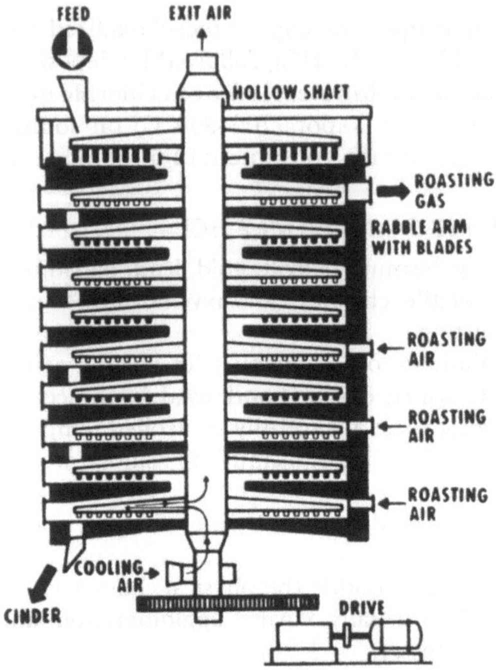

Figure 4.1-1. Multiple-hearth roasting furnace. Source: *Lurgi Manual*, Lurgi Gesellschaften, Frankfurt a. Main, June (1961), p. 192. (From Reference 1.)

This process is most appropriate for the roasting of sulfides which oxidize exothermally and require no additional fuel and hence the term *autogenous roasting*. In flash roasting the benefits of counterflow operation are partially lost.

4.1.2.3. Fluidized-Bed Roasting

The use of fluidized beds for roasting fine concentrate is obviously attractive[1] because of the very large mass- and heat-transfer coefficients between solid and gas. The gas flow rate must be large enough so that the pressure drop across the bed becomes independent of gas velocity. The bed under these conditions appears like a boiling liquid with bubbles bursting on its surface. This so-called *aggregative* fluidization is to be distinguished from *particulate* fluidization where the pressure drop across the bed decreases with increasing flow rate of the fluidizing gas. Particulate fluidization is approximately the condition found in ore-dressing equipment like hydraulic classifiers, jigs, etc., where the gas–solid contact is loose. By contrast, in aggregative fluidization the gas–solid contact is very complete. The gas flow is very turbulent so that the solids are thrown around in all directions with high energy but the relative velocity between gas and solid particle is not outside the laminar flow range. Both heat and mass transfer rates in fluidized beds are very high. Effective thermal conductivities of fluidized beds of about 100 times that of silver have been reported. Uniform temperatures in a bed are assured by convective transfer as the particles move about rapidly and randomly, and heat transfer between particles seems to be by conduction during collisions which break through the surface film of gas.

Figure 4.1-3 shows a typical fluidized-bed roaster. Either solid or gaseous fuels may be supplied in fluosolid reactors. The concentrate may be fed continuously and as a slurry, if necessary, displacing the column to discharge over a weir but some solids are always swept over with the gas and must be recovered with a cyclone. Because the reactions are so fast, both air requirements and heat losses are low and autogenous roasting is possible at a lower sulfur level than in other processes. The gas coming off from fluidized roasters is usually richer in SO_2 than that from conventional roasters.

The earliest applications of fluidized-bed techniques to extractive metallurgy were to sulfation roasting where close temperature control is necessary, particularly if differential sulfation is conducted. In differential sulfation, the temperature is controlled so that one metal present

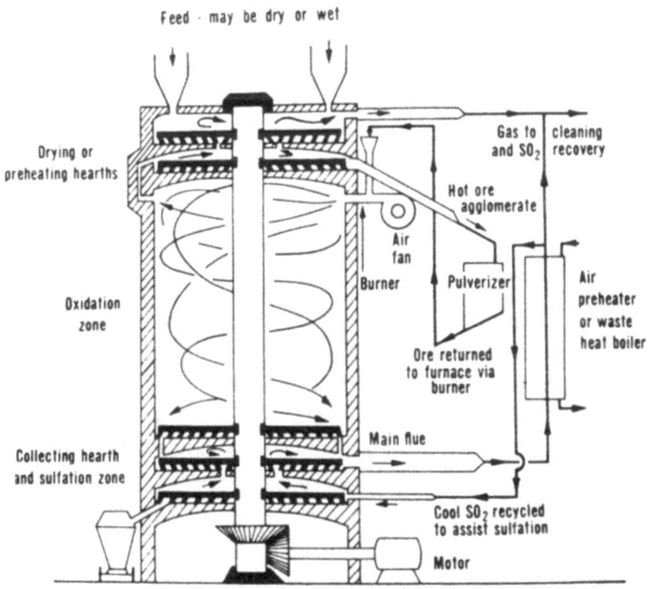

Figure 4.1-2. Schematic representation of a wedge-type multiple-hearth furnace modified for the flash roasting of sulfide concentrates. The upper hearths may be used for drying out slurry or for preheating concentrates. The lower hearths can be used for sulfating if required. (From Reference 3.)

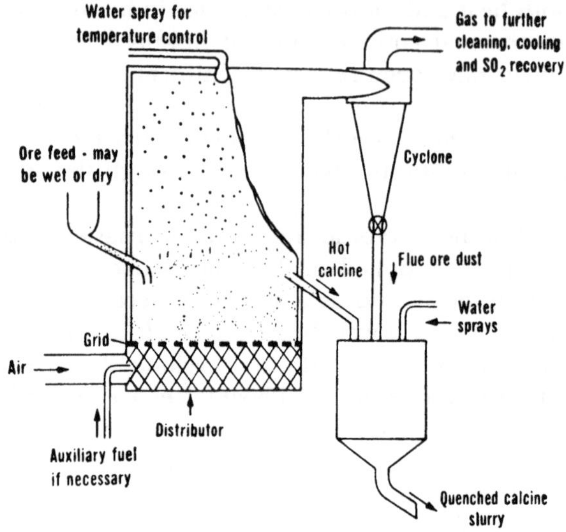

Figure 4.1-3. Schematic representation of a fluidized bed for roasting sulfide ore. Obviously a wide variety of feed and discharge arrangements are possible, appropriate to different operating conditions. (From Reference 3.)

forms the sulfate to the exclusion of all other metals which remain as insoluble oxides or silicates. Most sulfates, except those of lead, calcium, and barium are water soluble, but their decomposition temperatures to oxide and SO_3 (or a mixture of SO_2 and O_2) vary widely. Heavy-metal sulfates usually decompose at lower temperatures than those of alkaline or alkaline-earth sulfates.

Sulfate formation occurs under oxidizing conditions at high SO_2 pressure. In suspension roasting it follows oxidation in a special hearth below the main reaction chamber. In fluidized beds it apparently proceeds simultaneously with oxidation. Temperature must be controlled to be lower than the sulfate decomposition temperature of the metal to be extracted; e.g., about 770° C for $Al_2(SO_4)_3$, 850° C for $CuSO_4$ and $ZnSO_4$, 710° C for $Fe_2(SO_4)_3$, 900° C for $NiSO_4$, 980° C for $CoSO_4$, 1000° C for $MnSO_4$, and 300° C for $Cr_2(SO_4)_3$ when $P_{SO_2} = 1$ atm or rather lower if, as is usual, other gases are also present. In practice most sulfating is performed at about 600° C or at a lower temperature.

4.1.3. Thermodynamics of Roasting Reactions

A number of reactions take place during the roasting of sulfides. Using the roasting of NiS as an example, chemical changes can lead to either NiO or $NiSO_4$, depending on the roasting conditions. In addition, reactions involving formation of an intermediate oxysulfate (as in the cases of lead, copper, and zinc) can occur.[4]

Most roasting is carried to completion (so-called *dead* or *sweet* roasting) with elimination of all of the sulfur as SO_2 by the overall reaction

$$2\ NiS + 3\ O_2 = 2\ NiO + 2\ SO_2 \qquad (4.1\text{-}2)$$

Similar reactions can be written for zinc, copper, and iron. At lower temperatures, sulfates can be formed

$$2\ NiS + 4\ O_2 = 2\ NiSO_4, \qquad (4.1\text{-}3)$$

$$2\ NiO + 2\ SO_2 + O_2 = 2\ NiSO_4 \qquad (4.1\text{-}4)$$

The equilibrium constants of reactions (4.1-2) and (4.1-4) would be expressed in terms of the activities, a, of the solids and partial pressures, P, of the gases as

$$K_2 = \frac{a^2 NiO \cdot P_{SO_2}^2}{a^2 NiS \cdot P_{O_2}^3} \qquad (4.1\text{-}5)$$

$$K_4 = \frac{a\, NiSO_4^2}{a^2 NiO \cdot P_{SO_2}^2 \cdot P_{O_2}} \qquad (4.1\text{-}6)$$

Standard free-energy relationships for several of these equilibria are available in literature. The thermodynamics of sulfate systems can conveniently be presented in terms of predominance area diagrams ($\log P_{SO_2}$ versus $\log P_{O_2}$). For pure solids in their standard state at unit activity, equation (4.1-5) leads to

$$\log P_{SO_2} = \tfrac{1}{2} \log K_2 + \tfrac{3}{2} \log P_{O_2} \qquad (4.1\text{-}7)$$

Similarly, equation (4.1-6) leads to

$$\log P_{SO_2} = -\tfrac{1}{2} \log K_4 - \tfrac{1}{2} \log P_{O_2} \qquad (4.1\text{-}8)$$

A predominance area diagram for the nickel–oxygen–sulfur system is presented in Figure 4.1-4 for a temperature of 1000° K. The small square in Figure 4.1-4 encloses an area where the oxygen and sulfur dioxide pressures range from 0.01 to 0.1 atm. In operations at 1 atm total pressure, gas compositions from 3% (three times the lower limit of 0.01 atm) to 10% oxygen and 3–10% sulfur dioxide would yield nickel sulfate as the stable solid phase. At a gas phase composition of 1% oxygen and 1% sulfur dioxide, nickel oxide would be stable. At the univariant point A of Figure 4.1-4 ($P_{SO_2} = 2.5 \times 10^{-5}$ atm and $P_{O_2} = 5 \times 10^{-16}$ atm), the required conditions are not reducing enough to form metallic nickel in commercial roasting. The effect of temperature on sulfation equilibria can also be advantageously employed to produce a desired end product by shifting the location of the predominance areas.

Predominance area diagrams for copper and cobalt at 950° C are superimposed in Figure 4.1-5. From this diagram one can predict that sulfide concentrates of copper and cobalt can be sulfated in a fluidized bed operating at 950° K to produce water-soluble copper sulfate and cobalt sulfate with a roaster gas composed of 8% SO_2 and 4% O_2. The point representing this sulfation roasting of Cu–Co ores is shown to lie well inside an area where $CoSO_4$ and $CuSO_4$ are stable. If a separation of copper and cobalt is desired in the subsequent leaching step, a roasting operation in area A of Figure 4.1-5 would produce water-insoluble cupric oxide and soluble cobaltous sulfate.

In considering the thermodynamics of roasting, it should be noted that sulfur trioxide is formed in the gas phase according to the reaction

$$SO_2 + \tfrac{1}{2}O_2 = SO_3 \qquad (4.1\text{-}9)$$

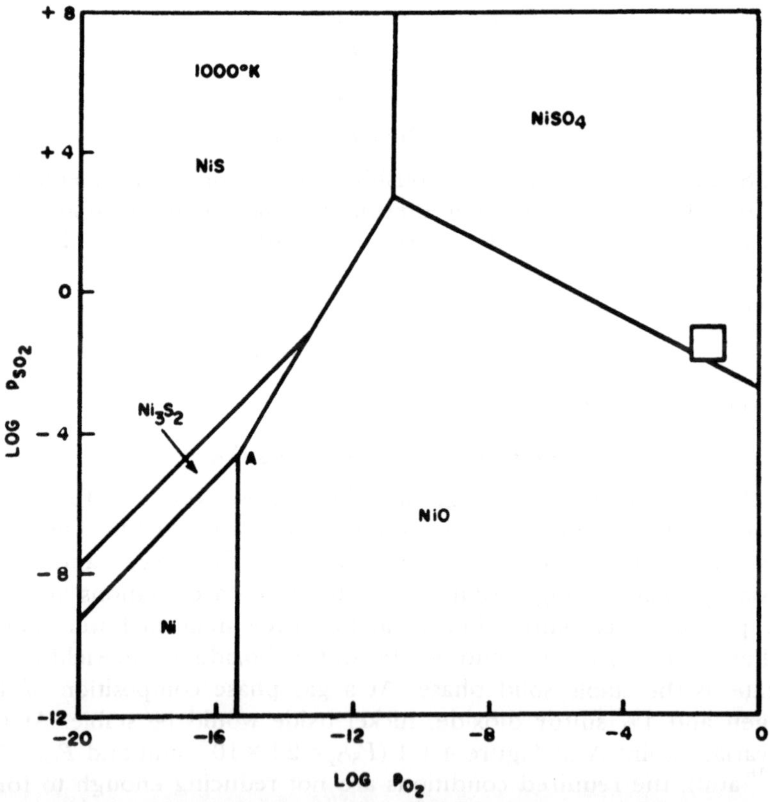

Figure 4.1-4. Predominance area diagram for the Ni–S–O System at 1000° K. Source: T. R. Ingraham, Sulphate stability and thermodynamic phase with particular reference to roasting, *Applications of Fundamental Thermodynamics to Metallurgical Processes*, Gordon and Breach, New York (1967), p. 187.

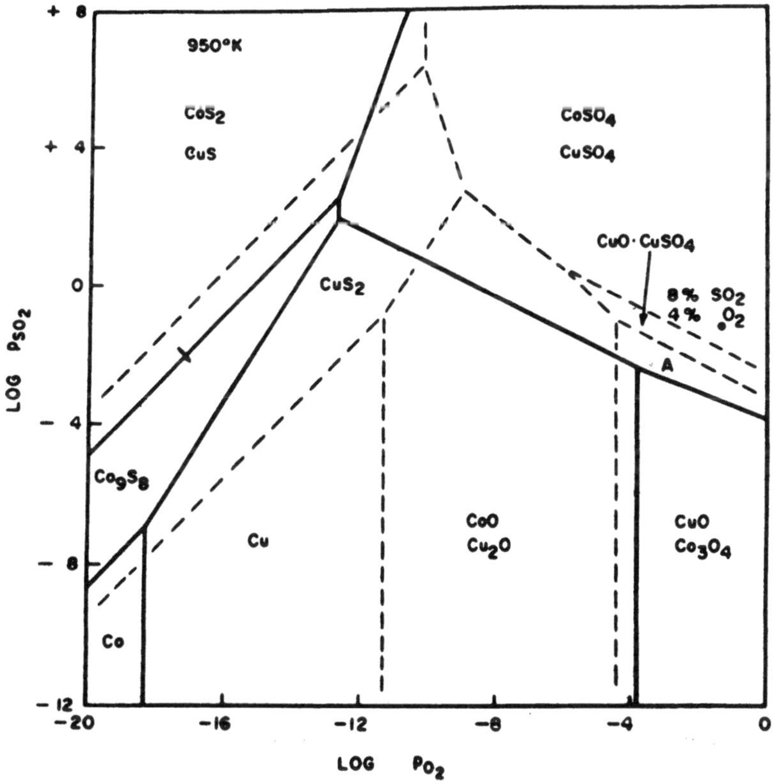

Figure 4.1-5. Superimposed predominance area diagrams for the Co–S–O and Cu–S–O Systems at 950° K. Source: T. R. Ingraham, Sulphate stability and thermodynamic phase with particular reference to roasting, *Applications of Fundamental Thermodynamics to Metallurgical Processes*, Gordon and Breach, New York (1967), p. 190.

Because it is exothermic, reaction (4.1-9) is favored at lower temperatures.

4.1.4. Gas–Solid Reactions in Roasting

4.1.4.1. Preliminary Energetics

Unlike the case of thermal decomposition or calcination, the overall chemical reaction in roasting is exothermic and hence not likely to be rate controlled by heat transfer or thermal diffusion.[4] In roasting the temperature is everywhere high enough to permit the desired chemical reactions to occur at a fairly early stage in the process, or at least that the reaction temperature isotherm is well ahead of the reaction front. Each particle is oxidized from the outside so that an inner unchanged core and an outer shell of solid product are in contact. At the interface between these zones the roasting reaction occurs, provided the ratio P_{O_2}/P_{SO_2} is locally higher than the equilibrium ratio for the reaction at the prevailing temperature. In the roasting of sphalerite for example,

$$2\,ZnS + 3\,O_2 \rightarrow 2\,ZnO + 2\,SO_2; \qquad K = P^2_{SO_2}/P^3_{O_2} \qquad (4.1\text{-}10)$$

At least three molecules of oxygen must come to the interface for every two of SO_2 that permeate away. Hence the gaseous flux, J, for each gas must satisfy[(3)] the condition

$$J_{O_2} \not< \tfrac{3}{2} J_{SO_2} \quad \text{or} \quad D_{O_2}\,\text{grad}\,P_{O_2} \not< D_{SO_2}\,\text{grad}\,P_{SO_2} \qquad (4.1\text{-}11)$$

The flux depends on the respective diffusion coefficient, D, and the partial pressure gradient down which the gas mass transfer takes place.

4.1.4.2. *Surface Reaction Steps*

At the interface, the reaction mechanism probably occurs in several stages. First, oxygen is adsorbed on to the surface of the sulfide. Secondly, it yields electrons and becomes incorporated in the lattice of the mineral. The electrons neutralize a sulfide ion on a nearby surface site as it unites with another oxygen molecule adsorbed beside it. The SO_2 molecule so formed desorbs and migrates away leaving a vacant sulfide ion site on the mineral surface. Another sulfur ion may diffuse from the interior to occupy this site and continue the reaction. When one surface layer is depleted, the interface advances into the mineral to reach more sulfur ions. The role of adsorption is probably to lower the chemical activation energy and thereby speed up the reaction.

An important side reaction in oxidative roasting is the formation of sulfate, most probably in a reaction zone in the outer and colder part of the oxide shell, because most sulfates decompose at a temperature below 900° C. By providing a "sink" for SO_2, this sulfate formation should permit a lower P_{O_2}/P_{SO_2} ratio to be operated without detriment to the main reaction rate. Sulfating of fine concentrate would be restricted to a separate part of the kiln where temperatures are uniformly lower and the SO_2 content of the atmosphere distinctly higher than in the oxidizing section.

A mass of fine concentrate must behave like a single large lump toward the reacting gases unless steps are taken to improve the gas–solid contact over that possible by diffusion alone, e.g., by raking or fluidization.

4.1.5. Kinetics of Roasting Reactions

Although thermodynamic free energy and enthalpy data are indispensable in predicting reaction feasibilities, calculating heat balances and evaluating maximum product yields, the exact reaction path and mechanism cannot be decided without kinetic implementation. Thermodynamic data, therefore, provide a necessary but not sufficient condition for predicting reaction feasibility. Hence, they guide but cannot decide the course and mechanism of metallurgical processes. Also, one cannot make money at equilibrium!

Metallurgical production rates obviously depend on chemical reaction rates. Significant progress has been made in the past few decades in using and applying the principles of thermodynamics, transport of mass, and heat and momentum, to quantitatively evaluate the overall behavior of metallurgical reactors. An understanding of the kinetics of metallurgical reactions is the only missing link to achieve dynamic simulation and computer control of the systems concerned. The time-honored and old-fashioned method of analyzing the metallurgical system in terms of elemental stoichiometry and overall mass and energy balances is now well established. Although this traditional approach will continue to be important, it constitutes only the first step in an effective system analysis. The proper and modern analysis of metallurgical systems should include the time element, and hence depends on application of the concepts of dynamic analysis (through its dynamic or kinetic parameters) to achieve feed-back and feed-forward compensation in reactor control.

Understanding of reaction mechanisms would also permit developing promoters and identifying the inhibitors of metallurgical processes. Promoters, by their very definition, speed up reactions by lowering their activation energy or barrier height. If the price of the promoter (which in general is used in small quantities) is less than that of the energy saved through suppressing the activation barrier, then it may be used to lower the overall energy requirement. In this era of energy shortages, nothing can more effectively minimize the energy requirement for metallurgical operations than a knowledge of their kinetic parameters, such as activation energies, half-life periods, rate coefficients, etc. Determination of reaction activation enthalpies will assist in estimating the minimum energy requirement for a given system in dynamic performance. Instead of dumping excess energy to conduct a given metallurgical operation, the minimum amount, as determined by the reaction activation energy, should be used in transient systems where equilibrium is seldom established.

4.1.5.1. *Reaction Rate Expressions*

Roasting is a gas–solid reaction of the general type

$$\text{solid(I)} + \text{gas(I)} \rightarrow \text{solid(II)} + \text{gas(II)} \qquad (4.1\text{-}12)$$

In oxidative roasting, gas(I) is oxygen and gas(II) is SO_2, whereas solid(I) and solid(II) are a metal sulfide and metal oxide, respectively. In sulfation roasting, gas(I) is SO_3, or a mixture of SO_2 and O_2, and gas(II) is absent, while solid(I) and solid(II) are metal oxide and metal sulfate, respectively.

The definition of a rate constant for a gas–solid reaction gives rise to some ambiguity. The main problem is that the concept of concentration has no significance, so that the rate constant cannot be defined in the same way as for a homogeneous reaction. Special procedures can be worked out in particular cases. If, for example, the product of a reaction is formed as a layer on the surface of the solid, one can consider the change with time of the thickness of that layer. In case of reactions involving solids, the symbol k, used for isothermal specific reaction-rate constants contains other physical properties of the solid, such as diffusion coefficients, densities, etc. The combination of these constants in the term k, which is to be termed *rate coefficient* for simplicity, will appear frequently below.

For the same reason, the activation energy of a gas–solid reaction does not have the significance that it has in a homogeneous reaction. It is only in rare cases that the apparent activation energy is a simple quantity; usually the temperature dependence of a diffusion coefficient or adsorption constant is involved, sometimes in a complicated way.

When solid(II) forms as a layer around the reactant solid(I), two possibilities may be distinguished:

1. The molar volume of solid(II) may be less than that of solid(I).
2. The molar volume of solid(II) may be greater than that of solid(I).

In Case 1, the product layer is probably porous so that the rate-determining step may be the chemical reaction occurring at the interface between the parent and daughter solids. Under these circumstances the rate is determined by the available surface area of the parent solid and such processes are referred to as *topochemical.* In Case 2 the product usually forms a protective layer around the solid(I), and the rate is probably diffusion controlled.

4.1.5.2. Kinetic Modeling

When the rate of flow of the gas is above the critical value for diffusion through the boundary layer to be rate controlling, different mechanisms of roasting can lead to different kinetic models. Spherical geometry of the reacting particles is assumed to remain as the roasting proceeds. The following cases have been reported in literature:

1. Contracting Sphere Model. If the reaction product is porous enough so that the process is controlled by chemical reaction at the interface, and the rate of roasting is proportional to the interfacial area, then

$$-\frac{1}{\rho}\frac{dw}{dt} = -\frac{dV}{dt} = kA \tag{4.1-13}$$

where ρ is the original density, w is the weight, V is the volume of a particle at time t, A its interfacial area, and k is the isothermal reaction rate constant in cm sec^{-1}.

For spherical particles, $A = 4\pi r^2$ and $V = \frac{4}{3}\pi r^3$, where r is the radius at time t; hence

$$-\frac{dV}{dt} = 4\pi k r^2 = (36\pi)^{1/3} k V^{2/3} \tag{4.1-14}$$

These topochemical reactions can, therefore, be regarded to be of *two-thirds order*. It can be seen from equation (4.1-14) that

$$-\frac{d}{dt}\left(\frac{4}{3}\pi r^3\right) = 4\pi k r^2 \tag{4.1-15}$$

so that

$$-\frac{dr}{dt} = k \tag{4.1-16}$$

The particle radius thus recedes at a constant rate.

In terms of the fraction reacted, $f = (r_3^3 - r^3)/r_0^3$, equation (4.1-15) can be integrated to give

$$1 - (1 - f)^{1/3} = kt \tag{4.1-17}$$

This equation was used by McKewan[5] in his study of the kinetics of iron oxide reduction. Adherence of roasting kinetics to equation (4.1-17) would suggest that the process is controlled by a phase-boundary chemical reaction.

2. Diffusion Models. The simplest treatment of a diffusion-controlled reaction between a solid and a gas can be made by assuming the rate of reaction to be inversely proportional to the thickness, x, of the product layer, thus

$$\frac{dx}{dt} = \frac{k}{x} \tag{4.1-18}$$

hence

$$x^2 = 2kt \tag{4.1-19}$$

This is known as the parabolic law, which is frequently observed in oxidation of metals. The constant k is proportional to the diffusion coefficient. Equation (4.1-19) is valid for thick protective layers on flat surfaces, assuming no product aging effect, no space charge layers, and that the phase boundary reaction is fast compared to the rate of ionic diffusion within the product layer.

For spherical geometry, equation (4.1-19) no longer applies. The assumption is still made that the reaction product is nonporous and consequently the process is controlled by diffusion through this product layer. In this case, two equations were found to be applicable.

Jander's equation

$$[1-(1-f)^{1/3}]^2 = \mathit{k}t \tag{4.1-20}$$

Crank, Ginstling, and Braunshtein's equation

$$1-\tfrac{2}{3}f-(1-f)^{2/3} = \mathit{k}t \tag{4.1-21}$$

Equation (4.1-20) applies for reactions controlled by ionic diffusion through a completely nonporous product layer, whereas equation (4.1-21) applies for control by gaseous diffusion through the very thin pores that may exist in an otherwise nonporous layer.

Valensi used Fick's first law of diffusion to express the kinetics of diffusion[6] through spherical and cylindrical shells and obtained

$$\frac{\varphi-[1+(\varphi-1)f]^{2/3}}{\varphi-1}-(1-f)^{2/3} = \frac{2P}{r_0^2}t = \mathit{k}t \tag{4.1-22}$$

and

$$\frac{1+(\varphi-1)f}{\varphi-1}\ln[1+(\varphi-1)f]-(1-f)\ln(1-f) = \frac{4P}{r_0^2}t = \mathit{k}t \tag{4.1-23}$$

where φ is the ratio of equivalent volumes of product to reactant. Equation (4.1-22) applies for spheres, whereas equation (4.1-23) applies for cylinders. Stewart *et al.*[6] took into consideration the geometrical factors as well as the direction of diffusion in their treatment of diffusion-controlled reactions. They tabulated the Valensi function for various values of φ and f, for cases of diffusion-in and diffusion-out, and for both spherical and cylindrical geometries.

3. Mixed Surface and Diffusion Control Models. Situations may arise in which the reaction product is nonporous and the rate of diffusion through this layer is of equal magnitude to the rate of chemical reaction at the interface. In this case, the rate equation has been shown to be

$$\frac{\mathit{k}}{2}\left[1-\frac{2}{3}f-(1-f)^{2/3}\right]+\frac{D}{r_0}[1-(1-f)^{1/3}] = \mathit{k}\frac{DP}{r_0^2\rho}t \tag{4.1-24}$$

where P is the partial pressure of the reacting gas, ρ is the density of the reacting solid, D is the diffusivity of the gas through the solid product layer, and k is the chemical reaction-rate coefficient at the solid–solid interface. This equation was independently derived by Lu,[7] Mori,[8] Seth and Ross,[9] St. Clair,[10] and Spitzer *et al.*[11]

4. Nucleation Growth Models. In some instances, the extent of conversion varies with time in a sigmoidal fashion. This was observed[12] in magnetic roasting of iron oxide at lower temperatures. The kinetic curve in these cases is characterized by an initial induction or incubation period, followed by an acceleratory stage in which the rate increases before falling off in the final stages. This behavior may be interpreted as arising from the production of nuclei at various places in the crystal, followed by the growth of these nuclei. An equation according to Prout and Tompkins is widely used to describe processes in which the rate of formation and growth of nuclei dominates the conversion of a solid. The Prout and Tompkins kinetic model assumes that the reaction starts at a series of active centers, the initial number of which is n_0. The rate of production of active sites, dn/dt, is given by

$$\frac{dn}{dt} = k_0 n_0 + k n - k_1 n \tag{4.1-25}$$

where k_0 is the nucleation constant, k is the probability of branching of the active centers, and k_1 is the probability of termination or disrupting the propagation of active sites (e.g., when the reaction front reaches an area that has already reacted). When the nucleation constant, k_0, is large, the original sites n_0 are soon exhausted, then

$$\frac{dn}{dt} = (k - k_1)n = k_2 n \tag{4.1-26}$$

Equation (4.1-26) is justified on the grounds that even if k_0 is not large enough, the branching process still predominates[13] and $(k - k_1)n \gg k_0 n_0$. At any instant, provided one considers strictly linear nuclei, the reaction rate, $\mathcal{R} = df/dt$, is proportional to the rate of production of active sites, thus

$$\frac{df}{dt} = k' \frac{dn}{dt} = k_3 n \tag{4.1-27}$$

Equations (4.1-26) and (4.1-27) are the basic foundation of the Prout–Tompkins theory. They cannot, however, be integrated until the functional dependence of f on k and k_1 is introduced. Prout and Tompkins considered the case of a symmetrical sigmoid for which the reaction is half completed at the inflection point, i.e., $f_i = 1/2$. Furthermore, at $t = 0$, $f = 0$, and k_1 must be zero because interference is not possible at this stage, whereas at f_i, dn/dt changes sign. Thus at $f = 1/2$, $k = k_1$, whence it is at least consistent to write $k_1 = k(f/f_i)$. Now k depends on crystal geometry and is normally constant.[13] Thus from equation (4.1-26)

$$\frac{dn}{dt} = k\left(1 - \frac{f}{f_i}\right)n = \frac{k}{k_3}\left(1 - \frac{f}{f_i}\right)\frac{df}{dt} \tag{4.1-28}$$

Therefore,

$$\frac{dn}{df} = \frac{k}{k_3}\left(1 - \frac{f}{f_i}\right) \tag{4.1-29}$$

Integrating equation (4.1-29) yields

$$n = \frac{k}{k_3}\left(f - \frac{f^2}{2f_i}\right) \tag{4.1-30}$$

Substituting from equation (4.1-27) and setting $f_i = 1/2$, one obtains

$$\mathcal{R} = \frac{df}{dt} = kf(1 - f) \tag{4.1-31}$$

which is the differential form of the Prout–Tompkins law.

5. Pore-Blocking Models. These models were found useful in the sulfation roasting of many minerals. The model suggests a rather unaffected silicate lattice in which the oxide of the metal to be sulfated is dissolved. When the mineral is subjected to a sulfating atmosphere, the metal is pulled out to the surface and generally the counterions, O^{2-}, must follow the metal ions to maintain electroneutrality. Probably, the rate-controlling step is the movement of these large oxygen ions. In some minerals, electroneutrality can be sustained by oxidation of Fe^{2+} to Fe^{3+} instead of movement of oxygen, in which case the rate-controlling step may be the chemical reaction on the surface. The transport of oxygen is supposed to take place through special routes called *pores*. The first step is from the initial position in the lattice to the nearest pore. The next is the diffusion out through the pore, and lastly the reaction with the gas on the surface

$$O^{2-} + SO_3 \rightarrow SO_4^{2-} \tag{4.1-32}$$

The metal ions can move easier in the lattice since they are usually much smaller than the oxygen ions and hence their transport cannot be rate determining. It is further supposed that pores can be blocked, thereby losing their ability to act as paths for oxide ion movement. The fewer open pores are left, the slower is the sulfation reaction. The differential pore blocking, dN is assumed[14] to be dependent on the differential degree of reaction df, the grain size r, and the number of open pores, N, hence

$$-dN - \alpha r N\, df$$

or

$$\frac{N}{N_0} = e^{-\alpha r f} \tag{4.1-33}$$

where N_0 is the initial number of pores, and α is a constant dependent on temperature and pressure. The rate, u, at which matter arrives at the pore in mole $\min^{-1}$, is assumed[14] to depend on f and N according to

$$u = u_0 \left(\frac{N}{N_0}\right)^{\beta} (1-f) \tag{4.1-34}$$

where u_0 is the value of u at the start of the reaction when $N = N_0$ and $f = 0$. The factor $(1-f)$ in equation (4.1-33) shows that the material transport is treated as a first-order process with respect to the solid component. The exponent β represents the order of the transport process with respect to the pore number. When every pore produces sulfates at a rate that is independent on how close the open pores are situated to each other, then $\beta = 0$.

The total production rate of sulfates is proportional to u, r, and N, hence

$$\frac{df}{dt} = urN = u_0 r N_0 (1-f) \exp\left[-(1+\beta)\alpha r f\right] = k_1 (1-f) e^{-k_2 f} \tag{4.1-35}$$

where k_1 is a constant that combines u_0, r, and N_0, and k_2 is another constant given by $(1+\beta)\alpha r$, and called the *pore blocking factor*. Equation (4.1-35) can be rearranged to

$$k_1\, dt = \frac{e^{k_2 f}}{1-f}\, df \tag{4.1-36}$$

and integrated to give

$$k_1 t = \int_0^f \frac{e^{k_2 f}}{1-f}\, df = -e^{k_2} \int_0^f \frac{e^{-k_2(1-f)}}{k_2(1-f)}\, d[k_2(1-f)]$$
$$= e^{k_2}\{E_1[k_2(1-f)] - E_1(k_2)\} \tag{4.1-37}$$

where the function $E_1(Z)$ is the exponential integral defined as $\int_Z^{\infty} (e^{-s}/s)\, ds$. The value of this integral depends on Z only, and is tabulated in mathematical handbooks.[15] If it had been supposed that the sulfation rate is independent of f, then

$$\frac{df}{dt} = k_1 N = k_1 N_0 e^{-k_2 f} \tag{4.1-38}$$

This would lead to the logarithmic law that Evans[16] used to explain the oxidation of some metals, according to which the fractional oxidation, f, of a metal is given by

$$f = k \log (k' t + 1) \tag{4.1-39}$$

This law was also used to model the sulfation of monazite ores. The foregoing models are chosen from the vast literature on the kinetics of roasting reactions. They by no means exhaust the kinetics of gas–solid reactions, for which many other models are known to apply. They merely serve to illustrate the mechanism of certain roasting processes by developing an applicable kinetic law.

4.1.6. Selected Cases in Metallurgical Roasting

The following sections are devoted to a discussion of some practical roasting operations:

1. sulfation roasting with special emphasis on the roasting of manganese and aluminum;
2. oxidative roasting, especially of copper sulfides; and
3. magnetic roasting of iron ores.

Some of these topics will be dealt with in depth, other important ones will be treated only briefly to keep the discussion within reasonable bounds.

4.1.6.1. Sulfation Roasting

This covers processes in which a mineral at elevated temperature reacts with SO_2 and O_2 or SO_3 to produce sulfates of one or more of the metals in the mineral. In some cases, the sulfating conditions (temperature, SO_3 pressure, etc.) are adjusted so that one metal is selectively sulfated to the exclusion of the other metals in the mineral, thereby achieving a beneficiation in addition to the extraction step.

1. Manganese Ores. The U.S. Bureau of Mines has conducted research up to pilot-plant experiments on sulfation roasting of manganese ore from the Cuyuna Range of Minnesota. In these experiments, Prasky *et al.*[17] roasted the carbonate slate that contains $FeCO_3$ at 700° C with 10% SO_2 in air. The $MnCO_3$ and $MgCO_3$ in the slate are transformed into sulfates but the iron is not attacked by SO_2 on account of the high temperature. The remaining Fe_2O_3 and silicates are not attacked by the sulfating gas. Experiments were conducted in a vertical-shaft furnace with the manganese ore moving countercurrently with the sulfating gas. The roasting solubilized 95% of the manganese in water, but only a few percent of the iron. Joyce and Prasky[18] were also able to combine sulfation with reduction roasting to recover manganese and iron, respectively, from manganiferous iron ore. First, the manganese mineral (7.5% manganese) was converted to water-soluble manganese sulfate by roasting in sulfur oxide gases at 700° C, then the hematitic iron minerals (31% iron) in the hot sulfatized charge were reduced to magnetite with simulated reformed natural gas in the temperature range of 350 to 425° C. Water leaching of the roasted product recovered over 90% of the manganese. An iron oxide concentrate (60% iron) with an iron recovery of 75% was obtained from the water-insoluble residue by magnetic separation.

Van Hecke and Bartlett[19] have recently studied the kinetics of sulfation of Atlantic Ocean manganese nodules using a thermogravimetric analysis (TGA) method. The concentrations of SO_2 were low, and typical of power-plant stack gases. Above 400° C, the rate of sulfation was found proportional to the SO_2 pressure and to the unreacted solid fraction. Thus the kinetic law was best represented by

$$\ln(1-f) = kt \tag{4.1-40}$$

The rate was independent of nodule particle size and the apparent activation energy was found to be low (1.6 kcal $mole^{-1}$). Leaching the completely sulfated nodules in boiling water provided Ni, Cu, and Co extractions above 80%. Much of the manganese but little of the iron was also extracted. At 300° C, the SO_2 sorption capacity was lower with most of the sulfation attributed to the manganese oxides.

2. *Copper in Pyrite Ash.* Leftover ashes after roasting of pyrite usually contain copper. Fletcher and Shelef[20] used sulfation roasting in a fluidized bed to extract this copper. In the temperature range of 680 to 700° C, copper was selectively sulfated but not iron. The reaction is rather slow and the reaction rate can be increased by addition of alkali sulfates. This additive also rendered Ni, Zn, and Co soluble after sulfation roasting of the pyrite ash.

3. *Nickel in Pentlandite.* Nickel can be extracted by sulfation roasting. The first step in the process is an oxidation roasting to form NiO and SO_2. In the next step, NiO reacts with the SO_2 from the first step and oxygen to form $NiSO_4$. The sulfation of pure NiO is very slow because the product $NiSO_4$ is impermeable to SO_2 since it forms a protective layer[21] around the unreacted NiO. The $NiSO_4$ can be made permeable[22] to the gas by the addition of Na_2SO_4. Fletcher and Shelef

assumed that alkali sulfates promote the sintering and agglomeration of $NiSO_4$. Thornhill[22] explained the beneficial effect of sodium sulfate by its ability to destroy the nickel ferrite formed from the interaction of NiO and Fe_2O_3 according to

$$NiO + Fe_2O_3 \rightarrow NiFe_2O_4 \quad (4.1\text{-}41)$$

$$Na_2SO_4 + NiFe_2O_4 \rightarrow Na_2Fe_2O_4 + NiSO_4 \quad (4.1\text{-}42)$$

A method for extracting copper and nickel from the Duluth Gabbro Complex by selective high-temperature sulfation has been reported recently by Joyce.[23] His tests demonstrated that over 95% of the copper and 80% of the nickel can be converted to water-soluble sulfates by roasting with 40% SO_2 in air at temperatures between 520 and 670° C.

4. Sulfation of Cobalt Oxide. Alcock and Hocking[24] found that $CoSO_4$ can be formed from CoO in a sulfating atmosphere with an intermediate layer of Co_3O_4 between CoO and $CoSO_4$. The parabolic law applies. If the CoO reactant was marked with gold, then after sulfation the gold markers were located between the Co_3O_4 and $CoSO_4$ layers. This showed that oxygen diffuses inward through the Co_3O_4 to produce additional Co_3O_4 as it meets the CoO interface.

5. Sulfation of Sphalerite. Zinc sulfide is usually roasted to ZnO at and above 800° C. At 700° C, the oxidation rate is too slow. At lower temperatures[25] ZnS reacts directly with SO_3 to produce $ZnSO_4$ according to

$$ZnS + 4\,SO_3 \rightarrow ZnSO_4 + 4\,SO_2 \quad (4.1\text{-}43)$$

or

$$ZnS + O_2 + 2\,SO_3 \rightarrow ZnSO_4 + 2\,SO_2 \quad (4.1\text{-}44)$$

No sulfation occurred when ZnS is treated with SO_2 and air under conditions where SO_3 is not formed.

6. Sulfation of Galena. Lead sulfide reacts with oxygen between 690 and 800° C to form lead sulfate[26] according to

$$PbS + 2\,O_2 \rightarrow PbSO_4 \quad (4.1\text{-}45)$$

The reaction follows the kinetic law, equation (4.1-21),

$$1 + 2(1 - \varphi f) - 3(1 - \varphi f)^{2/3} = \frac{k}{4r_0^2}t \quad (4.1\text{-}46)$$

where φ is the volume of $PbSO_4$ per unit volume of PbS, and r_0 is the radius of galena particles at the start of the reaction. Gold-marking experiments show that PbS diffuses from the inside out through the $PbSO_4$ layer to react on the surface.

7. Sulfation of Uranium. Swedish uranium shale was sulfated[14] at temperatures and SO_2 pressures that fulfill the following equation:

$$\frac{4228}{6.44-\log P_{SO_3}} < T < \frac{6540}{7.60-\log P_{SO_3}} \tag{4.1-47}$$

where T is the absolute temperature and P_{SO_3} is the pressure of SO_3 in mm Hg. The first term in equation (4.1-47) is the temperature at which $Al_2(SO_4)_3$ is decomposed. The operating temperature has to be higher than this temperature to prevent the production of aluminum sulfate. The last term represents the decomposition temperature of uranyl sulfate.

The addition of Na_2SO_4 or NaCl leads to the formation of the double sulfate of sodium and uranium. This double sulfate is more stable than uranyl sulfate so that a higher temperature can be chosen.

$$T < \frac{6670}{6.63-\log P_{SO_3}} \tag{4.1-48}$$

For practical SO_3 pressures, the temperature without sodium salt lies between 625 and 725° C. In the presence of sodium salt, the temperature can be raised up to 800° C. Uranium yields from the sulfation of this lean Swedish ore were about 60%. Because of the low uranium yield, sulfation roasting is regarded inferior to the conventional leaching with dilute sulfuric acid.

Asmund[14] sulfated uranium in the mineral steenstrupine from South Greenland. The sulfation temperature was 700° C and the gas mixture was 5% SO_2 in air containing 0.61 vol % water vapor. Metal and oxygen ions in the solid reactant are transported from the original position in the crystal lattice to the surface where they react with the gas. The transport rate for constant sulfating gas composition is proportional to the number of open micropores and the amount of unreacted solid component. During the reaction the pores are blocked thereby losing their ability to act as transportation paths for the oxide ions.

8. Beryllium in Silicates. Sulfation of beryllium in montmorillonite clay and in the silicate helvite has been patented by Kruse and Pitman.[27] Montmorillonite with 0.2% Be is roasted at 600° C in a fluidized bed with SO_3. After 30 min of sulfation, 85% of the beryllium was solubilized with small amounts of aluminum and magnesium. Helvite is roasted at 670° C in a fluidized bed for 30 min and yields 75% of the beryllium and very little aluminum and magnesium.

9. Aluminum in Clays. Aluminum sulfate can be produced by sulfation of gibbsite, boehmite, diaspore, kaolinite, or halloysite.[28] Sodium sulfate has to be added in more than catalytic quantities to achieve good yields of sulfation. The molar ratio of Na_2SO_4 to Al_2O_3 should be about 4 to get a yield of 80–90% from 1 hr of roasting at 500–600° C. The exact mechanism by which sodium sulfate promotes the sulfation of aluminum is somewhat controversial. Fletcher and Shelef[21] attributed the beneficial effect of Na_2SO_4 to improved permeability of $Al_2(SO_4)_3$ by creating discontinuities in its structure. Recent research[29] favors the concept that sodium sulfate acts as an SO_3 carrier via a sodium pyrosulfate intermediate according to

$$Na_2SO_4(s) + SO_3 \rightleftharpoons Na_2S_2O_7(liq) \qquad (4.1\text{-}49)$$

followed by the attack of alumina by the molten pyrosulfate phase, thus

$$3\ Na_2S_2O_7(liq) + Al_2O_3(s) \rightarrow Al_2(SO_4)_3(s) + 3\ Na_2SO_4(s) \qquad (4.1\text{-}50)$$

The effect of additives other than sodium sulfate was recently[29] investigated. Alkali and alkaline earth compounds were admixed with alumina in clay at a molar ratio of 1. The results in Table 4.1-1 indicate that the aluminum extraction was highest with a sulfate addition having the lowest melting point. Furthermore, only those metals capable of forming pyrosulfates exerted an accelerating effect on alumina sulfation.

This was substantiated by the high aluminum extractions with sodium and potassium additives which are known to form low-melting pyrosulfates (melting at about 400 and 300° C, respectively). Cations such as Cs, Mg, Ca, and Ba do not form pyrosulfates and hence cannot promote the sulfation of alumina. The aluminum extractions obtained with the sodium carbonate or sulfate additives were essentially the same because the former can be readily sulfated during the experimental run.

Table 4.1-1. Experimental Results Obtained with Different Additives at a Molar Ratio of 1

Additive		Al extraction with water, %
Type	Melting point, °C	
Na_2SO_4	884	24
Na_2SO_4	884	23
K_2SO_4	1076	6
Cs_2SO_4	1010	3
$BaSO_4$	1580	3
$MgSO_4$	1124[a]	2
$CaSO_4$	1450	1
Na_2CO_3	851	24
NaCl	801	6

[a] Sulfate decomposes at this temperature.

A considerably lower Al extraction was obtained with sodium chloride additive because of its difficulty to be sulfated under the present conditions. The requirement of a molten pyrosulfate phase to sulfate-alumina-bearing minerals can be understood on examining the surface acidity of the aluminum sites. Lozos and Hoffman[30] demonstrated that Al^{3+} of alumina is a poorer electron acceptor (Lewis acid) than that of silica–alumina. Since pure alumina was difficult to sulfate by a gas–solid reaction, a silica–alumina mineral would be even more difficult to sulfate and hence the need for Na_2SO_4.

Investigations were also conducted[29] on potential aluminum resource minerals other than clay. Briquets of these minerals with and without 30% w/w Na_2SO_4 added were calcined at 800° C for 6 hr and then sulfated. It is evident from column 8, Table 4.1-2, that the weight gain was small when the mineral was sulfated in absence of Na_2SO_4. However, a substantial weight gain was observed (column 9) when 30% w/w sodium sulfate was added. The highest weight gain occurred with the phyllosilicates which contain extended sheets of SiO_4 tetrahedras linked together by metallic cations. By contrast, tectosilicates (feldspar) with their framework structure were difficult to sulfate, even in presence of sodium sulfate promoter. The sulfation products of the minerals containing sodium sulfate additive were leached for 6 hr with hot water. The largest quantity of aluminum was extracted from halloysite and kaolinite both of which belong to the phyllosilicate group (column 10). Leaching with 0.2*N* H_2SO_4 increased the aluminum extraction (column 11). Halloysite and kaolinite appeared to be the most promising minerals for aluminum extraction by sulfation.

Despite their closed structure and the absence of alkali ions, orthosilicates (third block in Table 4.1-2) were moderately sulfated with better aluminum extractions than realized with tectosilicates. This is probably due to the lower Si/Al ratio in kyanite when compared to albite or orthoclase. This ratio is 0.5 in the former and 3 in the latter minerals. Increasing the Si/Al ratio in a mineral increases the surface acidity of the Al sites,[30] rendering them less amenable to sulfation. When this Lewis acidity is combined with the closed structure of tectosilicates, it becomes apparent why these minerals were poorly sulfated, even in presence of a sodium sulfate promoter (second block in Table 4.1-2).

Weathering of silicate rocks is known to give clays and sand. This process involves thawing and freezing of water in the rocks and the chemical action of water and carbon dioxide. For example, the weathering of feldspars to give kaolin is represented by

$$\underset{\text{feldspar}}{2\,KAlSi_3O_8} + 2\,H_2O + CO_2 \rightarrow K_2CO_3 + 4\,SiO_2 + \underset{\text{kaolin}}{Al_2Si_2O_5} \qquad (4.1\text{-}51)$$

Table 4.1-2. Experimental Data Obtained from Potential Aluminum Resources

Name	Formula	Partial analyses, %					Weight gain, %		Al extraction, %	
		Al_2O_3	SiO_2	Na_2O	K_2O	CaO	Mineral, *M*	M+30% Na_2SO_4	H_2O	0.2*N* H_2SO_4
Phyllosilicates (sheet structure)										
Halloysite	$Al_4Si_4O_{10}(OH)_8$	33.2	27.9	<0.1	<0.1	0.5	3.4	28.5	26	—
Kaolinite	$Al_4Si_4O_{10}(OH)_8$	36.9	45.5	0.2	0.3	<0.1	<0.5	31.5	23	43
Montmorillonite	$Al_4Si_8O_{20}(OH)_4 \cdot n\ H_2O$	37.0	48.0	—	—	—	1.8	34.9	11	16
Muscovite	$KAl_3Si_3O_{10}$	44.3	46.9	0.9	9.1	1.1	1.0	—	—	—
Tectosilicates (framework structure)										
Albite	$NaAlSi_3O_8$	15.6	65.7	10.0	0.3	1.8	<0.5	—	—	—
Orthoclase	$KAlSi_3O_8$	17.1	66.7	4.9	4.2	2.2	<0.5	9.9	2	8
Anorthite	$CaAl_2Si_2O_8$	18.1	51.6	<0.1	0.3	6.9	2.6	—	—	—
Anorthosite[a]	$[Na, Ca, Al, Si]_5O_8$	27.0	57.1	3.8	0.7	0.3	—	12.6	6	6
Orthosilicates (isolated SiO_4 tetrahedra)										
Kyanite	Al_2SiO_5	25.7	66.3	<0.1	0.2	0.8	—	15.4	13	33
Nonsilicates										
Dawsonite	$NaAl(OH)_2CO_3$	6.3	26.8	1.7	2.3	16.3	43.5	—	12[b]	—
Alunite	$K(AlO)_3(SO_4)_2 \cdot 3\ H_2O$	35.0	2.1	0.2	2.1	0.7	12.6	—	8[b]	15[b]

[a] A rock containing 90% or more anorthite.
[b] Data were obtained on minerals without Na_2SO_4 additive.

The soluble potassium carbonate formed is largely removed by water and the residue of sand and clay remains as soil. In that respect, a phyllosilicate may be regarded as weathered or partially weathered tectosilicate. Weathering, therefore, involves not only opening the silicate structure, but also a lowering of the Si/Al ratio.

In addition, two aluminum minerals, essentially free of lattice silicon, were subjected to dry sulfation. The results with dawsonite (a mineral in spent-oil shale) and alunite are given in the fourth block of Table 4.1-2. Reasonable degrees of sulfation were realized with these minerals without the need of Na_2SO_4 additions because of the absence of silicon and the presence of alkali metals in these minerals.

4.1.6.2. Oxidative Roasting

This category is designed to transform completely the sulfidic to an oxidic concentrate.

1. Oxidation of Sphalerite. The kinetics of oxidation of ZnS was investigated by Natesan and Philbrook using single-sphere samples[31] as well as fluidized-bed reactors.[32] In the temperature range from 740 to 1040° C, the kinetics of oxidation of ZnS pellets ranging from 0.4 to 1.6 cm in diameter was controlled by gaseous transport through the product layer of ZnO. For suspensions of individual ZnS particles in a fluidized-bed reactor in an oxygen–nitrogen mixture at 740–1000° C, the kinetics was controlled by a surface reaction at the ZnS–ZnO interface. The dissociation of ZnS into zinc and sulfur vapor complicated measurements of the kinetics above 1000° C in a fluidized bed.

4. Oxidation of Molybdenite. The oxidation of molybdenite, MoS_2, to MoO_3 involves the formation of MoO_2 as an intermediate. Coudurier *et al.*[33] investigated the oxidation of molybdenite in a multiple hearth furnace and found that the reaction rate was limited by gaseous diffusion on the upper hearth and surface reaction on the lower hearth. On the lower hearth where surface reaction controls, the rate varies with partial pressure of oxygen to the first power. Kinetic equations were developed for the upper and lower hearths and it was demonstrated that the optimum operating conditions could be determined by matching the heat balance with the observed kinetics. Coudurier *et al.* also observed that volatilization of rhenium, which was in solid solution in the molybdenite lattice, occurred concurrently with the formation of MoO_3. The kinetics of molybdenite oxidation was controlled by gaseous diffusion in the early stages and surface reaction in the latter stages of reaction.

Ammann and Loose[34] investigated the oxidation kinetics of molybdenite at 525–635° C for various oxygen pressures. The kinetics was monitored by measuring the rate of evolution of sulfur oxides from a thin layer of molybdenite powder arranged so that the temperature would not increase appreciably during oxidation. By using high gas velocities the geometric factors associated with the reaction interface were well correlated to the contracting sphere model,

$$1-(1-f)^{1/3}=\frac{kP}{2\rho r_0}t \tag{4.1-52}$$

where P is the pressure of oxygen, ρ is the density of the particle, r_0 is the particle radius, and k is an agglomeration of terms containing the reaction rate constant, stoichiometry factors, and the conversion units necessary to convert mole fractions to partial pressures. The experimental activation energy was determined to be 35.4 kcal $mole^{-1}$, and the rate was independent of the partial pressure of SO_2, indicating that back reaction kinetics was negligible.

3. Oxidation of Galena. The adsorption and incorporation of oxygen by lead sulfide was studied by Hillenbrand.[35] High-temperature experiments showed that some sulfur oxides are retained in the oxidized layer as, for example, in the formation of lanarkite, $PbO \cdot PbSO_4$. Hillenbrand demonstrated that the major part of oxygen was chemisorbed without incorporation into the lattice, and in some cases extensive surface coverages by oxygen were achieved with practically no evolution of sulfur dioxide.

4. Oxidation of Chalcocite and Covellite. Razouk *et al.*[36] showed that the oxidation of chalcocite leads to CuO as final product with the intermediate Cu_2O layer not built up to other than interfacial boundary dimensions of constant thickness. Their evidence was the fact that the overall weight remained constant during the course of oxidation, which proceeds according to:

$$Cu_2S+\tfrac{3}{2}O_2 \rightarrow Cu_2O+SO_2 \tag{4.1-53}$$

$$Cu_2O+\tfrac{1}{2}O_2 \rightarrow 2\,CuO \tag{4.1-54}$$

Ashcroft[37] claimed that oxide formation during the roasting of chalcocite resulted only from secondary decomposition of sulfate and oxysulfate that were formed as primary products. Wadsworth *et al.*[38] reported the formation of "puff-balls" involving trapped SO_2 and O_2 beneath a viscous film. They concluded that the presence of these entrapped gases would stabilize the sulfate formed.

A detailed study of the oxidation of covellite, CuS, was conducted by Shah and Khalafalla[39] to determine the sequence of reactions and intermediate phases during its roasting. The reactions were studied under a variety of conditions by means of DTA (differential thermal analysis), TGA, and X-ray diffraction of the reaction products. It was demonstrated, contrary to results reported by previous investigators, that tenorite, CuO, and dolerophanite, $CuO{\cdot}CuSO_4$, were formed prior to the formation of the copper sulfate phase and that they were physically located below the $CuSO_4$ layer. It was also noted that CuO formed at 450–460° C, which is well below the decomposition temperature of $CuO{\cdot}CuSO_4$, indicating that it must form in a sequence other than that requiring the decomposition of the basic sulfate. The proposed sequence is

$$CuS \rightarrow Cu_{1.8}S \rightarrow Cu_2O \rightarrow CuO \rightarrow CuO{\cdot}CuSO_4 \rightarrow CuSO_4 \rightarrow CuO{\cdot}CuSO_4 \rightarrow CuO \quad (4.1\text{-}55)$$

where $Cu_{1.8}S$ is a defect form of copper sulfide known as diginite. On the basis of these findings, Khalafalla and Shah[40] were able to develop a procedure for oxidative roasting of covellite with minimal retardation from the dolerophanite film by prior heating of the sulfide in nitrogen to about 650° C followed by oxidation. This energy-saving procedure for covellite roasting, designated PER (programmed environmental roasting) can form copper oxide at lower temperatures than normally required in aerobic roasting. Traditional oxidative roasting of chalcocite requires temperatures ranging from 800 to 850° C for conversion to tenorite. Unlike the situation with conventional roasting, $CuSO_4$ was not detected in the X-ray diffractogram of the product formed above 625° C with the PER technique. Furthermore, the quantity of $CuO{\cdot}CuSO_4$ decreased significantly as the halt temperature increased from 600 to 700° C with the new technique. Evidently the shell of oxysulfate is impervious to oxygen and/or sulfur dioxide and delays the formation of tenorite until the sulfate and oxysulfate are decomposed. If the oxysulfate stage is bypassed with an inert atmosphere, then, even if small quantities of this salt are formed upon introducing the oxidant, it would decompose at an appreciable rate and the impedance of its thin film to gaseous transport would be considerably diminished. The sudden shift of the roasting atmosphere from anaerobic to aerobic at 625–650° C results in easy transformation of sulfide to oxide unimpeded by transport resistances across the oxysulfate film. The initial liberation of elemental sulfur instead of SO_2 from covellite during the anaerobic stage, as well as the sudden evolution of the available SO_2 in large concentrations during the constant temperature aerobic stage, constitute added assets of the PER technique.

Metallurgical investigations of the sulfidic copper ores showed that pyritic concentrates were consistently more desulfurizable than chalcocitic concentrates when roasted under similar conditions.[41] The promoting effect of iron on the roasting reactions was, therefore, investigated by Khalafalla and Shah[40,42,43] in some detail. Small quantities of metallic iron powder or pyrite were found to accelerate the oxidative roasting of copper sulfides. The accelerating effect of externally added iron was confined to the low-temperature reaction steps, and hence it promoted the extent of sulfate formation. Iron did not, however, lower the thermal requirement for complete oxidation of CuS or Cu_2S because it has virtually no effect on the thermal decomposition of $CuO \cdot CuSO_4$.

Having established the sequence of reactions in the roasting of copper sulfide, Shah and Khalafalla[42,43] then proceeded to study the kinetics of the first and last steps in equation (4.1-55). The conversion of covellite to digenite follows linear kinetics, i.e., $f = kt$, with an apparent activation energy of 23 ± 3 kcal mole^{-1}. In a nitrogen atmosphere, the conversion proceeds at a slower rate but with the same activation energy. Externally added iron powder (0.6%) increased the rate of conversion, but had no effect on the apparent activation energy. Arrhenius diagrams depicting these observations are shown in Figure 4.1-6. These results are at variance with those reported by Pawlyuchenco and Samal[44] who fitted their data on thermal decomposition of CuS in vacuum in a nitrogen atmosphere at 310–338° C to the contracting sphere model

$$1-(1-f)^{1/3} = kt \tag{4.1-17}$$

Although their data suggest an oxygen attack at the covellite–digenite interface, the present TGA data on covellite pellets in the temperature range from 260 to 400° C suggest a rate-controlling reaction at the digenite–gas interface. Because the outermost surface area of the pellet remains essentially constant during oxidation, linear kinetics were repeatedly confirmed by Shah and Khalafalla[42] in their TGA data. This finding has important ramifications on the conversion mechanism, which is postulated by Shah and Khalafalla to be controlled by the dissociation $5\,Cu_{1.8}S \rightarrow 9\,Cu^+ + 9\,e^- + 5\,S$, at the digenite–gas interface. Reduction of CuS by Cu(I) ions and electrons presumably occurs by the rapid reaction $4\,Cu^+ + 4\,e + 5\,CuS \rightarrow 5\,Cu_{1.8}S$ at the CuS–$Cu_{1.8}S$ interface, which steadily contracts with time.

In the TGA method of kinetic analysis, the sample instantaneous weight, w_t, was recorded as a function of time and temperature. The active weight, w, was calculated from the instantaneous weight by normalizing for unreacted sample, i.e., by subtracting a multiple of the weight loss from the initial sample weight in accordance with the reaction stoichiometry. A Gerber derivimeter was used to obtain the slope of the (w-t) curve at selected points and hence to determine the reaction rate $R = -dw/dt$. Kinetic parameters were calculated from the TGA

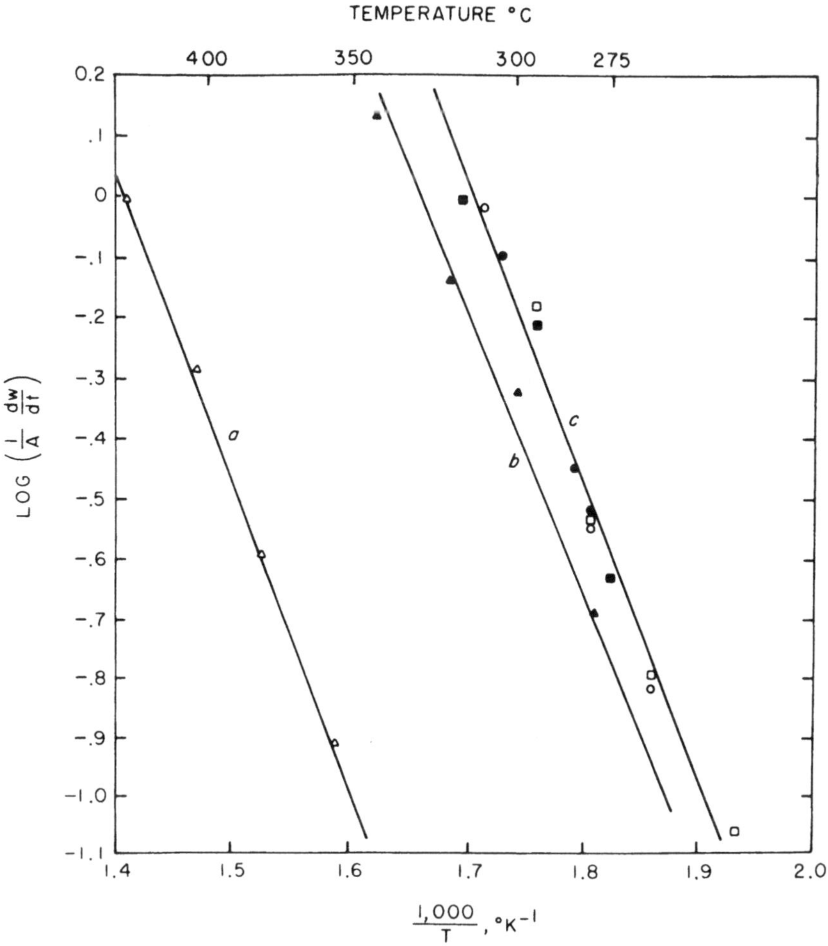

Figure 4.1-6. Arrhenius diagram for the conversion of covellite to digenite. (a) Nitrogen; (b) 20% O_2; (c) 20% O_2, pellet containing 0.6% iron powder.

thermogram by using the Chatterjee[45] method which requires at least two TGA determinations with different sample weights under similar reaction conditions. The reaction order, γ, can then be determined from the equation

$$\gamma = \frac{\log R_i - \log R_j}{\log w_i - \log w_j} \tag{4.1-56}$$

where the subscripts i and j refer to two points at which the active weights and the instantaneous rates are known. Usually a series of values of γ are determined at successive temperatures to see whether or not the value of γ remains constant during a particular range. Better accuracy for the reaction order can be achieved through increased redundancy. Thus, when four independent TGA curves are recorded at different sample weights, they would, if taken two at a time, yield six values of γ at a given temperature. The apparent activation energy, Q, was determined by plotting $(\gamma \log w - \log R)$ as a function of the reciprocal absolute temperature. The slope of the resulting Arrhenius line will give a value of $Q/2.303R$, where R is the molar gas constant.

The last two steps in equation (4.1-55) followed linear kinetics[43] with apparent activation energies of 55 ± 2 and 66 ± 3 kcal mole^{-1} for the thermal decomposition of $CuSO_4$ and $CuO{\cdot}CuSO_4$, respectively. The presence of oxygen decreased the rate of decomposition, whereas small quantities of iron did not produce significant changes in the kinetics. Neither oxygen nor iron had any appreciable effect on the experimental activation energy of the decomposition reaction. Ingraham and Marier[46] reported an activation energy of 57 ± 7 kcal for the isothermal decomposition of $CuSO_4$ and 67 ± 8 kcal for the decomposition of $CuO{\cdot}CuSO_4$ in dynamic nitrogen atmosphere.

Rao *et al.*[47] investigated the kinetics of oxidation of Cu_2S in oxygen at 750–950° C. They followed the kinetics by titrating SO_2 in standard iodine solutions. The initial nonisothermal rate was followed by an essentially isothermal rate. The experimental activation energy at the initial nonisothermal regime was 25 kcal mole^{-1}, and the kinetics was controlled by heat and mass transport. The apparent activation energy during the isothermal portion of the reaction was ~6 kcal mole^{-1} with mass transport being responsible for the observed kinetics. A detailed theoretical calculation involving models for both mass and heat transport showed an excellent correlation for the model based upon mass transport. Mass transport included diffusion through an outer-boundary layer, through the layer formed during the nonisothermal period, and through the layer of product formed during the continuing reaction.

5. Spontaneous Combustion of Sulfides. Kinetics of oxidation of metal sulfides has other important mining safety ramifications. It has been suggested that fires such as those that led to the disasters in 1972 and 1975 at the Sunshine Silver Mine may be the result of oxidation leading to spontaneous combustion of the sulfides in mill tailings. In the coal mining industry, burning of the coal rejects is often attributed to the oxidation of pyrite in the coal fines. It is apparent that the ultrafine sulfide particles, usually rejected in ore processing, are of colloidal dimensions and exhibit extreme reactivity in the presence of oxygen. Sulfide concentrates of many metals (copper, iron, lead, nickel, and zinc) have a tendency to undergo autogenous heating, which reportedly[48] occurs during the transport of concentrations, especially at sea. Combustion during storage prior to transport has also been reported.

In the usual pathway of metal sulfide oxidation, the mechanism involves many intermediate steps, some of which are exothermic, others endothermic. This tends to distribute the heat of reaction over a long reaction coordinate. By contrast, submicron-sized sulfide particles probably have no mechanism by which to dissipate the excessive enthalpies of reactions, and explosions may result. The standard enthalpies of combustion of pyrite (FeS_2), pyrrhotite (FeS), argentite (Ag_2S), and covellite (CuS) are −228, −196, −71, and −98 kcal per mole of sulfide converted to oxide. On exposure to oxygen, and in the presence of water vapor, a slow oxidation of sulfide to sulfate can occur in cases of zinc and lead according to

$$ZnS + 2\,O_2 + 7\,H_2O \rightarrow ZnSO_4 \cdot 7\,H_2O, \qquad \Delta H^\circ_{298} = -208\text{ kcal} \qquad (4.1\text{-}57)$$

$$PbS + 2\,O_2 \rightarrow PbSO_4, \qquad \Delta H^\circ_{298} = -197\text{ kcal} \qquad (4.1\text{-}58)$$

Especially with pyrite, sphalerite, and galena, the heat of reaction is comparable with that for the burning of methane, namely

$$CH_4 + 2\,O_2 \rightarrow CO_2 + 2\,H_2O, \qquad \Delta H^\circ_{298} = -213\text{ kcal} \qquad (4.1\text{-}59)$$

Identification of promoters and inhibitors for these sulfide oxidations can only be made through a detailed study of their kinetics and mechanisms, topics that are unfortunately still in their infancy. More work is clearly desirable in these areas.

4.1.6.3. Magnetic Roasting

Complete reduction of hematite to metallic iron is a complex process with several heterogeneous reactions occurring simultaneously. The reduction of hematite to magnetite is the first step in the overall reduction process which removes only one-ninth of the total oxygen, but results in the conversion of a nonmagnetic to a strongly magnetic ore that is amenable to metallurgical concentration.

Despite the extensive investigations reported on the kinetics of the overall reduction of iron oxides to the metallic state, very little work was done on the kinetics of the very first step, namely conversion of hematite to magnetite. The magnetic roast reaction can provide an important step in the beneficiation of low-grade iron ores. Nonmagnetic iron materials can be converted to a magnetic form using either reduction roasting to form magnetite (Fe_3O_4), or reduction followed by an oxidation roasting[49] to form maghemite (γ-Fe_2O_3).

The magnetic roast reaction in 2 CO–CO_2 atmosphere was studied by Hansen *et al.*[50] They found the reaction rate to be proportional to the hematite–magnetite interfacial area. At temperatures below 500° C, a definite nucleation effect was noted, amounting to as much as 120 min in some extreme cases. They also observed that as the temperature was increased, the reduction rate passed through a minimum at 675° C. Below 450° C, the apparent activation energy was approximately 13.6 kcal/mole of hematite converted to magnetite.

Nigro and Tiemann[51] studied the magnetic roasting reaction in CO–CO_2 mixtures over the temperature range of 500 to 900° C and found that the reaction rate increased when silica (20–80 mole %) was added to hematite. The increase in rate was attributed to silica preventing the sintering of hematite.

The taconites of Minnesota, Michigan, and Wisconsin represent the largest potential source of iron in the United States. Present mining activities are focused on the magnetic taconites because of their relative ease of beneficiation. As these ores become depleted, it will become mandatory to beneficiate nonmagnetic taconites with flotation and/or magnetic roasting techniques. Because of the large thermal requirements the latter techniques can be costly. Khalafalla *et al.*[12] studied the effect of various additives in an attempt to accelerate the magnetic roasting reaction to lower the overall energy requirement of the process. The reducing gas stream consisted of 5% H_2, 3.3% water vapor, and the balance nitrogen.

The low-temperature magnetic roasting of pure hematite exhibits a sigmoidal kinetic curve, characteristic of solid–solid transformations. The curve begins with an induction period whose duration increases with decreasing temperature. A typical kinetic curve for the reduction roasting of hematite at 550° C is shown in Figure 4.1-7, Curve *a*. The effect of temperature on the magnetic roasting reaction of pure hematite is illustrated by the curves in Figure 4.1-8. The induction period increases with decrease in temperature from about 1.4 min at 690° C to 9.1 min at 578° C and to 31.5 min at 510° C.

Although the addition of small quantities of alkali metal oxides and alkaline earth oxides to hematite accelerates its reduction to iron,[52] these additives did not accelerate the magnetic roasting reaction. For example, neither Na_2O, K_2O, nor CaO additives shortened the induction time (Figure 4.1-7).

Oxidized pellets containing 0.51 wt % CaO gave a kinetic curve at 550° C with an induction period of 160 min during which virtually no reduction took place, and after which the reduction reaction was slowly initiated. This anomalous behavior is shown in Curve *b* of Figure 4.1-7. When the pellet was reduced with pure hydrogen flowing at

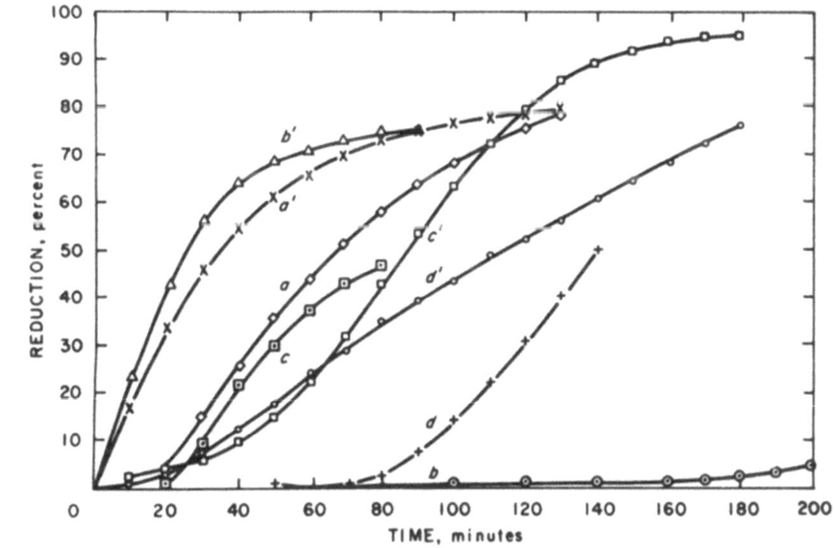

Figure 4.1-7. Kinetic curves of the magnetic roasting reaction at 550° C. Key: *a*, *a'*, pure hematite; *b*, *b'*, calcium-impregnated pellet; *c*, *c'*, sodium-impregnated pellet; *d*, *d'*, potassium-impregnated pellet (primed symbols indicate hydrogen pretreatment).

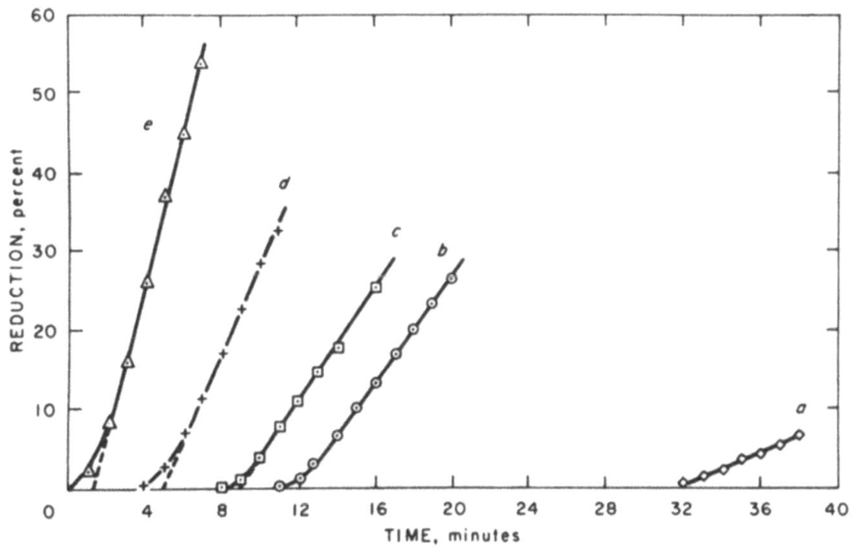

Figure 4.1-8. Effect of temperature on the magnetic roasting reaction of pure hematite pellets. Reduction temperature, °C: *a*, 510; *b*, 550; *c*, 578; *d*, 605; *e*, 690.

300 cm^3/min for 200 sec (1 liter of the gas) prior to introducing the magnetic roasting atmosphere, a vigorous reaction occurred with elimination of the induction period (Figure 4.1-7, Curve *b'*). At the same temperature, only a short induction time was observed with pure hematite, and this was also eliminated by hydrogen pretreatment as illustrated by Curves *a* and *a'* of Figure 4.1-7.

Curves *c* and *c'* were obtained when the additive in the hematite pellet was 0.28% Na_2O. Curves *d* and *d'* correspond to 0.43% K_2O in the pellet. It is concluded that the oxides of sodium and potassium, and particularly the oxide of calcium, prolong the induction time in the magnetic roasting curve. Hydrogen prereduction of the pellet, even when followed by reoxidation, appears to activate the surface for the reaction as evidenced by the significant shortening of the induction period.

Figure 4.1-9 shows the effect of low concentrations of silica and alumina on the magnetic roast reaction at 550° C. The induction time was decreased by about 22% when 8.4×10^{-4} g-atom Si as SiO_2 was mixed per gram of the hematite sample. With the same atom concentration of Al as Al_2O_3 and at the same temperature, the induction time was increased by 26%.

Below 550° C, pellets containing Al_2O_3 had longer induction times than the blanks, whereas samples containing SiO_2 had shorter induction

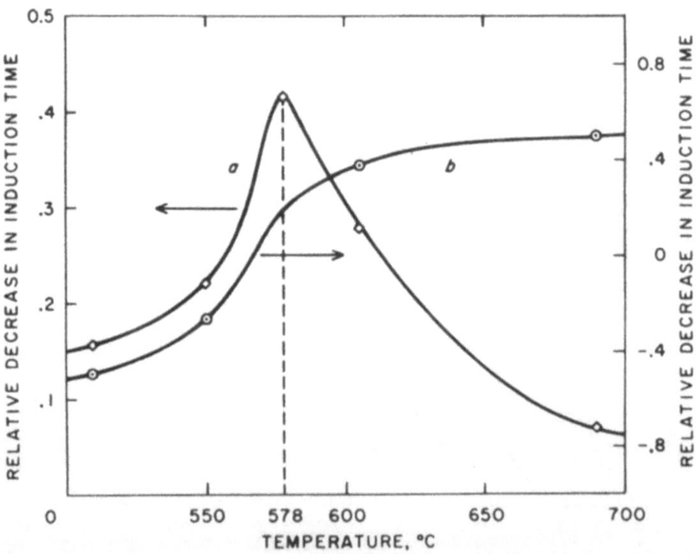

Figure 4.1-9. Effect of Al_2O_3 and SiO_2 on the induction period for the reduction of Fe_2O_3 to Fe_3O_4 at various temperatures. Hematite impurity level per gram mixture: *a*, 8.4×10^{-4} g-atom Si as SiO_2; *b*, 8.4×10^{-4} g-atom Al as Al_2O_3.

Table 4.1-3. Effect of Silica and Alumina on the Induction Time of the Magnetic Roasting Reaction

	Induction time, τ, min			Relative decrease in induction time	
Temp., °C	Fe_2O_3 (blank)	$Fe_2O_3+SiO_2$	$Fe_2O_3+Al_2O_3$	$\Delta(SiO_2)$	$\Delta(Al_2O_3)$
510	31.5	26.5	46.8	+0.2	−0.5
550	12.6	9.8	15.9	+0.2	−0.3
578	9.1	5.3		+0.4	
605	5.0	3.6	3.1	+0.3	+0.4
690	1.4	1.3	0.7	+0.1	+0.5

times. At temperatures above 600° C, both additives reduced the induction time. The average values of the induction times observed with untreated, silica-treated, and alumina-treated hematite pellets at various temperatures are given in Table 4.1-3.

To compare the relative effect of each additive, we define the degree of synergy,† Δ, as the relative decrease in induction time, thus,

$$\Delta = \frac{\tau_b - \tau_m}{\tau_b} \tag{4.1-60}$$

where τ_m represents the induction time for the mixed sample, and τ_b is the induction time for the blank. Values of Δ thus calculated are listed in Table 4.1-3 for both the silica- and the alumina-treated pellets. The variation of Δ with temperature is shown in Figure 4.1-9. For pellets containing SiO_2, a maximum in Δ was observed at 578° C (Curve *a*), whereas pellets containing Al_2O_3 showed a steady increase in Δ with temperature.

The shortening by quartz of the incubation time for transforming hematite to magnetite can be explained by a *secondary Hedvall effect*. The known (or primary) Hedvall effect[53] predicts enhanced reactivity when the reacting material undergoes a simultaneous phase transition. A secondary Hedvall effect is involved here whereby the dilatometrics and energetics associated with a phase transition may activate a contacting interface for chemical action. Silica exists in many crystallographic

† Synergy is the term used when the output from the combined effect of two inputs exceeds the summation of outputs from the separate inputs. An induction time, τ, of about 10 min at 570° C for reduction of hematite corresponds to a velocity constant, k, of 0.1 min^{-1}. For quartz, no reduction occurs, and hence $\tau = \infty$ and $k = 0$. For the two together one observes $\tau \sim 5$ min and $k \sim 0.2$ min^{-1}. If there was no synergy, one would expect a combined k of only 0.1 min^{-1} and τ of 10 min.

forms. Aside from the high-temperature tridymite and cristobalite, two modifications of quartz are known with a transition point of 575° C. Alpha- or low quartz (trigonal) is thermodynamically stable below this temperature, and β- or high quartz (hexagonal) is stable above it. The extensive research of Hedvall and his co-workers at Gothenburg University, Sweden, showed that increased solid-state chemical reactivity usually accompanies a crystallographic transformation.[53] Bond collapse and reconstruction at the transition point expose atoms in their nascent state. Solid surfaces thus assume a more or less defect structure which produces a more reactive state. Several investigators have reported an acceleration in reactivity at or near phase-transition temperatures.

Solid-state reactions are known to be strongly influenced by small quantities of selected impurities. The effect of structural factors on solid reactivity is also well established. For example, a newly formed solid phase on thermal dissociation usually possesses a particularly high reactivity,[53] e.g., CoO formed by the reaction $Co_3O_4 \rightarrow CoO + O_2$. Hedvall[53] referred to this as "reactivity in nascent state" and attributed it to defect or incompleteness of structures in the solid substance formed at the reaction. He also described a number of convincing experiments to prove the increased solid reactivity associated with crystallographic transformation.

The Hedvall effect has many applications in chemical metallurgy. Fishwick[54] has patented a process which describes the removal of lattice-bound iron from spodumene at the $\alpha \leftrightarrows \beta$ spodumene transition temperatures (1000–1150° C). At these temperatures, chlorine or hydrogen chloride directly attacks the lattice-bound iron and removes it as volatile iron chloride. Daellenbach *et al.*[55] found that maximum grindability for nonmagnetic taconite ores occurred near the quartz $\alpha \rightleftharpoons \beta$ phase transition temperature of 575° C. It has also been shown that the reaction of NiO with SiO_2 is intiated at the $\alpha \rightleftharpoons \beta$ phase transition temperature of quartz and that the reactivity of the system was greatly enhanced at the temperature of cristobalite formation.[53]

Direct observation of iron oxide reduction using high-voltage electron microscopy was recently reported by Tighe and Swann.[56] They showed that when hematite is heated in a reducing atmosphere, magnetite nucleates epitaxially at the surface in contact with the gas. The morphology and rate of growth of the lower oxides vary according to specimen orientation, temperature, and partial pressure of oxygen used for reduction. Magnetite, however, was found not to nucleate preferentially along or near grain boundaries or dislocations. Voluminous ions as those of alkali or alkaline-earth metals disturb the oxide lattice[52] and create more crystal imperfection. This follows from Volkenstein's theory of solid-state chemistry which proposes that lattice disturbances can be transmitted to the surface of a solid. Although these disturbances may activate the surface to chemisorption of a reducing gas and hence promote reduction with this gas, they would however, hamper the epitaxial nucleation and growth of magnetite on hematite. Hence, the antagonistic effect of Li_2O, Na_2O, and CaO reported in this work may be attributed to a strong retardation of the formation of elementary nuclei needed to initiate the surface reaction.

Primary and Secondary Hedvall Effects. The preceding data shows that the induction time in the low temperature magnetic roasting of hematite can be significantly shortened by the occurrence of phase transitions in certain additives. A secondary Hedvall effect is proposed to account for the observed results. The general Hedvall effect (primary or secondary) can be explained by an increase in information at the reaction phase boundary resulting from a phase transition. It is suggested that the increased reactivity associated with phase transitions be attributed to catalysts of the second kind (see below) to distinguish them from ordinary catalysts of the first kind (see below) which function by lowering the activation energy barrier. By contrast, catalysts of the second kind function by increasing the activation entropy and hence the information contained in the surface.

Below 700° C, the conversion of hematite to magnetite follows a sigmoidal kinetic curve with a finite incubation (induction) time which decreases with rise in temperature and increases in the presence of Li_2O, Na_2O, or CaO. It also disappears by flash reduction with hydrogen for a few seconds indicating that an incubation period is required to prepare and reorganize the hematite surface for an epitaxial transformation to magnetite. The presence of small quantities of quartz also shortens the induction period. A maximum in the relative decrease in induction time is observed at 578° C, the $\alpha \rightleftharpoons \beta$ transition point of quartz.

Synergistic effects on the magnetic roasting of hematite are also observed with Al_2O_3 at 690° C, but no maximum in the degree of synergy is observed up to 700° C because the induction period disappeared before the $\gamma \rightleftharpoons \alpha$ transition point[57] of Al_2O_3 (~1000° C). The effects observed with Al_2O_3 are even larger than those observed with silica. This is undoubtedly related to the larger heat of transition[57] of Al_2O_3 ($\Delta H_{tr} \approx 20.6$ kcal/mole) as compared with the heat of transition of quartz ($\Delta H_{tr} \approx 0.29$ kcal/mole). Presumably the more energetic the bond breakage is, the larger is the associated synergy because of the Hedvall effect.

The occurrence of an induction time in the low-temperature conversion of hematite to magnetite is ostensibly related to the period required to prepare an activated surface, according to

$$\text{Hematite} \rightarrow (\text{Hematite})^{\dagger} \rightarrow \text{Magnetite}$$

or

$$A \overset{k_1}{\rightarrow} B \overset{k_2}{\rightarrow} C \quad (4.1\text{-}61)$$

The reciprocal of the induction time can accordingly be symbolized by a rate process whose velocity increases with temperature rise. At elevated temperatures (above 700° C) the rate of surface preparation is large enough that no induction time can be observed experimentally. Surface activation can be achieved either by flash reduction with hydrogen or by mixing the reducible charge with a material capable of undergoing a phase transition.

The activation energies for the duration of the induction periods reported in Table 4.1-3 can be determined from the Arrhenius diagrams shown in Figure 4.1-10. For pure hematite and hematite-containing alumina, the apparent activation energies were 26.0 and 33.8 kcal/mole, respectively. For pellets containing silica, a break in the

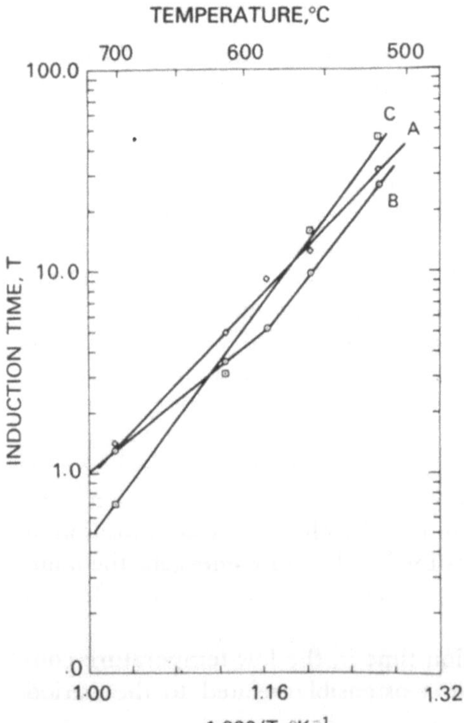

Figure 4.1-10. Arrhenius diagram for the induction time of the magnetic roast reaction. Key: *a*, blank Fe_2O_3, $Q = 26.0$ kcal/mole; *b*, $Fe_2O_3 + SiO_2$, $Q = 30.3$ and 20.2 kcal/mole; *c*, $Fe_2O_3 + Al_2O_3$, $Q = 33.5$ kcal/mole.

Arrhenius plot occurred at 578° C. Below 578° C the apparent activation energy was 30.3 kcal/mole, and above that temperature a value of 20.2 kcal/mole was obtained.

Kinetic Interpretation of the Variability in Activation Energy. In the consecutive reaction (4.1-61) where the active nuclei B are formed as the intermediate, classical reaction kinetics[58] shows that the time required to reach a maximum in the quantity of the active intermediate is given by

$$t_{max} = \frac{\ln (k_2/k_1)}{k_2 - k_1} \tag{4.1-62}$$

where k_1 is the specific rate constant for the "surface activating" reaction, and k_2 is the specific rate constant for the reaction of the active sites to form the final product "magnetite." Both reactions are assumed to be first order, or at least pseudo first order.

Because reorganization kinetics is usually much slower than reduction kinetics, k_1 would be much smaller than k_2; hence k_1 can be neglected in comparison to k_2 in the denominator of equation (4.1-62). Thus

$$t_{max} \approx \frac{1}{k_2} \ln\left(\frac{k_2}{k_1}\right) \tag{4.1-63}$$

Applying the Arrhenius equation for the two specific rate constants, we have

$$k_1 = A_1 e^{-Q_1/RT} \quad \text{and} \quad k_2 = A_2 e^{-Q_2/RT} \tag{4.1-64}$$

where A and Q are the frequency factor and the activation energy, respectively, and subscripts 1 and 2 refer to the pre- and postreactions for the activated surface. The activating or surface reorganization reaction is detectable only at low temperatures and becomes very fast at higher temperatures. This is reasonable since the induction time is observed only at lower temperatures and disappears at elevated temperatures. Hence the assumption can be made that $Q_1 > Q_2$. Let the ratio $A_2/A_1 = a$, and the difference $Q_1 - Q_2 = \Delta Q$. The induction time would be equal to t_{max}, or a certain fraction thereof, and hence will be given by

$$\tau \approx \frac{\ln (a e^{\Delta Q/RT})}{A_2 e^{-Q_2/RT}} \tag{4.1-65}$$

hence

$$\tau = \frac{1}{A_2}\left(\ln a + \frac{\Delta Q}{RT}\right) e^{Q_2/RT}$$

$$= \frac{\ln a}{A_2}\left(1 + \frac{\Delta Q}{RT \ln a}\right) e^{Q_2/RT} \tag{4.1-66}$$

Taking the logarithms in equation (4.1-66) and remembering that

$$\ln(1+x) = x - \frac{x^2}{2} + \frac{x^3}{3} + \cdots \approx x \tag{4.1-67}$$

when x is sufficiently small, i.e., when $\Delta Q < RT \ln a$, the above equation becomes more valid. Using the approximation of equation (4.1-67) in equation (4.1-66) one obtains

$$\begin{aligned} \ln \tau &\approx \ln\left(\frac{\ln a}{A_2}\right) + \frac{Q_2}{RT} + \frac{\Delta Q}{RT \ln a} \\ &= \ln\left(\frac{\ln a}{A_2}\right) + \frac{1}{RT}\left(Q_2 + \frac{\Delta Q}{\ln a}\right) \end{aligned} \tag{4.1-68}$$

Equation (4.1-68) predicts that the plot of $\ln \tau$ against $1/T$ should be approximately linear. The slope of the Arrhenius line for the induction time is determined not only by the activation energies of the pre- and postsurface activation reactions, but also by the ratio a of their entropies of activation. The phase transition of α- to β-quartz, or γ- to α-alumina, increases the activation entropy, and hence the preexponential factor, A_1, of the reorganization reaction. Since the ratio $a = A_2/A_1$ would decrease, the $1/\ln a$ would increase. This leads to the conclusion that the composite energy of activation, $\theta \simeq (Q_2 + \Delta Q/\ln a)$, will be larger in the presence of alumina and quartz. This is borne out by the experimental results (namely 33.8 and 30.3, respectively, compared to 26.0 kcal/mole in their absence). It is also clear that the Arrhenius line in the presence of quartz exhibits a break at 578° C, very close to the $\alpha \rightleftharpoons \beta$ transition point of 575° C. The activating effect of the phase transition appears to be less significant after the transition point is bypassed. No break in the Arrhenius line is noticed in the presence of alumina because the $\gamma \rightleftharpoons \alpha$ transition of alumina occurs at much higher temperatures[57] (~1000° C), well above the range of this study.

Equation (4.1-68) was derived under the constraint that $\Delta Q < RT \ln a$. The conclusion that the apparent energy of activation for induction time would increase when a phase transition occurs can nevertheless be deduced from equation (4.1-66) without any constraint on ΔQ other than that it must be positive. Letting ψ stand for the expression in the parenthesis in equation (4.1-66) then

$$\frac{\partial \ln \psi}{\partial a} = -\frac{\Delta Q}{a \ln a(RT \ln a + \Delta Q)} \tag{4.1-69}$$

The apparent activation energy, θ, as determined from equation (4.1-66) would be given by

$$\theta = Q_2 + R\left[\frac{\partial \ln \psi}{\partial(1/T)}\right] = Q_2 - RT^2\left(\frac{\partial \ln \psi}{\partial T}\right) \tag{4.1-70}$$

The directional change of θ with additives can be predicted from $d\theta/d\alpha$; thus

$$\frac{d\theta}{d\alpha} = -RT^2 \frac{\partial}{\partial \alpha}\left(\frac{\partial \ln \psi}{\partial T}\right) = -RT^2 \frac{\partial}{\partial T}\left(\frac{\partial \ln \psi}{\partial \alpha}\right) \tag{4.1-71}$$

Combining equations (4.1-69) and (4.1-71), one obtains

$$\frac{d\theta}{d\alpha} = RT^2 \frac{\partial}{\partial T}\left[\frac{\Delta Q}{\alpha \ln \alpha(\Delta Q + RT \ln \alpha)}\right]$$

$$= -\frac{R^2 T^2 \Delta Q}{\alpha(\Delta Q + RT \ln \alpha)^2} \tag{4.1-72}$$

Because both ΔQ and α are necessarily positive numbers, $d\theta/d\alpha$ is always negative regardless of whether ΔQ is larger or smaller than $RT \ln \alpha$, or whether A_1 is larger or smaller than A_2. The apparent activation energy, θ, should therefore increase when the ratio α decreases, or when A_1 increases by a phase transition.

The function of the silica or alumina surface activators is, therefore, not primarily to change the activation energy (as most catalysts do), but rather to increase the rates of reactions by increasing the activation entropy. Information is sometimes defined as *negentropy*,[59] and activation entropy has often been called *superinformation*.[60] When the activation entropy has a large negative value (which amounts to a small α term in equation (4.1-69), the "ignorance barrier" is very high and reactions are slow. These activators, therefore, lower the "ignorance barrier" by increasing the entropy of activation.

The preceding analysis attributes the Hedvall effect of solid-state chemistry to a catalytic effect associated with the phase transition. Catalysis produced by increase in activation entropy should be referred to as *catalysis of the second kind* to contrast it from ordinary catalysis (of the first kind) which function by lowering the energy of activation.

4.2. The Reduction of Metal Oxides (JWE and CHK)

4.2.1. Introduction

The reduction of oxides by gases, or by solids via gaseous intermediates, constitutes an important class of reactions in pyrometallurgy. By far the most significant of these reactions is the reduction of iron oxides by carbon monoxide, one of the two major reactions in the iron blast furnace process. An appropriate length of this section is therefore devoted to iron oxide reduction, although investigations of reduction by hydrogen appear to outnumber those concerned with reduction by carbon monoxide and the former predominate in our discussion. Perhaps the next most significant oxide reduction to the extractive metallurgist is the reduction of lead oxide. This oxide is given scant treatment here because of the paucity of published work on its reduction. The third oxide reduction reaction of major significance is that of nickel oxide. There have been many investigations of the reduction of this oxide by hydrogen and these are discussed at length below. Examples are also given of papers on the reduction of oxides of copper, molybdenum, tin, zinc, and tungsten.

All the reduction reactions under discussion are gas–solid reactions (or solid–solid reactions proceeding through gaseous intermediates) and in this section we will rely heavily on the earlier chapter on the fundamentals of the kinetics of heterogeneous reaction systems. It will be recalled that gas–solid reactions entail mass transfer of gaseous reactants (and product) through a "boundary layer" surrounding the reacting solid, pore diffusion of gaseous reactant (and product) within the solid, and the chemical step(s) within the solid. Depending upon conditions of temperature, gas flow, solid geometry, etc., any one of these steps (or a combination) may be rate determining. The pore structure of the solid product (and the solid reactant if it is porous) may affect the reaction rate. All these factors must be taken into account in interpreting experimental data or attempting extrapolation or generalizations concerning a particular reaction.

4.2.2. Iron Ore Reduction

The reduction of iron ores to yield metallic iron is perhaps the technologically most significant gas–solid reaction. At present the major portion of the ores being processed are being reduced in an iron blast furnace, although direct reduction processes are gaining in importance.[61]

The reduction of iron oxides involves one or more of the following steps:

1. Fe_2O_3 (hematite) $\rightarrow$ Fe_3O_4 (magnetite)
2. $Fe_3O_4 \rightarrow Fe$
3. $Fe_3O_4 \rightarrow FeO$ (wustite)
4. $FeO \rightarrow Fe$

Whether magnetite reduces directly to iron (Step 2) or via wustite (Steps

3 and 4) depends on the temperature. Wustite is unstable below 570° C and this is the approximate upper limit on the reduction of magnetite directly to iron. Wustite is a nonstoichiometric compound whose oxygen content varies between about 23 and 25%, depending on temperature and the atmosphere with which the solid is in contact.

4.2.2.1. *Thermodynamics*

The equilibrium diagrams for the reduction of iron oxides by hydrogen and by carbon monoxide are well established (Figure 4.2-1). The reduction of hematite to magnetite can be regarded as an irreversible reaction but the reductions of magnetite and wustite are reversible. For example, at 1000° C and 1 atm a 73% CO–27% CO_2 mixture is in equilibrium with wustite and iron, as is a 60% H_2–40% H_2O mixture.

In the reduction by carbon monoxide, Steps 1–4 are nearly isoenthalpic (having heats of reaction less than 5 kcal/g-atom of oxygen removed), except for the reduction of hematite to magnetite (Step 1) which is exothermic with a heat of reaction of 12 kcal/g-atom oxygen at 1000° C. Reduction by hydrogen is slightly endothermic at temperatures normally encountered in laboratory experiments and industrial plants,

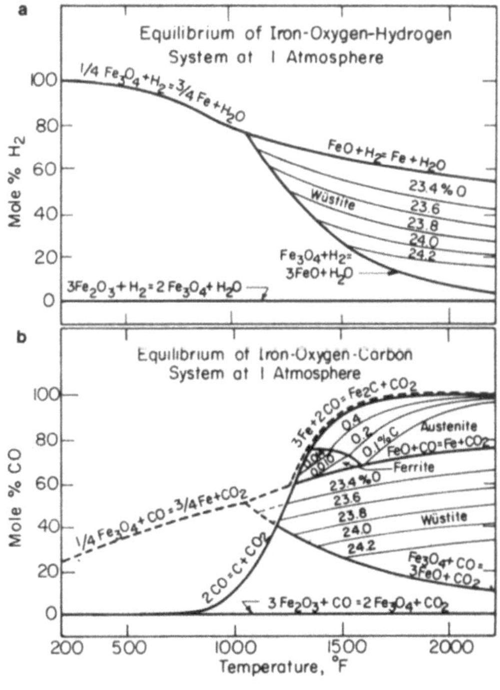

Figure 4.2-1. Equilibrium diagrams for the reduction of iron oxides by (a) hydrogen and (b) by carbon monoxide.

the heat of reaction being approximately 11 kcal/g-atom oxygen for the reduction of magnetite to wustite. The exception is the reduction of hematite to magnetite, which is slightly exothermic (roughly 2 kcal/g-atom oxygen).

4.2.2.2. Nature of Iron Produced

A simple material balance shows that the porosity of a solid produced by reduction of a nonporous oxide is given by

$$\varepsilon_p = 1 - N\frac{\rho_0}{\rho_p} \qquad (4.2\text{-}1)$$

where N is the number of moles of solid product formed by the reduction of one mole of oxide, and ρ_0 and ρ_p are the true (pore-free) molar densities of the oxide and product, respectively. This equation assumes no swelling or shrinking during reactions. The reactions in Steps 1–4 should, according to this equation, yield products of 2, 52, 15, and 44% porosity, respectively, starting with nonporous reactants and assuming no swelling or sintering takes place. We should therefore expect the reduction of nonporous natural ores or nonporous iron oxides to yield porous iron and intermediate products. This is illustrated in the micrographs (Figures 4.2-2 and 4.2-3) obtained by Turkdogan and co-workers revealing the pore structure of iron produced by the reduction of nonporous oxides. Turkdogan's group has carried out a very thorough examination of iron produced by gaseous reduction of iron oxides.[62–64] Figure 4.2-4 shows how the pore-size distribution of the iron obtained by the reduction of a Venezuelan hematite ore depended on the reduction temperature. It should be noted that the mean pore size increases with increasing temperature. This is also revealed in the scanning electron microscope pictures produced by this group (Figure 4.2-5). Pore surface area measurements were found by these investigators to show the same trend (Table 4.2-1). No further structural changes were observed on holding the iron at temperature once reduction had been completed. This indicates that the structural changes observed by Turkdogan *et al.* were not merely a sintering phenomenon.

Our attention thus far has been focused on reduction which is not accompanied by large changes in the gross dimensions of the ore pellet or lump, that is, in the absence of swelling or shrinking. Occasionally iron oxides exhibit swelling during reduction. This swelling has an effect on reduction kinetics through its effect on the structure of the iron produced (see below), but its effect on the operability of a blast furnace or direct reduction unit is probably more important. Swelling in excess of about 40% by volume has been found to lead to operating difficulties

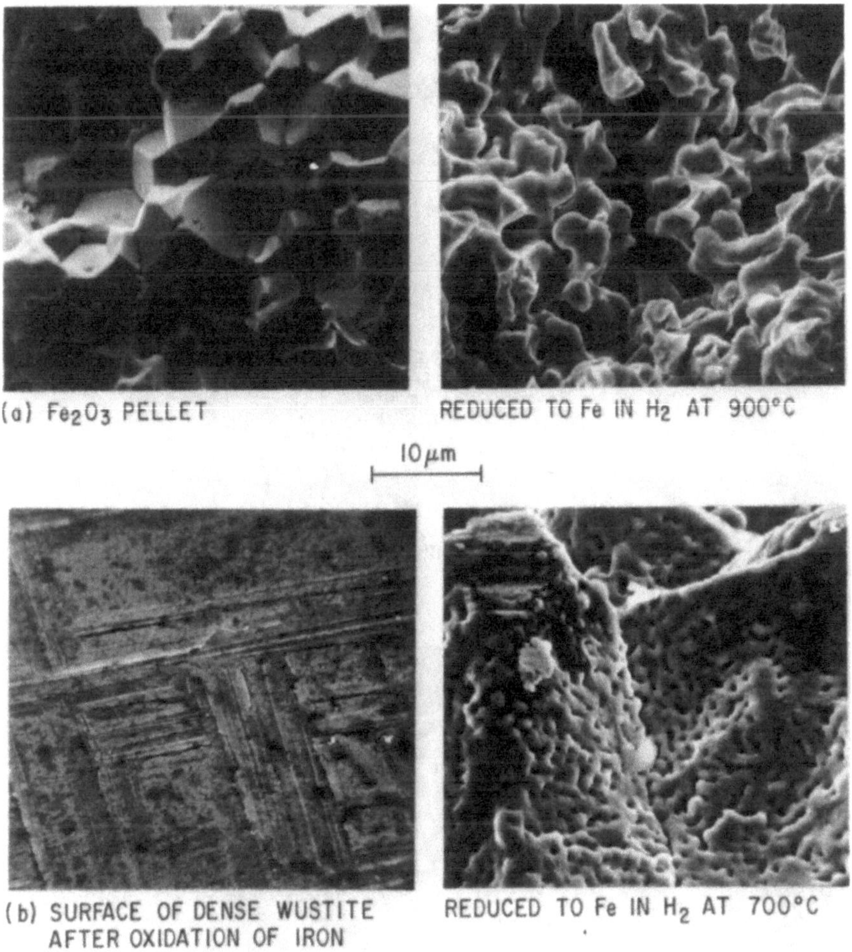

Figure 4.2-2. Scanning electron micrographs of Turkdogan.[87] (a) Fracture surface of sintered Fe_2O_3 pellet before and after hydrogen reduction. (b) Outer surface of dense wustite before and after hydrogen reduction.

Table 4.2-1. Trend of Pore Surface Area

Reduction temperature, C°	Pore area, m^2/g
200	38.6
400	23.2
600	13.7
800	4.2
1000	0.48
1200	0.10

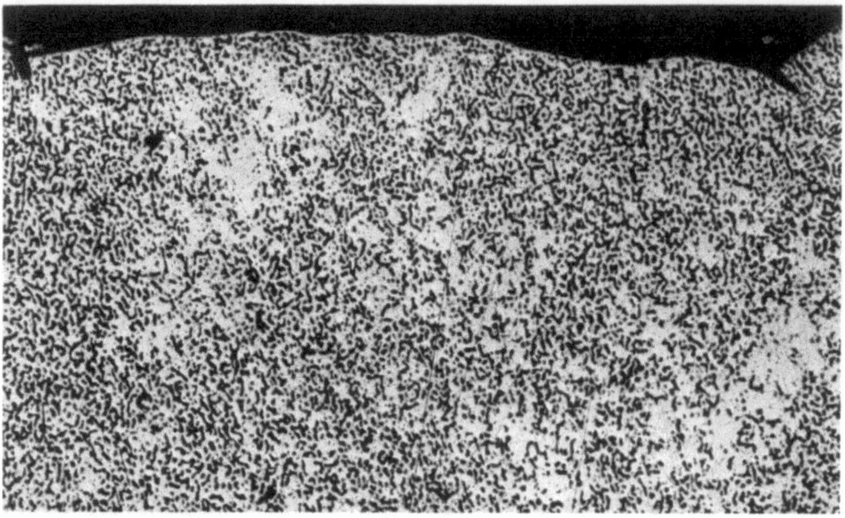

100% H_2

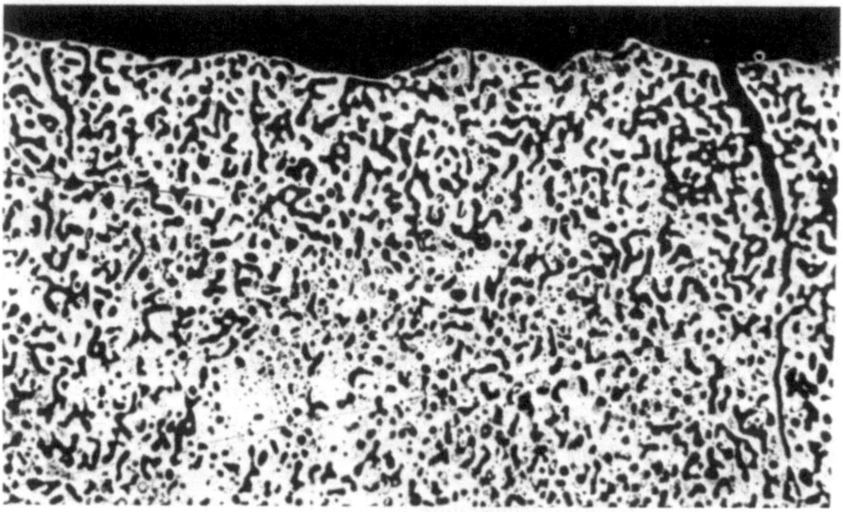

90% CO–10% CO_2

100μm

Figure 4.2-3. Scanning electron micrographs of Turkdogan and Vinters.[64] Polished sections showing network of pores in the iron formed by hydrogen and carbon monoxide reduction of dense wustite at 1200° C.

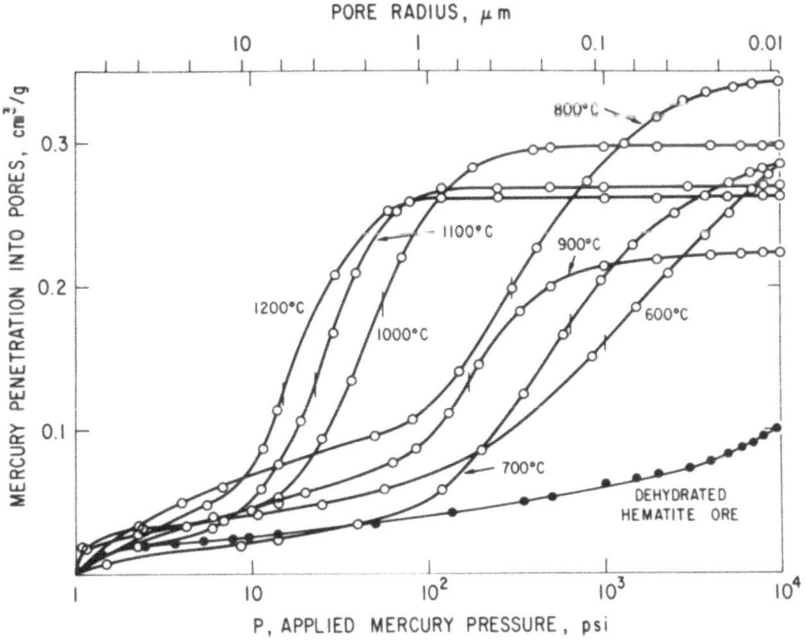

Figure 4.2-4. Effect of reduction temperature on pore size distribution in iron formed by hydrogen reduction of hematite ore.[63]

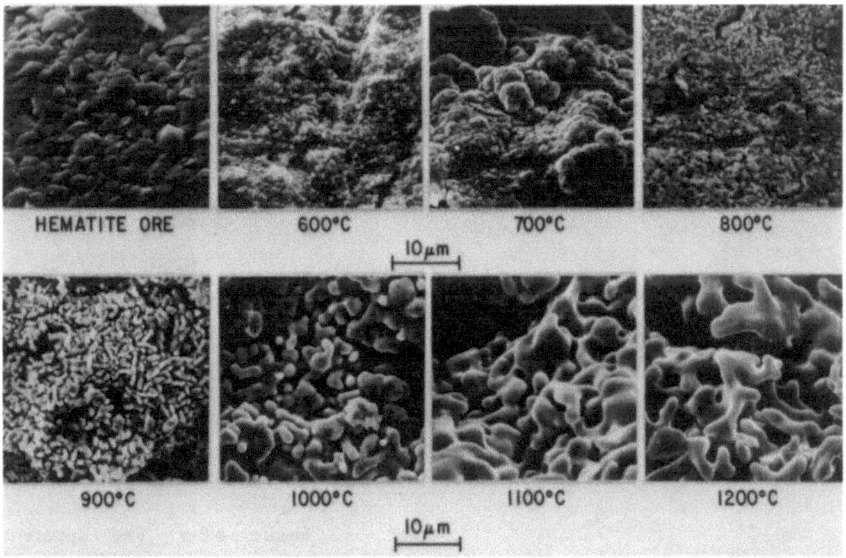

Figure 4.2-5. Scanning electron microscope pictures of fracture surface of lump hematite ore and porous iron obtained by hydrogen reduction at various temperatures.[62]

in the blast furnace; this swelling is often referred to as "catastrophic" swelling.[65] Chang and co-workers[66] studied the swelling of 17 production pellets from various steel plants and concluded that none exhibited catastrophic swelling. Swelling has been studied by several investigators.[67–80] Catastrophic swelling appears to be due to the formation of needle-like iron "whiskers" rather than the normal equiaxed iron grains. Wenzel *et al.*[80] published micrographs showing such whiskers.

4.2.2.3. Reduction Kinetics

The aim of most investigations into the kinetics of iron oxide reduction has been to determine the effect of various parameters (temperature, partial pressure of gaseous reactant, particle size, etc.) and to interpret and interpolate the experimental data in terms of a rate controlling step or perhaps a combination of such steps.

Many of the early investigations of iron ore reduction have to be regarded as incomplete. Variations of the solid structure of chemically identical oxides and of the product iron do not seem to have been recognized, although the effect of porosity on reducibility had been recognized by Joseph[81] in 1936. The shrinking core model was indiscriminately applied to both porous and nonporous oxides and sweeping claims as to the rate controlling step were made on fairly scant evidence.

Much of the early work on iron ore reduction carried out before 1970 is summarized in a book by Bogdandy and Engell.[82] Figure 4.2-6 shows plots of the data obtained by Bogdandy and Janke[83] from the reduction of Malmberg ore pellets using hydrogen and water vapor. It appears that under the conditions of these experiments the reduction is

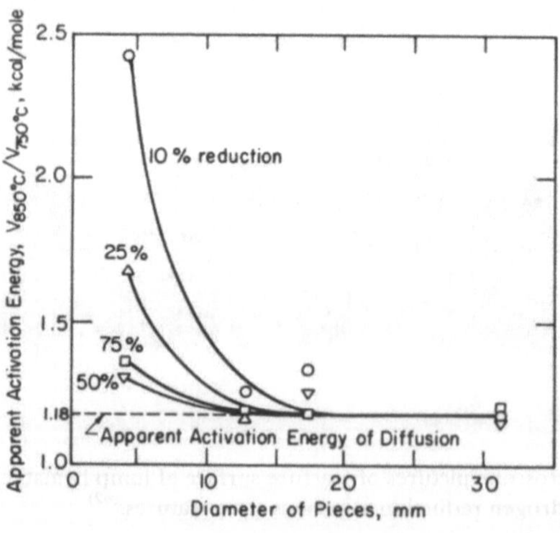

Figure 4.2-6. The apparent activation energy as a function of particle size and extent of reaction for Malmberg pellets.[83]

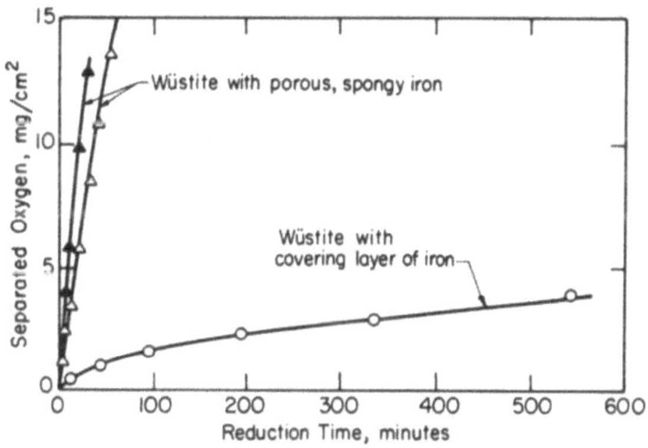

Figure 4.2-7. Progress of reduction with formation of porous spongy iron and with formation of a nonporous covering layer of iron.[64]

proceeding under diffusion control for pellets greater than about 15 mm in diameter, except for about the first 10% of reduction. Kohl and Engell[84] described the important role of the iron layer in some experiments summarized in Figure 4.2-7. Under conditions where the iron product forms a nonporous layer the progress of reaction is slowed down considerably, since now the mechanism of reaction entails solid-state diffusion.

Some investigators have reduced oxide samples of a geometry so that the interfacial area between oxide and metal (the reaction interface) does not change in size during the course of reaction, e.g., cylinders whose curved surfaces have been sealed. For such geometries (in contrast to the more usual spheres), valid conclusions concerning the rate controlling step can be drawn from a plot of extent of reaction versus time. A good example of such work is that of Quets *et al.*[85] the results of which are shown in Figure 4.2-8. The magnetite plates were reduced by hydrogen at a pressure of 0.86 atm and the linear nature of the reduction curves, coupled with the high activation energy, suggests control by chemical kinetics. According to these data, the first-order rate constant for hydrogen reduction of magnetite to iron was given by

$$k \simeq 2.0 \times 10^{-2} \exp(-14{,}600/RT) \text{ moles oxygen/cm}^2 \text{ sec atm hydrogen} \tag{4.2-2}$$

A similar approach was used by Lu and Bitsianes[86] to study the reduction of dense natural and synthetic hematites by hydrogen and carbon monoxide. The specimens used had the form of cylinders or rectangular parallelepipeds, and an impervious enamel coating was applied to some of the external surfaces, e.g., the curved surface in the case of cylinders. In this way, reaction was limited to a receding planar

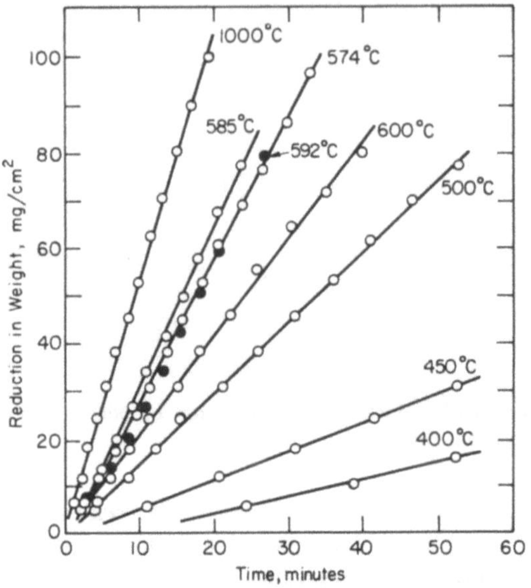

Figure 4.2-8. The linear relationship between extent of reaction and time for the reduction of magnetite plates by hydrogen.[85]

interface of constant area. Diffusion through the ash layer presented a significant resistance in these experiments as evidenced by the curvature of plots of reaction extent against time.

Most investigators have assumed that the multiple reactions taking place during iron ore reduction occur within a narrow region of the pellet or lump of ore. This is probably justified in the case of oxides of low porosity. Figure 4.2-9 illustrates the validity of this assumption.[87] From our earlier discussion of porosity it can be expected that the magnetite would have almost the same (low) porosity as the initial hematite, whereas that of the wustite would be only a little greater. This explains the inability of the hematite–magnetite and magnetite–wustite reaction interfaces to advance much beyond that of the wustite–iron interface; to do so would require diffusion of gaseous reactant and product through a thicker low-porosity layer. Spitzer *et al.*[88] presented an extension of the shrinking core model in which allowance is made for the existence of three reaction interfaces within a hematite pellet undergoing reduction to iron.

Seth and Ross[89] studied the reduction of ferric oxide compacts by hydrogen; their results are presented in Figure 4.2-10. It will be recalled that, for a nonporous spherical pellet, the time to achieve a given percentage reduction is proportional to the pellet size for chemical step control and to the square of the pellet size for diffusion control. The

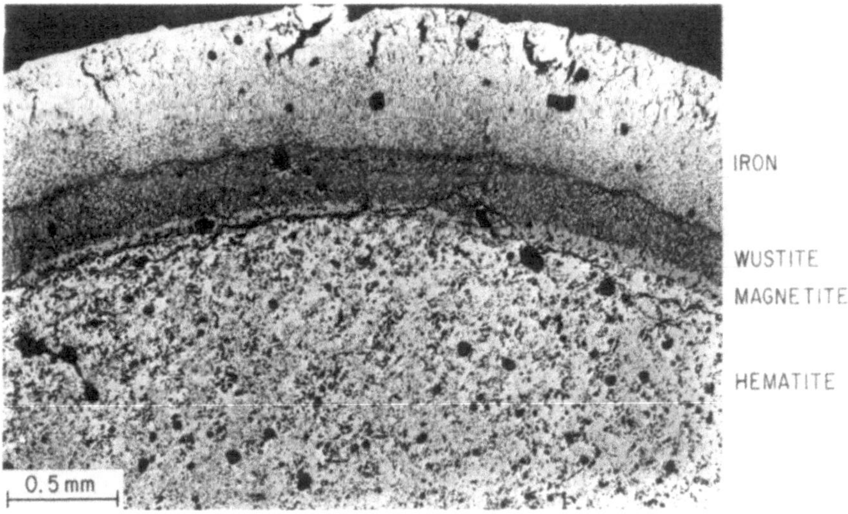

Figure 4.2-9. Polished section of a sintered hematite pellet which has been partially reduced in hydrogen.[87]

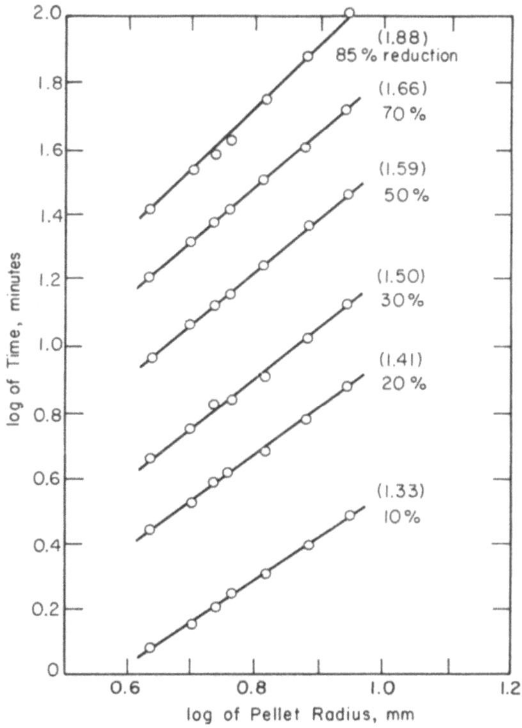

Figure 4.2-10. A plot of time to achieve a given degree of reduction as a function of pellet size.[78]

slopes of the log–log plot appearing in Figure 4.2-10 indicate mixed control under the conditions of the experiments.

Most of the investigations referred to so far have been of natural or synthetic ore of low porosity. Nowadays porous pellets represent a considerable portion of the feeds to blast furnaces in industrialized nations, whereas some of the natural ores used in direct reduction units have a high porosity. As a consequence, interest in the reduction kinetics of such porous oxides has recently increased considerably.

An extensive series of experiments has been carried out by Turkdogan and co-workers at U.S. Steel on the reduction of porous iron ore.[82–84] Attention was focused on the reduction of natural Venezuelan hematite of roughly 30% porosity and 18 m^2/g-mole pore surface area by hydrogen and by carbon monoxide. A wide range of temperatures (200–1200° C) and particle sizes (0.35–15 mm) was studied, enabling these workers to explore the diffusion-controlled and chemically controlled regimes. Most of the particles used were spherical but some "unidirectional" reduction experiments were also carried out. In these latter experiments the oxide was enclosed within a tube open at one end. Under conditions of chemical control the rate should stay constant in such experiments (reaction area stays constant) but for diffusion control the rate should decline with time (diffusion path lengths, and therefore diffusional resistance, increases with time). Like many other investigations, Turkdogan's noted "induction periods" at low temperatures (200–300° C). This is indicated in Figure 4.2–11 where the reduction passes through an initial accelerating phase when carried out at 200° C. Many

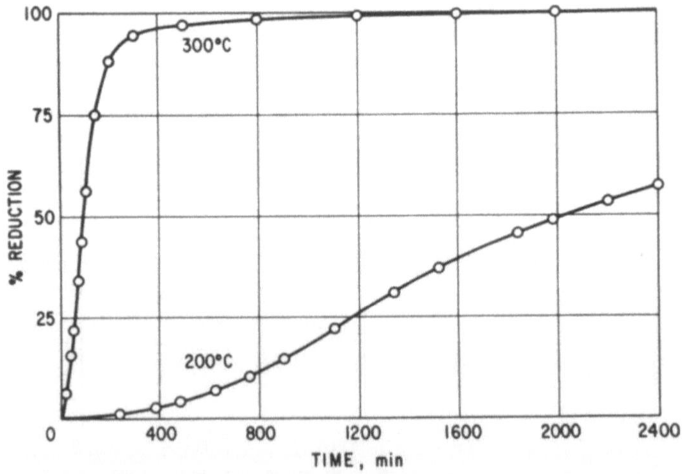

Figure 4.2-11. Reduction of hematite by hydrogen at low temperatures showing an induction period at 200° C.[82]

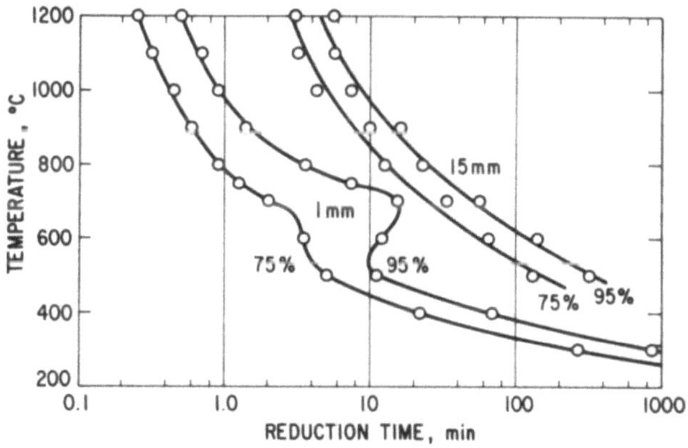

Figure 4.2-12. Time to achieve a given extent of reduction as a function of particle size and temperature.[82]

investigators (e.g., Strangway and Ross[90]) have noted a rate minimum in the vicinity of 700° C; this was also observed by Turkdogan *et al.*, but only for small particles (see Figure 4.2-12).

Turkdogan *et al.* found that the time to achieve a given extent of reduction of a sphere was *not* proportional to the particle diameter, as it would have been had reaction been proceeding according to a chemically controlled shrinking core model. Broad reaction zones were observed for the wustite–iron reduction and these zones increased in width with decreasing temperature and increasing porosity in conformity with ideas presented earlier in this volume. Reaction of a porous solid with a fluid takes place within a broad reaction zone within the solid. As the rate-controlling step is switched from pore diffusion to the chemical step (e.g., by lowering the temperature or increasing the porosity), the width of this reaction zone is expected to increase.

Beyond a particle radius of about one millimeter the *initial* rate of reaction was found to be inversely proportional to the particle size; this was interpreted as indicating mixed control during the initial stages of reaction. Remember that for chemical control in reduction of porous pellets, rate is independent of particle size; for diffusion control the rate is inversely proportional to particle size squared. Beyond the initial stages, however, these investigators believed that pore diffusion control was rate controlling for the particles >1 mm and were able to provide reasonable agreement between theory and experiment on this basis using the pore diffusivity as an adjustable parameter. This was also consistent with the results of the unidirectional experiments in which the rate declined with time in inverse proportion to the thickness of the product layer.

Turkdogan and co-workers calculated the ratio of effective gaseous reactant diffusivity to ordinary diffusivity, obtaining the former from their reduction experiments by curve fitting a shrinking core model (valid for porous pellets under conditions of pore diffusion control). This ratio is plotted versus reduction temperature in Figure 4.2-13. This figure is consistent with the solid structure determinations by porosimetry, surface area measurement, and scanning electron microscope referred to earlier. As the reduction temperature is increased, the iron takes on a more open structure of larger pores through which the gaseous reactant may readily diffuse. Furthermore, the diffusivity of gaseous reactant and product within the porous iron product was measured using a technique developed by Olson and McKewan.[91,92] The measured diffusivities were in close agreement with those obtained from the reduction studies and calculated from the measured pore size and porosity of the iron (see Figure 4.2-14).

In a further series of experiments, the group at U.S. Steel examined the reduction of granular Venezuelan ore in H_2–CO–CO_2 mixtures. Carbon deposition was observed and the reaction rate increased with increasing hydrogen content of the reducing gas. The surface area of the porous iron produced by reduction was again observed to decrease

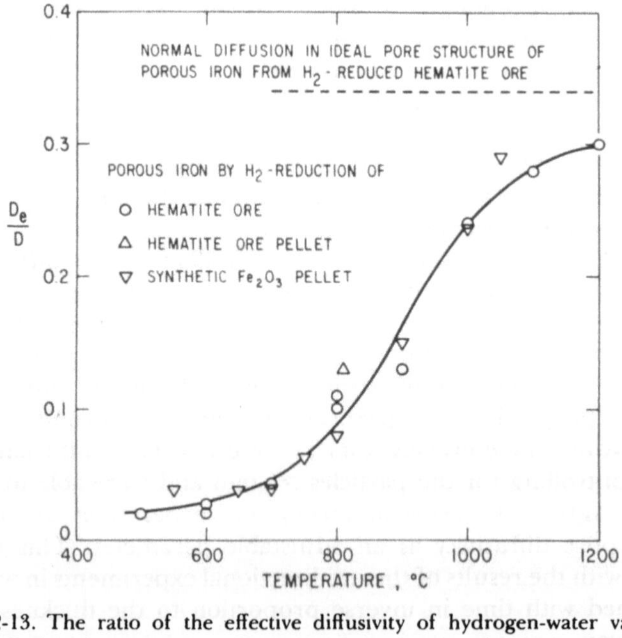

Figure 4.2-13. The ratio of the effective diffusivity of hydrogen-water vapor to the ordinary diffusivity for porous iron formed by reduction at various temperatures.[82]

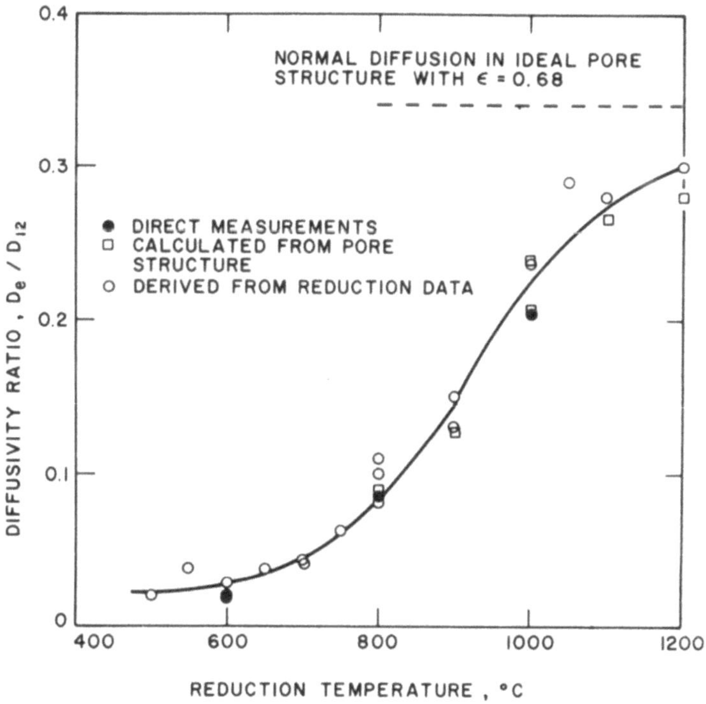

Figure 4.2-14. Comparison of calculated, measured, and derived diffusivity ratio for porous iron as a function of reduction temperature.[81,82]

with increasing reduction temperature. Typical experimental results are given in Figure 4.2-15 and show a weak dependence of rate on particle size below 1 mm. From this it can be deduced that the rate of reaction at these small particle sizes is controlled by the chemical step, which is the interpretation Turkdogan and co-workers placed on their results.

Papers on the reduction of iron oxides are legion, but some require special mention because of novelty. Koo[103] has examined the effect of heating on a Venezuelan Cerro Bolivar ore. Figure 4.2-16 illustrates the decline in surface area of the ore brought about by holding it at an elevated temperature under nitrogen. Figure 4.2-17 reveals that this decline is more severe the higher the temperature and that it parallels the decline in surface area of the reduced iron with increasing reduction temperature. Porosimetry measurements also reveal a coarsening of the ore structure on holding it under an inert gas at elevated temperatures (see Figure 4.2-18). This raises the possibility that the coarser pore structures observed in iron produced by reduction at higher temperatures may be due to this sintering of the ore. Certainly the change in

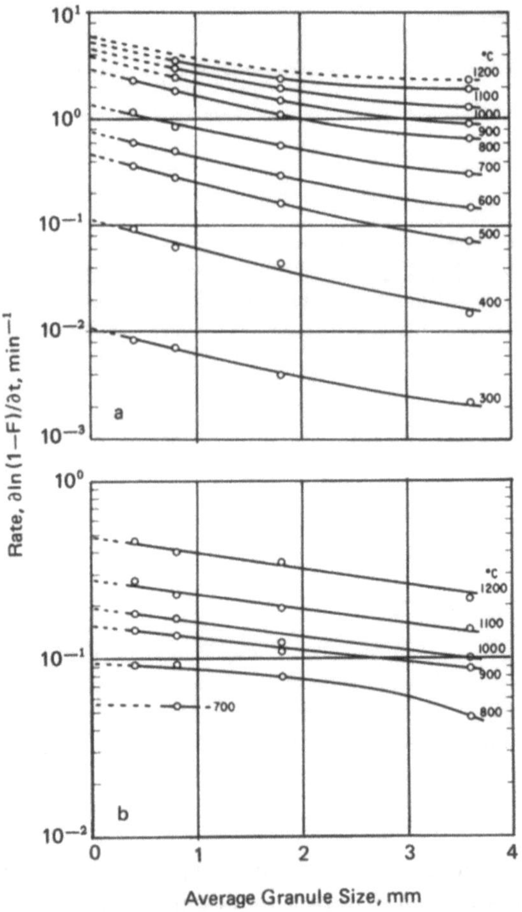

Figure 4.2-15. Effect of granule size on the rate of reduction of granular hematite, (a) in hydrogen and (b) in 90% CO-10% CO_2 mixture.[64]

surface area and pore size of the ore must be taken into account in interpreting data on reduction kinetics. Nabi and Lu[93] have carried out an investigation of the effect of various preparative methods and pretreatments on the reduction of hematite.

Nabi and Lu[104] also performed a particularly painstaking series of experiments on the reduction of hematite to magnetite by hydrogen–water vapor mixtures so that no reduction of the magnetite occurred. Chemisorption, surface reaction, and desorption were examined in detail. Landler and Komarek[104] performed experiments on the reduction of a wustite to a wustite of lower oxygen content using hydrogen–water vapor mixtures. Results were interpreted in terms of chemical reaction at the wustite–gas interface, coupled with solid-state diffusion of

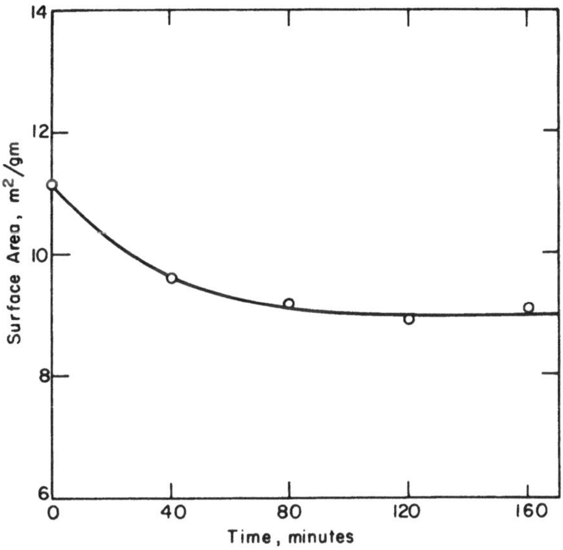

Figure 4.2-16. Effect of holding at 650° C under nitrogen on the pore surface area of a Venezuelan Cerro Bolivar ore.[103]

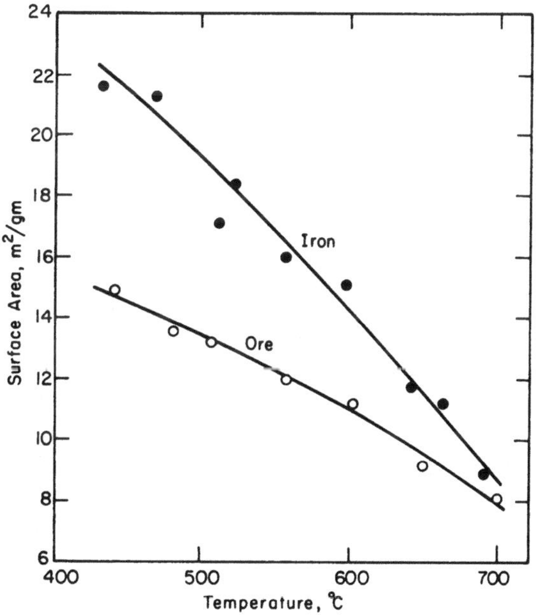

Figure 4.2-17. The pore surface area of iron produced by hydrogen reduction of Cerro Bolivar ore at various temperatures and the effect of holding for 80 min under nitrogen at various temperatures on the surface area of the ore.[103]

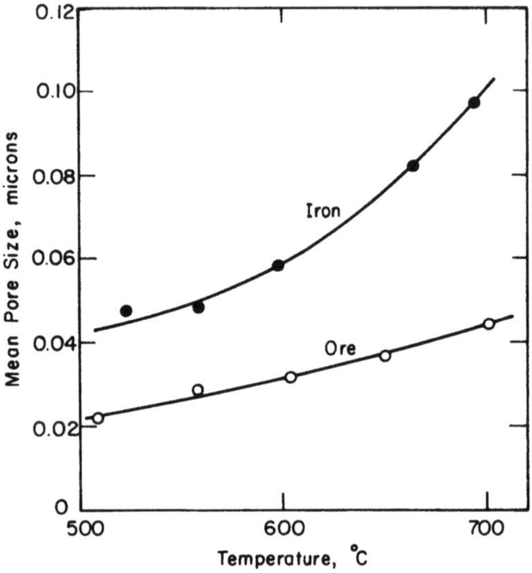

Figure 4.2-18. The mean pore size of iron produced by reduction at various temperatures and that of the ore after holding for 80 min under nitrogen at various temperatures.[103]

iron within the wustite. McKewan's investigation[95] of the reduction of magnetite by hydrogen at elevated pressure revealed a rate directly proportional to pressure up to 1 atm and approaching a rate independent of pressure above about 10 atm. This was interpreted by McKewan as a case of chemical step control with Langmuir–Hinshelwood kinetics; Evans and Haase[96] departed from the usual weight change method for following the progress of reaction and instead used a torsion technique in their study of the reduction of magnetite by hydrogen. Matyas and Bradshaw[97] and Tien and Turkdogan[98] employed models rather similar to the grain model described earlier in this volume to interpret experimental data on the reduction of iron oxides. The work of Cox and Sale[99] is an example of an investigation in which extensive use of scanning electron microscopy was made during a study of iron oxide reduction. Cech and Tiemann[106] have studied the influence of a corona discharge on the reduction of hematite. Szekely and El-Tawil[100] have recently studied the reduction of hematite pellets by hydrogen–carbon monoxide mixtures and pointed out that their experimental results should be interpreted by the application of the Stefan–Maxwell equation to the four-component gaseous pore diffusion. As examples of the study of iron oxide reduction in the presence of other oxides, one may point to the works of Tittle[101] on the reduction of calcined bauxite, McAdam[102] on the reduction of titanium rich ironsands, and Khalafalla *et al.*[105] on the reduction of hematite to magnetite in the presence of silica, alumina, lime, and lithia.

4.2.3. Nickel Oxide Reduction

Of the many commercially successful processes for nickel production several entail the gaseous reduction of nickel oxides. Lateritic nickel ores are usually reduced by CO–CO_2 mixtures prior to an ammoniacal leach, reduction of the oxide is an intermediate step in the Mond process, and both major North American nickel producers use hydrometallurgical processes wherein a final step is the reduction of high-purity oxide by hydrogen.

4.2.3.1. Thermodynamics

Reduction of nickel oxide by hydrogen or carbon monoxide can be considered irreversible, the equilibrium constants being 370 and 10,000, respectively, at 600° K, and 163 and 225, respectively, at 1000° K. Reduction by hydrogen is slightly exothermic, i.e., −1.89 kcal/g-mole at 600° K, −3.14 kcal/g-mole at 1000° K. Reduction by carbon monoxide is more exothermic (−11.2 kcal/g-mole at 600° K, −11.5 kcal/g-mole at 1000° K). These figures are calculated from the thermodynamic data in Elliott and Gleiser.[(107)]

4.2.3.2. Reduction Kinetics

Benton and Emmett[(108)] were amongst the first to investigate the reduction of nickelous oxide by hydrogen. Reaction was carried out in a packed bed of oxide powder through which hydrogen was passed at a flow rate which was probably too low for accurate kinetic measurements. The reactor was brought up to temperature in the range from 188 to 250° C under hydrogen over a period of 10–12 min. After the reaction, the water vapor formed was condensed. Like other workers in the field, Benton and Emmett observed "autocatalytic" behavior. For a short period after reaction temperature has been reached (or the hydrogen flow commenced, whichever is considered to be the start of reaction) no reaction is observable. A period in which a reaction which accelerates as time progresses may be observed follows this induction period, after which the rate remains constant or declines until reaction is complete. Benton and Emmett prepared their nickel oxide by roasting nickel nitrate in air or nitrogen for long periods. They observed that the higher the roasting temperature the longer the induction period during reduction and the lower the rate after this period. They established that water vapor increased the length of the induction period. The autocatalysis was ascribed to nickel, although adding nickel powder to the oxide before reduction had no effect.

It is noteworthy that Taylor and Burns[109] had earlier shown that considerable adsorption of hydrogen on nickel occurred in the temperature range of 25 to 218° C. Adsorption was found to be independent of pressure beyond 300 mm Hg at 25° C. It is therefore plausible to suggest that the induction period is associated with the growth of nickel nuclei. The chemical reaction may be controlled either by a surface reaction at the metal–oxide interface (an explanation put forward by Benton and Emmett) or by the adsorption of hydrogen on the nickel (presumed to be more rapid than adsorption on the oxide). In either case, the reaction accelerates as the nuclei grow and then passes into the declining rate period as the nuclei coalesce. Failure of added nickel to diminish the induction period and increase the rate is due to the inability of the nickel to achieve the intimate contact with the oxide enjoyed by the nickel formed during reduction. Presumably the nickel nuclei form around active sites which may be crystal defects. An annealing process would therefore explain the difficulty in reducing oxides roasted at a high temperature.

Kivnick and Hixson[110] studied the reduction of nickel oxide by hydrogen in a fluidized bed.

Parravano[111] studied the reduction of nickel oxide by hydrogen in a closed system of constant volume. Reduction was followed by the change in system pressure as water vapor condensed out of the recirculating gas stream. Granular nickel oxide was used in a small packed bed. The hydrogen partial pressure fell by at least 85% during reduction, calculated from Parravano's data. The rate of reduction during the constant rate period was found to be a linear function of the initial hydrogen pressure within the range (200–500 mm) examined. Parravano examined the effect of adding various foreign ions to the nickel oxide and related the changes in activation energy and rate to changes in the electronic properties of the oxide.

Simnad, Smoluchowski, and Spilners[112] studied the effect of proton irradiation upon hydrogen reducibility of nickel oxide. They produced their oxide samples by heating nickel sheet in oxygen at 1150° C for 3 or 24 hr, claiming that thereby a uniform solid layer of nickel oxide 15 or 60 μ thick, respectively, is developed. Samples were irradiated with 260-MeV protons for a total flux of 10^{16} protons/cm^2. The irradiated specimens had much shorter induction periods than the control specimens, which showed unusually long induction periods, i.e., 1200 min at 250° C for the 15-μ layer. In addition, the irradiated specimens showed much faster reaction during the constant rate period. The differences were relatively smaller for higher temperature reductions, the reductions of irradiated and nonirradiated 60-μ layers being almost identical at 400° C. These workers came to the conclusion that the reaction enhancement on proton irradiation was due to formation of lattice defects which either promoted the solid state diffusion process involved in the formation of nickel nuclei or increased the number of nuclei by provision of extra active sites. The unusually long induction periods of the samples which were not irradiated can be explained by the defect theory since the nickel oxide is formed at high temperature (1150° C), where presumably there is opportunity for annealing.

The experiments of Iida and co-workers[113–116] support these conclusions. A careful investigation of the variation in particle size, oxygen to nickel ratio, density, lattice parameter, electrical conductivity, and other properties, between samples of nickel oxide prepared by roasting the nitrate at various temperatures, was carried out.[113] Samples prepared by low temperature calcining showed the highest departure from stoichiometry, the highest electrical conductivity at 300° C (but the lowest activation energy for conductivity), and the highest number of vacancies per cm^3 (calculated from density measurements and Ni^{3+} atom %). Some samples prepared by calcining the hydroxide and the carbonate showed similar behavior. In another paper,[114] the effect of additives on the reduction of nickel oxide powder was examined. The nickel oxide was in the form of cylindrical pellets produced by compacting nickel oxide powder (from roasted nitrate). Several reductions were carried out in an environment of rapidly increasing temperature but some reductions were performed in an isothermal environment. The induction period was 2 hr for nickel oxide at 180° C but at 160° C a nickel oxide sample containing 5% CuO additive showed no induction period. For the CuO additive to be effective, it was necessary to heat the mixture prior to reduction, suggesting that a solid solution is needed before any effects are produced.

Iida and Shimada[115] also reduced single crystals of nickel oxide. In a constant temperature environment the single crystals showed rate maxima at about 350° C and a minimum at 500° C. Such behavior was not observed in the reduction of nickel oxide powder. This phenomenon was explained in terms of reduction in rate by sintering of the nickel (between 350 and 500° C) and the formation of pores in the nickel layer (above 500° C) by high-pressure water vapor. Some microscopic evidence was produced to support this theory. The reduction of cylindrical pellets (5–10 mm in diameter) of about 50% porosity was found to be topochemical at 250° C, but at 180° C reduction appeared to start at the center of the pellet and proceed outward.

These workers also published an important paper[116] on the effect of compacting and grinding on the hydrogen reduction of nickel oxide. Pellets were prepared by compacting oxide particles prepared by roasting the sulfate. It was found that the surface area, measured by the

BET (Brunauer–Emmett–Teller) method, of the compact did not diminish with compacting pressure, despite a reduction in porosity. The reductions were carried out at 200° C and it was found that increase in compacting pressure below 1000 kg/cm^2 caused an increase in the rate of reduction. An increase in rate also resulted by increasing the porosity of the 1000 kg/cm^2 compact by grinding. A single crystal reacted much more slowly than the powder but after grinding to 50 μ, particles reacted much more quickly. All this supports the hypothesis that the chemical step in the reduction of nickel oxide by hydrogen is governed by the formation and growth of nickel nuclei which in turn are dependent on the number of lattice defects. Deren and Stoch[(122)] have examined the effect of "biography" on the stoichiometry and chemisorptive properties of nickel oxide.

Bandrowski *et al.*[(117)] studied the reduction of pellets of nickel oxide in a packed bed. Reduction was followed by measuring the dew point and flow rate of the exit gas stream. The pellets were 4.0 mm long, 4.3 mm in diameter, and roughly cylindrical. Reduction was carried out by hydrogen and the data gathered were interpreted in terms of two simultaneous reactions. The first involved the reaction between nickel oxide and hydrogen adsorbed on nickel oxide and the second was the interfacial surface reaction between nickel oxide and hydrogen adsorbed on nickel.

Yamashina and Nagamatsuya[(118)] studied the hydrogen reduction of nickel oxide doped and mixed with cupric oxide. They were able to correlate the reduction velocity (during the constant rate period) of nickel oxide with the change in lattice parameter which resulted from doping with foreign ions. Ions increasing lattice parameter increased the reaction rate and decreased the induction period. Mixtures produced by mechanically mixing the oxide showed different results. Thus admixture of copper oxide extended the induction period and reduced the rate during the constant rate period. It was shown that this was largely a consequence of water vapor formed by reduction of cupric oxide.

Bicek and Kelly[(119)] and Levinson[(120)] studied the reduction of supported nickel oxide.

An important contribution to nickel oxide reduction by hydrogen was made by Charcosset *et al.*[(121)] Nickel oxide powder, produced by decomposition of nickel nitrate at 400° C and subsequent roasting in air at 950° C, was reduced in a closed system containing a large excess of hydrogen. Reaction was followed by measuring weight change on a recording balance. The oxide particles were mostly in the size range of 50 to 80 μ. The induction period increased and rate (during the constant rate period) decreased with increasing charge up to 100 mg. This was said to be due to traces of water produced. In keeping with the high temperature roasting, the induction periods were fairly long (15 min at 220° C). The rate showed an almost linear dependence on hydrogen pressure above 200 torr. These workers studied the effect of vacuum treatment prior to reduction at various elevated temperatures, with particular regard to the nickel produced (allegedly from reduction by stopcock grease) and subsequent reduction rate at 240° C. The rate was maximal after 350° C vacuum treatment, even though the nickel was formed more extensively at higher temperatures. Microscope examination showed the reduction of individual grains proceeding from few nuclei (frequently one) within a grain. Using a novel selective adsorption technique it was possible to measure both the total and the nickel surface at partial reductions. The nickel was abstracted from the grains of partly reduced oxide by treatment with bromine in methanol. Further adsorption measurements then gave the total oxide surface. The apparent oxide surface (that available for physical adsorption before bromine treatment) was found to decrease linearly with increasing extent of reaction. The nickel surface showed initial linear growth followed by a maximum and then a slight decline. The total oxide surface passed through an initial maximum and declined quickly thereafter. The difference between the total and apparent oxide surface is considered to be the reaction interface area. It rose during the first 25% of reduction and declined during the last 25%, showing a plateau in between. These remarks apply to experimental conditions under which the induction period extends to about 30% reduction. Charcosset and his fellow workers came to the conclusion that the number of nuclei depends on the hydrogen pressure and that vacuum treatment above 250° C is a way of generating artificial nuclei. They maintained that once the nuclei are established the advance of the interface is largely independent of the pressure of hydrogen and water vapor.

Despite the many investigations of nickel oxide reduction by hydrogen there is little agreement between the results. Activation energies ranging from 10.2 to 26.4 kcal/g-mole have been reported by authors referred to earlier. This may be due to the dependence of the reaction rate on the lattice structure of the oxide which has been discussed at length above. A further difficulty is an apparent change of the activation energy at the Néel temperature of the oxide (approximately 530° K) which has been reported by several authors.[123,124,127–130] Finally, the significance of the pore structure of the oxide and the role played by mass transport of gaseous reactant and product seems to have been ignored by most early investigators, although the high activation energies reported do indicate that experiments were being performed under conditions of chemical step control.

Mine *et al.*[125] studied the reduction of nickel oxide over a wide range of temperatures and examined the pore structure of the nickel product using scanning electron microscopy. The pore structure was found to depend on reduction temperature. Below 400° C, chemical reaction was considered to be rate controlling, whereas at higher temperatures pore diffusion became important.

Kurosawa *et al.*[126] carried out one of the few investigations in which the pressure of the hydrogen was varied (up to 31 atm). Experiments were carried out at 200, 220, and 240° C where reaction was likely to have been chemical step controlled. The dependency of the rate upon the hydrogen pressure indicated Langmuir–Hinshelwood kinetics but in the vicinity of 1 atm the dependence was such that an assumption of first-order kinetics would not be grossly inaccurate.

Chiesa and Rigaud[127] reduced powdered nickel oxide and nickel oxide compacts in the temperature range of 180 to 400° C. Above 250° C, water vapor strongly inhibited the reaction.

Szekely and Evans[128] studied the reduction of spherical nickel oxide compacts in the temperature range from 221 to 756° C. The oxide was characterized by mercury penetration porosimetry, scanning electron microscopy, and X-ray line broadening and the kinetic measurements were interpreted in terms of the grain model (see Figure 4.2-19). Table 4.2-2 shows the effect of pellet size and porosity. Subsequently, these authors[129] reported on reduction experiments at higher temperatures (up to 921° C) where sintering of the nickel product resulted in a dense outer shell of nickel which was nearly impervious to hydrogen) surrounding the remaining oxide. This dense shell brought the reaction to a near standstill (Figure 4.2-20).

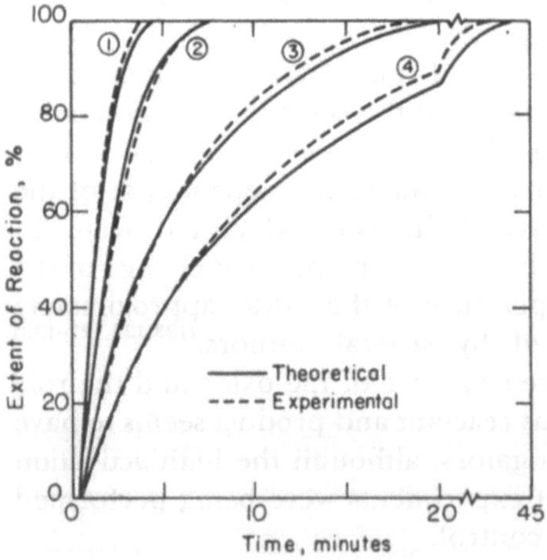

Figure 4.2-19. Plot of the extent of reduction of nickel oxide pellets by hydrogen at approximately 300° C,[131] and the effect of pellet size and initial porosity (cf. Table 4.2-2).

Table 4.2-2. Effect of Pellet Size and Porosity[131]

Curve[a]	Radius, cm	Porosity
1	0.77	0.73
2	1.18	0.73
3	0.80	0.38
4	1.13	0.38

[a] Figure 4.2-19.

Another application of the grain model to the interpretation of experimental data on nickel oxide reduction was that of Szekely *et al.*[129] They reduced nickel oxide pellets of various sizes and used the grain model to extrapolate the data to zero particle size where pore diffusion and external mass transfer effects would be zero. This is illustrated in Figure 4.2-21. Graph (a) shows reduction curves for three different pellet sizes (II, III, and IV). The slopes of these curves are plotted in

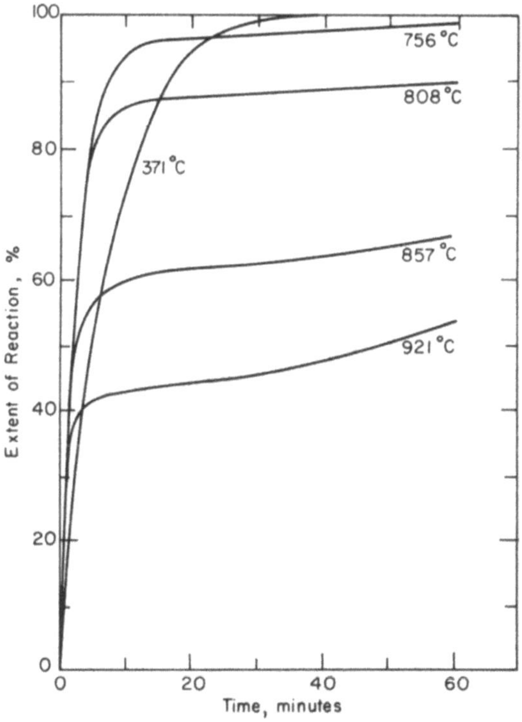

Figure 4.2-20. Plot of extent of reaction against time for high-temperature reduction of nickel oxide showing how the sintering of the nickel product impedes reaction.[131]

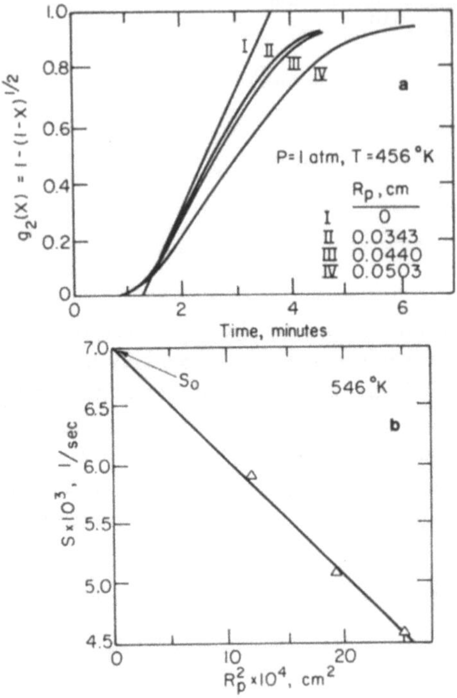

Figure 4.2-21. Illustration of the extrapolation of kinetic data to zero particle size in order to eliminate pore diffusion and external mass transfer effects.[129]

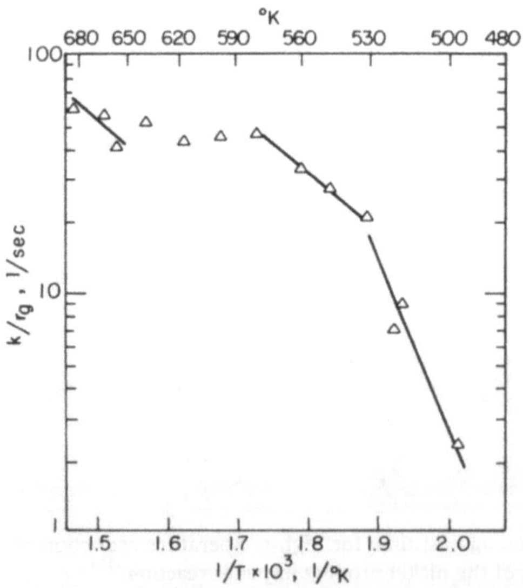

Figure 4.2-22. Arrhenius plot of kinetic data obtained by Szekely *et al.*[129] on the reduction of nickel oxide by hydrogen.

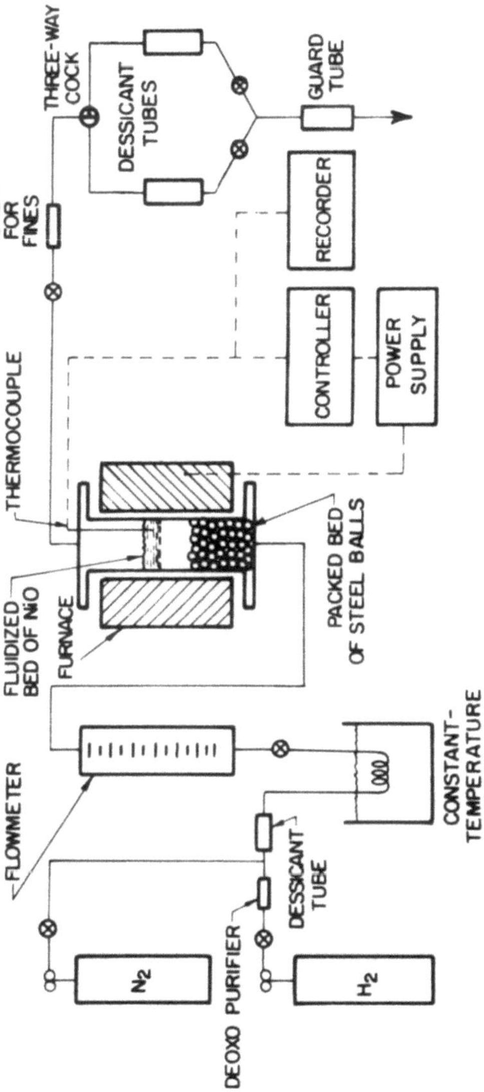

Figure 4.2-23. A batch fluidized-bed apparatus for studying the gaseous reduction of oxides.[132]

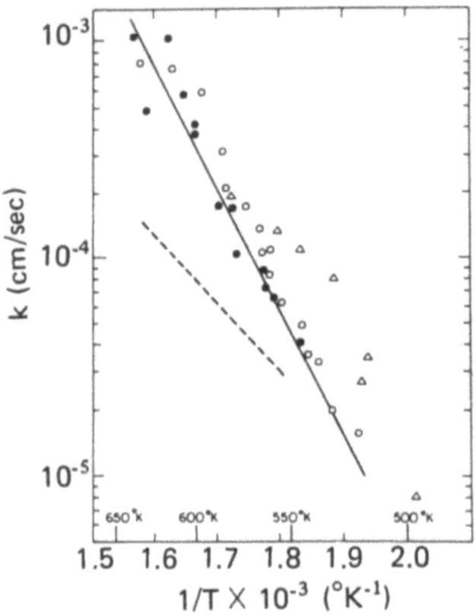

Figure 4.2-24. Comparison of the results of the fluidized-bed apparatus, the single-pellet (gravimetric) apparatus, and other investigations of the reduction of nickel oxide by hydrogen.[132] △, Szekely *et al.*[129]; ---, Szekely and Evans.[128,131] Present work: —●—, fluidized-bed results; ○, single-pellet results.

graph (b) and a slope at zero particle size (S_0) obtained by extrapolation. This slope appears as line I in graph (a). Figure 4.2-22 is an Arrhenius plot of the reaction rate constant divided by the grain radius obtained in this way.

An alternative technique for measuring reduction rates free of the influence of external mass transfer or pore diffusion was described by Evans *et al.*[132] Mass transport limitations were avoided by reducing small particles (−65 + 100 mesh) in a batch fluidized-bed reactor. As depicted in Figure 4.2-23, the progress of reaction was followed by determining the water vapor content of the exit gases using drying tubes.† The nickel oxide was characterized by BET surface area measurement and scanning electron microscopy. Figure 4.2-24 provides a comparison of the results of rate measurements using this apparatus and using the more conventional single pellet (gravimetric) apparatus. Also included in this figure are the results of two previous investigations[128,129] using a different nickel oxide.

Babushkin *et al.*[133] also employed a fluidized bed in a study of the reduction of nickel oxide using natural gas. They measured the progress of reaction by periodically discharging the bed and determining the extent of reaction by various experimental techniques.

Szekely and Hastaoglu[134] reduced mixtures of nickel oxide and hematite by hydrogen and interpreted the results using the grain model.

† An alternative procedure used with this apparatus in later studies, is to accurately measure the flow of the gaseous reductant to the bed and intermittently or continuously analyze the exit gas by gas chromatography or other suitable technique.

Selective reduction of the nickel oxide was found to be possible under conditions of chemical step control but not of pore diffusion control.

In the paper referred to earlier on hydrogen reduction, Minc *et al.*[125] also reported the results of experiments on carbon monoxide reduction of nickel oxide. Reduction below 600° C was prevented by the formation of a protective layer of Ni_3C. Above 600° C reduction by carbon monoxide would proceed but was much slower than in reduction by hydrogen.

Krasuk and Smith[135] performed rate measurements on the reduction of nickel oxide pellets of varying porosity by carbon monoxide in the temperature range from 566 to 796° C. Formation of Ni_3C was not observed; the composition of the CO–CO_2 mixture used for reduction was selected (partial pressure of CO less than 0.33 atm) so that the formation of Ni_3C was not expected on thermodynamic grounds. Reaction appeared to be first order in carbon monoxide and due allowance was made for intraparticle diffusion in interpreting the results obtained. The oxide pellets were characterized by mercury porosimetry and air pycnometry. An activation energy of 47 kcal/g-mole was determined over the temperature range from 566 to 682° C.

Szekely and Lin[136] reduced porous nickel oxide disks in the temperature range from 847 to 1100° C using carbon monoxide. The experimental results were analyzed in terms of the grain model. Although reaction was proceeding under mixed (pore diffusion and chemical step) control the authors were able to determine the reaction rate constant from initial reaction rates by correcting for the influence of pore diffusion.

Other papers on reduction of nickel oxide are those of Bielanski *et al.*,[137] Oates and Todd,[178] and Tumarev *et al.*[179]

4.2.4. Lead Oxide Reduction

Despite the importance of lead oxide reduction by carbon monoxide in the lead blast furnace, the authors were unable to find a publication on this topic although Chesti and Sircar[140] have investigated reduction by carbon, as have Ashin *et al.*[141]

Culver *et al.*[142] have reduced lead monoxide by hydrogen. The lead monoxide used was porous and reaction was observed to take place throughout the solid and to be insensitive to particle size indicating chemical step control. The rate appeared to be first order with respect to hydrogen. Water vapor inhibited the reduction. An activation energy of 39 kcal/g-mole was measured over the temperature range from 475 to 775° C.

Haertling and Cook[143] studied the same reaction and also the reduction of lead silicates, zirconate, and titanate. An activation energy of 18.4 kcal/g-mole over the temperature range from 300 to 540° C was reported.

4.2.5. Reduction of Other Oxides by Gases

Examples of investigations of the kinetics of reduction of other oxides are those of Taylor *et al.*[144,145] and Themelis and Yannopoulos[146] on copper oxide reduction; Vassilev *et al.*,[147] Hawkins and Worrell,[148] and Lavrenko *et al.*[149] on the reduction of molybdenum oxide by hydrogen; Decroly and Winand,[150] Bear and Caney,[151] and Rawlings and Ibok[152] on the reduction of tin oxides; Hougen *et al.*[153] on the reduction of tungsten oxides; and Guger and Manning[154] on the reduction of zinc oxide.

4.2.6. Reduction of Oxides by Solid Carbonaceous Materials

In the majority of industrial processes in which oxides are reduced to metal by carbon monoxide the reducing gas is produced in a separate part of the *reactor*, such as an iron blast furnace, or even in a separate reactor. However, it is possible to reduce oxides using coke (or other carbonaceous materials) which is more intimately mixed with the oxide, for example, in the same pellet. Reduction will then proceed by two coupled gas–solid reactions:

$$\text{Metal oxide} + CO \rightarrow \text{Metal} + CO_2$$

$$CO_2 + C \rightarrow 2\,CO$$

Examples of studies of such reactions are the papers by Baldwin[155] on the reduction of iron oxides by coke; Otsuka and Kunii[156] on the reduction of ferric oxide by graphite powder; Sharma *et al.*[157] on the reduction of nickel oxide by various solid carbonaceous materials; Kohl and Marinček[158,159] on the reduction of various metal oxides by graphite; Rao[160] on the reduction of hematite by various forms of carbon; Maru *et al.*[161] and Kulkarni and Worrell[162] on the reduction of chrome oxide by chrome carbide; and by El-Guindy and Davenport[163] on the reduction of ilmenite by graphite, in addition to papers already mentioned.[140,141]

Such reactions have recently been the subject of a detailed theoretical examination by Sohn and Szekely,[164] Rao,[165] and Rao and Chuang.[166,167]

4.2.7. Direct Reduction Processes

For centuries, the iron blast furnace has been the traditional method for producing iron. The modern blast furnace, has a production rate as high as 4000 tons per day. In the past two decades, alternatives to the blast furnace, which are economical on a small scale and do not require coking coal, have been developed. Natural gas and noncoking coal are used instead of coking coal either directly, or after conversion into hydrogen and carbon monoxide mixtures. These alternatives are *direct reduction processes.*

In some countries such as Venezuela, Iran, and Mexico, with abundant supplies of natural gas but limited supplies of coking coals, direct reduction with natural gas are the main iron production processes. In North America and Western Europe, direct reduction processes also produce significant tonnages of iron. In 1977, the worldwide direct reduction plants had a total production capacity of 11 million tons of sponge iron annually, as compared to a global blast furnace production of 650 million tons in 1975. The rapid development and growth of these direct reduction processes in recent years has proved these methods of iron production to be reliable from technical and economic viewpoints.[168]

Approximately 50 direct reduction processes have been proposed for the production of iron.[179] Roughly, these processes may be grouped into four categories according to their reduction reactors: fixed beds, shaft furnaces, fluidized beds, and rotary kilns. The product is usually a porous solid called *sponge iron* in the form of pellets, lumps, or particles which are then compressed into briquettes. Most of these iron products are used in electric arc furnaces for the production of steel. Some sponge iron is used as coolant in oxygen steel processes or as raw material in the open hearth furnace. The blast furnace may also use partially reduced ores to achieve further economies.

The selection of the process for the reduction depends mainly on the characteristics of the ore and reducing agent.

4.2.7.1. Fixed Bed Processes

The HyL Process[170] is one of the oldest direct reduction processes of commercial significance. Plant capacities are 600–1000 tons per day. The reduction is completed in the fixed-bed batch reactor with temperatures varying between 900 and 1100° C. The process flow diagram is shown in Figure 4.2-25. The reactors are charged from the top with high-grade lump iron ore or pellets. Gas flow is downward through the reactors. The reducing agent is a mixture of carbon monoxide and hydrogen gas

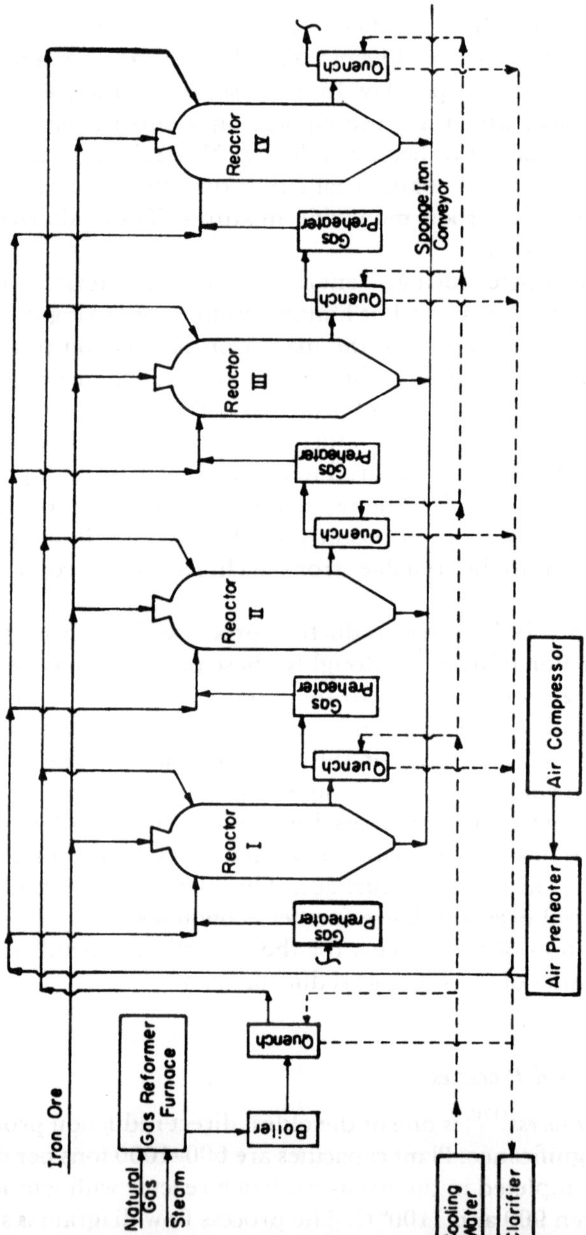

Figure 4.2-25. The HyL process.

Table 4.2-3. Operation of the HyL Process

Time elapsed, hr	Reactor I	Reactor II	Reactor III	Reactor IV
0–3	Cooling	Primary reduction	Secondary reduction	Discharging and charging
3–6	Discharging and charging	Cooling	Primary reduction	Secondary reduction
6–9	Secondary reduction	Discharging and charging	Cooling	Primary reduction
9–12	Primary reduction	Secondary reduction	Discharging and charging	Cooling

which is produced by catalytic steam reforming of natural gas or other hydrocarbons. The reducing gas has approximately the following composition: 74% H_2, 13% CO, 8% CO_2, and 5% CH_4. Each of the four reactors used passes through four 3-hr stages:

(1) discharging of final product and then charging of fresh ore,
(2) *secondary reduction* in which a partly used reducing gas is passed through the reactor,
(3) *primary reduction* in which a fresh reducing gas is passed through the reactor, and
(4) cooling of the product by cool fresh reducing gas.

Table 4.2-3 illustrates the operation. Typically, the iron ore charged has an iron content of 66%. The product has an iron content of around 85% and is not pyrophoric.

4.2.7.2. *Shaft Furnaces*

The Midrex process[171] is a continuous process in which the iron ore pellets or lump ore are continuously charged into the top of the furnace and the sponge iron product is discharged continuously from the bottom. The reduction zone is at the upper part of the reactor where the iron ore is heated and reduced to iron by countercurrent flow of the reducing gas at a temperature up to 875° C. The lower part of the reactor is the cooling zone for cooling the sponge iron before discharging. The process flow sheet is shown in Figure 4.2-26.

The reducing gas, containing 90–95% CO plus H_2, is generated by reforming recycled top gas. Natural gas, coke-oven gas, and methane-rich gases are used for this purpose.

The percent metallization of the product varies from 85 to 95% to meet desired specifications.

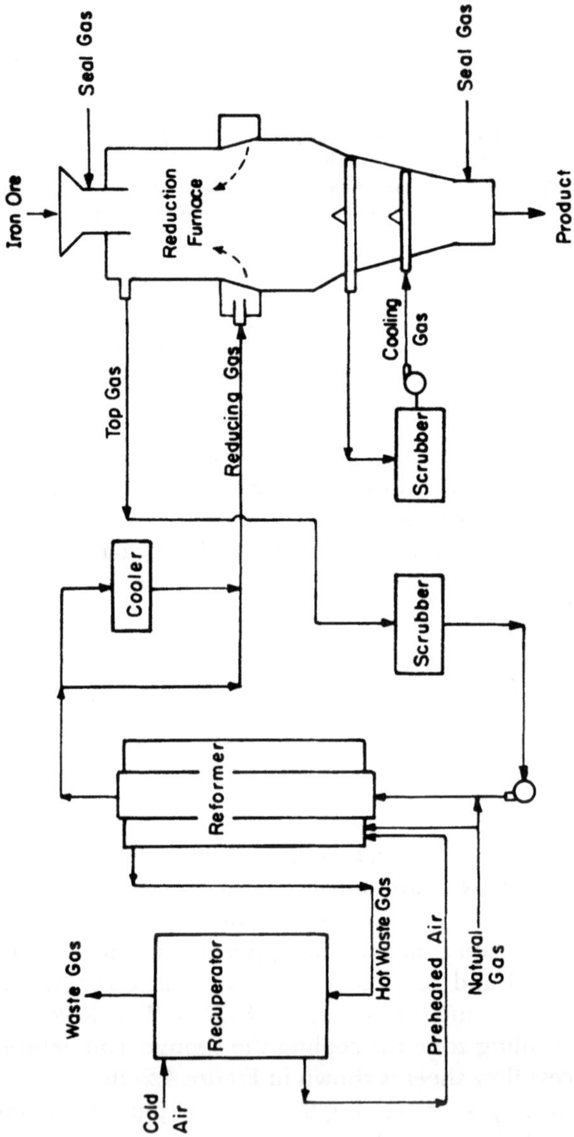

Figure 4.2-26. The Midrex process.

The Armco process[172] is similar to the Midrex process. The reducing gas consisting of 90% H_2 and CO is produced by a catalytic steam–methane reforming system. The iron ore pellets are reduced by H_2–CO gas mixture countercurrently in a shaft furnace at a temperature of 870° C. The cooling gas is not recycled but used as fuel in the reforming furnace. The product typically is 92% Fe.

The Purofer process[173] was developed in West Germany. It is similar to the two other processes except that the reducing gas is at a higher temperature (950° C). Cooling gas is dispensed with and the product iron is discharged hot (about 800° C) into a briquetting machine.

4.2.7.3. *Fluidized Bed*

The HIB process[174] is a two-stage fluidized-bed process for high-grade ore fines reduced with a H_2–CO gas mixture produced by steam reforming of natural gas at a temperature of 700 to 750° C and pressure of about 50 psi. This process was developed by U.S. Steel Corporation.

The dried ore with −10 mesh size is preheated in two preheater beds, then fed continuously into the top stage of the fluidized reactor. The partly reduced iron ore in the top stage flows continuously to the bottom stage and is then reduced to iron. The spongy iron particles (about 86.5% metallized) are briquetted. The top gas leaving the reactor is used as fuel in the reformer furnaces.

The FIOR process[175] (fluidized iron ore reduction process), uses four fluidized-bed reactors operating in series. The process flow chart is shown in Figure 4.2-27. The reducing agent is hydrogen gas preheated to 550° C. The ore fines (−12 mesh) are charged as batches into the top bed for preheating. The ore moves to the other three reactors sequentially for reduction at a temperature of 800° C. The hot product with 90–95% metallization is briquetted by double-roll briquetters.

4.2.7.4. *Rotary Kiln*

The SL–RN process[176] is a kiln process into which the ore, excess solid reductant (e.g., noncoking coal), and a small amount of flux (limestone or dolomite) are charged. The rotary kiln is fired with natural gas, oil, or solid fuel to a reduction temperature of 900–1100° C. The use of a wide variety of solid fuels is its advantage. Constant reduction temperature throughout the kiln is maintained by blowing air through shell burners for combustion. Either ore pellets or lumps could be fed into the kiln and move countercurrently to the flow of hot gas.

The Krupp Sponge Iron process[177] is a rotary kiln process used in Europe. High-grade lump ores or concentrates (with 65% Fe or more)

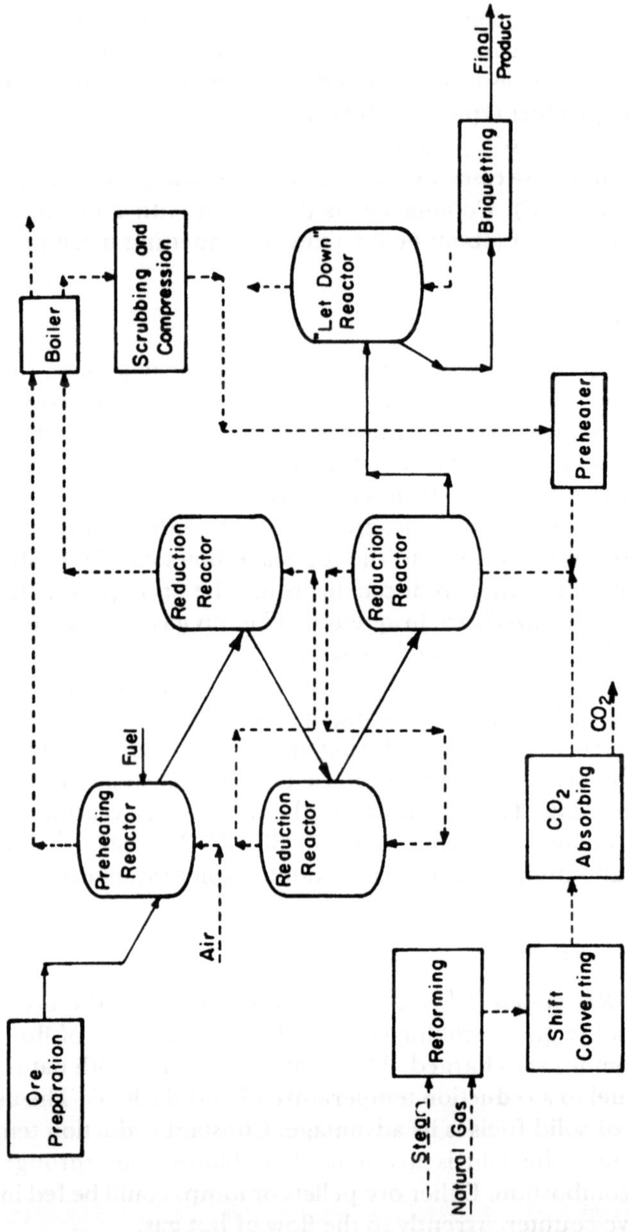

Figure 4.2-27. The FIOR process.

have been used as raw material in the recent years. The process is similar to the SL-RN process.

4.3. Calcination (HYS and ETT)

Calcination of limestone is an important reaction in many extractive metallurgical and chemical processes. Numerous investigations on this subject have been reported in the literature. The early ones have shown that the decomposition of limestone starts at the outside surface and proceeds toward the center. The reaction takes place within a thin layer between the undecomposed limestone core and the product (lime) layer. Although this picture is generally accepted for the progress of reaction, there have been conflicting claims regarding the controlling mechanism.

Thus, one or a combination of the component steps, namely the decomposition reaction, diffusion and heat conduction in the product layer, and external heat and mass transfer, has been reported to control the overall rate.[178–185] Upon closer examination of these investigations, it is apparent that the various controlling mechanisms hold for the specific reaction conditions under which the calcination was carried out, the most important parameters being the size of the limestone particles and the reaction temperature. Thus, the cases of different controlling steps represent various asymptotic regimes of the general scheme. The general rate expression may be obtained by incorporating all the component steps, as was discussed in Chapter 1 for nonisothermal reactions of porous or nonporous solids, depending on whether the limestone is initially porous or not. It is noted, however, that in calcination there is no gaseous reactant and thus the mass transport involves only the gaseous product, carbon dioxide. The transport of mass may involve the bulk flow of gaseous product toward the outside of particle. This phenomenon, which is significant at a higher rate reaction, has often been overlooked. One of the few analyses incorporating the bulk flow has been derived by Turkdogan *et al.*[184]

The derivation and discussion of the complete theory of calcination, including intrinsic decomposition kinetics, heat and mass transport, and transient effects is beyond the scope of this monograph. We shall, therefore, limit our discussion to the specific case most relevant to calcination encountered in metallurgical processes. The conditions under which the following analysis is valid are:

1. Particle size is sufficiently large so that the overall rate is controlled entirely by heat and mass transport, and at the interface equilibrium exists between the gas and solid.

2. Temperature is sufficiently high (and hence the rate of generation of carbon dioxide is fast enough) so that the gas at the reaction interface is essentially pure carbon dioxide.

3. The lime layer is sufficiently permeable so that the internal pressure is not much greater than the ambient pressure. Therefore the interface temperature remains constant at a value corresponding to a CO_2 partial pressure equal to the ambient pressure. Furthermore, the enthalpy of dissociation is rather large ($\Delta H = 38{,}550$ cal/g-mole), and thus a moderate change in the CO_2 partial pressure does not cause a significant change in the decomposition temperature.

4. The sensible heat is much smaller than the heat of decomposition.

Under these conditions, the temperature at the reaction interface, and also at the center, of limestone decomposing at 1-atm pressure is constant at 902° C, and the overall rate is controlled entirely by heat transfer. The relationship between conversion and time can be obtained by considering a heat balance including external heat transfer and the conduction through the product layer, as follows. (The procedure is entirely analogous to that in the case of the diffusion-controlled reaction of nonporous solids with gases discussed in Chapter 1.)

The energy balance in the product layer for a sphere is

$$\frac{d}{dr}\left(r^2\frac{dT}{dr}\right)=0 \tag{4.3-1}$$

with boundary conditions

$$T=T_d \qquad \text{at } r=r_c \tag{4.3-2}$$

and

$$-\lambda_e\frac{dT}{d\gamma}=h(T_b-T) \qquad \text{at } r=r_p$$

where T_d is the decomposition temperature; r_c and r_p are the radii of the unreacted core and the entire particle, respectively; λ_e is the effective thermal conductivity of lime layer; h is the external heat transfer coefficient; and T_b is the bulk temperature. The rate of heat transfer to the interface, obtained by solving equation (4.3-1) together with the boundary conditions, is related to the rate of decomposition of limestone by the following:

$$-\lambda_e\frac{dT}{d\gamma}\bigg|_{r=r_c}=\rho\Delta H\frac{dr_c}{dt} \tag{4.3-3}$$

where ρ is the bulk molar density of limestone, and ΔH is the molar enthalpy of dissociation of $CaCO_3$.

The relationship between r_c and time is then given by

$$1+2\left(\frac{r_c}{r_p}\right)^3-3\left(\frac{r_c}{r_p}\right)^2+\frac{2\lambda_e}{hr_p}\left[1-\left(\frac{r_c}{r_p}\right)^3\right]=\frac{6\lambda_e(T_b-T_d)}{\rho\Delta H r_p^2}t \qquad (4.3\text{-}4)$$

The limestone conversion, X, is related to r_c by

$$X=1-\left(\frac{r_c}{r_p}\right)^3 \qquad (4.3\text{-}5)$$

In general, for a slablike, a cylindrical, and a spherical particle of limestone, the relationship between X and time can be written as follows:

$$P_{F_p}(X)+\frac{2X}{\mathrm{Nu}^*}=\frac{2F_p\lambda_e(T_b-T_d)}{\rho\Delta H}\left(\frac{A_p}{F_pV_p}\right)^2 t \qquad (4.3\text{-}6)$$

where

$$P_{F_p}(X)\equiv\begin{cases} X^2 & \text{for } F_p=1 \text{ (infinite slab)} \\ X+(1-X)\ln(1-X) & \text{for } F_p=2 \text{ (long cylinder)} \\ 1-3(1-X)^{2/3}+2(1-X) & \text{for } F_p=3 \text{ (sphere)} \end{cases} \qquad (4.3\text{-}7)$$

$$\mathrm{Nu}^*\equiv\frac{h}{\lambda_e}\left(\frac{F_pV_p}{A_p}\right), \quad \text{the modified Nusselt number} \qquad (4.3\text{-}8)$$

Narsimhan[181] showed agreements between equation (4.3-6) and the experimental data by Satterfield and Feakes.[186]

For the case of negligible resistance due to external heat and mass transfer, Turkdogan *et al.*[184] obtained a solution which is valid in the case where only conditions 1 and 4 above are required. Their solution can be used even in the case where the surface temperature is less than 902° C. (Thus, CO_2 partial pressure at the interface is less than 1 atm.) Under these conditions, they showed that for given CO_2 partial pressure and temperature at the external surface, the partial pressure of CO_2 and thus the temperature at the discomposition interface remain constant regardless of the position of the interface. By considering the diffusion and bulk flow of CO_2 combined with the condition of heat through the product layer Turkdogan *et al.* obtained the following relationship:

$$\ln\left(\frac{P-P^b_{CO_2}}{P-P^c_{CO_2}}\right)=\frac{RT}{P\Delta H}\frac{\lambda_e}{D_e}(T_b-T_d) \qquad (4.3\text{-}9)$$

which is independent of r_c. The decomposition temperature at the interface, T_d, can be determined for a given T_b by combining equation

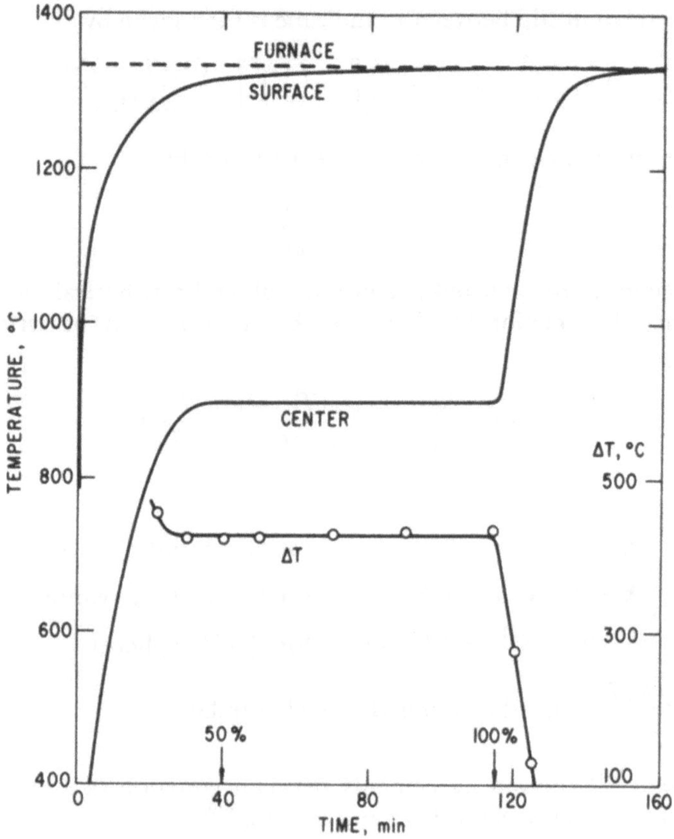

Figure 4.3-1. Variation of surface and center temperatures during calcination of an 8.3-cm-diam. limestone sphere at atmospheric pressure in argon at a furnace temperature of 1332° C.

(4.3-9) with the relationship between the equilibrium partial pressure of CO_2 and temperature. The experimental data of Baker[(187)] and Hills[(188)] can be expressed by the following:

$$\log p_{CO_2} = -\frac{8427}{T} + 7.169 \qquad (4.3\text{-}10)$$

where P_{CO_2} is in atm and T is in °K.

Turkdogan *et al*[(184)] measured surface and center temperatures during the calcination of spherical limestone samples in argon at atmospheric pressure. The results shown in Figure 4.3-1 are for an 8.3-cm-diam. sphere; times for 50 and 100% calcination are indicated. In this case, the center temperature remained unchanged at 897° C in the latter half of the calcination. At the end of the calcination the center temperature rose rapidly, approaching the surface temperature of 1330° C. After about 50% calcination, the surface temperature rose only a few degrees. The temperature difference $\Delta T = T_b - T_d$ was essentially constant at about 425° C for the latter half of the calcination. These experimental findings substantiate the reaction model assumed in the derivation of the foregoing equations.

It is evident from equations (4.3-9) and (4.3-10) that the temperature difference ΔT is a function of the surface temperature. That is, if the rate of dissociation of calcium carbonate at the limestone–lime interface is fast relative to other rate processes involved, the partial pressure of CO_2 in the pores at the interface is that coexisting in equilibrium with

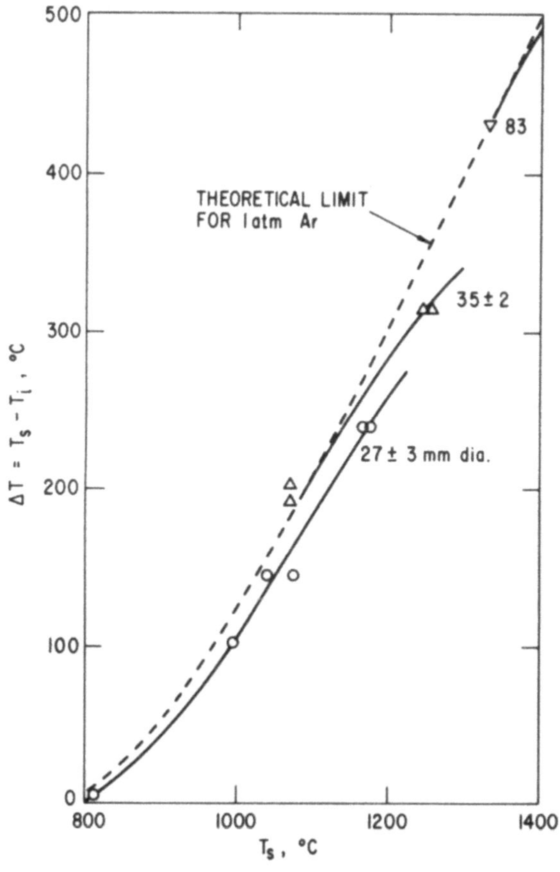

Figure 4.3-2. Steady-state temperature difference between surface and core temperatures as a function of the average surface temperature at 50–95% calcination.

CaO and $CaCO_3$ at the interface temperature $T_d \simeq$ center temperature. The dashed curve in Figure 4.3-2 depicts the relation ΔT vs. T_b at this theoretical limit, calculated by simultaneous solution of equations (4.3-9) and (4.3-10) for calcination in 1 atm argon. The curves drawn through the data points are those measured with limestone spheres of an average diameter of 27, 35, and 83 mm. Departure of the experimental points from the theoretical limit is a manifestation of excessive pressure build up at the interior of the stone or existence of nonequilibrium at the calcination front. For the equilibrium conditions to prevail with $T_d = 930°$ C, the excess pressure in the interior would have been about 0.48 atm which is much greater than that observed experimentally. Therefore, it was concluded by Turkdogan *et al.* that for the particle sizes less than 4-cm diam., the equilibrium conditions for the dissociation of calcium carbonate do not prevail at the calcination front.

The experimental results of Turkdogan *et al.*, reproduced in Figures 4.3-3 and 4.3-4 for limestone spheres of 1.8–14 cm diam., indicate that equation (4.3-6) with $Nu^* \rightarrow \infty$ fits the data well for conversion higher than 50%. From the slopes of the lines in Figures 4.3-3 and 4.3-4 and known values of r_p and ΔT, the effective thermal conductivity of burnt lime was found to be 1.26×10^{-3} cal/cm sec deg. In the early stages of calcination, however, chemical kinetics, external heat transfer, and/or transient effects appear to play an important role in

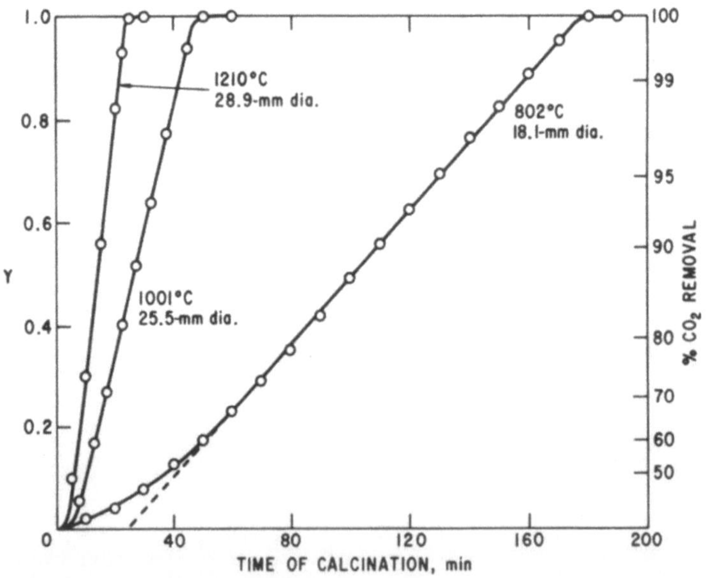

Figure 4.3-3. Calcination of limestone spheres in argon. $Y = 3 - 2F - 3(1-F)^{2/3}$, where F is fraction of CO_2 removed.

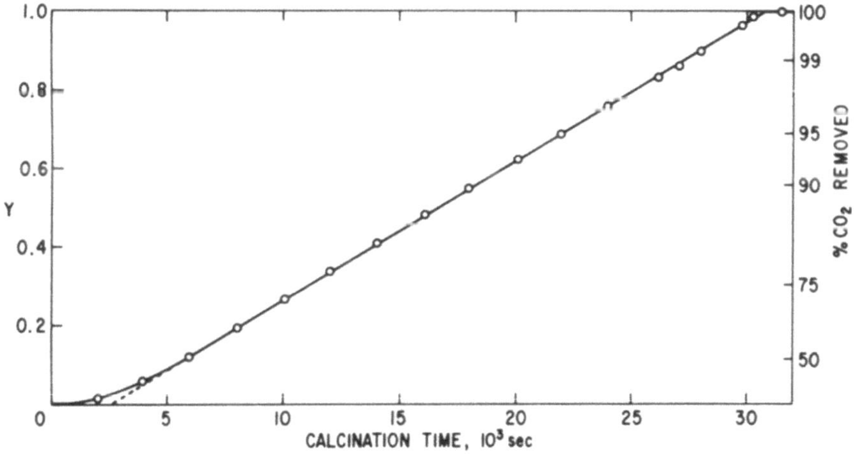

Figure 4.3-4. Calcination of 14-cm-diam. (3.58-kg) limestone spheroid in air at 1135° C. $Y = 3 - 2F - 3(1-F)^{2/3}$, where F is fraction of CO_2 removed.

determining the overall rate. These effects become more important at lower temperatures.

The reactivity of lime in various applications is much affected by its pore structure. The effects of time and temperature of calcination on the pore surface area of burnt lime are shown in Figure 4.3-5. The pore area of lime from crushed limestone calcined for 500 hr at 600° C is in accord with that of lime treated at higher temperatures for shorter periods. With increasing calcination temperature and soaking time the pore surface area decreases, owing to crystal growth and coarsening of the pore structure.

Mass flow in capillaries of porous media induces a balancing pressure gradient in the flow direction. Therefore, the diffusion or flow of CO_2 through the porous lime layer during calcination should be accompanied by a pressure buildup in the interior of the sample. If there are no additional pressure effects caused by diffusion, the pressure buildup can be approximated with the Hagen–Poiseuille law. On this basis Turkdogan *et al.* derived the following equation for the permeability of burnt lime:

$$\mathcal{P} = \frac{6RT\lambda_e\eta}{P\Delta H}\frac{\Delta T}{\Delta P} \tag{4.3-11}$$

where η is the viscosity coefficient of the gas mixture in the pores. The excess pressure ΔP measured in the calcination experiments was in the range from 10 to 60 torr with the corresponding permeability of 1×10^{-9} to 5×10^{-9} cm^2 for the calcination temperatures from 900 to 1175° C.

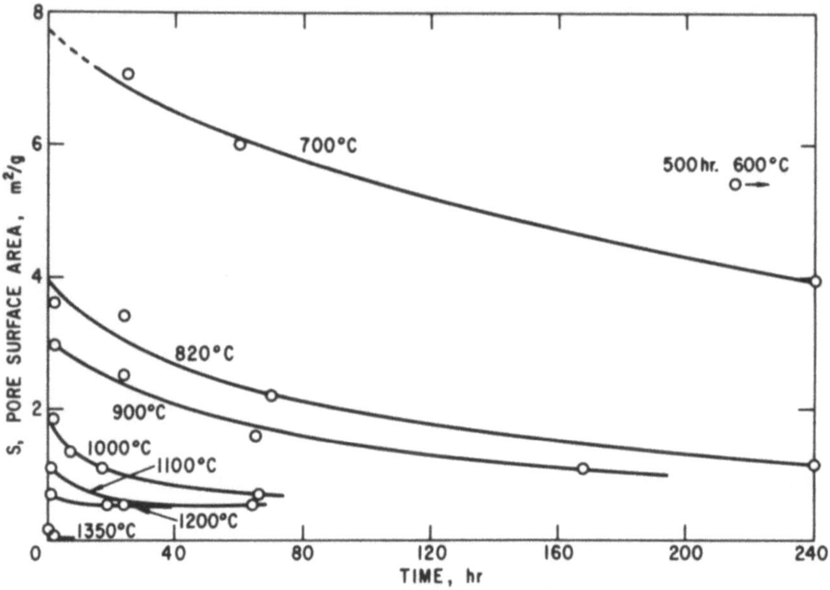

Figure 4.3-5. Effects of heat treatment time and temperature on pore surface area of burnt lime.[184]

In summary, the calcination reaction proceeds as, and only as, the latent heat is received by the limestone and converts it into CaO and CO_2. This heat flux, and thus the calcination rate, is determined by external conditions, the geometry of the sample, and the application of Fourier's law with appropriate boundary conditions. As already noted, with fixed furnace temperature, the temperature difference between the surface and center (interface) is essentially constant over a substantial range of conditions. The application of Fourier's law is thus rather simple. Furthermore, for given surface conditions and known effective thermal conductivity, effective mass diffusivity, density, latent heat, and equilibrium for the reaction, the rate of calcination is fixed and may be computed if the pressure buildup in the interior of the specimen is negligibly small and if equilibrium is essentially established at the interface between the limestone, lime, and CO_2. However, the experimental observations of Turkdogan *et al.* have indicated that these limiting conditions are not exactly fulfilled. A small pressure build-up was observed and the central temperature (under the conditions of steep gradient) was found higher than that calculated on the postulate of (local) equilibrium. Furthermore, the observed diffusiveness of boundary between limestone and lime indicates a temperature range of coexistence at essentially fixed partial pressure of CO_2, again indicating departure from equilibrium conditions. Thus, it appears that there is a small but observable effect of the finite chemical reaction rate. In some cases this may decrease the temperature gradient and hence the rate of calcination by perhaps 10%. However, when the temperature gradient is relatively small (low surface temperature or large specimen), this effect is so small, it cannot be observed.

4.4. *Sintering (IBC)*

4.4.1. *Introduction*

Understanding sintering is important in many different disciplines. Because of the great need to understand processes involved in powder metallurgy and in ceramics, sintering has been studied by scientists and engineers in these areas to a greater extent than elsewhere. However, sintering is just as important in many of the other fields of endeavor, ranging from extractive metallurgy to polymers and catalysis. To the powder metallurgists and ceramists sintering is vital, especially if it can be carried out to full densification of the materials involved. To the extractive metallurgist it is important that sintering produce a strong adherent compact. To the worker in catalysis densification is an undesirable property and to avoid loss of surface area would be most desirable. In reality, avoiding sintering is a much more difficult problem than obtaining sintering and densification.

4.4.2. *The Driving Force for Sintering*

From thermodynamics, it is easy to illustrate that the lowest state of free energy is the most stable state of the material. If a given material has a large surface area, it is in a higher free-energy state than the same material with minimum surface area. The difference in free energy or driving force for sintering is simply the surface free energy per unit area, γ, multiplied by the surface area of the material.

In terms of particle size, the driving force for sintering increases as the particle size decreases. Perhaps the most dramatic illustration of the effect of surface area can be obtained in describing the sintering of thoria (ThO_2), which has a melting point of 3220° C, but when obtained as a compact of particles from a colloidal sol where the particles are measured in terms of 100 Å, the consolidated gel derived from the colloidal suspension of thoria particles sinters at temperatures in the order of 1000° C.

Not every material can be sintered at such a small fraction of its melting point as in the case of thoria gel; but not all materials may be obtained in such a finely divided state of matter. For micron-sized particles of various metals and oxides, the sintering temperature is ordinarily about two-thirds of the melting point of the materials.

In the case of powder metallurgy, most metals can be consolidated so well by ductile deformation that it is not necessary to have them submicron size. For ductile metals, sintering strengthens the compacted particles that may be hundreds of microns in diameter. This is done by diffusion that once again eliminates the free surface and substitutes interfaces between the particles for the free surface that was originally a property of the powder compact.

For oxides, carbides, and the refractory metals, particles from 0.01 to 10 μm are employed. In general, the finer sizes are more difficult to compact and result in larger shrinkages during sintering.

In spite of small size, there are some materials that are very difficult to densify by sintering. Covalently bonded carbides are prime examples. To gain a better understanding of sintering, some of the recent advances will be reviewed.

4.4.3. Stages of Sintering

For purposes of coming to a better understanding of why and how sintering takes place, it is convenient to learn about the techniques for analytically describing sintering.[189,192] It has been convenient for many to describe a compact of particles in terms of spheres and spherical geometry. We realize that in reality very few systems approach spherical geometry, but it is difficult to describe nonspherical particles with equations. Most of us would prefer the simplicity of a spherical model. If we talk in terms of spheres, it is possible then to describe material transport in terms of spheres forming necks between them. This is illustrated in Figure 4.4-1 where a group of spheres is just touching each other. Up to 5–10% of linear shrinkage of this compact of spherical particles it is possible to describe sintering in terms of the neck formed between the particles. This is termed *the initial stage of sintering.*

Perhaps more shrinkage actually takes place during what has been called *the intermediate stage of sintering.* In this instance, the spheres touch each other to the point that the growth of necks between the particles actually interfere with each other. In this instance, it is more convenient to talk in terms of the *passageways of pore space* that penetrate the particle compact. One particular model describes this as a series of tubular pore space that decreases in diameter but is larger and smaller in diameter along its length, depending on the original placement of the particles in the particle compact. These tubular passageways through the particle

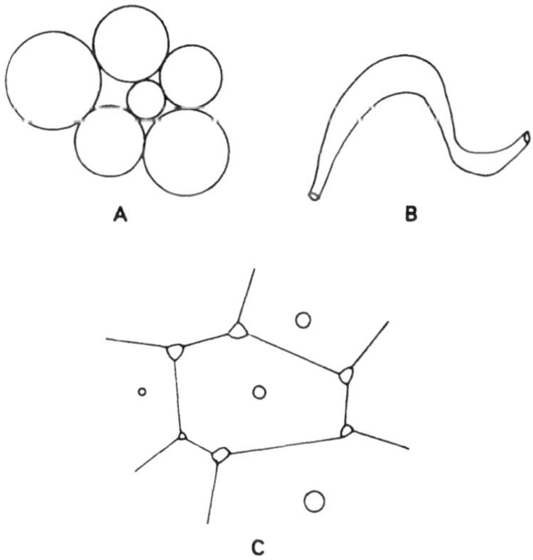

Figure 4.4-1. A. Initial stage sintering takes place as particles touch and form necks. B. Intermediate stage can be modeled with connected porosity. C. Final-stage sintering refers to elimination of isolated porosity.

compact decrease in diameter and pinch off as sintering proceeds toward the final stage.

Only about the last 5% of porosity in a particle compact is involved in the final-stage sintering. The pore space is no longer interconnected with the surface. It becomes isolated either as spherical pores within crystals or as nonspherical porosity at grain boundaries or three and four grain intersections. Of course in the case of glasses or liquid systems, pores are completely spherical; no grain boundaries exist. In most cases, the pores are gas filled and shrinkage only takes place to the point that pressure within the pore balances the surface tension forces attempting to shrink the pore. There are a few cases where pores devoid of gases or filled with a rapidly diffusing gas, can be completely eliminated from the powder compact. In these rare instances, full densification can be achieved if grain boundaries and pores remain together.

For the initial stage of sintering, there are some very simple equations that have been worked out by many investigators, starting with Frenkel in 1945, that describe the sintering of spherical particle compacts. These equations are represented in Table 4.4-1, along with mechanisms for material transport. The form of the equations listed in Table 4.4-1 is for shrinkage because this very readily differentiates between important methods or mechanisms of material transport. It is interesting to note that neither surface diffusion nor evaporation and

Table 4.4-1. Model Equations for Initial Isothermal Shrinkage of Spherical Powders

Mechanism of material transport	Shrinkage equation	Reference
Evaporation condensation	No shrinkage	
Surface diffusion	No shrinkage	
Plastic deformation	Not very significant	
Viscous diffusion	$\Delta L/L_0 = \dfrac{3\gamma t}{4a\eta}$	190
Volume diffusion	$\Delta L/L_0 = \left(\dfrac{5.34\gamma\Omega D_v t}{KTa^3}\right)^{1/2}$	190a
		190b
Grain boundary diffusion	$\Delta L/L_0 = \left(\dfrac{2.14\gamma\Omega b D_g t}{KTa^4}\right)^{1/3}$	190a
		190b
Liquid crystalline diffusion	$\Delta L/L_0 = \left(\dfrac{K\gamma\Omega\delta D_e t}{KTa^4}\right)^{1/3}$	190c

Key

ΔL change in length of linear dimension of a powder compact at any time, t

L_0 original length or effective length or length at which the equation becomes valid

γ surface tension

Ω volume of the unit or atom contributing to the transport of material, ordinarily the slowest-moving species

η viscosity of the liquid which is inversely related to the viscous diffusion coefficient

a equivalent spherical radius of the particle

k Boltzmann's constant

T absolute temperature

b grain-boundary width

D_v, D_g, D_e diffusion coefficients for volume, grain boundary, and liquid, respectively

δ separation distance between particles in a reactive liquid

K numerical constant containing configurational constants and solubilities for the reactive liquid system

condensation contribute to densification and shrinkage. These two processes do not take material from between the particles and deposit it in the neck as shown in Figure 4.4-2. This means that for a material that has a high vapor pressure or a large surface-diffusion coefficient, we should expect the material to become strong by forming necks between the particles when it is heated, but not become dense or have appreciable shrinkage. Sodium chloride and chromium oxide are examples of materials that have high vapor pressure and show little shrinkage,

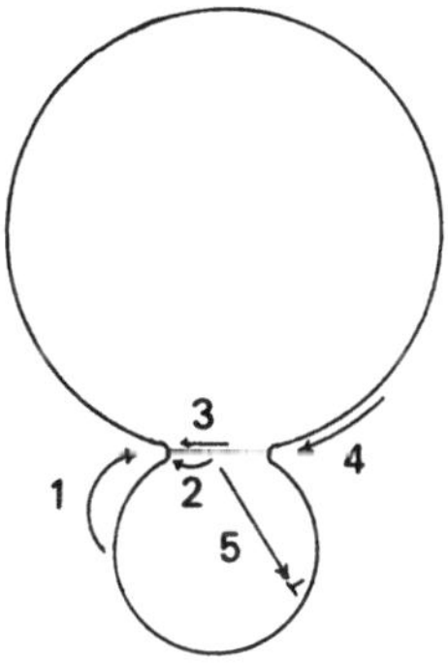

Figure 4.4-2. Material may be transported (1) through the vapor phase, (2) through the bulk by vacancies or by viscous flow, (3) along grain boundaries by vacancies or by thin liquid films, (4) over the surface by surface diffusion, or (5) through the bulk by dislocation motion (plastic deformation).

although they become somewhat stronger on annealing at high temperatures. The time dependence on shrinkage is shown in Figure 4.4-3.

The literature is not clear whether there are any materials that show sufficient surface diffusion to withstand shrinkage by some other transport process, such as grain boundary of volume diffusion. Surface diffusion coefficients obtained by thermal grooving are ordinarily misleading and would predict that shrinkage should never occur with many materials that we measure large amounts of shrinkage, even complete consolidation by sintering. As a result, we cannot rely on surface diffusion coefficients in the literature to tell us very much about the possibilities of sintering.

The simplest of the equations describing densification or shrinkage is the equation derived by Frenkel for viscous flow mechanism. Frenkel actually used this as a model for sintering metal, but since this time it has been shown to describe the sintering of glassy materials. The fact that diffusion coefficients are inversely proportional to viscosity simplifies the description of sintering of viscous and crystalline materials by diffusion

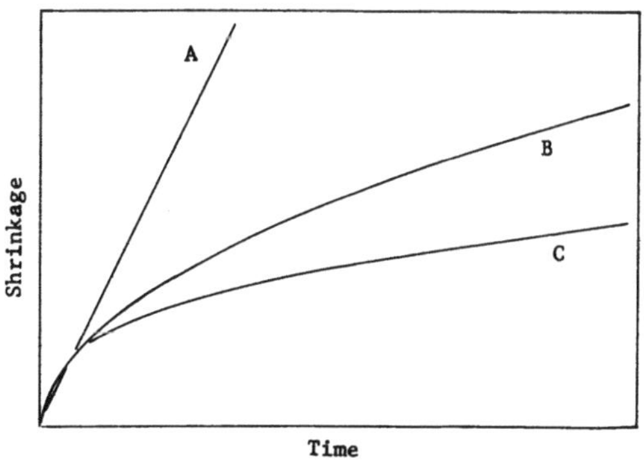

Figure 4.4-3. Initial isothermal shrinkage of powder compacts as a function of time illustrating: A, viscous flow; B, volume diffusion; C, grain boundary diffusion.

in much the same way. Note that the amount of shrinkage for a given time interval or the differential of shrinkage with respect to time, that is the rate of shrinkage, is inversely proportional to the particle size. This would be approximately true for volume and grain-boundary diffusion as well. This means that not only the driving force for sintering is increased by using small particles but also the rate at which consolidation takes place. This has great practical significance. In order to obtain densification at lower or more reasonable temperatures, ceramists have always recognized the importance of particle size and have attempted to obtain the smallest particle-size powders because they are the most active in their sintering characteristics. Extensive amounts of grinding often increase the rate of sintering by producing smaller particle sizes as described in the equations in Table 4.4-1.

Particle size of fine materials is often determined from line-broadening experiments or from surface-area measurements. Unfortunately, neither one of these techniques measure the particle size involved in sintering. Many fine particles are obtained by decomposition of minerals of various origins. The decomposition products are more often than not already partially sintered. Alpha aluminum oxide derived from the Bayer process by decomposition of the hydrate is a typical example. Magnesium oxide obtained by the decomposition of magnesium hydroxide is another example.[191] There are a multitude of other examples of powders utilized in the sintering process that are obtained by thermal decomposition of some mineral. In each case, during thermal decomposition, some sintering takes place. The decomposed product is often a pseudomorph of the original crystal. It is seldom possible to rely on either surface-area measurements or other techniques for measuring the particle for sintering. Most often, the particle size for sintering is that of the original crystal that was thermally decomposed to produce the powder.

One of the ways of altering this situation is to grind the decomposed powder in a ball mill or vibratory mill to liberate the individual crystals from their agglomerate. Another technique is to severely compress the powder in some compaction or briquetting operation. Each of these operations reduces particle size of the agglomerates and, hence, increases the rate of sintering.

The other important variable described in these equations is the diffusion coefficient or the viscosity. Both diffusion coefficient and viscosity can be altered with temperature. Both are exponential functions of temperature. The viscosity of a glass decreases exponentially as the temperature is raised and the diffusion coefficient increases exponentially with the temperature. Regardless of the mechanism of material transport, whether it be by liquid or viscous diffusion or along

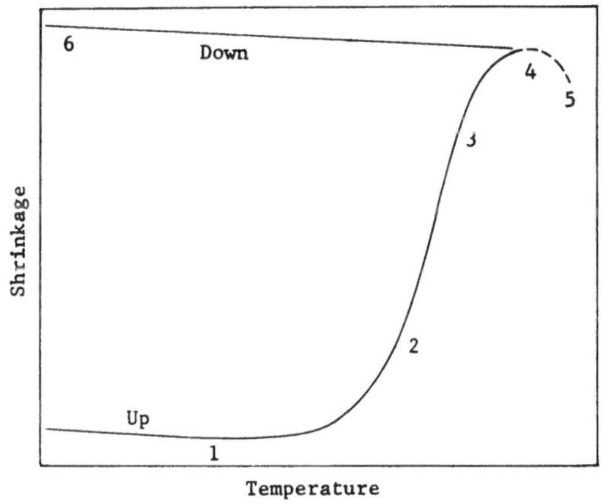

Figure 4.4-4. At constant rate of heating all mechanisms showing shrinkage have the same type of curve. From room temperature to 1 the powder compact expands by the normal thermal expansion processes. From 1 to 2 is the initial shrinkage portion of the shrinkage process; from 2 to 3 the intermediate stage sintering; from 3 to 4 final stage sintering; from 4 to 5 oversintering owing to expansion of trapped gases; from 4 to 6 thermal shrinkage similar to 0 to 1.

grain boundaries by grain boundary diffusion or through the bulk of the material by vacancy diffusion, all sintering increases exponentially with temperature. This is illustrated in Figure 4.4-4 where the two exponential features of the shrinkage as function of temperature curve can be seen for a constant rate of heating. Of course the exponential part of the curve cannot continue indefinitely inasmuch as the growths of necks begin to interfere with each other during the intermediate stage of sintering. Sintering slows down as the final stage of shrinkage takes place. In many cases where pores are filled with an insoluble gas, bloating or swelling can actually take place if the temperature is raised above the ideal temperature for complete consolidation of the material. This bloating or swelling is ordinarily avoided.

It should be noted in Table 4.4-1 that the very common practice of using a liquid to assist in the sintering process results in an equation very similar to grain boundary diffusion where material transport takes place in the liquid phase. This practice of adding a liquid-forming material that wets all of the grains is a practice often followed in powder metallurgy and ceramics and extractive metallurgy and is a very useful process if contamination by the liquid can be tolerated.

4.4.4. Variables That Change Sintering Characteristics of Materials

Table 4.4-2 lists the variables that are most prominent in changing sintering characteristics. The effect of temperature and particle size have already been described by the initial-stage equations listed in Table 4.4-1. Although those equations merely describe the initial stage, the intermediate and final stages are also appropriately altered by particle size and temperature. In other words, if we were dealing with a shrinkage curve as illustrated in Figure 4.4-4, using a larger particle size would

Table 4.4-2. Sintering Variables

Variable	Effect on rate of densification
Temperature	An exponential function of temperature
Particle size	Inversely proportional to particle size
Diffusion coefficient	Can be increased or decreased with appropriate additions (see Temperature)
Diffusion path	Some alterations possible by atmosphere control or additives
Powder compaction	No effect on rate

shift the curve to the right. A smaller particle sized material would shift the curve to the left. It is also interesting to note that a change in the rate of heating, since this curve describes a constant rate of heating process, has a somewhat similar effect, although the shift is not as large as for changes in particle size. A higher rate of heating shifts the curve to the right and a slower rate shifts the curve to the left. This particular fact illustrates another principle that ceramists have long understood, at least intuitively, that is, that the time for sintering at the sintering temperature is not nearly as important as the temperature itself. Increasing the sintering time by a factor of two or ten is not nearly as significant as changing the sintering temperature by 50° C.

The diffusion coefficient, of course is the variable that is being changed by changing the sintering temperature, but there are ways of changing diffusion coefficients and viscosities by means other than temperature. Unfortunately, most diffusion coefficients are tabulated for pure materials. Very seldom are they ever listed for impure materials and, of course, viscosities are notoriously lacking for the important materials we would like to use. How can we change a diffusion coefficient? In the case of metals, it is difficult to change the diffusion coefficient very much by additions of other materials. This is not the case in compounds or nonmetallic types of materials. The diffusion coefficient is altered thermally by generating more vacancies. However, vacancies can be chemically generated in the case of compounds. In Figure 4.4-5, we have illustrated the insertion of vacancies in a simple ionic compound, magnesium oxide, by the addition of titania that forms a limited solid solution in MgO and upsets the stoichiometry of the material. Incorporation of vacancies by chemical means may change sintering temperatures by hundreds of degrees.

In the case of liquid-phase sintering, chemical additions can be very instrumental in lowering viscosity. The use of fluorides to decrease the viscosity of slags is very well known among metallurgists. Where a silicate liquid is utilized in sintering as a second-phase addition, the same concept applies; a small amount of fluoride may be extremely useful in sintering. Hydroxides act almost identical to fluoride and hence, water vapor is likewise an effective means of reducing viscosity of silicates in many industrial applications.

Perhaps the most useful technique is to alter the path through which diffusion takes place. This might be the addition of a liquid so that solid-state diffusion would not be the limiting process but diffusion through the liquid could be relied upon for transporting material. When evaporation and condensation, because of high vapor pressure is the limiting step in densification, some means of changing the vapor pressure must be incorporated. One possibility is to lower the temperature for sintering by obtaining smaller particle sizes. This may not be entirely satisfactory. Sometimes it is possible to change the atmosphere and change the vapor pressure. This is true in the case of chromium oxide which is rather outstanding for its high vapor pressure, but by altering the partial pressure of oxygen it is possible to minimize the vapor transport and maximize the opportunity for volume diffusion and consolidation by sintering.

Although powder compact density has little or no effect on the rate of shrinkage and does not even come into the equations for the initial

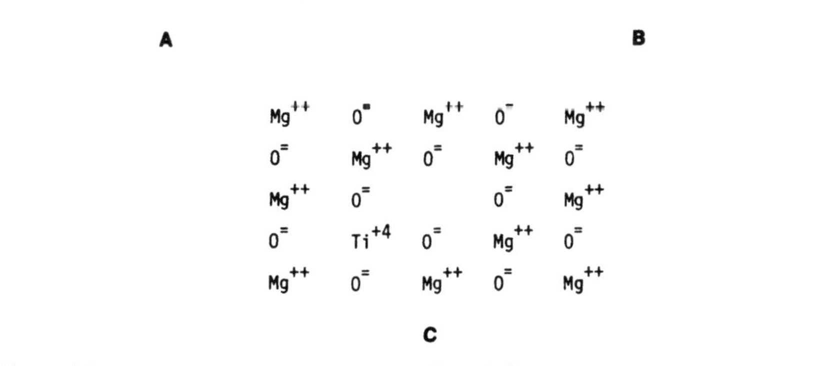

Figure 4.4-5. Schematic representation of MgO at: A, low temperature; B, high temperature with thermally induced vacancies; and C, low temperature with cation vacancy induced by solution of Ti^{4+}.

stage of sintering, it must be kept in mind that the final stages of sintering are largely dependent upon the number and sizes of pores that were initially incorporated into the powder compact. As a result, powder compaction is extremely important in sintering, but it is most important in determining the amount of shrinkage that will take place up to and through the final stages of sintering. This determines the strength of the partially sintered compact and, in particular, is extremely important when complete consolidation is not desirable or necessary. The degree of compaction determines the number of particle contacts and the strength of the partially sintered material.

4.4.5. Interaction between Grain Growth and Densification

It is well established that grain growth of crystalline materials accompanies the densification process. This is particularly true in the intermediate and final stages of sintering. Some grain growth processes, such as exaggerated grain growth, actually interfere with sintering during the later two stages and may terminate densification. Exaggerated grain growth has been noted in several different systems, including aluminum oxide, ferrites, silicon carbide, and β- and β''-alumina. This exaggerated grain growth, sometimes referred to as *abnormal grain growth*, takes place most readily in the presence of a small and disappearing amount of liquid phase.

Additives that contribute to complete densification are thought to inhibit rapid grain growth and maintain grain boundaries in the proximity of pores. If the grain boundaries remain in the vicinity of the pores, it is possible for the pores to shrink and disappear unless there is an interfering gas atmosphere inside the pores.

Pores themselves will interfere with grain growth as a matter of fact. During the initial stage of sintering the particles themselves have such small neck areas that the grain boundary cannot very well migrate out from between the particles. As densification proceeds, the neck area becomes a significant portion of the cross-sectional area of the particles themselves; it is then possible for the grain boundaries to migrate out from between particles and the grains may grow to larger sizes. Because porosity continues to occupy a significant part of the grain boundary area it will inhibit grain growth unless the porosity migrates as the boundary migrates.

4.4.6. Relationship of Strength to Densification by Sintering

Although there is a minor amount of strength in a particle compact as pressed (green strength), appreciable strength is obtained as necks

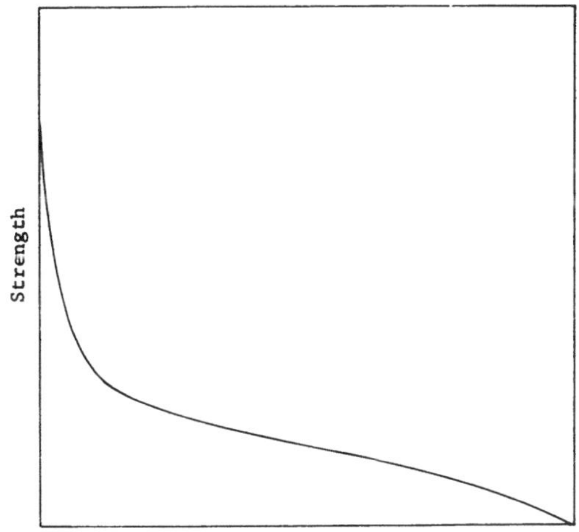

Figure 4.4-6. Schematic of development of strength as a function of densification. Sintering proceeds from right to left or from high toward low porosity.

form in between the particles during sintering. This means that the more particle-to-particle contacts, the higher the density will be for a green compact and the more rapid will be the development of strength during sintering. Few scientists are much concerned about strength being developed during the initial stage of sintering but are rather concerned about the effect of porosity on strengths during the final stages. A small amount of porosity decreases the strength by a rather large amount. It is well known that for a given grain size, hot-pressed samples that have full densification are much stronger than sintered samples with 1–2% porosity. The effect of porosity on strength is an exponential degradation of the strength properties. This is well recognized in powder metallurgy, as well as in ceramics.

In Figure 4.4-6, we have a schematic of strength as a function of porosity. Sintering begins with a particle compact, hopefully pressed to a density greater than 60% of theoretical (or with somewhat less than 40% porosity). As sintering takes place, shrinkage also takes place and porosity decreases. The strength increases as the neck area is increased. As pores become isolated one from another at 5% porosity, strength increases rapidly. Hence, we end up with a curve that has a rather complicated shape, and a change of curvature.

4.4.7. Application of Principles of Sintering to Extractive Metallurgical Operations

In a majority of applications to extractive metallurgy the purpose of sintering is to obtain an agglomeration of small particles with sufficient strength for subsequent processing. On the basis that strength is the goal of sintering and that a certain amount of porosity is acceptable, the important factors appear to be first of all, the number of particle–particle contacts. This is controlled by the green density of the particle compact. The higher the green density or the higher the pressure during compaction, the larger will be the sintered strength derived from the compact during the early stages of sintering.

For the same sintering temperature, strength and the amount of sintering and densification increases with decreasing particle size. Particle size then plays an important feature in the sintering process. It must be kept in mind that particle size is not the same as crystal size derived from surface area measurements but is determined by the actual size of the particles or agglomerates that are being compacted into a pellet for sintering.

The kinetics of sintering indicate the amount of liquid as well as its viscosity influences the rate of sintering. Quite often the amount of liquid must be held to a minimum because it becomes a contaminant in the further processing of the sinter. It is well recognized that viscosity of liquids, and in particular silicate liquids, may be altered by various minor additions. Fluorides, in particular, are effective in reducing viscosity of liquids.

4.5. Smelting and Refining (ETT)

4.5.1. Introduction

The present state of knowledge of rate phenomena in smelting and refining processes is presented in this critical review. Selected properties of gases, liquid metals, mattes, and slags that are encountered in pyrometallurgy are presented in the form of equations or graphs. The discussion of the rate phenomena is divided into three groups: (1) kinetics of interfacial reactions, (2) effects of gas blowing and gas evolution, and (3) reactions in ladle treatment and during solidification. Whenever possible, an attempt is made to demonstrate the usage of rate theories or concepts in the understanding and development of pyrometallurgical processes.

For historical reasons, most of the research effort in pyrometallurgy has been in areas that are of particular interest to iron and steelmaking processes; technical information on the processing of nonferrous metals is less abundant. Consequently, the numerous examples of metallurgical reactions rates given in this presentation are mostly on iron-base systems. However, the basic theories and concepts of rates and mechanisms of reactions discussed should be applicable to various types of pyrometallurgical processes involving gases, metals, slags, or mattes.

Discussion of reaction equilibria in metallurgical systems is almost completely excluded. With the exception of a few pertinent special cases, the theoretical and experimental model studies of reactions and heat and mass transfer are not discussed.

4.5.2. Selected Properties of Gases, Liquid Metals, Mattes, and Slags

Most metallurgical reactions are heterogeneous reactions occurring at interfaces between gas, metal, matte, or slag phases. The transport of reactants to and from interfaces plays an important role in the rates of interfacial reactions. Also the chemisorption of surface-active solutes can have a decisive influence on the kinetics of interfacial reactions. For better understanding of the rate phenomena, due consideration should be given first to a few selected properties of metallurgical systems relevant to heat and mass transfer and reaction kinetics.

4.5.2.1. Transport Properties of Gas Mixtures

The transport properties of gases, e.g., mass diffusivity, thermal diffusivity, thermal conductivity, and viscosity, can be predicted with sufficient accuracy from the kinetic theory of gases developed by Enskog and Chapman.[193,194] For the present purpose it is adequate to give some basic equations (without discussing their derivations) which are often used in computing the transport properties of gases.

Force Constants. For binary systems, the maximum energy of attraction, ε, between unlike molecules 1 and 2 is approximated by the geometrical mean of the pure components

$$\varepsilon_{1,2} = (\varepsilon_1 \varepsilon_2)^{1/2} \tag{4.5-1}$$

and the collision diameter, σ, is taken as the arithmetical mean of the components

$$\sigma_{12} = \frac{\sigma_1 + \sigma_2}{2} \tag{4.5-2}$$

The force constants σ and ε/k, where k is the Boltzmann constant, of selected gases are given in Table 4.5-1. The dynamics of binary collisions between the nonpolar spherical molecules is accounted for by the relative collision integral Ω^*. The omega integral designated by $\Omega^{*''}$ is used in computing viscosity and thermal conductivity, and $\Omega^{*'}$ for mass diffusivity. These are given in Table 4.5-2.

Table 4.5-1. Force Constants for the Lennard-Jones (6–12) Potential[a]

Gas	ε/k, °K	σ, Å
He	10.22	2.58
Ne	35.7	2.79
Ar	124	3.42
H_2	33.3	2.97
O_2	113	3.43
N_2	91.5	3.68
Air	97	3.62
CO	110	3.59
CO_2	190	4.00
CH_4	137	3.88
SO_2	252	4.29
H_2S	221	3.73
H_2O	380	2.65

[a] From Hirschfelder *et al.*(194)

Diffusivity. The self-diffusivity, D, and binary interdiffusivity, D_{12}, are given by

$$D = 1.8583 \times 10^{-3} \frac{T^{3/2}}{P\sigma^2 \Omega^{*\prime}} (2/M)^{1/2} \tag{4.5-3}$$

$$D_{12} = 1.8583 \times 10^{-3} \frac{T^{3/2}}{P\sigma_{12}^2 \Omega^{*\prime}} (1/M_1 + 1/M_2)^{1/2} \tag{4.5-4}$$

where M is the molecular weight in grams, P is the total pressure in atmospheres, and σ is the collision diameter in Å giving D in cm^2/sec.

Viscosity. The viscosity for a single component gas is given by

$$\eta = 2.6693 \times 10^{-5} \frac{(MT)^{1/2}}{\sigma^2 \Omega^{*\prime\prime}} \tag{4.5-5}$$

where η is in poise, i.e., dyne sec/cm$^2 \equiv$ g/cm sec.

Thermal Conductivity. The thermal conductivity of gases is derived from the viscosity, thus

$$\text{for monatomic gases} \qquad \kappa = \frac{15R}{4M}\eta \tag{4.5-6}$$

$$\text{for polyatomic gases} \qquad \kappa = \left(c_p + \frac{9R}{4M}\right)\eta \tag{4.5-7}$$

where c_p is the specific heat at constant pressure and R is the gas constant.

Table 4.5-2. Functions for Prediction of Transport Properties of Gases at Low Densities[a]

kT/ε or kT/ε_{12}	$\Omega^{*''}$ (for viscosity and thermal conductivity)	$\Omega^{*'}$ (for mass diffusivity)	kT/ε or kT/ε_{12}	$\Omega^{*''}$ (for viscosity and thermal conductivity)	$\Omega^{*'}$ (for mass diffusivity)
			2.50	1.093	0.9996
0.30	2.785	2.662	2.60	1.081	0.9878
0.35	2.628	2.476	2.70	1.069	0.9770
0.40	2.492	2.318	2.80	1.058	0.9672
0.45	2.368	2.184	2.90	1.048	0.9576
0.50	2.257	2.066	3.00	1.039	0.9490
0.55	2.156	1.966	3.10	1.030	0.9406
0.60	2.065	1.877	3.20	1.022	0.9328
0.65	1.982	1.798	3.30	1.014	0.9256
0.70	1.908	1.729	3.40	1.007	0.9186
0.75	1.841	1.667	3.50	0.9999	0.9120
0.80	1.780	1.612	3.60	0.9932	0.9058
0.85	1.725	1.562	3.70	0.9870	0.8998
0.90	1.675	1.517	3.80	0.9811	0.8942
0.95	1.629	1.476	3.90	0.9755	0.8888
1.00	1.587	1.439	4.00	0.9700	0.8836
1.05	1.549	1.406	4.10	0.9649	0.8788
1.10	1.514	1.375	4.20	0.9600	0.8740
1.15	1.482	1.346	4.30	0.9553	0.8694
1.20	1.452	1.320	4.40	0.9507	0.8652
1.25	1.424	1.296	4.50	0.9464	0.8610
1.30	1.399	1.273	4.60	0.9422	0.8568
1.35	1.375	1.253	4.70	0.9382	0.8530
1.40	1.353	1.233	4.80	0.9343	0.8492
1.45	1.333	1.215	4.90	0.9305	0.8456
1.50	1.314	1.198	5.0	0.9269	0.8422
1.55	1.296	1.182	6.0	0.8963	0.8124
1.60	1.279	1.167	7.0	0.8727	0.7896
1.65	1.264	1.153	8.0	0.8538	0.7712
1.70	1.248	1.140	9.0	0.8379	0.7556
1.75	1.235	1.128	10.0	0.8242	0.7424
1.80	1.221	1.116	20.0	0.7432	0.6640
1.85	1.209	1.105	30.0	0.7005	0.6232
1.90	1.197	1.094	40.0	0.6718	0.5960
1.95	1.186	1.084	50.0	0.6504	0.5756
2.00	1.175	1.075	60.0	0.6335	0.5596
2.10	1.156	1.057	70.0	0.6194	0.5464
2.20	1.138	1.041	80.0	0.6076	0.5352
2.30	1.122	1.026	90.0	0.5973	0.5256
2.40	1.107	1.012	100.0	0.5882	0.5170

[a] From Hirschfelder *et al.*[194]

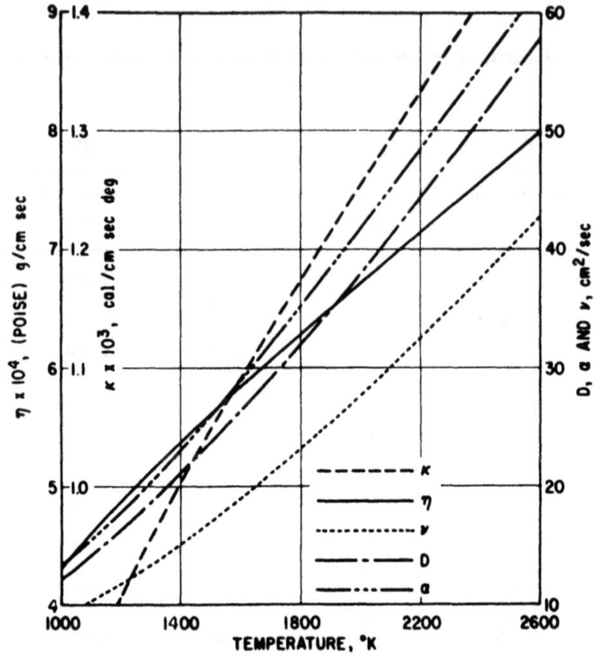

Figure 4.5-1. Transport properties of helium.

Thermal Diffusivity. The thermal diffusivity, α, is defined by the ratio

$$\alpha = \frac{\kappa}{\rho c_p} \tag{4.5-8}$$

where ρ is the density.

The transport properties calculated for helium and argon are given as examples in Figures 4.5-1 and 4.5-2. The argon–metal vapor (Mn, Cr, Co, Ni, and Fe) interdiffusivities measured by Grieveson and Turkdogan[(195)] are compared in Figure 4.5-3 with those calculated from equation (4.5-4) and the estimated force constants.

4.5.2.2. *Diffusion in Metals, Mattes, and Slags*

In the technical literature, we often come across references to diffusion coefficients identified as mutal diffusivity, volume diffusivity, interdiffusivity, chemical diffusivity, intrinsic diffusivity, etc. For binary systems there is only one coefficient of diffusion. As shown by Stark,[(196)] the mutal diffusion coefficient, or simply diffusivity, is an invariant of binary diffusive motion and, therefore, independent of the frame of

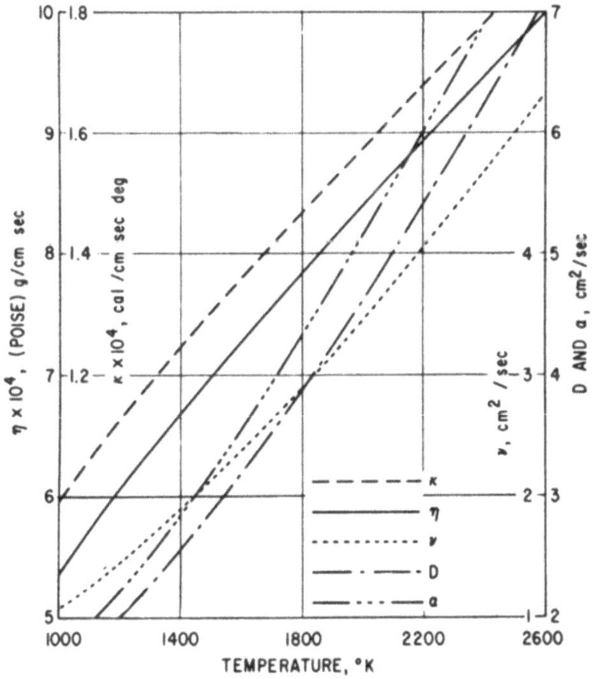

Figure 4.5-2. Transport properties of argon.

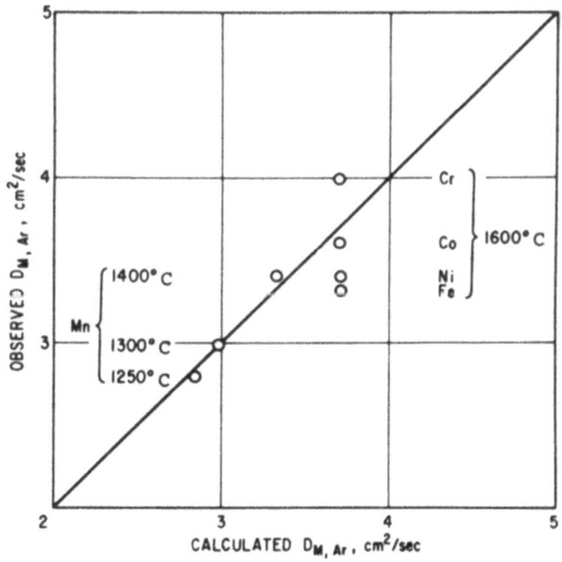

Figure 4.5-3. Comparison of estimated and experimental interdiffusivities of metal-vapor–gas binary mixtures at 1 atm pressure.

reference of the diffusion process. The diffusivity, thus defined, is given by the following expression in terms of fluxes J_1 and J_2, concentration gradients $\partial c_1/\partial x$ and $\partial c_2/\partial x$, and the partial molar volumes $\bar{V}_1$ and $\bar{V}_2$:

$$D = -\frac{c_2 \bar{V}_2 J_1}{\partial c_1/\partial x} - \frac{c_1 \bar{V}_1 J_2}{\partial c_2/\partial x} \tag{4.5-9}$$

where subscripts 1 and 2 refer to solvent and solute, respectively, and c is the molar concentration per unit volume.

For dilute solutions, i.e., $c_2 \to 0$ and $c_1 \bar{V}_1 \to 1$, equation (4.5-9) reduces to the form of Fick's first law, thus

$$J_2 = -D\frac{\partial c_2}{\partial x} \tag{4.5-10}$$

For an ideal gas, equation (4.5-9) simplifies to the following form, which is often called the Maxwell–Stefan equation:

$$J_1 = -\frac{D}{RT}\frac{\partial p_1}{\partial x} + \frac{p_1}{P}(J_1 + J_2) \tag{4.5-11}$$

where $P = p_1 + p_2$ is total pressure and T is the absolute temperature. Equation (4.5-11) is solved for known interrelation between the fluxes J_1 and J_2 for a given binary diffusion process.

For a binary mixture 1+2, the thermodynamic effect on the diffusivity is given by Darken's† equation[(197)]

$$D = (c_1 \bar{V}_1 D_2^* + c_2 \bar{V}_2 D_1^*)\frac{\partial \ln a_2}{\partial \ln c_2} \tag{4.5-12}$$

where D_1^* and D_2^* are the self (tracer) diffusivities of components 1 and 2 and a_2 is the activity of component 2.

For ternary and multicomponent systems the interpretation of diffusional fluxes becomes more complex; the phenomenological definition of the mass flux in a multicomponent system is discussed, for example, by Bird *et al.*[(198)] Although metallurgical systems are complex, for a solute at low concentrations the diffusivity may be defined simply by invoking Fick's first law, equation (4.5-10).

† Darken gave the simplified form of this equation for the special case of constant and equal molar volumes; thus in terms of atom fractions, N_1 and N_2

$$D = (N_1 D_2^* + N_2 D_1^*)\frac{\partial \ln a_2}{\partial \ln N_2}$$

or in terms of the activity coefficient γ_2,

$$D = (N_1 D_2^* + N_2 D_1^*)\left(1 + \frac{\partial \ln \gamma_2}{\partial \ln N_2}\right)$$

In liquid metals, the diffusivities of many solutes are in the range from 10^{-5} to 10^{-4} cm^2/sec with heats of activation 8–12 kcal. The hydrogen diffusivity in liquid metals is in the range from 6×10^{-4} to 10^{-3} cm^2/sec with heats of activation 1–5 kcal.

The temperature dependence of the self-diffusivity of O, Ca, S, Al, and Si in calcium–alumino–silicate melts is shown in Figure 4.5-4.[199–201] The relative order of self-diffusivities of Ca, Al, and Si is as would be expected from the ionic and covalent bond structure of silicates and aluminates. However, it is hard to explain the observed high self-diffusivity of oxygen from structural considerations. There is no simple way of calculating oxide interdiffusivities from the self-diffusivities of ions. A few measurements made indicate that the oxide interdiffusivities in silicates are similar to the self-diffusivities of cations, i.e., in the range from 2 to 20×10^{-6} cm^2/sec.

The iron–oxygen interdiffusivity in liquid iron oxide decreases with increasing Fe^{3+}/Fe^{2+} ratio[202,203] in the oxide, i.e., at 1550° C, $D = 4 \times 10^{-4}$ cm^2/sec for iron oxide in equilibrium with liquid iron, and $D = 5 \times 10^{-5}$ cm^2/sec for iron oxide saturated with magnetite. For liquid silicates of composition Fe_2SiO_4 and $CaFeSiO_4$, the iron–oxygen interdiffusivity is 1.5×10^{-5} cm^2/sec at 1535° C.[204]

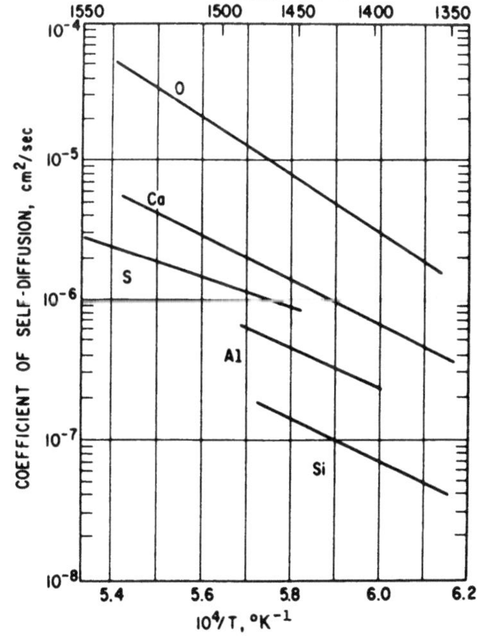

Figure 4.5-4. Self-diffusivities in calcium aluminosilicate melts.

Self-diffusivities in mattes are in the range from 2 to 10×10^{-5} cm^2/sec.[205] For example, in a 50:50 mixture of FeS and Cu_2S (by weight), at 1160° C the self-diffusivity of iron is 2.9×10^{-5} cm^2/sec and of copper, 5.5×10^{-5} cm^2/sec.

4.5.2.3. *Surface Tension*

Metals have high surface tensions, and as would be expected, the higher the heat of vaporization the higher is the surface tension.[206] For example, at the melting temperatures of metals, for heat of vaporization $\Delta H_v = 0.25$ kcal/cm^3, the surface tension is $\gamma = 56$ dyne/cm and for $\Delta H_v = 22$ kcal/cm^3, $\gamma = 2600$ dyne/cm. Table 4.5-3 lists a few examples of the surface tension of some liquid metals in inert gases at 1550° C.[207]

The surface tension of metals is lowered appreciably by small amounts of surface-active solutes such as oxygen, sulfur, selenium, and tellurium. Effects of these solutes on the surface tension of liquid iron (1550° C)[207] and liquid copper (1150° C)[208] are shown in Figures 4.5-5 and 4.5-6. The fact that the lowering of the surface tension is a function of the solute activity instead of concentration is illustrated in Figure 4.5-7 for the Fe–C–S system at 1450° C.[207] Carbon has no effect on the surface tension of iron. However, since carbon raises the activity of sulfur in iron, in the presence of carbon the sulfur has a greater effect on the surface tension of iron, as evidenced from the data in Figure 4.5-7.

The extent of surface coverage at any given solute activity can be calculated from the data on surface tension vs. solute activity by invoking the Gibbs adsorption equation

$$\Gamma_i = -\frac{1}{kT}\left(\frac{\partial \gamma}{\partial \ln a_i}\right)_T \qquad (4.5\text{-}13)$$

where Γ_i is the excess surface concentration of the adsorbed species, atom/cm^2. A plot of surface tension of liquid iron against ln (% O) is shown in Figure 4.5-8.[209] At high oxygen contents (≡activity) the curve approaches a straight line asymptotically. The slope of this line gives

Table 4.5-3. Surface Tension of Metals in Inert Gases at 1550 °C

Liquid metal	γ, dyne/cm
Iron	1788
Cobalt	1886
Nickel	1934
Copper	1280

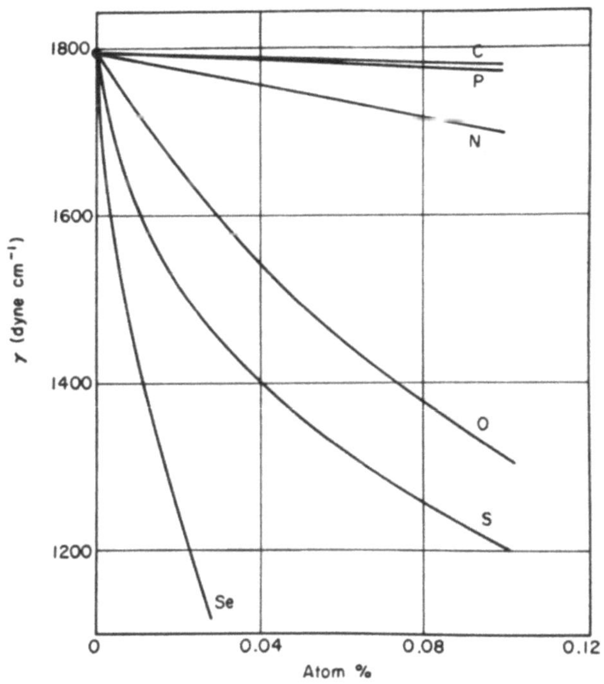

Figure 4.5-5. Surface tensions for solutions of C, P, N, O, S, and Se in liquid iron at 1550° C.

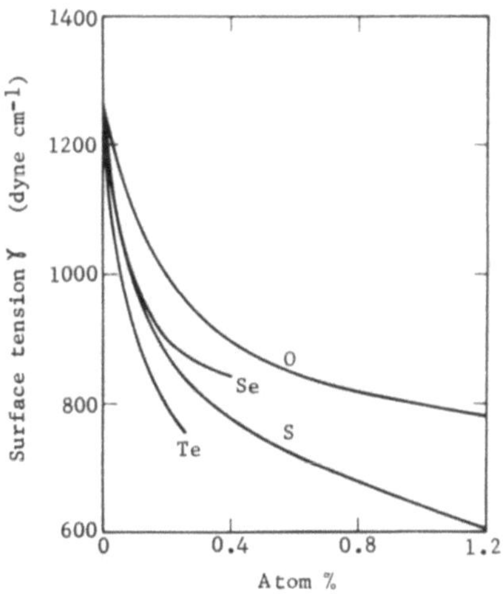

Figure 4.5-6. Surface tensions for solutions of O, S, Se, and Te, in liquid copper at 1150° C.

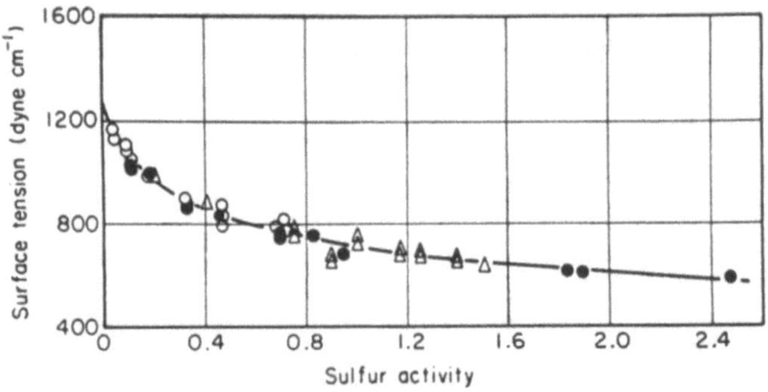

Figure 4.5-7. Surface tensions for Fe–C–S solutions at 1450° C shown as a function of sulfur activity. The standard state for sulfur is such that the activity equals wt % in pure iron at high dilution. Key: ○ 1.25% C; ● 2.5%; △ 4.0%.

$\Gamma = 1.05 \times 10^{15}$ oxygen atoms/cm^2. This value of Γ is a reasonable estimate of the total number of surface sites, denoted by Γ_0.

From statistical mechanics a relation can be derived for the fraction of sites, $\theta_i = \Gamma_i/\Gamma_0$, occupied by the adsorbed species in terms of its activity, thus

$$a_i = \varphi_i \frac{\theta_i}{1-\theta_i} \tag{4.5-14}$$

where the temperature-dependent proportionality factor φ_i is an activity coefficient-like term for the species i in the adsorbed layer with fixed number of sites. This relation is known as the *Langmuir adsorption isotherm* for an ideal monolayer for which φ_i is a function of temperature only. As in the case of three-dimensional solutions, adsorbed layers (two-dimensional solutions) may not be ideal. That is, φ_i may vary with the surface coverage.

The fractional coverage of surface by adsorbed oxygen, derived from Figure 4.5-8, is shown in Figure 4.5-9 as a function of the oxygen content of iron. As $\theta_0 \to 0$, the value of $\varphi_i = 0.014$ and decreases to $\varphi_i = 0.0013$ as $\theta_0 \to 1$. The curve for an ideal monolayer with $\varphi_i = 0.0013$ is the dotted line showing the extent of departure from the ideal solution law. It should be noted that the chemisorption isotherm derived from the surface tension data and the Gibbs adsorption equation as described above is subject to error, because some judgment is involved in drawing tangents to the γ vs. $\ln a_i$ curve.

Combining equations (4.5-13) and (4.5-14), and integrating gives

$$\gamma_0 - \gamma = kT\Gamma_0 \ln\left(1 + \frac{a_i}{\varphi_i}\right) \tag{4.5-15}$$

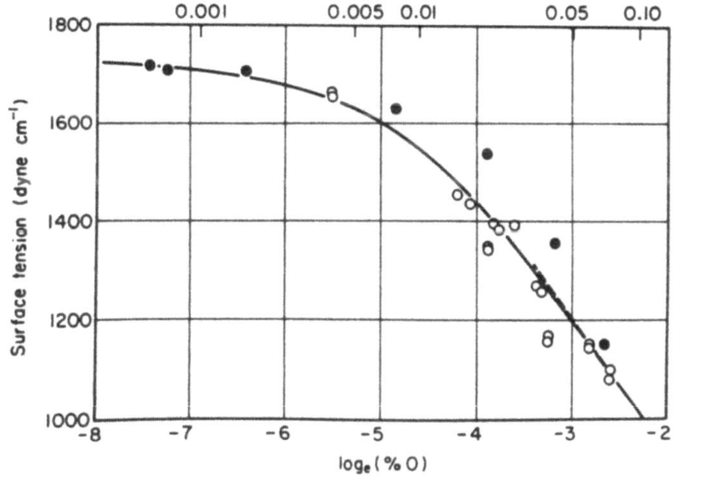

Figure 4.5-8. The surface tension of liquid Fe–O solutions at 1550° C.[209]

where γ_0 is the surface tension of the pure metal at $a_i = 0$. Belton[210] used this form of the adsorption equation to estimate the value of φ_i for various systems by assuming that an ideal monolayer is formed, i.e., φ_i is assumed to be independent of the fractional surface coverage. The extent of departures from ideality in chemisorbed layers cannot be resolved at present with the limited data available. Although the calculation of adsorption isotherms from the available surface tension data are approximate, such information provides guidance to the relative effects of adsorbed species on the kinetics of surface reactions.

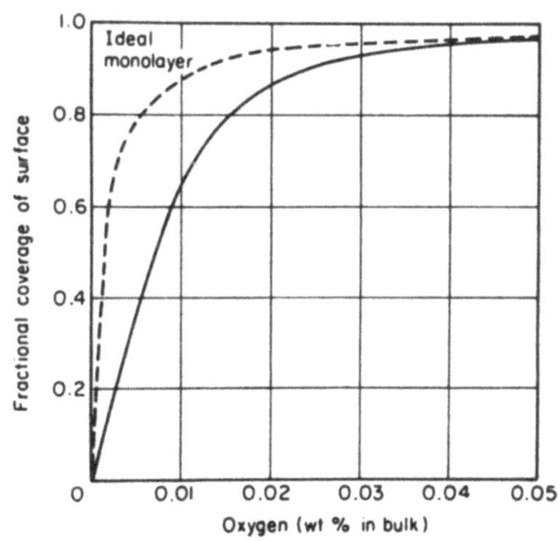

Figure 4.5-9. Oxygen adsorption isotherm for liquid iron at 1550° C, as calculated by Swisher and Turkdogan.[209]

The heat of adsorption, ΔH_a, for a given coverage is obtained from the temperature dependence of φ_i; thus

$$\left[\frac{\partial \ln \varphi_i}{\partial(1/T)}\right]_{\theta_i} = \frac{\Delta H_a}{R} \tag{4.5-16}$$

As in most chemical reactions, chemisorption is an exothermic reaction (ΔH_a has a negative value); therefore, φ_i increases with increasing temperature, i.e., the surface coverage decreases and surface tension increases with increasing temperature. However, with pure liquids in inert atmospheres the surface tension decreases with increasing temperature. For an ideal monolayer (Langmuir type), φ_i is a constant; hence, the heat of adsorption is independent of coverage. In many systems, the adsorbed layer is nonideal; consequently, φ_i and ΔH_a vary with θ_i; several such examples for systems at relatively low temperatures are given in Ref. 211.

The surface tensions of liquid oxides, slags, and mattes of metallurgical interest are in the range from 300 to 700 dyne/cm.[212] Sulfides, fluorides, and carbides lower the surface tension of slags; in addition, several oxides are surface-active when dissolved in liquid iron oxide.

Little is known about the slag–metal and slag–matte interfacial tension. Generally speaking, the slag–metal interfacial tension is between those for gas–metal and gas–slag interfaces. For liquid iron–silicate melts, the interfacial tension is in the range from 900 to 1200 dyne/cm.[213] The energy of adhesion at the interface is given by the relation

$$W(\text{metal–slag}) = \gamma(\text{gas–metal}) + \gamma(\text{gas–slag}) - \gamma(\text{metal–slag}) \tag{4.5-17}$$

When a surface-active solute is transferred from metal to slag, the interfacial tension decreases; hence the adhesion between metal and slag increases. Such a phenomenon has been observed in the transfer of sulfur from a carbon-saturated iron to an alumino–silicate slag by Kozakevitch.[214]

4.5.2.4. Viscosity

The viscosities of liquid metals, mattes, and halides are in the range from 0.005 to 0.05 poise (dyne sec/cm^2). The viscosity of iron, cobalt, nickel, and copper measured by Cavalier[215] is given in Figure 4.5-10. There are no discontinuities in the viscosity-vs.-temperature lines extending below the melting temperatures of the undercooled metals.

The viscosities of silicates, aluminates, and slags have been studied extensively. References to most of the previous work are given in a

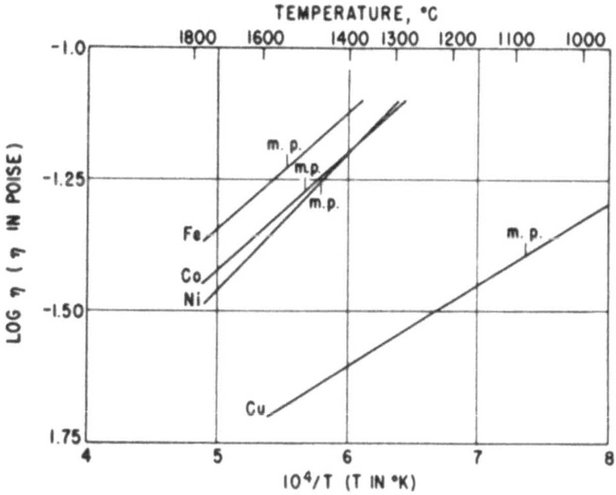

Figure 4.5-10. Temperature dependence of viscosity of liquid iron, cobalt, nickel, and copper.

review paper by Turkdogan and Bills.[216] As examples, viscosity data are given in Figures 4.5-11 and 4.5-12 for $CaO–Al_2O_3–SiO_2$ at 1800° C and $CaO–FeO–SiO_2$ in equilibrium with solid iron at 1400° C, respectively.[217,218] Up to about 2%, sulfur has no effect on the viscosity of blast furnace slags. However, the addition of calcium fluoride has a

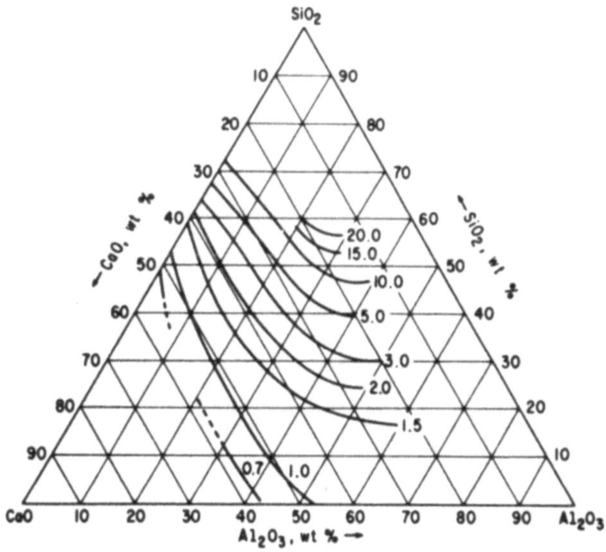

Figure 4.5-11. Isokoms (in poise) of $CaO–Al_2O_3–SiO_2$ melts at 1800° C.

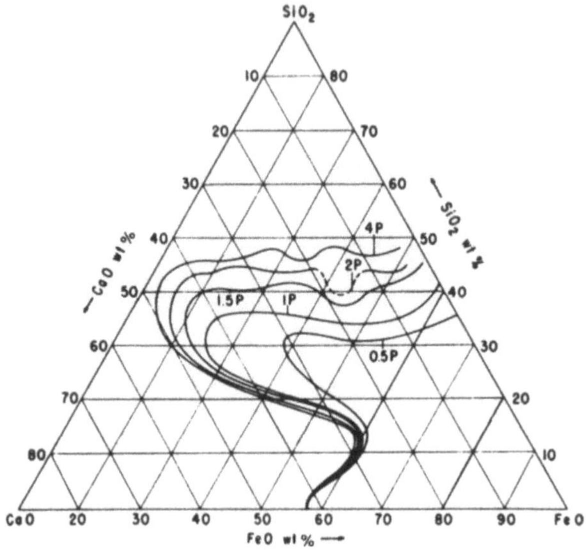

Figure 4.5-12. Isokoms (in poise) of $CaO–FeO–SiO_2$ melts at 1400° C.

pronounced lowering effect on the viscosity of slags,[219] as shown in Figure 4.5-13.

4.5.2.5. Slag Foam

When gas is bubbled through a liquid of low surface tension and high viscosity, motion of small gas bubbles and the liquid film around

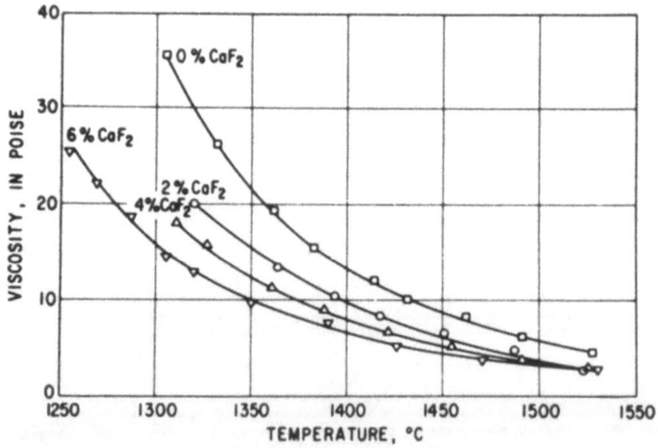

Figure 4.5-13. Effect of calcium fluoride on the viscosity of alumino–silicate melts containing 44% SiO_2, 12% Al_2O_3, 41% CaO, and 3% MgO.

them is retarded; such a system is called *foam*. In pneumatic metal refining processes, the fast rate of gas injection and gas evolution brings about slag foaming as in oxygen top- or bottom-blowing steelmaking processes. The energy required to create foam increases with increasing surface tension of the liquid. The rate of drainage of liquid film between the bubbles increases with decreasing viscosity of the liquid. Consequently, liquids with high surface tension and low viscosity, such as metals and mattes, cannot foam. The metallurgical slags have relatively lower surface tensions and much higher viscosities, hence the slag foam is readily formed during refining accompanied by extensive gas evolution.

Cooper *et al.*[220,221] made an extensive study of the foaming characteristics of metallurgical slags. A few examples of their experimental results are shown in Figure 4.5-14. The relative foam stability increases with decreasing temperature, decreasing CaO/SiO_2 ratio, and increasing P_2O_5 content of the slag; a similar effect is observed with the addition of B_2O_3 and Cr_2O_3 to the slag.[222] Silica adsorption at the bubble surface as a monomolecular film of SiO_4^{4-} tetrahedra results in high surface viscosity. Incorporation of phosphate, borate, or chromate anion groups in this adsorbed layer increases the surface elasticity of the viscous film, hence increases the foam stability.

Small quantities of suspended solid matter in the slag also stabilize the foam by increasing the apparent viscosity of the medium and by

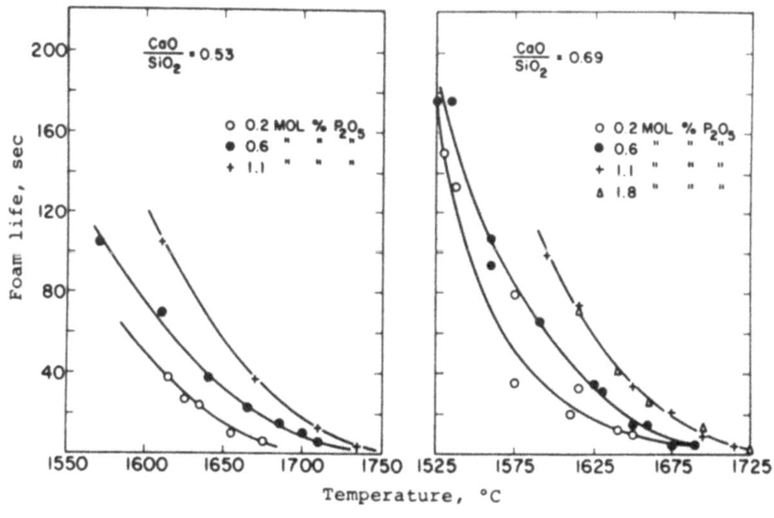

Figure 4.5-14. Variation of foam stability with temperature, CaO/SiO_2 ratio, and P_2O_5 content of $CaO–SiO_2–P_2O_5$ slags.

retarding the coalescence of the bubbles on which the solid particles are attached by surface tension.

The rate processes in pneumatic metal refining are much affected by the foaming characteristics of slags. In fact, in oxygen top blowing, a stable slag foam suppresses the extent of metal splash from the converter and increases the extent of metal emulsification which brings about a faster rate of refining.

4.5.3. Kinetics of Interfacial Reactions

The rate phenomena considered in this section are confined to the phenomenological concepts of reaction kinetics with particular emphasis on the role of adsorption of surface-active elements on the high-temperature reaction rates.

4.5.3.1. Theory

Unlike the laws of thermodynamics, reaction-rate theory is far from being rigorous. However, much progress has been made in the understanding of this subject during the past four decades. Of the numerous contributions made to the development of the reaction rate theory, those by Eyring[223,224] are the most outstanding. The theory is based essentially on two principal concepts: (a) the formation of an activated complex in equilibrium with the reactants adsorbed on the surface, and (b) the universal specific rate for the decomposition of the activated complex.

For the case of fast rate of transport of reactants and products to and from the reaction site and rapid rate of nucleation of a second phase, the rate is controlled by a chemical reaction occurring in the adsorbed layer at the interface. The reaction between adsorbed species L and M on the surface producing Q via an activated complex $(LM)^{\ddagger}$ is represented by

$$L + M = (LM)^{\ddagger} \rightarrow \text{product } Q \qquad (4.5\text{-}18)$$

For a given temperature the equilibrium constant, $K_{\ddagger}$, is stated by

$$K_{\ddagger} = \frac{a_{\ddagger}}{a_L a_M} \qquad (4.5\text{-}19)$$

where $a_{\ddagger}$ is the activity of the activated complex, a_L and a_M are the activities of the reactants. The equilibrium constant $K_{\ddagger}$ may be represented in a more general form by

$$K_{\ddagger} = \frac{a_{\ddagger}}{\prod a_i^{\nu_i}} \qquad (4.5\text{-}20)$$

where $\prod a_i^{\nu_i}$ is the product of the activities of the reactants in equilibrium with the activated complex.

Next to be considered is the specific rate of decomposition of the activated complex to the overall reaction product represented by

$$\mathcal{R} = \frac{kT}{h}\Gamma_{\ddagger} = \frac{kT}{h}\Gamma_0\theta_{\ddagger} \tag{4.5-21}$$

where

$\dfrac{kT}{h}$ = universal rate (k is the Boltzmann constant = 1.380×10^{-16} erg/deg, and h is the Planck constant = 6.626×10^{-27} erg sec),

$\Gamma_{\ddagger}$ = concentration of the activated complex in the adsorbed layer, moles/cm^2, and

Γ_0 = total number of adsorption sites at the surface, hence the fractional coverage by the activated complex $\theta_{\ddagger} = \Gamma_{\ddagger}/\Gamma_0$.

For single-site occupancy by the activated complex† in the adsorbed layer, the activity of the complex is represented by

$$a_{\ddagger} = \varphi_{\ddagger}\frac{\theta_{\ddagger}}{1-\theta} \tag{4.5-22}$$

where $\theta = \sum \theta_i$ is the total fractional occupancy of the sites by the adsorbed species i, and $\varphi_{\ddagger}$ is the activity coefficient of the complex in the adsorbed layer.

Assuming that the activated complex is the same for the forward and reverse reactions, the isothermal net reaction rate is given by

$$\mathcal{R} = \frac{kT}{h}\Gamma_0\frac{K_{\ddagger}}{\varphi_{\ddagger}}(1-\theta)\left(\prod a_i^{\nu_i} - \prod\nolimits_{\text{eq}} a_i^{\nu_i}\right) \tag{4.5-23}$$

where $\prod_{\text{eq}} a_i^{\nu_i}$ is the value that would prevail at equilibrium with the products and is found from the equilibrium constant and the activities of the reaction products.

At low-site fillage, the term $(1-\theta)$ is approximately one, and since $\theta_{\ddagger}$ is very small, $\varphi_{\ddagger}$ is essentially constant, and equation (4.5-23) reduces to

$$\mathcal{R} = \frac{kT}{h}\Gamma_0\frac{K_{\ddagger}}{\varphi_{\ddagger}}\left(\prod a_i^{\nu_i} - \prod\nolimits_{\text{eq}} a_i^{\nu_i}\right) \tag{4.5-24}$$

At high-site fillage, the term $(1-\theta)$ approaches zero. For this limiting case of almost complete surface coverage by a single species, p,

$$1-\theta = \frac{\varphi_p}{a_p} \tag{4.5-25}$$

† When the activated complex occupies x number of sites, $a_{\ddagger} = \varphi_{\ddagger}\theta_{\ddagger}/(1-\theta)^x$.

This solution is usable only when the high-site fillage is dominated by a single species. It is only in this case that each site has essentially the same surroundings in the adsorbed layer and hence the activity coefficients φ_p and $\varphi'_\ddagger$ are constant. Therefore, for the limiting case of $(1-\theta)\to 0$, the rate equation is

$$\mathcal{R} = \frac{kT}{h}\Gamma_0 \frac{K_\ddagger \varphi_p}{\varphi'_\ddagger} \frac{(\prod a_i^{\nu_i} - \prod_{\text{eq}} a_i^{\nu_i})}{a_p} \tag{4.5-26}$$

where $\varphi'_\ddagger$ is the value of $\varphi_\ddagger$ at $(1-\theta)\to 0$; for the Langmuir-type ideal monolayer $\varphi_\ddagger \simeq \varphi'_\ddagger$.

The heat of activation of the rate-controlling reaction is obtained from the slope of the plot of the $\ln \mathcal{R}$ vs. the reciprocal of the absolute temperature.† The true heat of activation of the reaction is that associated with the coefficient of temperature dependence of $\log K_\ddagger$. However, the temperature dependence of the rate constant determined experimentally involves three parameters, $K_\ddagger$, φ_p, and $\varphi_\ddagger$, which are all temperature dependent. Hence, the temperature dependence of the rate constant obtained experimentally gives only the apparent heat of activation, which may change with the value of $(1-\theta)$, or the activity of strongly adsorbed species.

For an ideal monolayer involving a single adsorbed species, p, equation (4.5-14) is transformed to

$$1-\theta = \frac{\varphi_p}{a_p + \varphi_p} \tag{4.5-27}$$

Inserting this in equation (4.5-23) gives for the net reaction rate in an ideal monolayer,

$$\mathcal{R} = \frac{kT}{h}\Gamma_0 \frac{K_\ddagger \varphi_p}{\varphi_\ddagger} \frac{(\prod a_i^{\nu_i} - \prod_{\text{eq}} a_i^{\nu_i})}{a_p + \varphi_p} \tag{4.5-28}$$

Several examples of rate measurements were given in a review paper by Darken and Turkdogan[225] showing that chemisorption plays an important role in the kinetics of many metallurgical reactions. Here, we shall briefly review the state of the present understanding of the kinetics of metallurgical reactions pertinent to smelting and refining. The examples given are limited to cases where the rate is controlled primarily by an interfacial reaction.

Unless the metal surface is poisoned with a strongly absorbed species, the rates of most gas–metal chemical reactions are relatively fast compared to the diffusional fluxes. Therefore, there are but a few good examples of chemical reaction control pertinent to smelting and refining

† Over a wide temperature range it is more correct to plot $\ln(\mathcal{R}/T)$ against $1/T$.

processes. Furthermore, most of the systems studied are in the area of ferrous metallurgy. In addition, it is well to point out that the rates of chemically controlled interfacial reactions cannot always be interpreted satisfactorily in terms of the phenomenological concepts of the reaction kinetics as outlined above.

4.5.3.2. Gas–Metal Reactions

Nitrogenation and Denitrogenation of Iron. Several investigators have demonstrated that the nitrogen dissolution rate in, or evolution from, liquid or solid iron is markedly retarded by the presence of small amounts of surface-active solutes in the metal, e.g., oxygen, sulfur, and selenium.[226–234] From the studies made in the early 1960s,[226–229] it was concluded that the transfer of nitrogen across the surface of iron follows first-order-type reaction kinetics. That is, the absorption rate is proportional to the square root of the partial pressure of nitrogen, and the desorption rate is proportional to the nitrogen content of the iron. Therefore, when the transport processes are not rate limiting, the reaction controlled rate of nitrogen transfer across the gas–metal interface is represented by

$$\frac{d\%\mathrm{N}}{dt} = -\frac{q}{l}(\%\mathrm{N}_e - \%\mathrm{N}) \tag{4.5-29}$$

where

l = depth of liquid,
$\%\mathrm{N}$ = nitrogen content at time t,
$\%\mathrm{N}_e$ = nitrogen content in equilibrium with p_{N_2} in the gas, and
q = rate constant $\propto (1-\theta)$.

As is shown in Figure 4.5-15a, the rate constant q for nitrogen reaction with liquid iron, derived from the available experimental data,[227,228] is inversely proportional to the oxygen content of the metal. As was pointed out previously (Figure 4.5-9), at oxygen contents above 0.04% the limiting case of $(1-\theta) \rightarrow 0$ is satisfied. Therefore, the linearity of the plot in Figure 4.5-15a is consistent with the interpretation of the surface tension data for the Fe–O system. At low oxygen activities, the nitrogen reaction rate increases to such an extent that the reaction is ultimately controlled by a transport process. This is demonstrated in Figure 4.5-15b. As the fraction of surface vacant sites increases, the rate parameter q reaches a constant value, indicating that diffusion through the boundary layer of inductively stirred melt ultimately becomes the rate-controlling step.

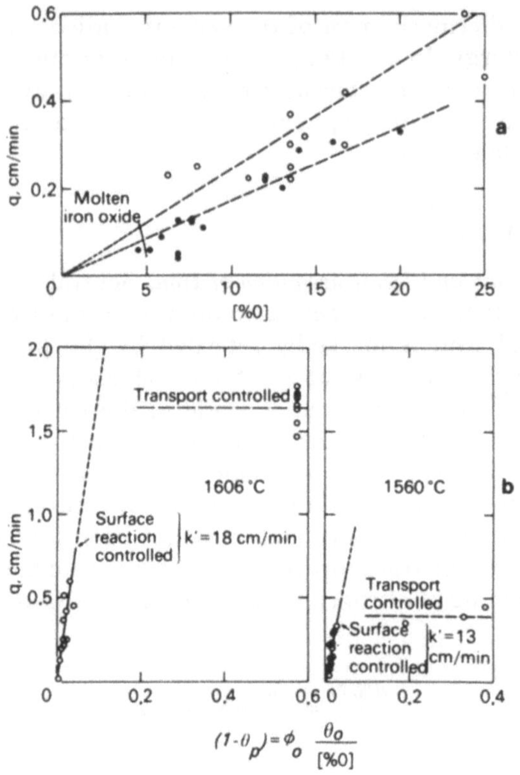

Figure 4.5-15. Rate parameter q for nitrogenation of liquid iron as a function of (a) oxygen activity at $(1-\theta)\to 0$ and (b) at $0<(1-\theta)<0.6$ from data of Pehlke and Elliott[227] (O, 1606° C) and Schenck *et al.*[228] (●, 1560° C).

In later studies, Fuwa *et al.*,[230] Mori and Suzuki,[231] and Narita *et al.*[232] measured the rate of denitrogenation of iron containing oxygen in a stream of argon impinging on the surface of the melt. They are of the opinion that, contrary to the findings of the previous investigators,[227,228] the rate is a second-order type, thus

$$\frac{d\%\mathrm{N}}{dt} = \frac{g'}{l}[(\%\mathrm{N}_e)^2 - (\%\mathrm{N})^2] \tag{4.5-30}$$

where $q' \propto (1-\theta)$ is in units of cm/min %N. As is seen from the results in Figure 4.5-16, the rate parameter is directly proportional to 1/%O at oxygen levels above about 0.02%, similar to the findings for the absorption rate. Because of the microscopic reversibility of the reaction, the rate constant q' can be calculated from q for absorption, at least for 1 atm N_2, as shown in Figure 4.5-16. Although there is an apparent agreement between the results for the absorption rate of nitrogen at 1 atm and the initial desorption rate, the differences in the order of the forward and reverse reactions observed by different investigators remains to be resolved.

Our understanding of the kinetics of the reaction of nitrogen with liquid iron is further confused with the results of Inouye and Choh.[233] They noted that at oxygen concentrations below 0.01%, the nitrogenation rate is proportional to $p_{N_2}^{1/2}$, but at higher oxygen concentrations the rate is proportional to p_{N_2}. Yet, at all levels of oxygen, the time dependence of increase in the nitrogen content follows a first-order-type relation. Inouye and Choh proposed a reaction model to account for this unusual behavior; their reasoning, however, is hard to justify in terms of our present understanding of the absolute reaction-rate theory. Despite these inconsistencies, their absorption rates for 1 atm N_2 are in reasonable accord with those of Pehlke[227] and Schenck,[228] who found first-order reaction kinetics at all levels of oxygen in iron and various reduced N_2 pressures in the gas.

However, one feature common to all the experimental findings is that the absorption and desorption rates of nitrogen in, or evolution from, liquid or solid iron is retarded by oxygen chemisorbed on the surface of the metal. Similarly, the rate of nitrogen transfer across the iron surface is retarded by other surface active elements such as sulfur[227,233] and selenium.[234]

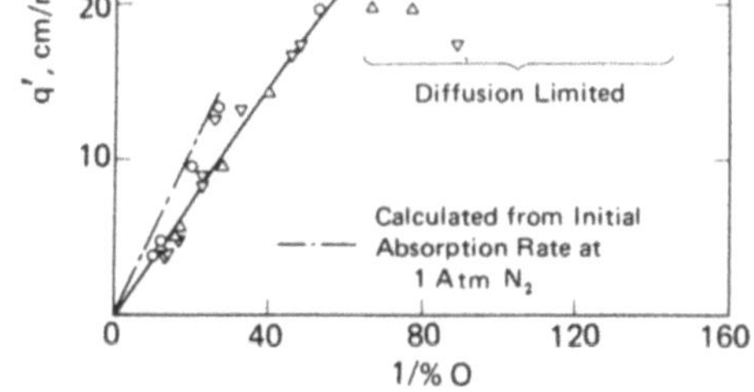

Figure 4.5-16. Rate parameter q' for denitrogenation of liquid iron at 1600° C as a function of oxygen content. ○, Fuwa *et al.*[230]; △, Mori and Suzuki[231]; ▽, Narita *et al.*[232]

Decomposition of CO_2 on Iron or Iron Oxide Surfaces. The decomposition rate of CO_2 on the surface of iron and iron oxides has been studied by numerous techniques. Whether the iron is oxidized or decarburized, the decomposition rate of CO_2 is found to be proportional to the partial pressure of CO_2 in the gas. That is, the rate is represented by

$$\mathcal{R} = \Phi_{CO_2}(1-\theta)p_{CO_2} \tag{4.5-31}$$

where Φ_{CO_2} is the rate constant which is a function of temperature only. The data compiled by Turkdogan and Vinters[(235)] show the variation of $\Phi_{CO_2}(1-\theta)$ with temperature as illustrated in Figure 4.5-17. The rate constant for the surface of iron is about an order of magnitude greater than for the surface of wustite. The rate constants were obtained over a

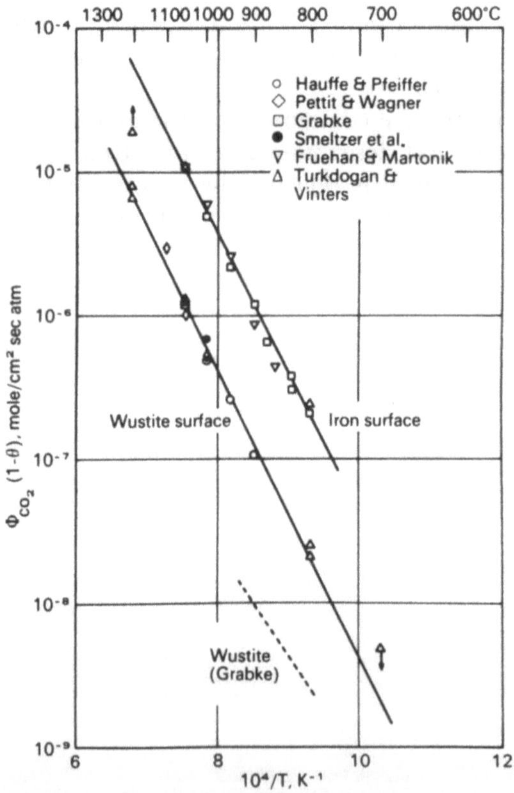

Figure 4.5-17. Temperature dependence of the specific rate constant for the dissociation of carbon dioxide on the surface of iron and wustite.

wide range of oxygen potentials, e.g., those corresponding to iron saturated with wustite to iron saturated with graphite. Yet, in all cases the same value of $\Phi_{CO_2}(1-\theta)$ is obtained, despite the anticipated variations in $(1-\theta)$ for the stated wide range of oxygen potentials. In other words, the kinetics of dissociation of carbon dioxide on the surface of iron cannot be explained satisfactorily in a way that would be consistent with the rate theory involving strong chemisorption of oxygen only at relatively high oxygen activities.

Extrapolation of the top line in Figure 4.5-17 to higher temperatures indicates that the rate of decarburization of liquid iron at 1600° C may be in the range from 2×10^{-3} to 10^{-2} mole/cm^2 sec atm. This is a relatively fast chemical reaction rate which may not be achieved by ordinary experimental techniques, because of the difficulty of maintaining rapid transport of reactants and products to and from the interface. In fact, many studies have shown that the rate of decarburization of liquid iron in CO_2 gas streams is controlled by gas diffusion near the surface of the melt.[236-237]

In a recent study, Sain and Belton[238] employed high gas-flow rates to overcome gas diffusional problems in measuring the rate of decarburization of liquid iron–carbon alloys in CO_2. They found that in melts containing more than 1% carbon, the rate was independent of the carbon concentration and directly proportional to the partial pressure of CO_2. The temperature dependence of the rate constant is shown in Figure 4.5-18 for sulfur-free and 0.009% sulfur melts. A small amount of sulfur in the melt appears to have a marked effect on the decomposition rate of CO_2. The extension of the dot–dash line in Figure 4.5-18 for solid iron, reproduced from Figure 4.5–17, lies in the area of the iron–carbon melts. Sain and Belton propose that the reaction on sulfur-free melts is controlled by the rate of chemisorption of CO_2 on the surface of the melt with an apparent heat of activation of about 23 kcal.

Forster and Richardson[239] investigated the oxidation rate of liquid copper and nickel in CO_2–CO gas mixtures using the levitation technique. As is shown in Figure 4.5-18, the rates are much lower than those observed for liquid iron. At present, it is difficult to account for large variations in the observed rate of decomposition of CO_2 on different metal surfaces.

4.5.3.3. Gas–Slag–Metal Reactions

Study of the kinetics of slag–metal reactions are often complicated by gas evolution. For example, in the reaction of a silicate slag with carbon-saturated iron, the silicon transfer to the metal is accompanied

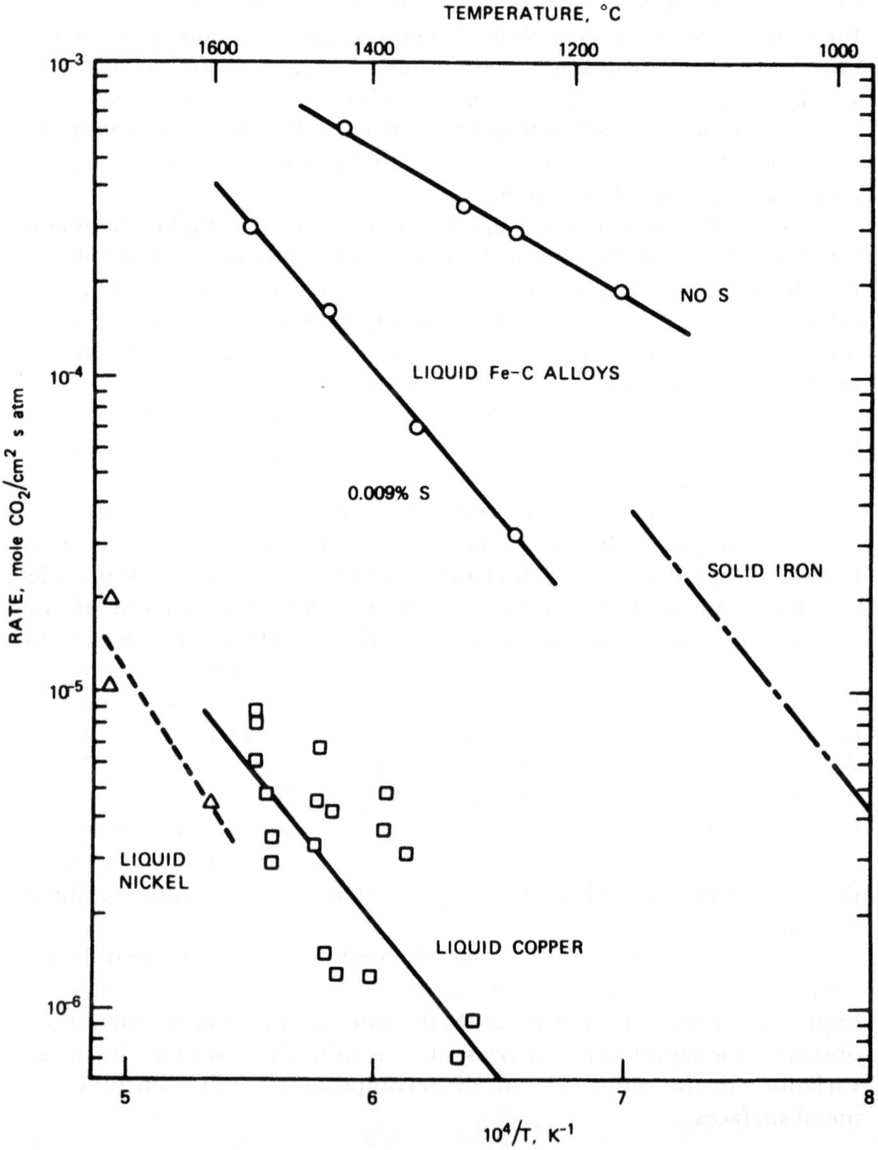

Figure 4.5-18. Rate constants for oxidation of liquid nickel, copper (Forster and Richardson[239], and liquid iron–carbon alloys (Sain and Belton[238]).

by CO evolution; thus the overall reaction is

$$SiO_2 + 2\,\underline{C} = \underline{Si} + 2\,CO \tag{4.5-32}$$

where underlines are used for elements dissolved in the molten metal. When considering the kinetics of heterogeneous reactions, a three-phase reaction at an interface cannot be realized. This overall reaction may be subdivided into two series of reactions

$$SiO_2 = \underline{Si} + 2\,\underline{O} \tag{4.5-33a}$$

$$2\,\underline{C} + 2\,\underline{O} = 2\,CO \tag{4.5-33b}$$

In an attempt to measure the kinetics of reaction (4.5-33a), Turkdogan *et al.*[240] carried out experiments with graphite-saturated iron and CaO–SiO_2 and CaO–BaO–SiO_2 melts in such a way that bubbles of CO were injected into the slag just above the metal, so that the interface would be swept by CO bubbles without breaking it up. The rate of reduction of silica was found to increase with increasing silica activity of the slag. Assuming equilibrium adsorption of silica on the metal surface as a silicate ion and forming an ideal monolayer, the fraction of the vacant sites is represented by

$$(1-\theta) = \frac{\varphi}{a_{SiO_2} + \varphi} \tag{4.5-34}$$

where φ is the activity coefficient of the silicate ions adsorbed on the surface of iron. From equation (4.5-28), the isothermal rate of reduction of silica is

$$\mathcal{R} = \Phi_{Si} \frac{a_{SiO_2}}{a_{SiO_2} + \varphi} \tag{4.5-35}$$

where Φ_{Si} is the specific rate constant and a_{SiO_2} the activity of silica relative to solid silica. Rearranging this equation gives

$$\frac{\mathcal{R}}{a_{SiO_2}} = \frac{\Phi_{Si}}{\varphi} - \frac{\mathcal{R}}{\varphi} \tag{4.5-36}$$

The experimental results of Turkdogan *et al.*[240] in Figure 4.5-19 are in general accord with this rate equation from which the following rate parameters are obtained for 1600° C: $\Phi_{Si} = 2.08 \times 10^{-3}$ mole SiO_2/cm^2 hr and $\varphi = 0.128$, relative to pure solid silica.

Because of the ionic nature of molten slags and the nonpolar nature of metals, the transfer of an element from a molten metal to a molten slag is accompanied by exchange of electrons between the reacting species. This is well demonstrated by King and Ramachandran[241] and by Nilas and Frohberg[242] who investigated the transfer of sulfur from

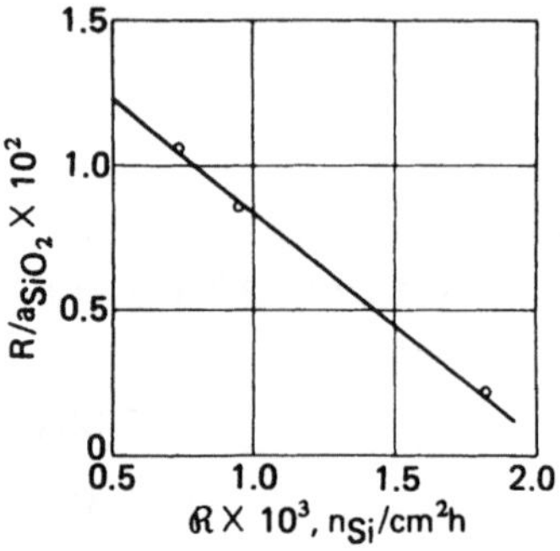

Figure 4.5-19. Effect of silica activity in slag on its rate of reduction by graphite-saturated iron at 1600° C and 1 atm CO.

graphite-saturated iron to slags. When sulfur is transferred from metal to slag, electrons must be provided, thus

$$\underline{S} + 2e = S^{2-} \tag{4.5-37}$$

Since electroneutrality is maintained in the slag and metal, in the absence of an electric field across the slag–metal interface, the solutes in the metal become oxidized to provide the electrons needed for reaction (4.5-37); thus

$$\underline{Fe} = Fe^{2+} + 2e \tag{4.5-38a}$$

$$\underline{Mn} = Mn^{2+} + 2e \tag{4.5-38b}$$

$$\underline{Al} = Al^{3+} + 3e \tag{4.5-38c}$$

$$\underline{Si} = Si^{4+} + 4e \tag{4.5-38d}$$

$$\underline{C} + O^{2-} = CO + 2e \tag{4.5-38e}$$

Therefore, in the presence of these solutes in iron the transfer of n_S moles of sulfur from metal to slag should be accompanied by the transfer of an equivalent total number of moles of solutes to satisfy the stoichiometric requirements:

$$2\, n_S = 2\, n_{Fe} + 2\, n_{Mn} + 3\, n_{Al} + 4\, n_{Si} + 2\, n_C \tag{4.5-39}$$

The sign of n is positive for metal→slag transfer and negative for slag→metal transfer. Typical examples of the results of King and Ramachandran[241] are given in Figure 4.5-20a, where silicon and sulfur contents of iron, sulfur and iron oxide contents of slag, and the amount of carbon monoxide evolved are plotted against the time of reaction. In Figure 4.5-20b, the change in iron and silicon contents and carbon monoxide evolution are given in terms of sulfur equivalence. It is seen that the amount of sulfur transferred from metal to slag calculated from

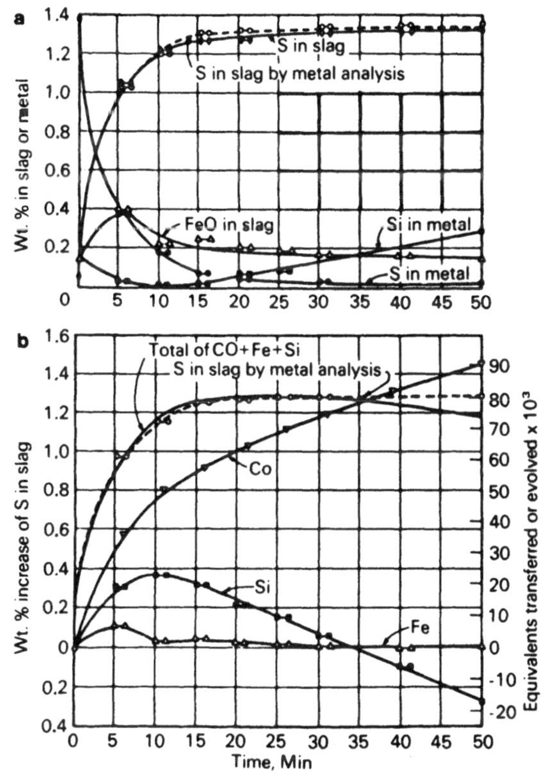

Figure 4.5-20. Equivalent of S, Fe, and Si transferred from metal to slag and equivalents of CO evolved for graphite-saturated iron reacting at 1505° C with a slag (48% CaO, 21% Al_2O_3, 31% SiO_2).

n_{Fe}, n_{Si}, and n_{CO}, as in equation (4.5-39), agree well with those observed experimentally. In the example considered, as slag and metal move toward equilibrium with respect to the slag–metal–sulfur partition, silicon and iron initially move away from equilibrium. These coupled reactions are considered as electrochemical. The relative rates of these electrochemical reactions are determined by their relative potential–current relations. It follows from these considerations that, if the initial concentrations of manganese, silicon, aluminum, etc., in the metal are higher than the ultimate equilibrium values for a given metal and slag system, the rate of sulfur transfer from metal to slag increases. This conclusion is borne out also from the results of Goldman, Derge, and Philbrook.[243]

Another example of an electrochemical coupled reaction is provided by the studies of Turkdogan and Pearce,[244] investigating the

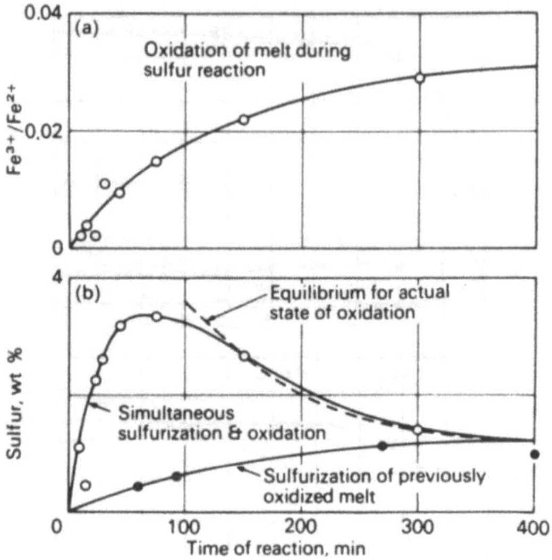

Figure 4.5-21. (a) Oxidation and (b) sulfurization of iron silicate melts (40% SiO_2) in a CO–CO_2–SO_2 gas mixture at 1550° C (ingoing gas composition $P_{CO}/P_{CO_2} = 3.0$ and $P_{SO_2} = 0.028$ atm).

reaction between iron silicate melts with gases containing CO, CO_2, and SO_2. The two following reactions take place simultaneously:

$$\tfrac{1}{2} S_2 + O^{2-} = S^{2-} + \tfrac{1}{2} O_2 \qquad (4.5\text{-}40a)$$

$$\tfrac{1}{2} O_2 + 2\, Fe^{2+} = O^{2-} + 2\, Fe^{3+} \qquad (4.5\text{-}40b)$$

As is indicated by the results in Figure 4.5-21a, reaction (4.5-40a) is faster than reaction (4.5-40b) with the result that the sulfur content of the melt reaches a maximum at some stage of the reaction when the following equilibrium is approached:

$$\tfrac{1}{2} S_2 + 2\, Fe^{2+} = S^{2-} + 2\, Fe^{3+} \qquad (4.5\text{-}41)$$

However, the system as a whole is not at equilibrium. As the oxidation of iron in the melt continues, the concentration ratio Fe^{3+}/Fe^{2+} increases, and since p_{S_2} in the gas is maintained constant, the sulfur content of the melt must decrease as dictated by reaction (4.5-41). As the whole system is approaching equilibrium, reaction (4.5-41) remains at a state of pseudoequilibrium. When the ferrous silicate melt is first equilibrated with the prevailing oxygen partial pressure of the gas, sulfurization proceeds in a normal manner as shown by the lower curve in Figure 4.5-21b.

Richardson and co-workers and others[(245)] have shown that at low oxygen activities, sulfur enters a slag by displacing oxygen by reaction (4.5-40a). Therefore, a slag which does not contain elements with more than one type of valency, such as Fe^{2+} and Fe^{3+}, can be used in principle as a medium to transfer sulfur from a low chemical potential to a high one. This has been demonstrated by Turkdogan and Grieveson[(246)] by using a soda-glass-coated Mn–MnS–MnO pellet and Fe–S–O melts contained in an evacuated silica capsule (Figure 4.5-22a). Upon heating, there was an equilmolar countercurrent sulfur and oxygen transfer across the thin film of soda-glass which acted as an ionic membrane (Figure 4.5-22b).

4.5.3.4. *Vaporization of Metals and Alloys*

Vaporization plays a role to varying degrees in smelting and refining processes. For example, in a blast furnace the sulfur in coke is transferred to iron and slag partly via the formation of volatile species such as S_2, COS, and SiS. Also silicates and phosphates in the gangue matter are reduced to volatile species as SiO and PO_x which react subsequently with the reduced iron. In the vacuum refining of metals, impurities are removed either in the elemental form or as volatile oxides or sulfides. Vaporization may also play some role in the transfer of impurities from metal to slag in pneumatic steelmaking processes, and in the conversion of copper matte to blister copper. Mechanisms of reactions involving vaporization depend much on the composition, temperature, and pressure of the system.

The maximum rate of vaporization, or free vaporization, which requires no energy of activation is derived from the kinetic theory of gases

$$\mathcal{R} = \frac{p_i}{(2\pi M_i R T)^{1/2}} \qquad (4.5\text{-}42)$$

where

$\mathcal{R}$ = mole/cm^{2}sec,
p_i = vapor pressure of the vaporizing species at the surface of the melt, dyne/cm^2 (1 atm = 1.0132×10^6 dyne/cm^2),
M_i = molecular weight, g, and
$R = 8.1344 \times 10^7$ erg/mole deg.

Inserting the appropriate units in equation (4.5-42) gives

$$\mathcal{R} = 44.3 \frac{p_i(\text{atm})}{(M_i T)^{1/2}} \text{ mole/cm}^2\text{sec} \qquad (4.5\text{-}43)$$

Studies made by Ward[(247)] of vaporization of manganese, copper, and chromium from iron alloys at 1580° C, using 10-kg melts heated

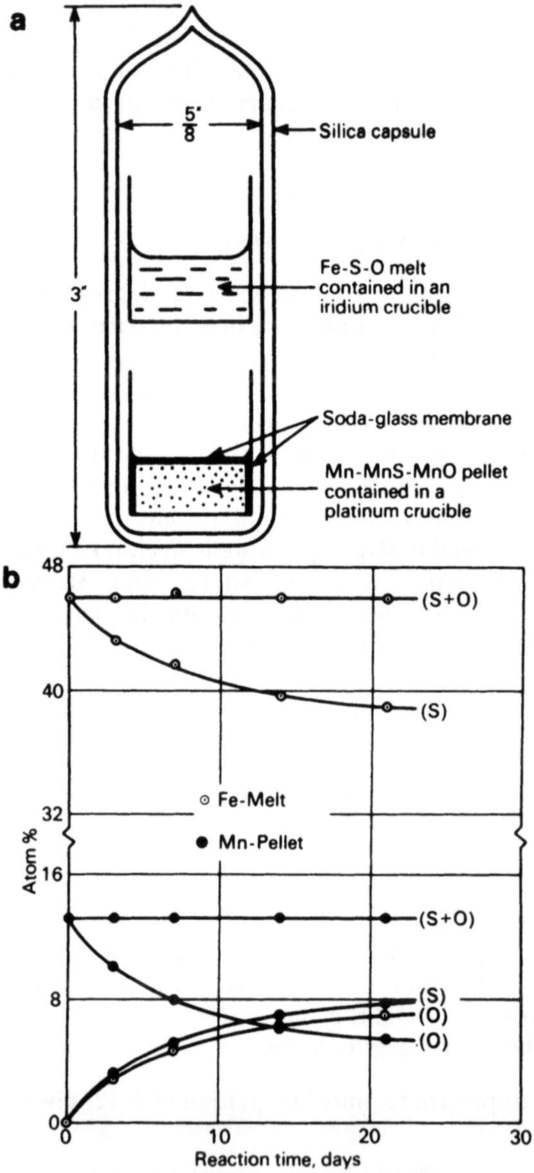

Figure 4.5-22. Stoichiometric displacement of sulfur and oxygen in Fe melt and Mn pellet separated by an ionic membrane at 1127° C.

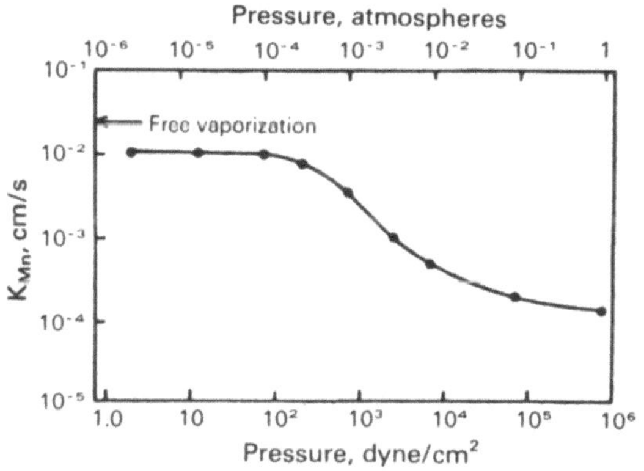

Figure 4.5-23. Effect of pressure (argon) on the rate of vaporization of manganese from inductively stirred iron–manganese melts at 1580° C.

inductively, illustrate the effect of liquid- or gas-phase diffusion on the vaporization rate. The rate constant shown in Figure 4.5-23 for the vaporization of manganese approaches a plateau below 10^{-4} atm. At these low pressures the vaporization rate is about 70% of that for free vaporization, relative to the bulk manganese content of the melt, indicating a slight depletion of manganese near the surface of the melt. In melts stirred by induction heating, the mass transfer of solute to the metal surface accompanying free vaporization can be estimated using a mass transfer equation derived by Machlin,[(248)] and by Tarapore and Evans.[(249)] With increasing pressure of argon in the system, the vaporization rate decreases. This is due to a decrease in the diffusive flux of the manganese vapor away from the surface with increasing gas pressure. The higher the temperature the greater is the transport effect on the rate of vaporization.[(250)] At lower temperatures and/or with solutes of lower vapor pressures, the maximum rate of vaporization can be achieved in inductively stirred melts, as, for example, in the desulfurization of graphite-saturated iron at low pressures ($\sim 10^{-4}$ atm) measured by Fruehan and Turkdogan[(251)]; this is shown in Figure 4.5-24.

If two solutes react in the metal and form a volatile species of higher vapor pressure, the rate of vaporization increases. For example, Fruehan and Turkdogan[(251,252)] showed that desulfurization of liquid iron alloys *in vacuo* is markedly enhanced by silicon, because of the formation of volatile SiS. Their results suggested that the equilibrium with the activated complex might be represented by

$$\underline{Si} + 2\,\underline{S} = (SiS_2)^{\ddagger} \rightarrow \text{product SiS} \qquad (4.5\text{-}44)$$

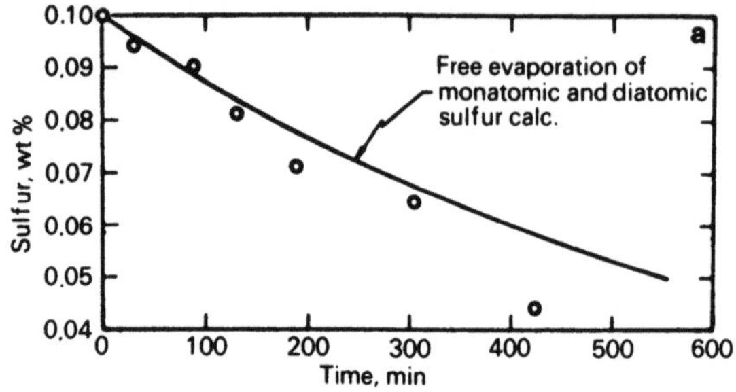

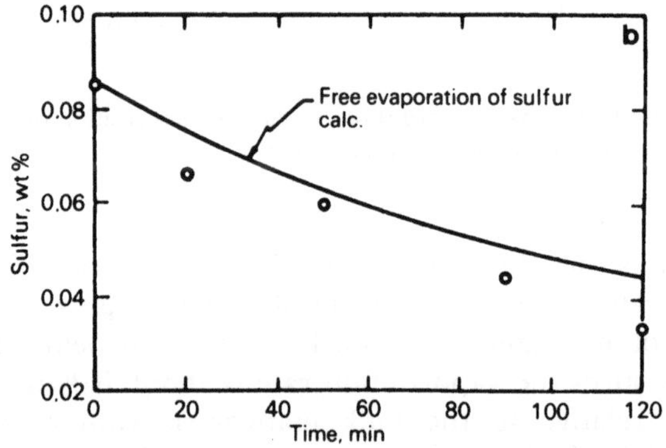

Figure 4.5-24. Calculated and experimental rate of vaporization of sulfur from graphite-saturated iron at 10 μm pressure and (a) 1600° C and (b) 1500° C.

for which the rate equation is

$$\mathcal{R} \propto a_{Si} a_S^2 (1-\theta) \tag{4.5-45}$$

In these experiments the sulfur activity in the melt was high enough to satisfy the limiting case of $(1-\theta) \to 0$. Therefore, inserting $(1-\theta) \propto 1/a_S$ for high-site fillage by adsorbed sulfur and integrating gives

$$\ln \frac{\%S}{(\%S)_0} = -k_1 a_{Si} f_S t \tag{4.5-46}$$

where k_1 is the rate constant, a_{Si} is the silicon activity in the melt, and f_S is the effect of silicon on the activity coefficient of sulfur in iron. The proportionality of the rate to silicon activity in iron was confirmed experimentally.[251] As is shown in Figure 4.5-25, the measured rate constant k_1 is about an order of magnitude smaller than the rate of free evaporation of SiS from liquid iron. That is, in the desulfurization of iron–silicon alloys, the rate of evolution of SiS vapor is controlled by the transfer of SiS from the bulk metal phase to the adsorbed layer via the formation of a complex $(SiS_2)^‡$. However, below 0.01% S the rate of desulfurization was retarded by slow sulfur transfer to the surface of the inductively stirred melts.

In reactions between two condensed phases involving a volatile reacting species, the gas phase plays an important role as a transfer medium. This is demonstrated well by the experiments of Boyd *et al.*[253] in the desulfurization of carbon-saturated iron *in vacuo* by lime crystals. It was found that the rate of sulfur pick up by lime suspended over the melt was the same as with the lime crystal immersed in the metal. Since the diffusivity of sulfur vapor is high at low pressures, desulfurization by lime is just as efficient even when the lime is not in direct contact with the melt. In this system the rate of desulfurization is controlled by diffusion through the CaS layer surrounding the lime particle.

Another important vaporization phenomenon is that occurring in reactive atmospheres. Turkdogan, Grieveson, and Darken,[254] have shown that when a metal vaporizes in a stream of a reactive gas, such as argon–oxygen mixtures, there is a counterflux of metal vapor and oxygen, and at some short distance from the surface of the metal a metal oxide mist forms (Figure 4.5-26). Increasing oxygen partial pressure decreases the distance δ through which the metal vapor is diffusing, i.e., the rate of vaporization increases. On further increase in oxygen partial pressure, the flux of oxygen toward the surface of the metal eventually becomes greater than the equivalent counterflux of metal vapor, resulting in the oxidation of the metal surface and cessation of vaporization. Just before this cutoff the rate of vaporization is close to the maximum rate of free vaporization as given by equation (4.5-42). As is seen from

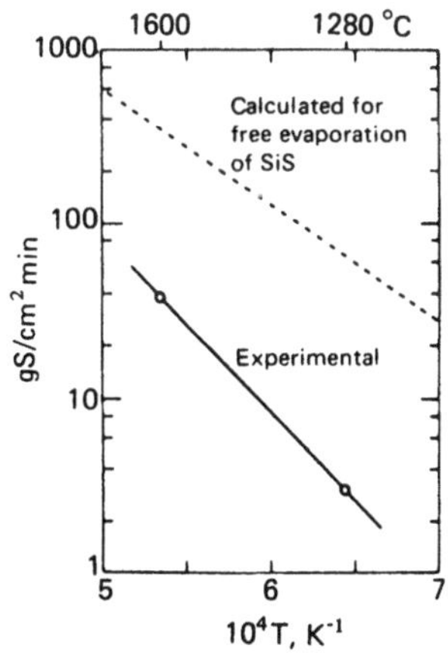

Figure 4.5-25. Measured rate constant for vaporization of SiS from iron alloy is compared with free evaporation rate.

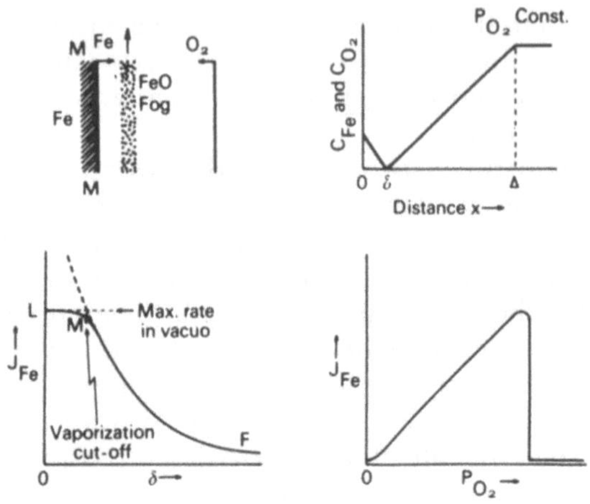

Figure 4.5-26. Schematic representation of gas–vapor phase reaction, counterflux of iron vapor and oxygen, and resulting enhanced vaporization of iron.

some examples of the experimental results of Turkdogan *et al.*[(254)] given in Figure 4.5-27, the rates of vaporization of iron and copper increase with increasing oxygen partial pressure until a maximum rate is obtained, beyond which vaporization almost ceases. The results summarized in Figure 4.5-28 show that the measured maximum rates of vaporization agree well with those calculated for the free rates of vaporization. This mechanism of enhanced vaporization is believed to be the cause of the formation of metal oxide fumes in smelting and refining processes. A quantitative analysis of these observations in terms of mass transfer under conditions of laminar gas flow over the metal surface has been given by Turkdogan *et al.*[(254)] Further similar studies with iron were made by Distin and Whiteway.[(255)]

Impurities in metals can also be vaporized at faster rates by again flowing a reactive gas over the surface of the alloy. For example, Kor and Turkdogan[(256)] showed that the rate of vaporization of phosphorus from liquid iron can be increased appreciably by flowing an oxygen–argon mixture over the surface of the melt without oxidizing the surface. In this case, the vaporizing species are Fe, PO, and PO_2 which react with oxygen in the gas, forming an iron phosphate fog close to the surface of the melt. Enhanced vaporization of impurities from metals in oxidizing atmospheres may provide the mechanism for rapid rate of refining in oxygen- or air-blowing processes. Thoughts on this subject are presented later.

The rate of vaporization increases also by a temperature gradient. When there is a temperature gradient between a heated object and its immediate fluid surroundings, two processes occurring simultaneously

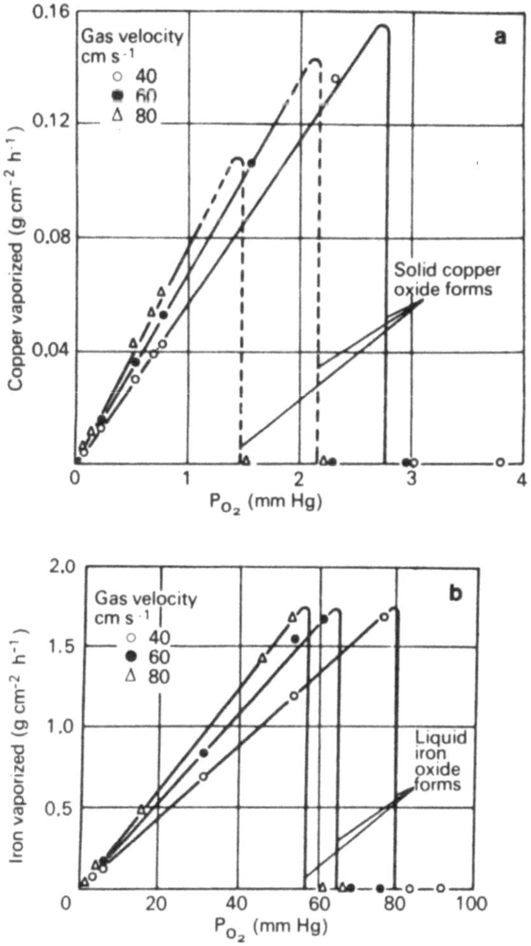

Figure 4.5-27. The effects of oxygen at various partial pressures in argon at a total pressure of 1 atm and at various flow rates on the rate of vaporization of (a) copper at 1200° C and (b) iron at 1600° C.

bring about enhancement of the rate of vaporization of the heated object: (1) natural convection, which is a transport process, and (2) condensation of the vapor, which is a reaction process. As shown by Turkdogan,[257] this is a complex phenomenon involving supersaturation for nucleation of metal droplets in a steep temperature gradient and the effect of the latent heat of condensation on the temperature profile and growth rate of droplets. Rate equations have been derived by Turkdogan,[257] Rosner,[258] and Hills and Szekely.[258] Turkdogan and Mills[259] measured the vaporization rate of iron–nickel alloys in a steep temperature gradient in a stream of helium using electromagnetic levitation. The results, shown in Figure 4.5-29 are in reasonable accord with those calculated from an approximate rate equation involving homogeneous nucleation of the condensate at a cooler temperature in the gas phase. Enhancement of the vaporization rate caused by a temperature gradient is less extensive than that obtained in a stream of reactive gas.

We see from numerous examples given that the vaporization rate of an alloy is affected by (1) diffusion in the alloy or gas above it, (2) total gas pressure and velocity parallel to the surface of the heated object, (3) reaction in the gas phase close to the surface, (4) a temperature gradient close to the surface of the vaporizing substance, and (5) a chemical reaction at the surface. At sufficiently low pressures, the gas or vapor molecules do not collide; under these conditions the vaporizing species do not return back to the surface of the condensed phase, hence free vaporization occurs at a maximum rate. The maximum vaporization rate can be obtained also at high pressures in a gas stream by lowering the partial pressure of the vaporizing species by a reaction in the gas phase. High-speed motion pictures of a zirconium drop falling in an oxidizing atmosphere and in a temperature gradient, taken by Nelson and Levine,[261] illustrate well the condensation occurring in the gas phase around the metal drop.

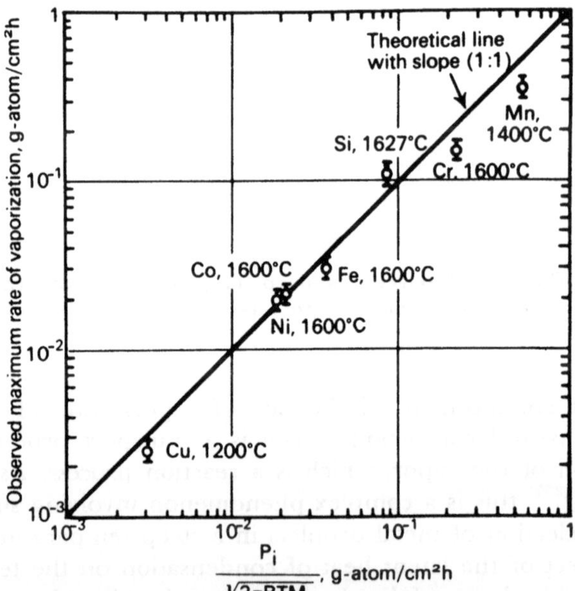

Figure 4.5-28. Experimentally observed maximum rate of vaporization of metals in flowing argon–oxygen mixtures is compared with the rate of free vaporization.

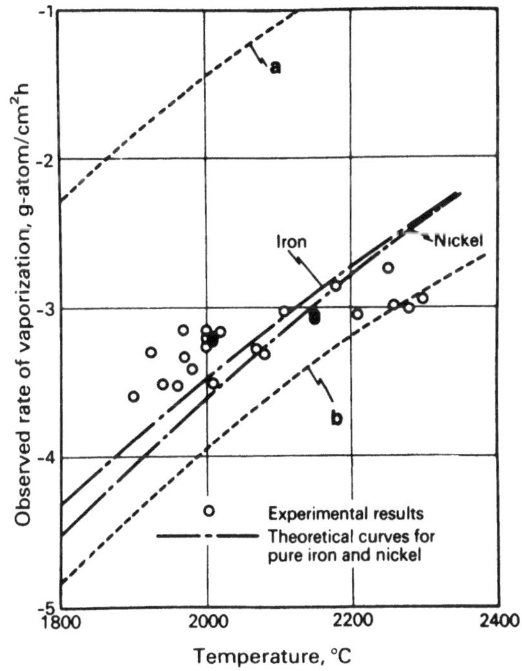

Figure 4.5-29. Experimental data on rate of vaporization of Fe–Ni alloys, levitated in an atmosphere of helium, are compared with the theoretical curves for pure iron and nickel. Curve a gives the maximum rate of vaporization of iron, *in vacuo*; curve b gives the minimum rate, due to diffusion in helium at 1 atm, under isothermal conditions.

Several aspects of vaporization phenomena briefly outlined here provide guidance for better understanding and control of some of the pyrometallurgical processes and even the burning and ablation of projectiles falling in the earth's atmosphere.

4.5.4. Effects of Gas Blowing and Evolution on Rates of Metal Refining

In almost all types of pyrometallurgical processes, gas blowing and evolution play vital roles in the smelting and refining of metals. Rates of reactions are much influenced by interplay of gas bubbles with metal and slag phases. Basic aspects to be discussed here are the slag–metal mass transfer affected by gas evolution, metal and gas emulsion in slags, gas bubbles in metals, refining by air or oxygen blowing, and refining *in vacuo*.

4.5.4.1. Mass Transfer across Bubble-Stirred Interfaces

Most of the smelting and refining processes involve gas evolution which brings about effective mixing in the metal and slag phases, and also enhances mass transfer across the slag–metal interface. Several model studies and experimental work done on the mass transfer across two-liquid interfaces stirred by bubbles have been collated in a recent review paper by Robertson and Staples.[262]

Richardson and co-workers[263–265] studied the effect of gas bubbling on the mass transfer in several two-liquid systems. For the simple case of no dispersion of the two liquid phases, no solute-induced interfacial turbulence, and assuming the interfacial reaction equilibria, the mass transfer of a solute from one phase to another is represented approximately by the following relation:

$$m^2 = bD\dot{Q} \tag{4.5-47}$$

where

m = average mass transfer coefficient, cm/sec,
b = proportionality factor, cm^{-1},
D = solute diffusivity, cm^2/sec, and
$\dot{Q}$ = volume flux of gas bubbles across the interface, $ml/cm^2 sec$.

The proportionality of m^2 to the product $D\dot{Q}$ is also predicted from the theories proposed by Machlin[248] and by Davies.[266] For liquid depths of more than about 5 cm, the proportionality factor b is about 120 cm^{-1} for metals and ~50 cm^{-1} for slags. On the basis of this mass transfer relation, Subramanian and Richardson[264] demonstrated, at least in principle, the primary rate-controlling mechanism of decarburization in open hearth steelmaking during the period of gentle carbon boil.

In open hearth steelmaking, oxygen is transferred from slag to metal and carbon monoxide bubbles are nucleated in crevices of the refractory furnace lining. For the overall reaction

$$\underline{C} + \underline{O} = CO \tag{4.5-48}$$

the rate is represented by

$$\dot{n}/A = m_A(C^i_M - C_M) = m_{sl}(C_{sl} - C^i_{sl}) \tag{4.5-49}$$

where

$\dot{n}$ = rate of oxygen transfer to metal ≅ −rate of decarburization, g-atom/sec,
A = static slag–metal interfacial area, cm^2,
m = oxygen mass transfer coefficient across the interface (M in metal, sl in slag), cm/sec,
C^i = iron oxide content at the interface (M in metal, sl in slag), g-atom/cm^3, and
C = iron oxide content of bulk metal (M) and slag (sl), g-atom/cm^3.

Assuming that there is equilibrium distribution of iron oxide in iron and slag at the interface, equation (4.5-49) is transformed to

$$\dot{n}/A = m'(k_O C_{sl} - C_M) \tag{4.5-50}$$

where

k_O = equilibrium ratio, $C^i_M/C^i_{sl} = 1/26$ for basic slags at steelmaking temperatures,

m' = overall mass transfer coefficient, $(1/m_M + k_O/m_{sl})^{-1}$

Noting that $\dot{n}/A \propto \dot{Q}$, equations (4.5-47) and (4.5-50) give

$$\dot{n}/A \propto D_O(k_O C_{sl} - C_M)^2 \tag{4.5-51}$$

where D_O is the diffusivity of oxygen in liquid iron. This relation indicates the autocatalytic nature of decarburization in the open hearth steelmaking. That is, the mass transfer coefficient for iron oxide is a function of the rate of CO evolution such that a doubling of the oxygen concentration driving force increases the rate of decarburization by a fourfold.

The autocatalytic decarburization of liquid iron by iron oxide in the slag, as indicated by the relation in equation (4.5-51), is also substantiated by the experiments of Philbrook and Kirkbride.[267] They measured the rate of reduction of iron oxide in blast-furnace-type CaO–Al_2O_3–SiO_2 slags by carbon-saturated iron, and found that the rate is proportional to the square of the iron oxide content of the slag. The rate constant calculated using equation (4.5-51) is about a third of the values obtained in their experiments.

According to a study of open hearth plant data by Brower and Larsen,[268] during decarburization of a 34-cm-deep metal bath from about 1.0 to 0.1% C at a rate of 0.15 wt % C per hour, the oxygen content of steel is about 0.03 wt % below that for the bulk slag–metal equilibrium. That is, the concentration difference $(k_O C_{sl} - C_M)$ equals 1.5×10^{-4} g-atom/cm^3 and the observed rate of decarburization corresponds to $\dot{n}/A = 8.5 \times 10^{-6}$ g-atom/cm^2 sec. Inserting these values in equation (4.5-50) gives for the overall mass transfer coefficient $m' = 0.057$ cm/sec. Let us now compare this with the mass transfer coefficient derived from equation (4.5-47) based on room temperature experiments with amalgams and aqueous solutions.[263–265]

Assuming for liquid steel, $b = 120$ cm^{-1} and $D_O = 4 \times 10^{-5}$ cm^2/sec; for slag, $b = 50$ cm^{-1} and $D_O = 10^{-5}$ cm^2/sec; and noting that $\dot{n}/A = 8.5 \times 10^{-6}$ g-atom/cm^{2}sec $\equiv \dot{Q} = 1.3$ ml/cm^{2}sec of CO at 1600° C, we obtain, equation (4.5-50), $m_M = 0.079$ cm/sec and $m_{sl} = 0.021$ cm/sec. Combining these values gives for the calculated overall mass transfer coefficient $m' = 0.069$ cm/sec which compares well with m' derived from the open hearth data.†

† Numerical values used in these calculations are somewhat different from those given originally by Subramanian and Richardson.[264]

4.5.4.2. Metal Emulsion in Slags

When the interface between two liquids is disturbed by chemical or mechanical means to create strong eddy currents, it becomes unstable and brings about emulsification of the liquid phase of lower viscosity in the liquid phase of higher viscosity. In relation to the study of metallurgical systems, we shall consider only the metal emulsion in slags brought about by surface gas injection and by gas bubbles crossing the interface.

In steelmaking with oxygen top blowing, as in the BOF process, entrainment of metal droplets in the slag is caused by the impact of the oxygen jet and rapid CO evolution from the metal bath. One of the effects of gas jet impact is illustrated in Figure 4.5-30.[269] The schematic drawings next to the photographs illustrate the observed liquid motion imparted by a gas jet impinging on the surface of oil overlaid on water. A negative pressure brought about by the motion in the lower liquid

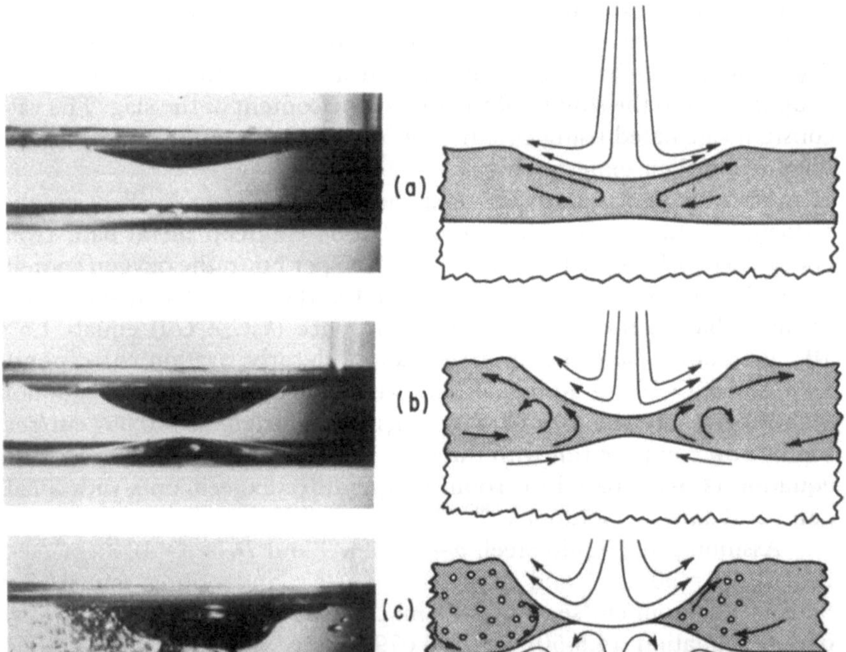

Figure 4.5-30. Photographs and drawings demonstrating the dome formation at the oil–water interface beneath the depression in the oil layer caused by impinging argon jet stream. Photograph (c) shows the beginning of water emulsification in oil when the water surface is exposed to the jet stream.

causes the formation of a dome at the two-liquid interface below the depression in the overlying liquid. On further increase in jet momentum, the depression and the dome meet and, subsequently, the water surface is exposed to the central part of the jet stream. This situation immediately brings about water emulsification in the oil layer. Beyond a certain critical jet momentum, splashing occurs and gas bubbles become entrapped in the oil–water emulsion. This is, in fact, what happens in pneumatic metal-refining processes.

The effect of viscosity on metal emulsion has been demonstrated by Poggi *et al.*[(270)] by bubbling argon through (a) mercury covered with a layer of water containing glycerin, (b) lead at 520° C covered with a fused salt, and (c) copper at 1200° C covered with a slag. The amount of metal dispersed in the overlying liquid was found to increase greatly with increasing viscosity of the upper liquid phase and with increasing gas-flow rate.

The extent of metal emulsion in slags was first brought to light by Kozakevitch and co-workers.[(271–273)] During decarburization in the basic Bessemer process, the slag contains 20–40% metallic iron ranging in size from 1 to 16 mm. Meyer and co-workers[(274)] and Trentini[(275)] found that in a 200-ton BOF steelmaking furnace, about 30% of the metallic charge is dispersed in the slag during the period of high rate of decarburization. Figure 4.5-31a shows the surface skin of a slag sample exhibiting small gas pockets and blow holes, and metal droplets and undissolved lime particles. A metal droplet with its CO bubble embedded in the slag is shown in Figure 4.5-31b. The metal droplets are in the range from 0.1- to 5-mm diameter, and the metal–slag interfacial area in commercial BOF converters is about 2000 ft^2 per ton of metal in the furnace. For a 200-ton converter, this corresponds to about 10^8 cm^2 slag–metal interfacial area in the emulsion during the period of high decarburization rate. Assuming that the decarburization rate is controlled by diffusion of carbon in metal droplets, the average residence time of metal droplets in the slag foam is estimated to be about 1–2 min.[(275)] The decarburization rate in the metal emulsion is about 0.3–0.6% C per minute and that in the entire system, about 0.15% C per minute. These relative rates of decarburization depend on the extent of metal dispersion which is determined by blowing conditions, slag composition, and the foaming characteristics of the slag.[(273,276,277)]

It is difficult to identify the mechanisms of rate-controlling reactions in the gas–metal–slag emulsions. However, the oxygen activity of the slag in BOF steelmaking, as determined from its iron oxide content, is much higher than that of the bulk metal phase. Therefore, it is expected that decarburization in the metal emulsion occurs by oxygen transfer from the slag phase and also directly from the oxidizing jet stream. Near the

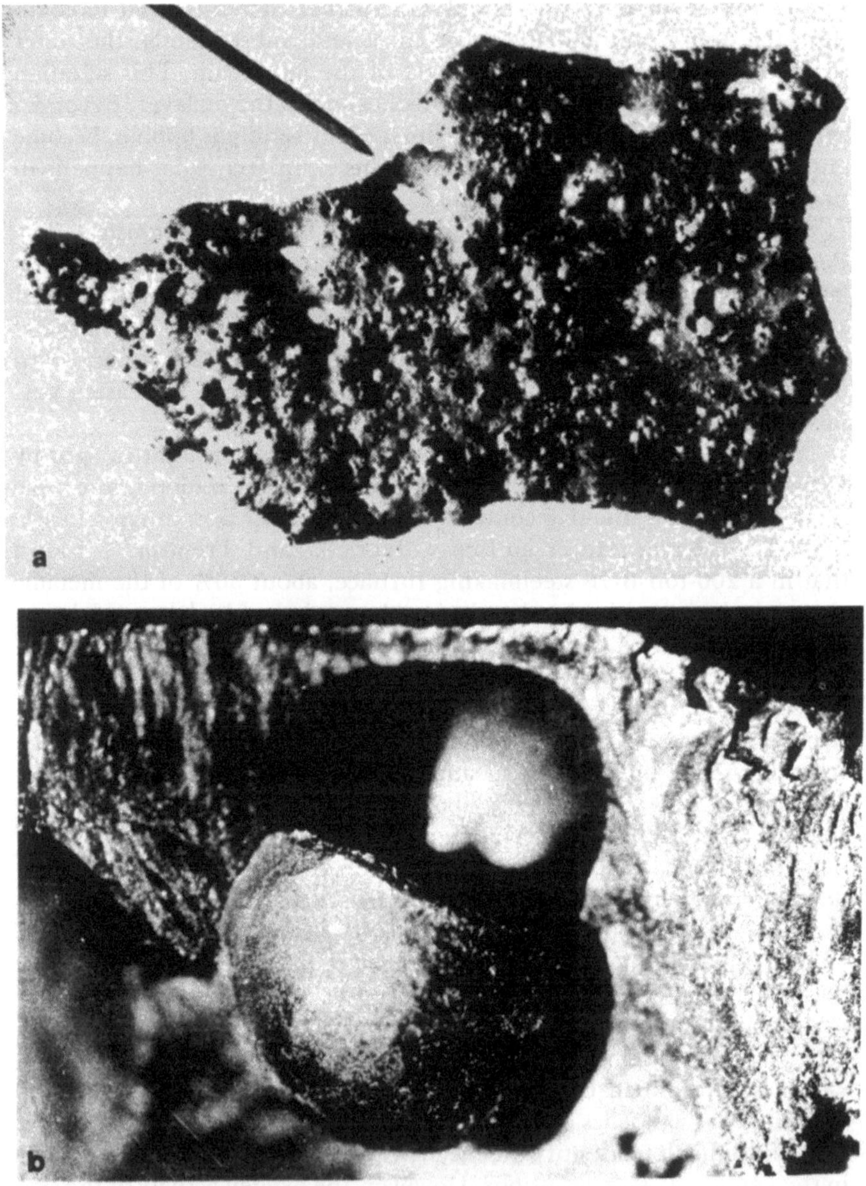

Figure 4.5-31. (a) Surface skin of solidified emulsion sample showing blow holes and minute metal droplets. (b) Photograph of a metallic droplet and its CO bubble (magnification 50 times, reduced 10% for reproduction).

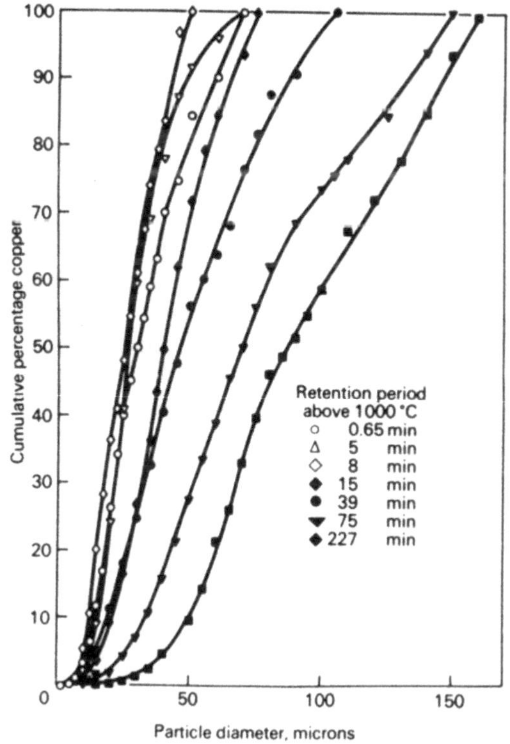

Figure 4.5-32. Size distribution of copper oxide and sulfide particles in Noranda process slags as a function of cooling rate.

end of decarburization, the slag foam subsides and most of the metal droplets return to the bulk metal phase.

There is also metal and matte emulsion in copper- and nickel-smelting processes. For example, in the Noranda process for continuous smelting and converting of copper concentrates, the tapped slag contains 8–12% Cu. However, only one third of the copper is in the form of entrained matte droplets, the remainder being in solution in the slag; upon cooling, copper oxide and sulfide precipitate. To obtain a high-grade copper concentrate from the slag by milling and flotation, the slag has to be heat treated to precipitate and coalesce the copper particles. The effect of the cooling rate on the size distribution of copper oxide and sulfide particles in Noranda process slags is shown in Figure 4.5-32.[278] By controlled cooling of the slag, most of the copper is recovered in the concentrate containing about 50% Cu as oxides and sulfides and 0.5% Cu in the tailing.

4.5.4.3. Bubbles in Liquid Metals

To facilitate the discussion of rates of reactions of gas bubbles with liquid metals, a brief mention will be made here of some characteristic features of gas bubbles in liquids with particular emphasis on the swarm of bubbles.

Gas bubbles rising in liquids acquire shapes that depend primarily on their size. Bubbles of equivalent diameter below 0.4 cm are spherical, between 0.5 and 0.9 cm in diameter they are ellipsoidal and above 1 cm diameter they are spherical-cap shape. From a theoretical analysis, Davies and Taylor[279] derived the following equation for the terminal velocity of spherical-cap-shape bubbles, given here in terms of the diameter of a sphere with the same volume as the spherical-cap bubble (the so-called *equivalent sphere*):

$$u_t = 0.72(gd)^{1/2} \tag{4.5-52}$$

where

g = gravitational acceleration, and
d = equivalent sphere diameter.

On the basis of this relation, Baird and Davidson[280] derived a theoretical equation for the liquid-phase mass transfer coefficient across single spherical-cap bubbles

$$m_l = 0.975D^{1/2}\left(\frac{g}{d}\right)^{1/4} \tag{4.5-53}$$

where D is the solute diffusivity in the liquid. Experimental observations, including those with liquid metals, are in general accord with these theoretical relations. For an average diffusivity of 5×10^{-5} cm/sec and for d from 1 to 2 cm, m_l is about 0.04 cm/sec.

In most metallurgical processes we are concerned mainly with the swarm of bubbles for which equation (4.5-52) does not apply, except perhaps at low-volume fractions of gas bubbles in liquids. Mass transfer phenomena in the swarm of bubbles have been studied extensively using aqueous solutions; this subject is well documented in a review paper by Calderbank.[281] For an average bubble velocity u_b, the liquid-phase mass transfer coefficient is given by

$$m = 1.28\left(\frac{Du_b}{d}\right)^{1/2} \tag{4.5-54}$$

In the swarm of bubbles, the bubble velocity u_b is a function of the superficial gas velocity, u_s, and the fractional holdup of gas, ε; thus

$$u_b = \frac{u_s}{\varepsilon} \tag{4.5-55}$$

where ε = (volume of gas)/[volume of (liquid + gas)].

Yoshida and Akita[282] have measured the increase in the fractional holdup of gas in nonfoaming aqueous solutions with increasing superficial gas velocity up to about 60 cm/sec. From their experimental results the following relation is obtained for $u_s > 3$ cm/sec:

$$\log \frac{1}{1-\varepsilon} = 0.146 \log (1 + u_s) - 0.06 \tag{4.5-56}$$

At high superficial gas velocities in nonfoaming liquids, the limiting value of ε in most practical situations is about 0.5. Yoshida and Akita also measured the volumetric liquid-phase mass transfer coefficient in the swarm of bubbles. Noting that the bubble interfacial area per unit volume† of the aerated liquid, a, is

$$a = \frac{6\varepsilon}{d} \text{ cm}^2/\text{cm}^3 \text{ of aerated liquid} \tag{4.5-57}$$

the volumetric mass transfer coefficient from equations (4.5-54) and (4.5-55) is

$$m_l a = 7.68(\varepsilon D u_s)^{1/2} d^{-3/2} \tag{4.5-58}$$

When equation (4.5-58) is applied to the results of Yoshida and Akita, it is found that the calculated bubble diameter is about 0.45 cm and independent of u_s over the range studied. Leibson *et al*[283] also found that gas injected into water from single orifices at high flow rates gave an average bubble diameter of 0.45 cm which is independent of the orifice diameter and gas-flow rate. We see from this analysis that equation (4.5-54), which was initially derived for single spherical-cap bubbles,[284] is applicable also to the swarm of bubbles. With $d = 0.45$ cm and taking $\varepsilon = 0.5$ for high superficial gas velocities, the limiting liquid-phase mass transfer coefficient for gas-aqueous solutions is simplified to

$$m_l = 2.7(D u_s)^{1/2} \tag{4.5-59}$$

The results of Yoshida and Akita do, in fact, indicate that this proportionality is approached at $u_s > 50$ cm/sec.

† Sometimes it is convenient to use bubble interfacial area S relative to the unit mass of the quiescent liquid, i.e.,

$$S = \frac{a}{(1-\varepsilon)\rho_l} = \frac{6\varepsilon}{1-\varepsilon}\frac{1}{\rho_l d} \quad \text{cm}^2/\text{g liquid}$$

In principle, these mass transfer relations for the swarm of bubbles should also apply to the gas–liquid metal systems. For the case of single bubbles in liquid silver, Guthrie and Bradshaw[285] found that the mass transfer coefficients of oxygen measured experimentally were only 10–30% lower than those predicted from equation (4.5-53). However, an important parameter to be evaluated, or estimated, for the swarm of bubbles in liquid metals is the average bubble size. Because of the high surface tension of liquid metals, the bubble size achieved is expected to be larger than that in water.

Relatively large bubbles in motion are subject to deformation and ultimately to fragmentation into smaller bubbles. The drag force exerted by the liquid on a moving bubble induces rotational and probably turbulent motion of the gas within the bubble. This motion creates a dynamic pressure on the bubble surface. When this force exceeds the surface tension, bubble breakup occurs. Because of the large difference between the densities of the gas and liquid, the energy associated with the drag force is much greater than the kinetic energy of the gas bubble. Therefore, the gas velocity in the bubble is similar to the bubble velocity. On the basis of this reasoning, Levich[286] derived the following equation for critical bubble size as a function of bubble velocity:

$$d_c = \left(\frac{3}{C_D \rho_g \rho_l^2}\right)^{1/3} \frac{2\gamma}{u_b^2} \tag{4.5-60}$$

where C_D is the drag coefficient which is taken as unity.[279] Combining this with equation (4.5-52) gives the critical bubble diameter at its terminal velocity.

$$d_c = \left(\frac{3}{\rho_g \rho_l^2}\right)^{1/6} \left(\frac{3.86\gamma}{g}\right)^{1/2} \tag{4.5-61}$$

On the basis of a similar concept of dynamic equilibrium, Hu and Kintner[287] expressed the critical size of liquid droplets in free fall in terms of a critical Weber number:

$$\mathrm{We}_c = \frac{d_c^2(\rho_l - \rho_g)g}{6\gamma} = 2.4 \tag{4.5-62}$$

They also found experimentally that We_c equals 2.4 for the breakup of water drops during free fall in air.

The critical bubble sizes at terminal velocities in liquids calculated from equation (4.5-61) are given in Table 4.5-4 for water, liquid iron, liquid copper, and liquid matte.

Table 4.5-4. Critical Bubble Sizes at Terminal Velocities

Liquid	Temperature, °C	Pressure, atm	γ, dyne/cm	d_c, cm
Steel	1600	1.5	1500	6.0
Blister copper	1300	1.5	900	4.5
Matte	1300	1.5	500	3.8
Water	20	1.0	72	1.96

In stirred melts, the bubble velocity is much greater than the terminal velocity; therefore the bubble size should be smaller. However, as the bubble size becomes smaller, the extent of gas circulation diminishes and equation (4.5-61) would no longer be applicable. Evidently, such a limit is reached in water stirred by injected gas for which the limiting bubble diameter is 0.45 cm, which is about one-fourth the value for free-rising bubbles. From observations of bursting of bubbles in open hearth furnaces, Richardson[288] estimated the average bubble diameter as 3.3 cm, which is about half that calculated from equation (4.5-61) for free-rising bubbles. This is consistent with the case for gas–water systems.

The rate of reaction controlled by the liquid-phase mass transfer to the swarm of bubbles, with a concentration driving force $\Delta\% x_i$, is represented by

$$\frac{dw}{dt} = m_l S W \rho_l \frac{\Delta\% x_i}{100} \tag{4.5-63}$$

where m_l is the liquid-phase mass-transfer coefficient for a single bubble given by equation (4.5-54), W is the liquid mass, and the rate of solute transfer, dw/dt, is in units of mass per unit time. Noting that the total bubble surface area per unit mass of the melt is

$$S = \frac{6\varepsilon}{1-\varepsilon} \frac{1}{\rho_l d} \tag{4.5-64}$$

the rate equation can be given in terms of ε and d as

$$\frac{dw}{dt} = \frac{6\varepsilon}{100d(1-\varepsilon)} m_l W(\Delta\% x_i) \tag{4.5-65}$$

For large superficial gas velocities with $\varepsilon = 0.5$ and the limiting relation for m_l in equation (4.5-59) for $d = 0.45$ cm, the rate equation for gas–

water systems simplifies to

$$\frac{d\% x_i}{dt} = 36(Du_s)^{1/2}(\Delta\% x_i) \tag{4.5-66}$$

where the units for D and u_s are cm^2/sec and cm/sec, respectively.

For pneumatic metal refining processes with $\varepsilon = 0.5$, an average solute diffusivity of $D = 5\times 10^{-5}$ cm^2/sec and an average bubble diameter of $d = 3$ cm, the approximate rate equation for diffusion-controlled mass transfer to and from the swarm of bubbles is

$$\frac{d\% x_i}{dt} \simeq 0.01 u_s^{1/2}(\Delta\% x_i) \tag{4.5-67}$$

In oxygen bottom blowing of a 200-ton steel with $\varepsilon = 0.5$ and $d =$ 3 cm, the total bubble surface area calculated from equation (4.5-64) is 0.6×10^8 cm^2 which is similar to the slag–metal interfacial area in the emulsion during the period of high rate of decarburization of a 200-ton steel by oxygen top blowing.

4.5.4.3. *Metal Refining by Oxygen or Air Bottom Blowing*

In steelmaking by oxygen bottom blowing, as in Q-BOP, the conditions in the bath differ from those in oxygen top blowing as in BOF. In bottom blowing there is extensive gas dispersion in the metal, whereas in top blowing there is extensive metal and gas dispersion in the slag. Some aspects of top blowing have already been discussed; here we shall present the results of some recent studies and thoughts on the rates and mechanisms of steelmaking reactions in Q-BOP and other metal-refining processes.

Fruehan[289] studied the rates of several gas–metal reactions occurring in the Q-BOP, using plant data obtained from 30-ton experimental heats. For most cases involving gas injection into liquids the mass transfer in the gas phase is relatively fast. Therefore, in the decarburization and argon flushing for hydrogen removal and deoxidation, the rates of reactions were shown to be determined either by gas starvation at high solute contents or by liquid-phase mass transfer at low solute contents. In the case of nitrogen absorption or desorption, the rate is also affected by the oxygen (or carbon) content of the metal, as discussed earlier. From two plant data points and the rate constant for nitrogen reaction at high and low oxygen levels (Figure 4.5-15), Fruehan derived the following rate parameters for nitrogen absorption controlled by liquid-phase mass transfer when nitrogen is blown at a rate of 1500 ft^3/min (STP):

$m_l = 0.03$ cm/sec, and
$S = 3.7\times 10^4$ cm^2 bubble area per ton of melt.

Similar values were obtained from four data points on the deoxidation of the bath by argon flushing at 2000 ft^3/min (STP). Practical observations indicate that in bottom blowing at superficial velocities from 70 to 250 cm/sec, the fractional hold up of gas is about 0.5. With $S = 3.7 \times 10^4\ cm^2$/ton of steel and $\varepsilon = 0.5$, Fruehan calculated an average bubble diameter to be about 8 cm. Furthermore, the mass transfer coefficient calculated using equation (4.5-54) was found to be close to that derived from the Q-BOP data.

If, as discussed above, the average bubble diameter were 3 cm, for $\varepsilon = 0.5$ the bubble surface area calculated from equation (4.5-64) would be $S = 30 \times 10^4\ cm^2$ per ton of steel, which is about eight times the value estimated from the plant data. It appears desirable to conduct mass transfer experiments with the swarm of bubbles in liquid metals at high superficial velocities for better evaluation of m_l and S. Nevertheless, it is difficult to reconcile the large bubble diameter derived by Fruehan from the analysis of Q-BOP plant data with the smaller bubble diameter anticipated from the concept of bubble fragmentation.

Another interesting aspect of steelmaking reactions in the oxygen bottom-blowing process is the dephosphorizing reaction. In Q-BOP steelmaking with lime bottom blowing, the steel is dephosphorized to levels much below those expected from the composition of the slag low in iron oxide. In an attempt to explain this observation, Turkdogan[290] proposed a reaction mechanism involving vaporization of phosphorus as PO_x which then reacts with the lime particles in the oxidizing jet stream, forming a relatively stable calcium phosphate which subsequently is absorbed by the slag. This reaction mechanism is based on the concept of enhanced vaporization of metals and alloys in oxidizing gas streams. The same principle would also explain the rapid oxidation of silicon in the early stages of the blow at high carbon levels. In this case, the intermediate reaction product is thought to be SiO.

In AOD stainless steelmaking in which argon–oxygen mixtures are blown instead of oxygen alone, steel is decarburized to low carbon levels with lesser chromium oxidation and lesser additions of ferro alloys for chromium recovery from the slag. The results of laboratory experiments and plant trials indicate, however, that the chromium oxidation occurs at carbon levels much higher than those predicted from equilibrium considerations. As is shown in Figure 4.5-33, Fulton and Ramachandran[291] found from 150-lb experimental melts that the carbon content at which significant chromium oxidation initiated increases with increasing gas-flow rate, and that the carbon levels are much higher than those predicted from equilibrium considerations for a given argon–oxygen mixture. Similar observations were made also by Choulet and coworkers.[292]

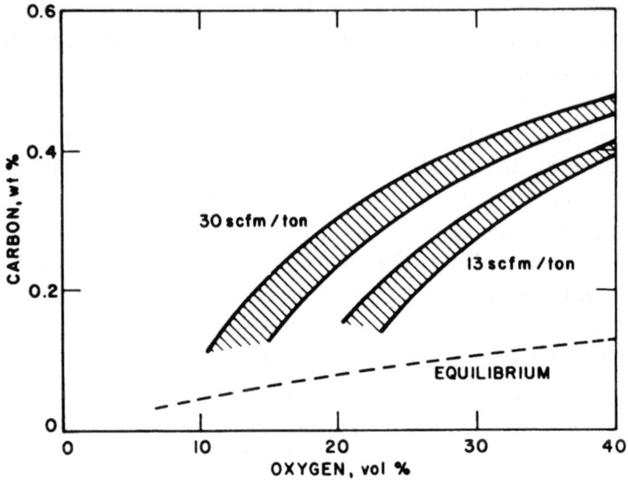

Figure 4.5-33. Approximate relation for carbon content of the melt at which significant chromium loss initiates for various argon–oxygen mixtures and two gas-flow rates at about 1650° C in Fe–14 Cr–C melts (the results of Fulton and Ramachandran[99]) as compared with that calculated for $Cr_2O_3(s) + 3\,C = 2\,Cr + 3\,CO(g)$ equilibrium.

These observations and also the results of laboratory experiments of Fruehan[293] suggest that as oxygen enters the melt, chromium and carbon are oxidized independently of each other. Subsequently, as the chromic oxide floats out of the melt, it reacts with carbon and results in partial recovery of chromium. That is, the overall reaction is:

1. *Near oxygen entrance*

$$2\,\underline{Cr} + \tfrac{3}{2}O_2 = Cr_2O_3 \tag{4.5-68a}$$

$$\underline{C} + \tfrac{1}{2}O_2 = CO \tag{4.5-68b}$$

2. *In metal bath*

$$Cr_2O_3 + 3\,\underline{C} = 2\,\underline{Cr} + 3\,CO \tag{4.5-69}$$

Experimental results of Fruehan[293] indicate that the rate of reduction of chromic oxide by carbon is controlled primarily by diffusion of carbon. On the assumption that local equilibrium prevails for reaction (4.5-69) at the bubble surface and that the rate is controlled by diffusion of carbon, Fruehan derived from the plant data the decarburization curves for various O_2/Ar ratios in the gas blown at 1800 ft^3/min(STP). As is shown in Figure 4.5-34a, the O_2/Ar ratio in the gas blown has little effect on the decarburization rate at carbon levels above 0.2%. The carbon contents calculated from the curve in Figure 4.5-34a are in accord with those observed in two trial heats (Figure 4.5-34b).

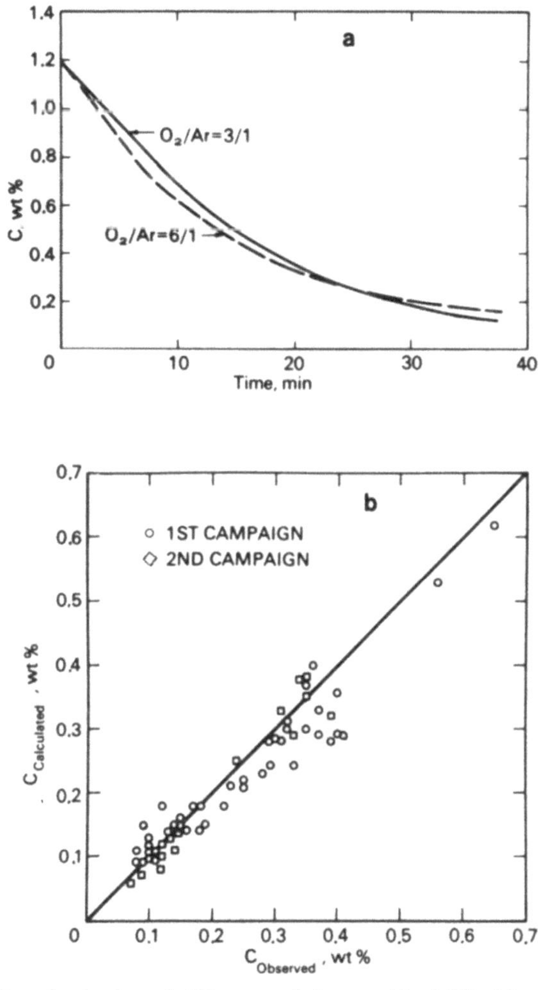

Figure 4.5-34. Decarburization of 409-type stainless steel in AOD: (a) calculated rate of decarburization, and (b) comparison of calculated and observed carbon contents.

In all pneumatic metal-refining processes, the oxidizing conditions prevailing in the gas jet stream are expected to bring about rapid vaporization of the solvent metal and some of the solutes, except of course carbon in solution. Finely dispersed particles of the oxides of iron, chromium, or manganese thus formed react subsequently with carbon in the bath. This cycle of oxidation and reduction continues during the decarburization of the steel. At lower carbon levels, e.g., $<0.2\%$ C, the rate of reduction becomes slower than the rate of vaporization and oxidation. It is for this reason that in air or oxygen bottom-blowing practices, copious iron oxide fumes are evolved only during the later stages of refining.

In the conversion of matte to blister copper by air blowing, the rate is controlled primarily by the rate of air blow for which the practical limitations are dictated by the movement of the melt in the converter. In the second stage of the blow to convert white metal to blister copper, the following sequence of reactions is usually envisaged:

$$Cu_2S + \tfrac{3}{2}O_2 = Cu_2O + SO_2 \tag{4.5-70a}$$

$$Cu_2S + 2\,Cu_2O = 6\,Cu + SO_2 \tag{4.5-70b}$$

$$6\,Cu = 6\,\underline{Cu} \tag{4.5-70c}$$

Little is known about the kinetics of these reactions. In a recent study of the rate of reaction of liquid copper oxide with liquid copper sulfide, Byerley *et al.*(294) interpreted their data for a single-liquid phase of an oxide and sulfide mixture. However, their melt compositions indicate that liquid copper must have been formed during the progress of the reaction for which their interpretation of the reaction rate would not apply.

In the presence of copper, the cuprous oxide solubility in liquid cuprous sulfide is low; at 1 atm SO_2 the solubility is about 3 mole % Cu_2O.(295) However, such a low oxide solubility in the sulfide does not exclude the formation and decomposition of cuprous oxide in the conversion of white metal to blister copper. In fact, since the overall reaction involves three phases, i.e., matte, copper, and gas, there should be a sequence of 2-phase reactions such as (4.5-70a, b, c). In reaction (4.5-70a) the matte becomes supersaturated with the oxide; the matte then becomes supersaturated with copper via reaction (4.5-70b). Ultimately the metallic copper separates from the supersaturated matte in reaction (4.5-70c). The same end result is achieved also by considering only two sequences of reactions

$$Cu_2S + O_2 = 2\,Cu + SO_2 \tag{4.5-71a}$$

$$2\,Cu = 2\,\underline{Cu} \tag{4.5-71b}$$

Although some of the metallic copper settles at the bottom of the converter, smaller copper particles exposed to the oxidizing atmosphere of the jet stream vaporize and oxidize and subsequently are reduced by the sulfide. In addition, there is direct oxidation of sulfur vaporizing from the matte in the gas-jet stream.

In copper conversion, the efficiency of oxygen utilization is close to 100%; in gaseous poling with CO injection, the rate is proportional to the gas flow rate above 0.1% oxygen in the metal.(296) Therefore, it appears that in these processes gas starvation controls the rate of refining.

4.5.5. Metal Refining in Vacuo

Vacuum refining of liquid metals and alloys has been used to a varying extent in the metals industry, and most of the large-scale applications are in the steel industry. Hydrogen, carbon monoxide, and to lesser extent nitrogen can be removed from steel by an appropriate method of exposing the metal to a sufficiently low-pressure environment. The effectiveness of the vacuum treatment depends on the growth of gas bubbles in the bulk of the liquid, diffusion of solute to the surface of the melt and, in certain cases, the reaction kinetics at the surface of the melt.

The excess pressure, ΔP, in a bubble is determined by the surface tension of the liquid and the bubble radius, r, thus

$$\Delta P = \frac{2\gamma}{r} \tag{4.5-72}$$

The theory of homogeneous nucleation of bubbles has been substantiated experimentally with aqueous solutions of low surface tension. The application of this theory to liquid metals of high surface tension indicates that an excess pressure of the order of 10^4 atm is needed for bubble nucleation.[297] Such a prediction is not consistent with experimental observations. In levitated liquid iron–carbon–oxygen alloys, not in contact with a refractory surface and presumably free from inclusions, the burst of gas evolution has been observed at carbon and oxygen contents which correspond to the CO supersaturation of the melt to a pressure of 20–100 atm.[298] If this is a manifestation of homogeneous bubble nucleation, the supersaturation required is some orders of magnitude less than that predicted by the theory.

In practice, the bubble nucleation occurs primarily in crevices of the refractory lining of the metal container, provided the solute content is sufficiently high to sustain the required excess pressure in the bubble under a given head of liquid metal. This is demonstrated in Figure 4.5-35 showing the concentrations of hydrogen, oxygen (for 0.05% C), and nitrogen in steel at 1600° C in equilibrium with indicated H_2, CO, and N_2 pressures and the corresponding head of steel above which there is vacuum (or a very low gas pressure).

The higher the head of liquid, the higher will be the solute concentration for the bubble to grow. For example, for bubble growth under a 10-cm head of steel, there should be more than 2 ppm hydrogen or more than 40 ppm nitrogen. Partially deoxidized steel containing, for

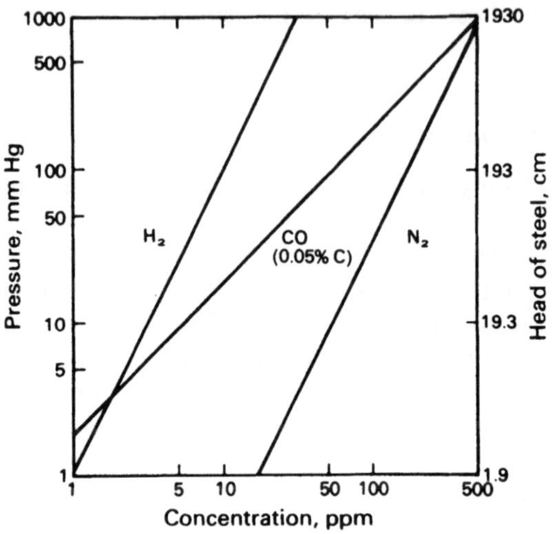

Figure 4.5-35. Concentrations of hydrogen, oxygen (for 0.05% C) and nitrogen in steel at 1600° C in equilibrium with H_2, CO, and N_2 pressures and the corresponding head of steel (in cm).

example, 0.05% C will have about 120 ppm oxygen in solution and, consequently, CO bubbles will grow even under a much higher head of steel. Once the growth of gas bubble has been established, other solutes such as hydrogen and nitrogen will be flushed out of the metal. Degassing by purging of the metal with an inert gas such as argon is frequently used in the metals industry. Combination of argon purging and vacuum treatment as in the R–H (Ruhrstahl–Heraus) process further enhances the efficiency of degassing.

In a study of the thermodynamic and kinetic aspects of vacuum degassing, Bradshaw and Richardson[299] computed the quantity of purge gas required to remove hydrogen, nitrogen, and carbon monoxide from steel at total surface pressures 0.01; 0.1, and 1.0 atm. The calculations were based on the assumption that the purge gas leaving the metal is in equilibrium with the residual solute in the metal. The results of their calculations are reproduced in Figures 4.5-36 and 4.5-37. The relative gas partial pressure is defined as

$$p_i^* = \frac{p_i}{P_{\text{tot}}} \tag{4.5-73}$$

where p_i is the gas partial pressure in equilibrium with the initial solute concentration; i.e., in terms of % of solutes,

$$p_i = \left(\frac{\% X_i^0}{K_i}\right)^2 \tag{4.5-74}$$

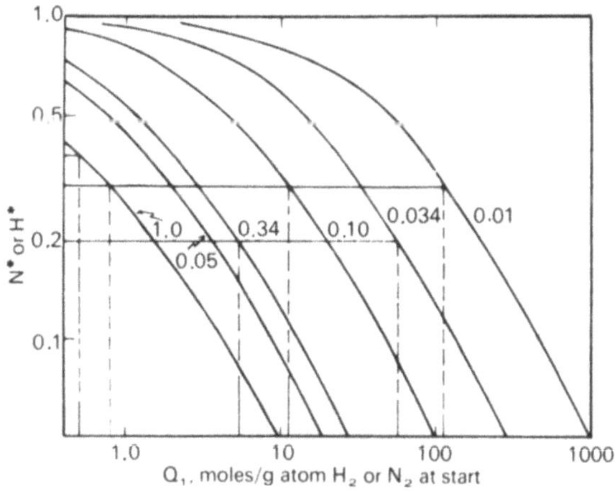

Figure 4.5-36. Ratio of final to initial hydrogen or nitrogen contents, with varying quantities of purge gas at different values of p^*; purge gas is assumed to leave in equilibrium with the metal.

where K_i is the equilibrium constant for solution of H_2 or N_2 in steel. Similarly for carbon deoxidation,

$$p_{CO} = \frac{(\%\,O^0)(\%\,C^0)}{K_{CO}} \tag{4.5-75}$$

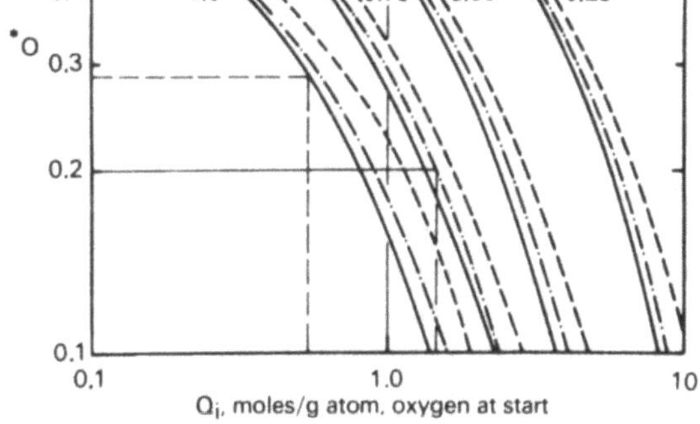

Figure 4.5-37. Ratio of final to initial oxygen contents with varying quantities of purge gas at different values of p^* and for various concentration ratios of oxygen to carbon (R); the purge gas is assumed to leave in equilibrium with the metal.

The values of the equilibrium constants, with pressure in atm, for 1600° C are: $K_{H_2} = 2.74 \times 10^{-3}$, $K_{N_2} = 4.57 \times 10^{-2}$, and $K_{CO} = 2 \times 10^{-3}$. For known values of $\% X_i^0$ and total surface pressure P_{tot}, the quantity of purge gas, Q_i, required to achieve a desired level of degassing can be derived from the calculated data in Figures 4.5-36 and 4.5-37.

For example, a steel containing 5×10^{-4} $\% H^0$ is purged with argon until the concentration is reduced to 1×10^{-4} $\% H$. For 1-atm total-surface pressure, $p^*_{H_2} = 0.034$; therefore, for $H^* = \% H/\% H^0 = 0.2$, the required quantity of purge gas is $Q = 58.3$ mole/g-atom of hydrogen present at start of purging. The total volume of purge gas is 234 ft^3 (STP) per ton of steel.

For 0.1-atm total-surface pressure: $p^*_{H_2} = 0.34$, $H^* = 0.2$, and $Q_i =$ 5.43 mole/g-atom H^0; total volume of purge gas = 21.8 ft^3 (STP) per ton of steel.

Several other numerical examples are given by Bradshaw and Richardson.[299] With decreasing total pressure over the steel surface, a lesser volume of purge gas is needed to achieve the same level of degassing; this also minimizes the cooling effect of the purge gas. It is well to point out, however, that the equilibrium conditions assumed in the calculation of the data in Figures 4.5-36 and 4.5-37 give the maximum degassing values which have a limited range of applicability. In fact, with this in mind, Bradshaw and Richardson also discussed in their paper the liquid-phase mass transfer control in the degassing of metals by purge gas. As discussed earlier, there is no clear understanding of mass transfer to the swarm of bubbles in liquid metals. Speculative discussion of this subject will not be pursued further here.

In addition to the mass transfer control, a slow chemical reaction takes place at the bubble surface, as with the nitrogen reaction already discussed.

To overcome diffusional effects, degassing is accomplished by exposing the molten metal in a vacuum chamber over as great a surface area as possible. The commercial processes used in degassing of steel have been compiled in a comprehensive review paper by Flux.[300]

In the present commercial stream-degassing processes, a stream of steel enters a vacuum chamber, breaks into a spray of droplets, and falls into a collecting vessel at the base of the chamber. Degassing takes place from the droplets in a contact time of less than one second, and from the liquid pool in the collecting vessel. In present practices the chamber pressure is maintained at about 1 mm Hg. However, as demonstrated by Olsson and Turkdogan,[301] effective vacuum carbon deoxidation can be achieved even at much higher chamber pressures, provided there is good stream breakup to give droplet sizes of the order of 100-μm diameter. This was achieved by injecting inert gas bubbles into the liquid

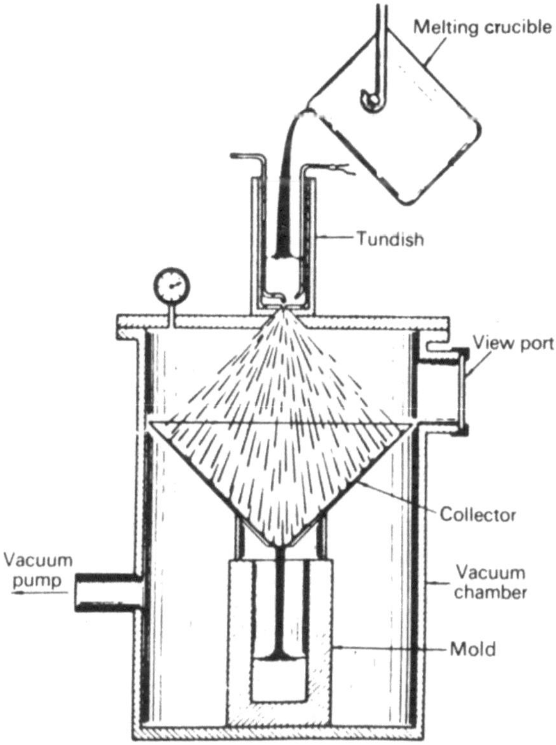

Figure 4.5-38. Stream breakup by gas injection.

stream just before passage through a nozzle to a chamber at a lower pressure (see Figure 4.5-38). The oxygen removal from 0.1% carbon steel is shown in Figure 4.5-39a as a function of chamber pressure. The extent of oxygen removal to about 70% is essentially independent of chamber pressure up to about 200 mm Hg. In these experiments the average time of droplets flight was about 0.08 sec. The diffusion-controlled carbon deoxidation of 100-μm-diameter steel droplets in 0.08 sec (for oxygen diffusivity 4×10^{-5} cm^2/sec) would give about 80% oxygen removal, which is close to the value obtained experimentally. Using this method of stream break up in a commercial unit, where the residence time of droplets would be about 0.3 sec, about 95% oxygen removal is anticipated (Figure 4.5-39b).

In vacuum degassing of metals, there is some loss of certain alloying elements such as manganese and chromium which have relatively high vapor pressures. The rate of loss of such elements will be less than that calculated for free vaporization [equation (4.5-42)] because of the diffusional effects. We should also keep in mind the reduction of some

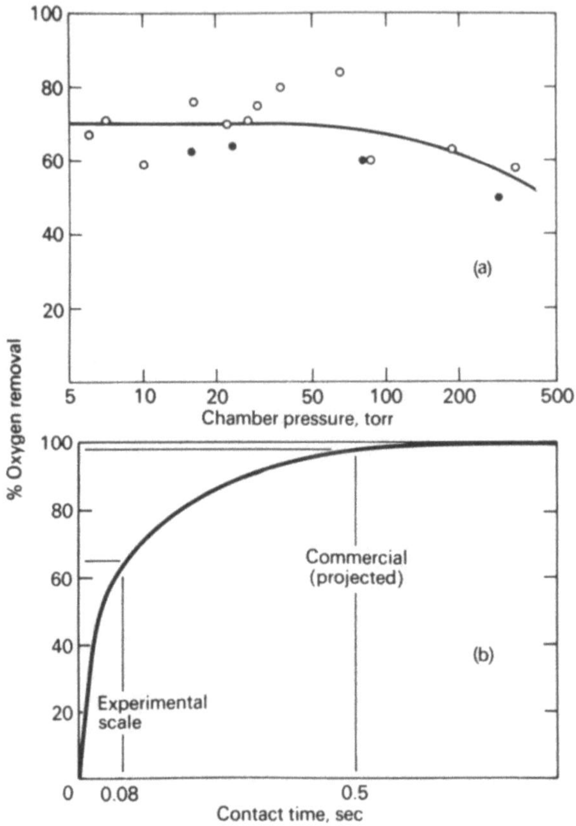

Figure 4.5-39. "Vacuum" carbon deoxidation from a broken-up liquid steel stream: (a) effect of chamber pressure, and (b) effect of droplet residence time and projection to commercial-size operation. Amount of Al added in mold: ○, none; ●, 0.1%.

refractory oxides and inclusions such as MgO and SiO_2 by carbon in steel during vacuum treatment. In fact, Philbrick[302] has observed that with stream degassing, the carbon loss is about twice that accounted for by the decrease in the oxygen content.

In some steelmaking practices vacuum treatment is applied to fully killed steel to improve its purity. Effective stirring achieved during the repeated recycle of metal from the ladle to the vacuum chamber, as in the D–H and R–H processes, apparently aids the coalescence of small inclusions and facilitates their separation from the melt prior to casting.

4.5.6. Reactions in Ladle Refining and during Solidification

Ladle treatment of refined metals, particularly in steelmaking and many foundry applications, is an important final phase of the refining.

In addition, there are several reactions that occur during the solidification of metals and alloys which affect the quality of castings. We should, therefore, discuss some salient aspects of the rate phenomena involved in the ladle treatment of metals, in particular of steel, and reactions occurring during the freezing of metals.

4.5.6.1. *Deoxidation*

Aside from gaseous deoxidation and degassing, ladle treatment in steelmaking achieves deoxidation and control of inclusions in castings. Thermodynamic and kinetic aspects of deoxidation of steel have been studied on numerous occasions; references to most of the recent studies are given in a critical review paper by Turkdogan.[303] The brief discussion of the subject here is confined to the kinetic aspects.

Although elements such as aluminum interact strongly with oxygen in steel which is deoxidized to a few parts per million residual oxygen in solution, the total oxygen content of the steel is often much higher (20–100 ppm oxygen). This is due to sluggish separation of oxide inclusions from the bulk of the steel in the ladle treatment. There are three basic consecutive steps involved in the deoxidation reaction: (1) formation of critical nuclei of the deoxidation product in a homogeneous medium, (2) progress of deoxidation resulting in growth of the reaction products, and (3) their flotation from the melt.

Since any deoxidation reaction yields a second phase as a reaction product, oxide particles must first be nucleated for the reaction to proceed. The extent of supersaturation of the melt with reactants needed for homogeneous nucleation of the reaction product is estimated from the following relation derived from the theory of homogeneous nucleation[304–306]:

$$\frac{K_s}{K} = e^{\Delta F/kT} \tag{4.5-76}$$

where K is the deoxidation solubility product (in terms of activities of the reactants), K_s is the corresponding product for the actual concentrations in supersaturated solution, and ΔF is the free-energy change accompanying the nucleation process derived from the oxide–metal interfacial energy.† Turpin and Elliott[307] gave several examples of estimating the

† The rate formation of I number of nuclei/cm^3 sec is given by

$$I = A_0 e^{-\Delta F^*/kT}$$

where A_0 is a constant ($\sim 10^{27}$/cm^3 sec) and ΔF^* is the free energy of activation defined by

$$\Delta F^* = \frac{16\pi\gamma^3}{3(\Delta F)^2} \quad \text{erg/nucleus}$$

The value of ΔF^*, hence ΔF, is insensitive to the choice of a value of I; for spontaneous nucleation $I = 10^3$ is a good approximation.

supersaturation needed for homogeneous nucleation of various types of inclusions in liquid steel. For liquid iron–alumina or iron–zirconia systems the interfacial energy is probably in the range from 1500 to 2000 erg/cm^2 for which K_s/K is of the order of 10^5–10^{10}. In fact, experimental observations of Bogdandy *et al.*[308] show that the homogeneous nucleation of these oxides in liquid iron occurs only when $K_s/K > 10^8$. The interfacial energy for liquid silicates and liquid iron is of the order of 800–1000 erg/cm^2, for which the value of K_s/K for spontaneous nucleation is estimated to be within the range from 500 to 4000. With an electrochemical cell technique, Sigworth and Elliott[309] investigated the formation of silica from the supersaturated solution in liquid iron. Their results and those of Forward and Elliott[310] give $K_s/K \simeq 100$ for spontaneous nucleation of silica. This supersaturation ratio is much lower than the value that might have been anticipated from the interfacial energy of liquid silicates and liquid iron. In any event, it is certain that during the dissolution of added deoxidizers in liquid steel, numerous regions in the melt are very rich in the solute content and, consequently, the oxide particles nucleate readily at the time of addition of the deoxidizers.

The rate of progress of deoxidation and growth rate of oxide inclusions controlled by diffusion is strongly affected by the number of nuclei. Furthermore, because of the buoyancy effect, growing deoxidation products float out of the melt. Unless additional new oxide particles are nucleated, deoxidation reaction may cease in regions of the melt depleted of oxide particles. Based on a simplified model, Turkdogan[311] calculated the rate of deoxidation controlled by diffusion of oxygen to growing oxide particles and their separation from the melt by flotation. Such calculations lead to the relation in Figure 4.5-40, where the average

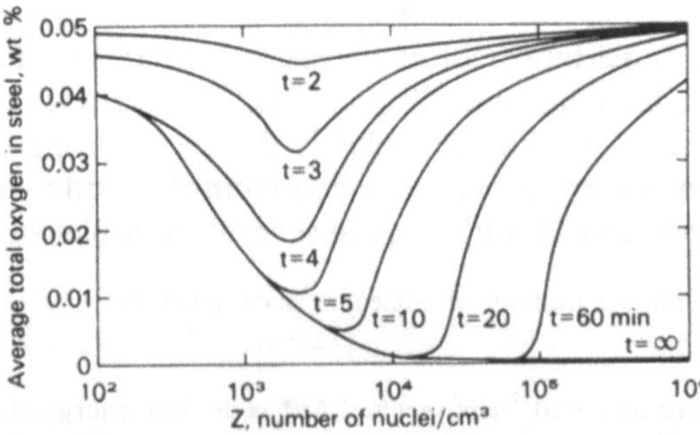

Figure 4.5-40. Calculated data on average total oxygen content of steel as function of number of nuclei/cm^3 in liquid steel and holding time after deoxidizer additions. $[\%O]_0$ initial = 0.05; $[\%O]_i$ equilibrium = 0, and depth of liquid steel $l = 200$ cm.

total oxygen content of steel is plotted against the number of nuclei assumed to be present immediately after the addition of deoxidizers. In these calculations, the depth of liquid steel is taken to be 2 m, the initial uniform oxygen content as $[\%O]_0 = 0.05$, and that in equilibrium with the oxide $[\%O]_i$ = zero, for convenience. If the number of nuclei Z is less than the critical value Z_m, the relatively large inclusions float out of the melt quickly leaving behind unreacted high-residual oxygen in solution, provided no additional nuclei are formed. For $Z > Z_m$, the inclusions attain their full growth in a relatively short time, i.e., the residual oxygen in solution approaches the equilibrium value in a short time. Under these conditions, the inclusions are small in size and hence their separation from the melt takes a long time, as indicated by the curve for infinite time in Figure 4.5-40. Therefore, the curve to the right of the minimum point gives the average oxygen present as oxide inclusions in steel where oxygen in solution is at the equilibrium value. At the time of addition of deoxidizers, the number of nuclei formed is probably of the order of 10^5–10^6 for which the diffusion-controlled deoxidation reaction is essentially complete within a few minutes, and the overall oxide removal is controlled primarily by flotation of the inclusions.

Under practical conditions of deoxidation during filling of the ladle, there is sufficient turbulence that some inclusion growth may take place by collision and coalescence. Furthermore, stirring by purge gas or electromagnetic force field as in ASEA-SKF ladle, may bring about a motion in the melt such that the inclusions could become attached to the surface of the ladle lining and caught by the slag layer. Another consequence of stirring is that some of the large inclusions may be prevented from floating out of the melt. Some aspects of these complications have been demonstrated by Iyengar and Philbrook,[(312)] Lindskog and Sandberg,[(313)] and Emi.[(314)]

Despite rough approximations involved in estimating the rate of deoxidation in terms of solute diffusion to the Z number of growing inclusions and their flotation, this rate phenomenon may also explain apparent discrepancies in the available data on the solubilities of oxides and oxysulfides of rare earths in liquid steel. As indicated by Turkdogan[(315)] in a study of sulfide formation in liquid steel, at low chemical potentials of the reactants as, for example, in the deoxidation by cerium, the diffusion-controlled reaction may not reach equilibrium within a practical period of reaction time. Since the sulfur and oxygen reactions in steel are coupled, the sulfide solubility product would be high when the metastable deoxidation product is high. Therefore, for incomplete deoxidation of steel by cerium within a practical limit of reaction time, the corresponding CeS and Ce_2O_2S solubility product would be greater than that expected from the reaction equilibrium.

Another example of rate phenomena in ladle treatment is that involved in the coalescence of oxide and matte particles in slowly cooled copper reverberatory slags. As already mentioned, Subramanian and Themelis(278) were able to coarsen the particles of entrained copper oxides and sulfides in slags by heat treatment above 1000° C. This is another example of the Ostwald ripening effect which brings about the growth of large particles at the expense of small particles by a diffusion process; the driving force for diffusion is the lowering of interfacial energy with increasing particle size. The equation derived by Wagner(316) gives the time dependence of the mean radius $\bar{r}$ of an assembly of particles;

$$\bar{r} = \left(\frac{8\gamma D c V^2}{9RT}\right)^{1/3} t^{1/3} \tag{4.5-77}$$

where D is the diffusivity, c is the bulk concentration of the diffusing species, and V is the molar volume of the growing particle (oxide or sulfide) per mole of the diffusate. We see from the data of Subramanian and Themelis(278) in Figure 4.5-32 that the mean particle size of 25 μm after a 5-min heat treatment above 1000° C increases to 87 μm after 277 min. This is in close agreement with the theory [equation (4.5-77)] which states that the mean particle size increases in proportion to the cube root of time.

Copper dissolves in slags primarily in the oxidized or sulfided state; according to the studies of Billington and Richardson(317) the solubility of metallic copper in calcium-alumino silicates is only 0.055 atom % at 1530° C. Nagamori *et al.*(318) measured the solubility of copper oxide in iron-alumino silicates at 1200 and 1300° C and at cuprous oxide activities up to 0.2. At $a_{Cu_2O} = 0.2$ the copper oxide solubility is in the range from 3 to 4% Cu. On the basis of these equilibrium measurements, it is anticipated that even partial reduction of the copper oxide in the slag by CO purge may enhance the growth of copper particles during the heat treatment of the slag.

4.5.6.2. *Solidification*

Soundness of castings and distribution of oxide, silicate, and sulfide inclusions in the cast structure are much affected by the state of deoxidation in the ladle. Oxygen and sulfur in steel solution are the two main elements which may react with the other solutes when their concentrations become sufficiently high in the enriched interdendritic liquid. That is, microsegregation accompanying dendritic solidification is responsible for the precipitation of inclusions along the grain boundaries of the cast structure and for the formation of blow holes.

Microsegregation resulting from solute rejection during dendritic freezing is represented by the following equation derived by Scheil(319) for the simple case of no diffusion in the solid phase and complete diffusion in the interdendritic liquid:

$$C_l = C_0(1-g)^{k-1} \quad (4.5\text{-}78)$$

where C_0 is the initial uniform solute content, C_l is the solute content of the liquid at a given fractional solidification g, and $k = (C_s/C_l)$ is the solute distribution ratio. When two or more solutes interact, such as oxygen reacting with silicon and manganese, due account must be taken also of the solute depletion caused by reactions in the enriched liquid. The concept of reactions occurring during solidification of alloys developed by Turkdogan(320) is briefly outlined below.

An example of calculations is given in Figure 4.5-41 showing the oxygen content of interdendritic liquid, controlled by Si–Mn deoxidation reaction, as a function of percentage of solidification. The dashed curve gives the oxygen enrichment in the absence of any deoxidation reaction computed from equation (4.5-78). In the case of steel containing, for example, 0.015% Si, the Si–Mn deoxidation starts in the enriched liquid at about 60% solidification, and the residual oxygen as a function of the

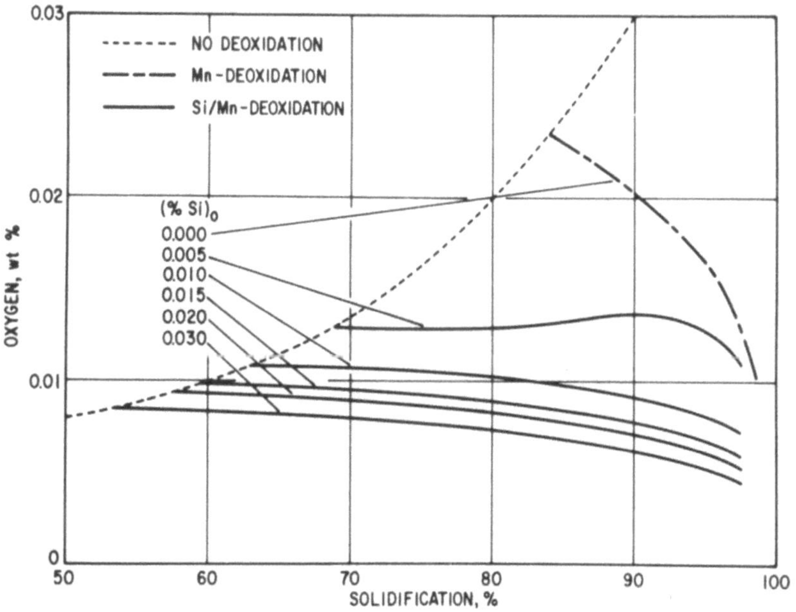

Figure 4.5-41. Change in the oxygen content of entrapped liquid during freezing of steel; $(\%O)_0 = 0.004$; $(\%Mn) = 0.50$; freezing temperature = 1525° C.

extent of solidification follows the path indicated by the corresponding curve in Figure 4.5-41.

The blow hole is another form of inclusion in steel. When the carbon content of the interdendritic liquid becomes sufficiently high, it reacts with oxygen forming carbon monoxide. Similarly, at sufficiently high concentrations of hydrogen and nitrogen in the enriched liquid, there will be gas evolution. This gas formation results in the displacement of liquid to neighboring interdendritic cells; hence the formation of blow holes. The equilibrium oxygen content of steel for the carbon–oxygen reaction at 1 atm CO at various stages of solidification for several initial carbon concentrations is superimposed in Figure 4.5-42 on the curves given in Figure 4.5-41. If the concentrations of silicon, manganese, or other deoxidizers are sufficiently high, the oxygen in solution in the enriched liquid is maintained at low levels during solidification so that the carbon–oxygen reaction would not take place; thus the absence of blow holes (if the hydrogen and nitrogen contents are sufficiently low). As shown in Figure 4.5-42, a particular carbon line tangential to the Si–Mn deoxidation curve gives the critical carbon content of steel below which blow holes do not form. In a composition plot in Figure 4.5-43, the sections of laboratory ingots which showed

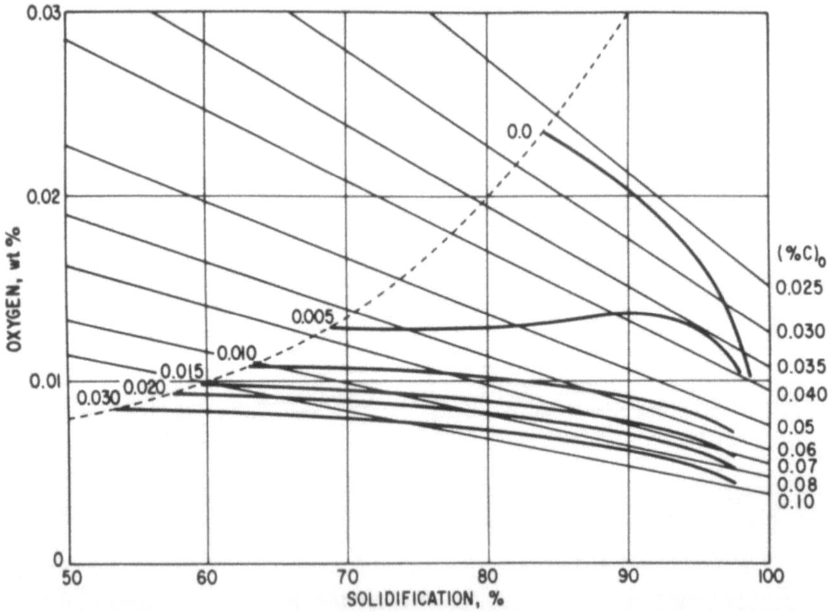

Figure 4.5-42. Carbon lines superimposed on silicon curves of Figure 4.5-41 for 0.5% Mn and $P_{CO} = 1.0$ atm.

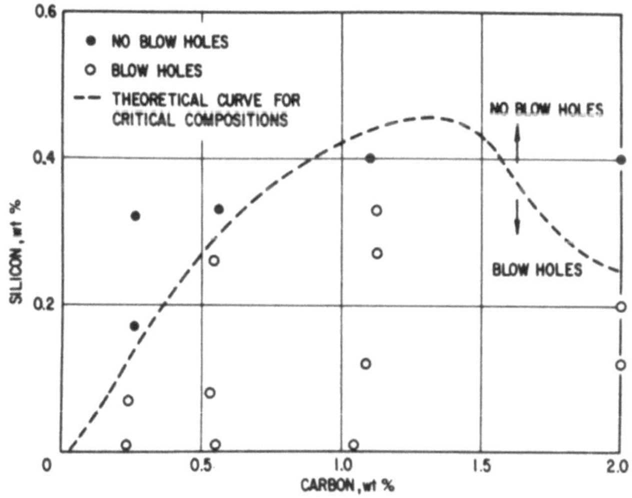

Figure 4.5-43. Critical Si–C curve for suppression of blow holes in steel containing 0.5–0.6% Mn and 25–65 ppm O_2.

blow holes are indicated by open circles and the castings free of blow holes by filled circles. The critical curve derived from a theoretical analysis,[320] briefly mentioned here, is well supported by the experimental observations.[322]

If the nitrogen and hydrogen contents of the steel are high, the blow hole formation is enhanced, and the position of the critical Si–C curve in Figure 4.5-43 moves to lower carbon levels. When the partial pressure of carbon monoxide resulting from the carbon–oxygen reaction is at a pressure less or greater than 1 atm, the position of the curve in Figure 4.5-43 for 1 atm CO is shifted to lower or higher carbon levels, respectively, as shown in Figure 4.5-44. This finding has a particular significance in relation to the deoxidation of semikilled steel for which the bottom of the ingot is sound and the blow holes are confined to the upper part of the ingot, thus eliminating piping. In continuous casting the steel composition should be above the curve for atmospheric pressure so that the blow holes do not form close to the surface of the casting where the total pressure is only slightly above atmospheric during the early stages of freezing. These concepts of reactions during solidification were helpful in the development of deoxidation practices for special casting applications, particularly for continuous casting.

The sulfide and oxysulfide inclusions are formed in the interdendritic liquid during the last stages of solidification in a manner similar to that discussed above for the formation of interdendritic oxide and silicate inclusions. This has been demonstrated by several studies

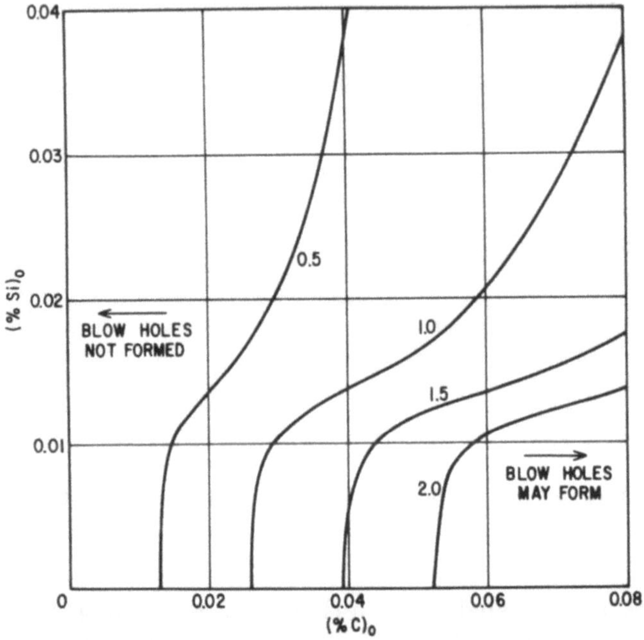

Figure 4.5-44. Effect of carbon monoxide partial pressure on critical %Si and %C for the formation of blow holes.

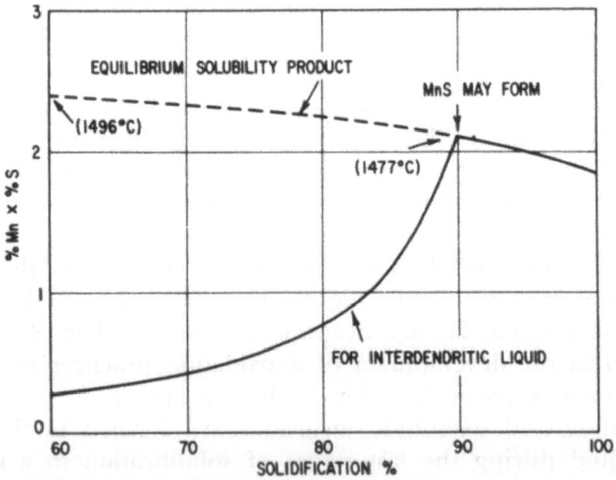

Figure 4.5-45. Formation of MnS during dendritic freezing of steel containing 1.5% Mn, 0.25% C, and 0.05% S.

with complex steels[321,323] and with the Fe–S–O ternary alloys.[324] The product [% Mn] [% S] calculated for the enriched interdendritic liquid is plotted in Figure 4.5-45 against the percentage of solidification of steel containing 1.5% Mn, 0.25% C, and 0.05% S. At about 90% solidification this product equals the MnS solubility product, indicating the beginning of MnS precipitation, if little or no supersaturation is needed. Since the sulfide formation occurs at the later stages of freezing, sulfide inclusions are accumulated in the interdendritic regions which ultimately become the grain boundaries in the cast structure.

In steels treated with rare earth alloys, which interact strongly with oxygen and sulfur, oxysulfides precipitate continuously during cooling of liquid steel and throughout the progress of solidification. What is usually called *the improved sulfide morphology by rare earths* is, in fact, the dispersion of fine particles of rare earth oxysulfides throughout the casting and suppression of the sulfide accumulation along the grain boundaries.[315]

4.5.7. Concluding Remarks

We see from this overview of the accumulated knowledge that major advances have been made during the past two decades in the understanding of pyrometallurgical reaction rates. It is reassuring to note that numerous isolated pieces of research findings, much like the pieces of a three-dimensional jig-saw puzzle, are now being put together to consolidate in more absolute terms the fundamental concepts of the rate phenomena. It is even more reassuring that such knowledge is now being used to a greater extent in the metals industry.

It appears that most of the gas–liquid metal reactions are fast relative to the rates of heat and mass transfer. There are but a few examples of metallurgical reactions that are controlled by a chemical reaction at the surface, particularly when there is strong chemisorption.

Chemisorption on matte or slag surfaces and on slag–metal interfaces is less well understood. Indications are that the reduction of silicates, and presumably aluminates, by metals is controlled by an interfacial reaction kinetics. However, the transfer of simple cations and anions such as Mn^{2+}, Cu^{+}, O^{2-}, and S^{2-} across the metal–slag (or matte) interface, via an electrochemical reaction step, is controlled by mass transfer to and from the interface. The effect of interface stirring on the rate of mass transfer is reasonably well understood. With dispersion of metal droplets in slags the rates of reactions are enhanced by several orders of magnitude, because of the vast increase in the interfacial surface area and rapid diffusion in minute metal particles, as in pneumatic metal refining processes.

Gas bubbles in metals and slags play a vital role in metal refining. The present state of knowledge of the dynamics of the swarm of bubbles in liquid metals, mattes, and slags is by no means adequate. Although the bubble size and mass transfer relations for the swarm of bubbles are known for gas–water systems, to what extent they can be applied to gas–metal systems remains to be resolved by future research. Model experiments with gas–water systems may not necessarily simulate gas–liquid metal systems. There is need for more sophisticated theoretical and experimental work in this area.

There are many examples of coupled reactions in gas–slag and slag–metal systems. In fact, in the early stages of the transfer of a reactant from one phase to another, other accompanying reactions may move away from equilibrium initially. Furthermore, we have seen that vaporization has a decisive role in the transfer of reactants from one condensed phase to another. In reactions involving the formation of a dispersed condensed phase, as in the deoxidation of metals, the overall rate of reaction is markedly influenced by nucleation, growth, and flotation.

Although a great deal more analytical research is needed to resolve the complexities of metallurgical reaction rates, there is even a greater need for the development of new technology for the smelting and refining of metals by applying the already well-stocked wealth of accumulated knowledge of pyrometallurgy.

4.6. Zone Refining (CHP)

4.6.1. Background

Zone refining is a specialized purification technique based upon passing a series of molten zones through the charge in one direction.[325–328] Zone refining is actually a modification of the classical purification technique of fractional crystallization. Purification with this technique results from the difference in solubility of a solute in the solid and liquid phases of the solute. The earliest example of fractional solidification occurred during solidification of the earth's molten layer or magma. The distribution of minerals and elements in the earth's crust can be explained on the basis of fractional solidification processes. One of the earliest examples of the use of the principle of fractional crystallization was in the early metallurgical process of liquation. Agricola describes a 14th-century process whereby lead and silver are separated from a copper–lead–silver alloy by gradually heating and allowing the lead and silver to melt, leaving the copper behind. Another early exam-

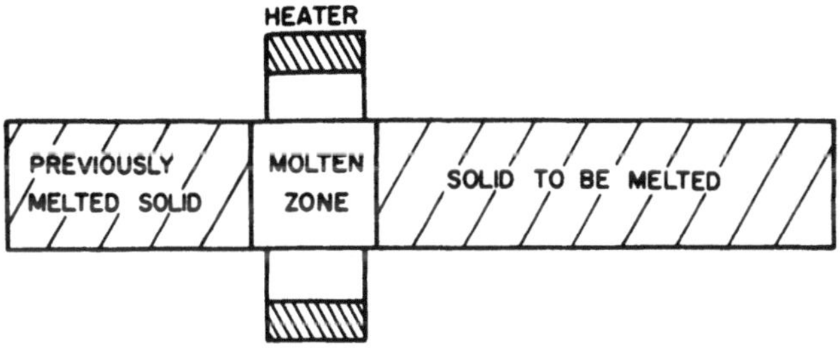

Figure 4.6-1. Illustration of zone melting technique.

ple of the use of fractional solidification was when the whalers of the 19th century melted sea ice to use for drinking water. The relationship between segregation during freezing and the thermodynamics of phase equilibria was slowly realized during the 1920s and 1930s. Slow freezing was utilized in 1929 by Straumanis to partially remove bismuth from zinc. As the use of solidification for separation or purification developed, better techniques were devised. One of these was the Bridgeman–Stackbarger technique of slowly lowering a tube of molten metal from a furnace so that freezing started at one end and gradually progressed through the entire liquid. This method produced not only high-purity materials but also single crystals. Many applications of this technique have been made since about 1945 with both organic and inorganic materials being purified. Part of the ingot has to be discarded after each solidification to carry out repeated purification in batches since remelting the solid redistributes the impurities. The need for discarding part of the ingot was eliminated in 1952 by Pfann's invention[325] of zone melting where only a thin zone of the solid is melted at one time as shown in Figure 4.6-1. The molten zone moves slowly through the solid; it can be passed through many times achieving further purification with each zone pass. Nearly all the common metals have been purified by zone refining techniques. A fairly complete list of organic and inorganic materials which have been zone refined is given in the appendix of Ref. 327. Zone refining has also been used in preparing semiconductors and intermetallic compounds as well as ionic compounds and oxides.

Since zone refining involves movement of impurities from one position in an ingot to another over a period of time it may be considered a rate or mass transport process.

4.6.2. *Equilibrium Solidification*

Solidification of most metals and inorganic nonmetals from their melts is closely approximated by the assumption of equilibrium at the

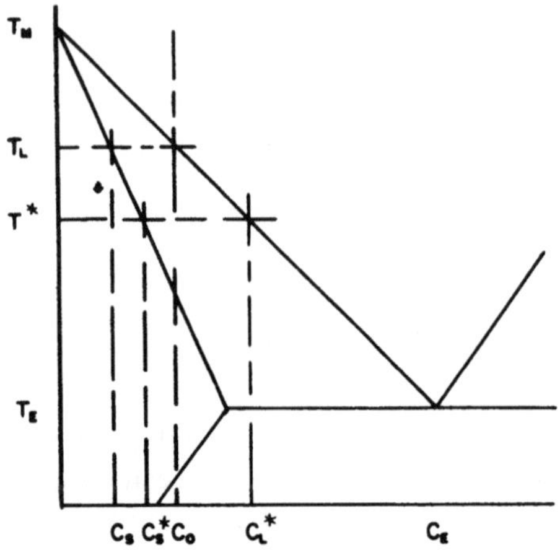

Figure 4.6-2. Idealized phase diagram illustrating relations in zone melting.

interface during growth. Large concentration gradients may occur in the solid and liquid during solidification but there is very little barrier to atom transport across the surface. Consider an idealized phase diagram as shown in Figure 4.6-2 and an alloy of initial composition *Co* at a temperature high enough to be completely liquid. As the temperature declines at one end of an ingot of this alloy, it eventually reaches the liquidus temperature where solid of composition *Cs* begins to form. The ratio of the solute composition in the solid to the solute composition in the liquid is known as the equilibrium partition ratio,

$$k = C_s/C_L \tag{4.6-1}$$

It should be noted that if the composition of the liquid alloy changes at the interface, say, to C_s^*, the equilibrium partition ratio still is the same or

$$k = C_s^*/C_L^* = C_s/C_L \tag{4.6-2}$$

Under conditions of extremely slow solidification the solute could diffuse through the solid and liquid phases to maintain equilibrium conditions throughout the freezing process. Consider an ingot of length L and initial composition C_0 freezing from one end. The first solid begins to form at T_L and has a composition, kC_0, lower in solute than the original melt. The solute is rejected from the solid–liquid interface and diffuses into the liquid. During subsequent cooling and solidification both the liquid and solid become enriched in solute. Because under equilibrium conditions diffusion in the liquid and solid is assumed complete, the entire solid at any given time is of a uniform composition C_s^* and the entire liquid is of a uniform composition C_L^*. At temperature

T a materials balance may be written

$$C_s f_s + C_L f_L = C_0 \tag{4.6-3}$$

where f_s and f_L are weight fractions of the liquid and solid. Owing to the equilibrium nature of the solidification just described a large amount of solute redistribution must occur by diffusion to maintain equilibrium.

4.6.3. Mass Transfer in Fractional Solidification

Mass transfer may be defined as a diffusive process that arises from a concentration gradient. During solidification segregation takes place at the freezing interface. The resulting change in liquid and solid composition at the interface causes diffusion. As the solid forming increases in solute composition because of increasing solute concentration in the liquid, diffusion occurs in the solid but at a much slower rate than in the liquid. Consequently one approximation useful in practical systems is to assume no solid diffusion with complete and rapid diffusion occurring in the liquid phase. Consider a crucible of liquid alloy of length L and initial composition C_0 freezing from one end. As in equilibrium solidification, the first small amount of solid to form is of composition kC_0 at temperature T_L. During subsequent cooling and solidification the liquid becomes richer in solute and so the solid that forms is of higher solute content at later stages of solidification. Since there is no diffusion in the solid state the composition of the solid formed initially remains the same. At temperature T^*, solid of composition C_s^* is freezing from liquid of composition C_L^* (see Figure 4.6-2). A quantitative expression for the relation between C_s^*, f_s, C_0, and k is obtained by equating the amount of solute rejected when a small amount of solid forms to the resulting increase of solute in the liquid.

$$(C_L^* - C_s^*)df_s = (1 - f_s)dC_L \tag{4.6-4}$$

or

$$\frac{dC_L^*}{C_L^* - C_s^*} = \frac{df_s}{1 - f_s} = \frac{dC_L^*}{C_L^*(1-k)} = \frac{df_s}{1 - f_s}$$

$$\frac{1}{1-k} \ln C_L^* = -\ln(1 - f_s) + \ln K' \qquad \text{at} \quad f_s = 0, \qquad C_L^* = C_0$$

so

$$\ln K' = \frac{1}{1-k} \ln C_0$$

then

$$\ln\left(\frac{C_L^*}{C_0}\right)^{1-k} = \ln(1-f_s)^{-1}$$

$$C_L^* = C_0(1-f_s)^{k-1}$$
$$C_s^* = kC_0(1-f_s)^{k-1} \tag{4.6-5}$$

In terms of liquid composition

$$C_L = C_0 f_L^{k-1} \tag{4.6-6}$$

These equations closely describe solute redistribution in solidification under a wide range of experimental conditions.

Another important set of limiting conditions that can be applied is where no diffusion occurs in the solid and limited diffusion without convection occurring in the liquid. In this case the solute rejected into the liquid is transported only by diffusion so an enriched boundary layer forms and gradually increases in solute. If the ingot is long enough, a steady-state situation arises in which the composition of the solid is the same as that of the initial alloy. This is illustrated diagrammatically in Figure 4.6-3. Equilibrium at the interface requires that the composition in the liquid at the interface by C_0/k and that solidification be occurring at the corresponding solidus temperature, T_s. In setting up the differential equation which describes this situation one considers the moving interface to be at the origin of the coordinate system. Because the interface is actually moving, it is as though liquid is flowing into the interface with this coordinate system. The following equation represents

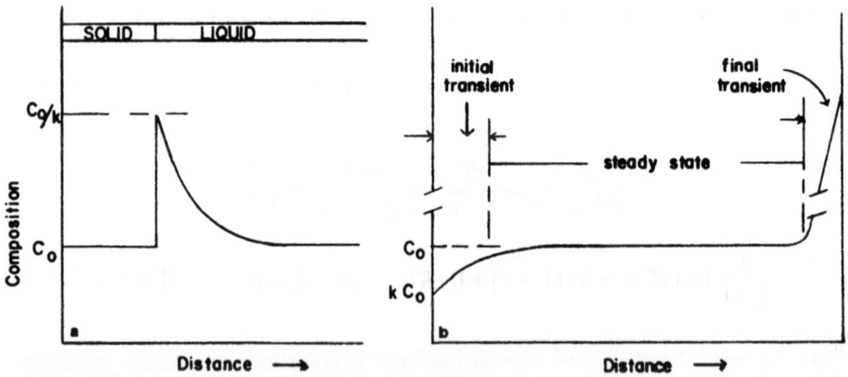

Figure 4.6-3. Composition distribution in solidification: (a) steady state, and (b) after solidification.

the assumed conditions:

$$D_L\frac{d^2C_L}{dX'}+R\frac{dC_L}{dX'}=0 \tag{4.6-7}$$

where X' is the distance from the interface, D_L is the diffusion coefficient of solute in the liquid, R is the rate of liquid flow into the interface, and C_L the liquid composition. The solution to this equation is

$$C_L = C_0\left(\frac{1-k}{k}\right)e^{-(R/D_L)X'} \tag{4.6-8}$$

using the boundary conditions that at steady state

$$k=\frac{C_0}{C_a-C_0}$$

where C_0 is the original alloy composition and C_a is the increase in solute composition at the liquid–solid interface. The quantity D_L/R is called the characteristic distance or the distance at which the value of $C_L - C_0$ falls to $1/e$ of the maximum $C_a + C_0$. The characteristic distance depends only on D_L and R, the diffusion coefficient and the rate of interface movement. Equation (4.6-8) applies in many real cases of single crystal growth even though some convection is usually present. This type of solidification results in a crystal of nearly uniform composition except for the initial and final transients as shown in Figure 4.6-3b. The initial transient is formed while the solute boundary layer builds up to its maximum steady-state value. Solute redistribution is calculated during this period using the time-dependent form of equation (4.6-7)

$$D_L\frac{d^2C_L}{dX'^2}+R\frac{dC_L}{dX'}=\frac{dC_L}{dt} \tag{4.6-9}$$

For small values of k an approximate solution to this equation is

$$C_s^* = C_0[1-(1-k)e^{-kRx/D_L}] \tag{4.6-10}$$

The characteristic distance for the initial transient from equation (4.6-10) is D_L/kR. The final transient is much smaller than the initial transient since it arises from the impingement of the boundary layer on the end of the crucible. The characteristic length of the final transient is thus of the order of D_L/R for small k and increases with decreasing k. Another type of transient of importance in engineering applications occurs with a change in velocity of the interface. This is shown schematically in Figure 4.6-4 which indicates a band of increased solute content caused by a velocity increase. Likewise, a decrease in velocity would cause a solute-poor band.

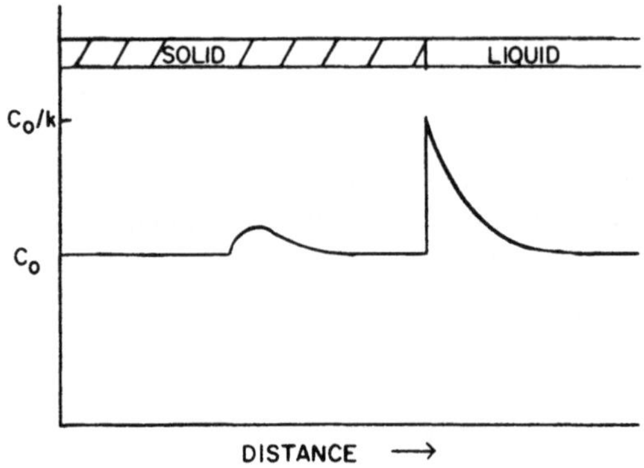

Figure 4.6-4. The effect of an increase of velocity in fractional solidification.

In most cases of crystal growth some convection is present. The analytical treatment in this case assumes a boundary layer at the interface of thickness, δ, in which diffusion plays the major role. Outside of this layer the liquid composition is assumed to be made uniform by convection. When steady state is reached during solidification, equation (4.6-7) applies with the boundary conditions that $C_L = C_L^*$ at the liquid–solid interface ($x' = 0$) and $C_L = C_0$ at $x' = \delta$. The solution to equation (4.6-7) with these boundary conditions is

$$C_L^* - C_s^* = (C_0^* - C_s^*)e^{-R\delta/D_L} \qquad (4.6\text{-}11)$$

dividing by C_s^*

$$\frac{1}{k'} - 1 = \left(\frac{1}{k} - 1\right)e^{-R\delta/D_L}$$

solving for k'

$$k' = \frac{k}{k + (1-k)e^{-R\delta/D_L}} \qquad (4.6\text{-}12)$$

This expression is useful since it relates the composition of the solid forming in crystal growth to alloy composition and growth conditions. It can be used to describe solute redistribution in crucibles of finite extent, provided that the thickness of the boundary layer δ is small compared with the length of the crucible. When this is true, a dynamic equilibrium is maintained between the bulk melt and growing solid, and equations identical to (4.6-5) and (4.6-6) are readily derived with the effective partition ratio k' replacing the equilibrium partition ratio

$$\begin{aligned} C_s^* &= k'C_0(1-f_s)^{k'-1} \\ C_L &= C_0 f_L^{k'-1} \end{aligned} \qquad (4.6\text{-}13)$$

When $R\delta/D_L \ll 1$, that is, at a slow growth rate, high liquid diffusivity, and maximum stirring to make δ small, the situation approaches the case in which infinite diffusivity in the liquid was assumed and $k' \to k$. The maximum value of k', $k = 1$, is achieved when $R\delta/D \gg 1$. The limit is described by the special case discussed earlier where no convection occurs.

4.6.4. Theory of Zone Melting

Consider a cylindrical charge of uniform composition consisting of two components that can form a solid solution as shown in Figure 4.6-2. A molten zone of length l is caused to traverse the charge slowly. The distribution coefficient k is less than one. Passing the zone through the ingot distributes the solute as shown in Figure 4.6-3b which is similar to the distribution obtained in normal freezing. To determine an analytical equation describing the distribution of solute in zone melting one may consider the increment of solute frozen out of the zone as

$$kC_L\, dx$$

where

C_L = solute concentration in the molten zone,
$C_0\, dx$ = the increment solute entering the zone,
$C_L = s/l$, and
l = zone length.

If s = amount of solute in the molten zone, the net change of solute in the molten zone is then

$$ds = (C_0 - kC_L)\, dx$$

and

$$\frac{ds}{dx} + k\frac{s}{l} = C_0 \qquad (4.6\text{-}14)$$

Using the appropriate boundary conditions this equation integrates to

$$\frac{C}{C_0} = 1 - (1-k)e^{-kx/l}$$

where x is the distance along the ingot.

This equation is valid in all but the last zone length where the solute distribution is given directly by the normal solidification equation. If convection is present and the boundary layer δ is much smaller than l, equation (4.6-14) can be expressed in terms of the effective distribution coefficient. There is no simple way to calculate solute distribution after

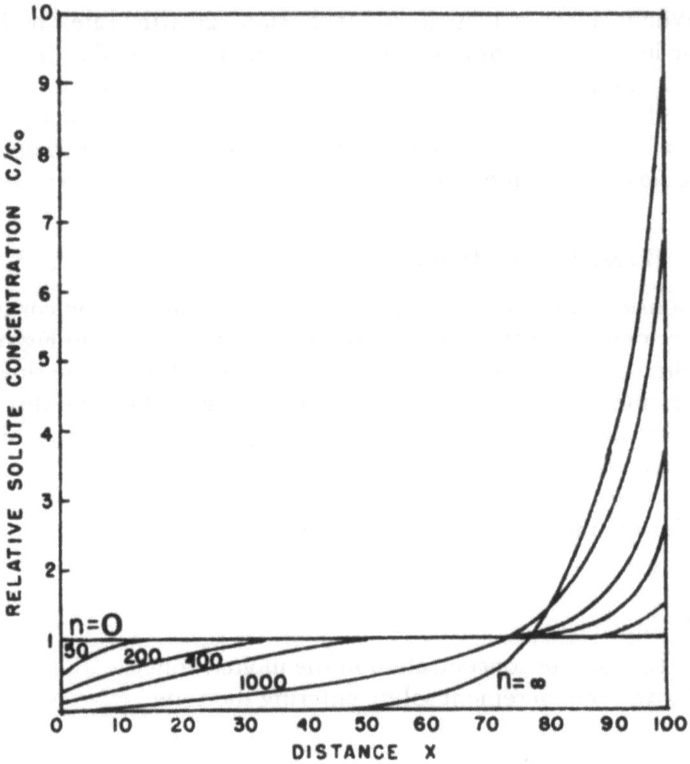

Figure 4.6-5. Relative solute concentration C/C_0 vs. distance x, with number of passes n as a parameter for $K = 0.9724$.

multiple pass zone melting but a number of computed curves have been given by the inventor of the process, W. S. Pfann. One example is shown in Figure 4.6-5 for a k value close to one. After 2000 passes only a small fraction of the solute initially present remains in the first half of the bar. When only one or a few passes are conducted in zone melting and the zone is small compared with the ingot length, the bulk of the ingot solidifies at steady state with the solid material in this region being of uniform composition C_0. When zone solidification is conducted in this manner to obtain uniformity, it is called *zone leveling*. When it is conducted to obtain maximum transport of solute it is called *zone purification* or *zone refining*. An interesting variation of zone melting is *zone freezing* in which case a solid zone is caused to pass through a liquid melt. In zone freezing substances which lower the melting point accumulate ahead of the liquid–solid interface.

When there are volatile impurity elements present, enhanced purification can result from the volatilization of these elements. Conversely, special precautions must be taken to retain volatile alloy elements if they are desired in the alloy. Solute distribution equations for the various types of solidification presented earlier can now be rederived including the effect of volatilization. The applicable equation for volatilization is

$$J = \alpha \rho_L (C_L^e - C_L) \tag{4.6-15}$$

where

J = flux across the liquid–vapor interface, g/cm^2 sec
α = evaporation coefficient, cm/sec
ρ_L = liquid density,
C_L^e = liquid composition that would be in equilibrium with the gas phase, and
C_L = actual liquid composition.

A simple example would be that of vertical normal solidification (Bridgeman method) in which the top surface is exposed to zero partial pressure of solute. Solute volatilization occurs by transfer across the liquid–vapor interface. The material balance is

$$(C_L C_s^*) df_s \rho_L V = (1 - f_s) dC_L \rho V J A \, dt \tag{4.6-16}$$

$$\begin{bmatrix} \text{g solute rejected} \\ \text{from solid} \end{bmatrix} = \begin{bmatrix} \text{g solute added} \\ \text{to liquid} \end{bmatrix} - \begin{bmatrix} \text{g solute} \\ \text{vaporized} \end{bmatrix}$$

where V is the crystal volume and A is the crystal cross section.

Assuming constant liquid and solid densities

$$\frac{df_s}{dt} = \frac{A}{V} R \tag{4.6-17}$$

Substituting equations (4.6-15) and (4.6-17) into (4.6-16) with $C_L^e = 0$ and integrating from $C_L = C_0$ at $f_s = 0$ gives

$$C_s^* = kC_0(1 - f_s)^{(k-1)+(\alpha/R)} \tag{4.6-18}$$

Equation (4.6-18) is the simple equation for normal solidification with no solid diffusion modified by the α/R term in the exponent. If diffusion is limited in the liquid but convection maintains the bulk liquid at uniform composition C_L, then the effective partition ratio k' can be substituted for k. If the volatile component is a desirable alloying element in the crystal to be grown, several techniques can be used to retain it in solution. One method is to encapsulate the melt so that the vapor pressure builds up to its equilibrium and no further volatilization occurs. Another technique is to cover the melt with a slag impervious to the volatile species.

In most analyses of solute redistribution in crystal growth diffusion of solute in the solid is neglected. This is generally an excellent assumption for the usual types of crystal growth. If can be shown that if the dimensionless parameter $D_s k/RL \ll 1$, no appreciable diffusion occurs in the solid.

4.6.5. Matter Transport during Zone Refining

When a molten zone is passed through an ingot in an open horizontal boat, the ingot usually becomes tapered.[329] The taper may only be slight after one pass but it can become prominent after repeated passes and can even cause an overflow of matter at the end of the boat. This matter transport by a molten zone arises from the change of density on melting. The magnitude and direction of matter transport correspond to the magnitude and sign of the density change. The mechanism of matter transport is basically similar to the mechanism of solute transport analyzed in the preceding sections and the phenomena are described by similar equations. Contraction on melting causes forward transport, whereas expansion on melting causes reverse transport. Analysis of the phenomenon can be accomplished with the aid of Figure 4.6-6, where the solid is shown to contract on melting. The height of the liquid is αh_0 where α is the ratio of solid density to liquid density. If the zone advances a distance dx, the volume of solid melted is $h_0\,dx$ and the volume of solid frozen is $h\,dx = h_0\,\alpha dx$ since the solid freezes at the level of the liquid. Since αh_0 is less than h_0 the zone increases in height as it travels until the height entering and leaving become equal. Let dv be the change in liquid volume as the zone advances a distance dx. Then

$$dv = l\,dh = \alpha(h_0 - h)\,dx$$

solving for h gives

$$\frac{h}{h_0} = 1 - (1-\alpha)e^{-\alpha x/l} \tag{4.6-19}$$

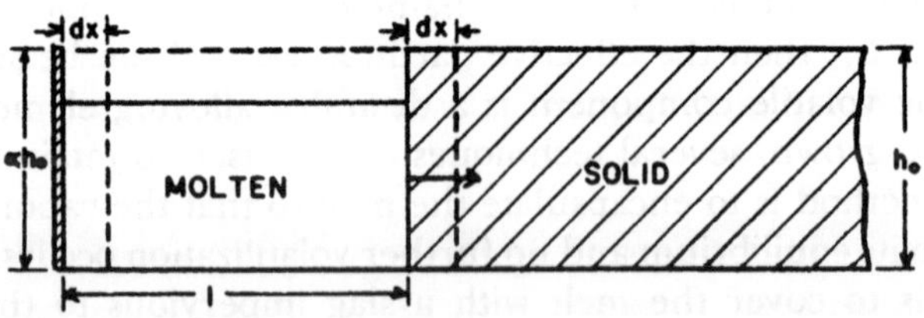

Figure 4.6-6. Formation and advance of a molten zone in a level uniform ingot.

This equation is valid in all but the last zone length where the normal solidification equation applies or

$$\frac{h}{h'_0} = (1 - f_s)^{\alpha - 1}$$

where h'_0 is the height of liquid at a beginning of normal freezing. To prevent undesirable matter transport the ingot can be tilted to the proper angle. Tilting effectively changes the values of α to unity. The critical angle of tilt is given by

$$\theta_c = \tan^{-1} \frac{2h_0(1-\alpha)}{l}$$

Matter transport can also be prevented by making the height to the solid in the molten zone at the beginning of the ingot h_0/α so that after melting the height of the liquid is h_0.

4.6.6. Summary

Mass transfer for purification or other purposes can be achieved by the techniques of progressive solidification or zone refining. The purification phenomena is caused by rejection of solute at the liquid–solid interface. This phenomena has been used throughout history for purification of materials. The resulting solute concentration profile can be calculated using various limiting assumptions. Derivation of concentration profile equations are given for no solid diffusion, with complete liquid diffusion and convection combined. Volatilization of impurities is considered along with matter transport during zone refining.

References

1. R. D. Pehlke, *Unit-Processes of Extractive Metallurgy*. American Elsevier Publ. Co., New York (1973), pp. 7–23.
2. I. Iwasaki, A thermodynamic interpretation of the segregation process for copper and nickel ores, *Minerals Sci. Eng.* **4** (2), 14–23 (1972).
3. J. D. Gilchrist, *Extraction Metallurgy*, Pergamon Press, London (1967), pp. 169–171.
4. H. H. Kellogg, A critical review of sulfation equilibria, *Trans. TMS-AIME* **230**, 1622–1634 (1963).
5. W. M. McKewan, Kinetics of iron oxide reduction, *Trans. TMS-AIME* **212**, 791–794 (1958).
6. R. W. Stewart, A. P. Linden, I. B. Cutler, and M. E. Wadsworth, Geometric factors in solid-state reactions, in *Energetics in Metallurgical Phenomena*, W. M. Mueller, ed., Vol III, Gordon and Breach, New York (1967), pp. 19–51.
7. W. K. Lu, The general rate equation for gas–solid reactions in metallurgical processes, *Trans. TMS-AIME* **227**, 2303–2306 (1963).

8. K. Mori, A rate equation for the reduction of iron ore, *Tetsu to Hagane*, **50**, 2259–2260 (1964); *Chem. Abstr.* **63**, 14,429e (1965).
9. B. B. L. Seth and H. U. Ross, The mechanism of iron oxide reduction, *Trans. TSM-AIME* **233**, 180–185 (1965).
10. H. W. St. Clair, Rate of reduction of an oxide sphere in a stream of reducing gas, *Trans. TMS-AIME* **233**, 1145–1152 (1965).
11. R. H. Spitzer, F. S. Manning, and W. O. Philbrook, Mixed control reaction kinetics in the gaseous reduction of hematite, *Trans. TMS-AIME* **236**, 726–742 (1966).
12. S. E. Khalafalla, G. W. Reimers, and M. J. Baird, Effect of low concentrations of silica, alumina, lime, and lithia on the magnetic roasting of hematite, *Met. Trans.* **5**, 1013–1018 (1974).
13. D. A. Young, *Decomposition of Solids*, Pergamon Press, New York (1966), pp. 49–52.
14. G. Asmund, *Chemistry and Kinetics of the Sulfating Roasting of Uranium-Bearing Silicates*, Danish Atomic Energy Commission, Report No. 253, Kopenhagen, (Oct. 1971), 125 pp.
15. W. Gautschi and W. F. Cahill, Exponential integrals and related function, in *Handbook of Mathematical Functions*, M. Abramowitz and I. A. Stegun, eds., U.S. Dept. of Commerce, NBS AMS55, Washington (1964), pp. 227–251.
16. U. R. Evans, *The Corrosion and Oxidation of Metals. Scientific Principles and Practical Applications*, Edward Arnold Publishers, Ltd., London (1961), pp. 819–859.
17. C. Prasky, F. E. Joyce, Jr., and W. S. Swanson, Differential sulfatizing process for the recovery of ferrograde manganese, *U.S.B.M.* 6160 (1963), 30 pp.
18. F. E. Joyce, Jr. and C. Prasky, Sulfatization–reduction of manganiferous iron ore, *U.S.B.M. RI* 7749 (1973), 17 pp.
19. M. C. Van Hecke and R. W. Bartlett, Kinetics of sulfation of Atlantic Ocean manganese nodules, *Met. Trans.* **4**, 941–947 (1973).
20. A. W. Fletcher and M. Shelef, A study of the sulfation of a concentrate containing iron, nickel, and copper sulfides, *Trans. TMS-AIME*, **230**, 1721–1724 (1964).
21. A. W. Fletcher and M. Shelef, Symposium on unit processes in hydrometallurgy, *Met. Soc. Conf.* **24**, 946–970 (1963).
22. P. G. Thornhill, Roasting sulfide concentrates in fluidized beds, U.S. Pat. 2,813,015 (1957), Falcon Bridge Nickel Mines, Ltd.; Roasting nickeliferous sulfide concentrates in fluidized beds, U.S. Pat. 2,813,016 (1957), Falcon Bridge Nickel Mines, Ltd.
23. F. E. Joyce, Jr., Extraction of copper and nickel from the Duluth-Gabbro complex by selective high-temperature sulfatization, *U.S.B.M. RI* 7475 (1971), 15 pp.
24. C. G. Alcock and M. G. Hocking, Kinetics and mechanism of formation of cobalt sulfate on cobalt oxide, *Inst. Min. Met. Trans. C.* **75**, (712), 27–36 (1966).
25. A. W. Sommer and H. H. Kellogg, Oxidation of sphalerite by sulfur trixoide, *Trans. TMS–AIME* **215**, 742–744 (1959).
26. N. B. Gray, N. W. Stump, W. S. Boundy, and R. V. Culver, The sulfation of lead sulfide, *TMS–AIME* **239**, 1835–1840 (1967).
27. J. M. Kruse and N. J. Pitman, Extracting beryllium from very lean nonpegmate ores, U.S. Pat. 3,148,022 (1964), E.I. du Pont de Nemours and Co.
28. T. Chao and S. C. Sun, Study of sulfatization of alumina with gaseous sulfur trioxide, *Trans. TMS-AIME* **238**, 420–429 (1967).
29. L. A. Haas and S. E. Khalafalla, Formation of water-soluble aluminum compounds by dry sulfation of clay, Presented at Annual Meeting, AIME, New York, February 1975.
30. G. P. Losos and B. M. Hoffman, Electron paramagnetic resonance of a nitroxide adsorbed on silica, silica–alumina, alumina, and decationated zeolites, *J. Phys. Chem.* **78**, 2110–2116 (1974).

31. K. Natesan and W. O. Philbrook, Oxidation kinetic studies of zinc sulfide pellets, *Trans. TMS–AIME* **245**, 2243–2250 (1969).
32. K. Natesan and W. O. Philbrook, Oxidation kinetic studies of zinc sulfide in a fluidized-bed reactor, *Met. Trans.* **1**, 1353–1360 (1970).
33. L. Coudurier, I. Wilkomirsky, and G. Morizat, Molybdenite roasting and rhenium volatilization in a multiple hearth furnace, *Inst. Min. Met. Trans. C* **79**, 34–40 (1970).
34. P. R. Ammann and T. A. Loose, The oxidation kinetics of molybdenite at 525° to 635° C, *Met. Trans.* **2**, 889–893 (1971).
35. L. J. Hillenbrand, The adsorption and incorporation of oxygen by lead, *J. Phys. Chem.* **73**, 2902–2908 (1969).
36. R. I. Razouk, M. Y. Farah, R. S. Mikhail, and G. A. Kolta, The roasting of precipitated copper sulfide, *J. Appl. Chem.* **12**, 190–196 (1962).
37. E. A. Ashcroft, Sulfate roasting of copper ores and economic recovery of electrolytic copper from chloride solutions, *Trans. Electrochem. Soc.* **23**, 23–44 (1933).
38. M. E. Wadsworth, K. L. Leiter, W. H. Parter, and J. R. Lewis, Sulfating of cuprous sulfide and cuprous oxide, *Trans. TMS–AIME* **218**, 519–525 (1960).
39. I. D. Shah and S. E. Khalafalla, Chemical reactions in roasting of copper sulfides, *U.S.B.M. RI* 7549 (1970), 21 pp.
40. S. E. Khalafalla and I. D. Shah, Oxidative roasting of covellite with minimal retardation from the $CuO{\cdot}CuSO_4$ film, *Met. Trans.* **1**, 2151–2155 (1970).
41. C. Goetz, Obtaining nonferrous metals such as copper from iron containing ores, U.S. Pat. 2,082,284 (June 1, 1937).
42. I. D. Shah and S. E. Khalafalla, Kinetics and mechanism of the conversion of covellite (CuS) to digenite ($Cu_{1.8}$), *Met. Trans.* **2**, 2637–2643 (1971).
43. I. D. Shah and S. E. Khalafalla, Kinetics of thermal decomposition of copper(II) sulfate and copper(II) oxysulfate, *U.S.B.M. RI* 7638 (1972), 21 pp.
44. M. M. Pavlyuchenko and G. I. Samal, Kinetics and mechanism of the thermal decomposition of copper sulfide, *Geterogennye Reakts. i. Reakts. Sposobnost* (*Minsk: Vyssh. Shkola*), *Sb.* 85–94 (1964) (in Russian); *Chem. Abstr.* **64**, 15048c (1966).
45. P. K. Chatterjee, Application of thermogravimetric techniques to reaction kinetics, *J. Polymer Sci. A* **3**, 4253–4262 (1965).
46. T. R. Ingraham and P. Marier, Kinetics of the thermal decomposition of cupric sulfate and cupric oxysulfate, *Trans. TMS-AIME* **233**, 363–367 (1965).
47. V. V. Rao, V. S. Ramakrishna, and K. P. Abraham, Kinetics of oxidation of copper sulfide, *Met. Trans.* **2**, 2463–2470 (1971).
48. N. W. Kirshenbaum, Transport and handling of sulfide concentrates, problems and possible improvements, Dept. of Mineral Engr., Stanford University, Ca. (1967), 218 pp.
49. P. W. Scott, A review and appraisal of iron ore beneficiation–Part II. *Min. Cong. J.* **49**, 78–83 (June 1963).
50. J. P. Hansen, G. Bitsianes, and T. L. Joseph, A study of the kinetics of magnetic roasting, *Blast Furnace, Coke Oven, Raw Materials Conf.* **19**, 185–205 (1960).
51. J. C. Nigro and T. D. Tiemann, Kinetics and mechanism of the gaseous reduction of hematite to magnetite and the effect of silica, Ph.D Thesis, University of Wisconsin, Madison (1970).
52. S. E. Khalafalla and P. L. Weston, Jr., Promoters for carbon monoxide reduction of wustite, *Trans. TMS-AIME* **239**, 1493–1499 (1967).
53. J. A. Hedvall, *Solid-State Chemistry*, Elsevier Publ. Co., New York (1966), pp. 28–34.
54. J. H. Fishwick, Treatment of spodumene, U.S. Pat. 3,394,988 (July 30, 1968), Foote Mineral Co.

55. C. B. Daellenbach, R. A. Vik, and W. M. Mahan, Influence of reduction and thermal shock on nonmagnetic taconite grindability, *Trans. TMS-AIME* **250**, 212–217 (1971).
56. N. J. Tighe and P. R. Swann, Direct observation of iron oxide reduction using high-voltage microscopy, Final Scientific Report TR 72-1285 to Air Force Office of Scientific Research (AFOSR) through the European Office of Aerospace Research (OAR), U.S. Air Force Grant E00AR-69-0067, Washington (1972), 32 pp.
57. F. Habashi, *Extractive Metallurgy*, Vol. 1, Gordon and Breach, New York (1969), p. 258.
58. S. W. Benson, *The Foundations of Chemical Kinetics*, McGraw-Hill Book Co., New York (1960), p. 35.
59. P. I. Richards *Manual of Mathematical Physics*, Pergamon Press, London (1959), p. 197.
60. D. H. Andrews and M. L. Boss, The transference of information in growth processes, *Yale Sci. Mag.* **25** (8) (May 1971).
61. J. R. Miller, *Sci. Am.* **235**, 68 (1976).
62. E. T. Turkdogan and J. V. Vinters, *Met. Trans.* **2**, 3175 (1971).
63. E. T. Turkdogan, R. G. Olsson, and J. V. Vinters, *Met. Trans.* **2**, 3189 (1971).
64. E. T. Turkdogan and J. V. Vinters, *Met. Trans.* **3**, 1561 (1972).
65. S. Watanbe and M. Yoshinaga, *Sumitomo J.* **77**, 323 (1965).
66. M. C. Chang, J. Vlanty, and D. W. Kestner, Ironmaking Conf., AIME, 26th, Chicago, (1967).
67. W. Wenzel and H. W. Gudenau, *Stahl Eisen* **90**, 689 (1970).
68. W. Wenzel and H. W. Gudenau, *Germ. Offen.* 440 (1970).
69. L. Granse, Proc. Int. Conf. Sci. and Tech. of Iron and Steel, *Trans. Iron Steel Inst. Japan* **11**, 45 (1971).
70. J. T. Moon and R. D. Walker, *Ironmaking Steelmaking*, No. 2, 30 (1975).
71. E. T. Turkdogan and J. V. Vinters, *Can. Met. Q.* **12**, 9 (1973).
72. O. Burghardt, H. Kostmann, and B. Grover, *Stahl Eisen* **90**, 661 (1970).
73. E. E. Hoffman, H. Rausch, and W. Thum, *Stahl Eisen* **90**, 676 (1970).
74. H. Vom Ende, J. Grege, S. Thomalla, and E. E. Hofmann, *Stahl Eisen* **90**, 667 (1970).
75. P. Lecomte, R. Vidal, A. Poos, and A. Decker, *CNRM Metall. Rep.* No. 16 (Sept. 1968).
76. A. Poos and R. Balon, *J. Metals* **19**, 93 (1967).
77. R. L. Bleifuss, *Trans. Met. Soc. AIME* **247**, 225 (1970).
78. B. B. L. Seth and and H. U. Ross, *Trans. Met. Soc. AIME* **233**, 180 (1965).
79. W. Wenzel, H. W. Gudenau, and M. Ponthenkandath, *Aufbereitungs-technik* **11**, 154, 1970.
80. *Ibid*, 492.
81. T. L. Joseph, *Trans. AIME* **120**, 72 (1936).
82. L. von Bogdandy and H.-J. Engell, *The Reduction of Iron Ores*, Springer-Verlag, Berlin and New York (1971).
83. L. von Bogdandy and W. Janke, *Z. Electrochem. Ber. Bunsenges. Phys. Chem.* **61**, 1146 (1957).
84. H. K. Kohl and H.-J. Engell, *Arch. Wiss.* **34**, 411 (1963).
85. J. M. Quets, M. E. Wadsworth, and J. R. Lewis, *Trans. Met. Soc. AIME* **218**, 545 (1960).
86. W.-K. Lu and G. Bitsianes, *Can. Met. Q.* **7**, 3 (1968).
87. E. T. Turkdogan, *Ironmaking Conf. Proc. AIME* **31**, 438 (1972).
88. R. H. Spitzer, F. S. Manning, and W. O. Philbrook, *Trans. Met. Soc. AIME* **236**, 726 (1966).

89. B. B. L. Seth and H. U. Ross, *Trans. Met. Soc. AIME* **233**, 180 (1965).
90. P. K. Strangway and H. U. Ross, *Trans. Met. Soc. AIME* **242**, 1981 (1968).
91. R. G. Olsson and W. M. McKewan, *Trans. Met. Soc. AIME* **236**, 531 (1966).
92. R. G. Olsson and W. M. McKewan, *Met. Trans.* **1**, 1507 (1970).
93. G. Nabi and W. K. Lu, *J. Iron Steel Inst.* **211**, 429 (1973).
94. P. F. J. Landler and K. L. Komarek, *Trans. Met. Soc. AIME* **236**, 138 (1966).
95. W. M. McKewan, *Trans. Met. Soc. AIME* **224**, 387 (1962).
96. J. W. Evans and K. Haase, *High Temp. Sci.* **8**, 167 (1976).
97. A. G. Matyas and A. V. Bradshaw, *Ironmaking Steelmaking*, No. 3, 180 (1974).
98. R. H. Tien and E. T. Turkdogan, *Met. Trans.* **3**, 2039 (1972).
99. T. G. Cox and F. R. Sale, *Ironmaking Steelmaking*, No. 4, 234 (1974).
100. J. Szekely and Y. El-Tawil, *Met. Trans.* **7B**, 490 (1976).
101. K. Tittle, *Proc. Aust. Inst. Min. Met.* **243**, 57 (1972).
102. G. D. McAdam, *Ironmaking Steelmaking*, No. 3, 138 (1974).
103. C.-H. Koo, Ph.D. Dissertation, University of California, Berkeley, 1977.
104. G. Nabi and W.-K. Lu, *Ind. Eng. Chem. Fundam.* **13**, 311 (1974).
105. S. E. Khalafalla, G. W. Reimers, and M. J. Baird, *Met. Trans.* **5**, 1013 (1974).
106. R. E. Cech and T. D. Tiemann, *Met. Trans.* **3**, 590 (1972).
107. J. F. Elliott and M. Gleiser, *Thermochemistry for Steelmaking*, Addison-Wesley, Reading, Mass., (1960).
108. A. F. Benton and P. H. Emmett, *J. Am. Chem. Soc.* **46**, 2728 (1924).
109. H. S. Taylor and R. M. Burns, *J. Am. Chem. Soc.* **43**, 1273 (1921).
110. A. Kivnick and A. N. Hixson, *Chem. Eng. Prog.* **48**, 394 (1952).
111. G. Parravano, *J. Am. Chem. Soc.* **74**, 1194 (1952).
112. M. T. Simnad, R. Smoluchowski, and A. Spilners, *J. Appl. Physics* **29**, 1930 (1958).
113. Y. Iida, K. Shimada, and S. Ozaki, *Bull. Chem. Soc. Jap.* **33**, 1372 (1960).
114. Y. Iida and K. Shimada, *Bull. Chem. Soc. Jap.* **33**, 8 (1960).
115. Y. Iida and K. Shimada, *Bull. Chem. Soc. Jap.* **33**, 790 (1960).
116. Y. Iida and K. Shimada, *Bull. Chem. Soc. Jap.* **33**, 1194 (1960).
117. J. Bandrowski, C. R. Bickling, K. H. Yang, and O. A. Hougen, *Chem. Eng. Sci.* **17**, 379 (1962).
118. T. Yamashina and T. Nagamatsuya, *J. Phys. Chem.* **70**, 3572 (1966).
119. E. J. Bicek and C. J. Kelly, *Am. Chem. Soc.*, Chicago Meeting, 57, (1967).
120. G. S. Levinson, *Am. Chem. Soc.*, Chicago Meeting, 47 (1967).
121. H. Charosset, R. Frety, Y. Trambouze, and M. Prettre, *Proc. Int. Symp. React. Solids*, 6th, 1968, J. W. Mitchell, ed., Wiley-Interscience, New York, (1969).
122. J. Deren and J. Stoch, *J. Catal.* **18**, 249 (1970).
123. A. Roman and B. Delmon, *Compt. Rend. Ser. B* **269**, 801 (1969).
124. A. Roman and B. Delmon, *Compt. Rend. Ser. B* **271**, 77 (1970).
125. H. Mine, M. Tokuda, and M. Ohtani, *J. Jap. Inst. Metals* **34**, 814 (1970).
126. T. Kurosawa, R. Hasegawa, and T. Yagihashi, *J. Jap. Inst. Metals* **34**, 481 (1970).
127. F. Chiesa and M. Rigaud, *Can. J. Chem. Eng.* **49**, 617 (1971).
128. J. Szekely and J. W. Evans, *Met. Trans.* **2**, 1699 (1971).
129. J. Szekely, C. I. Lin, and H. Y. Sohn, *Chem. Eng. Sci.* **28**, 1975 (1973).
130. H. Forestier and G. Nury, *Proc. Int. Symp. React. Solids*, p. 189 (1952).
131. J. Szekely and J. W. Evans, *Chem. Eng. Sci.* **26**, 1901 (1971).
132. J. W. Evans, S. Song, and C. E. Leon-Sucre, *Met. Trans.* **7B**, 55 (1976).
133. V. N. Babushkin, A. I. Tikhonov, and V. I. Smirnov, *Tsvet. Metal.* (*Engl. Transl.*), **12**, 13 (1971).
134. J. Szekely and A. Hastaoglu, *Trans. Inst. Min. Met.* **85**, C78 (1976).
135. J. H. Krasuk and J. M. Smith, *AIChE J.* **18**, 506 (1972).

136. J. Szekely and C. I. Lin, *Met. Trans.* **7B**, 493 (1976).
137. A. Bielanski, R. Dziembaj, and H. Urbanska, *Bull. Acad. Pol. Sci. Ser. Sci. Chim.* **19**, 447 (1971).
138. W. A. Oates and D. D. Todd, *J. Aust. Inst. Met.* **7**, 109 (1962).
139. A. S. Tumarev, L. A. Panyushin, and V. A. Pushkarev, *Izv. Vysshikh. Uchebn. Zavedenii. Tsvetn. Met.* **8**, 39 (1965); *Chem. Abstr.* **63**, 9527C (1965).
140. A. R. Chesti and S. C. Sircar, *Indian J. Tech.* **9**, 339 (1971).
141. A. K. Ashin, S. T. Rostovtsev, and O. L. Kostelov, *Russian Metallurgy* (*Engl. Transl. S.I.C. London*) **1**, 18 (1971).
142. R. V. Culver, I. G. Matthew, and E. C. R. Spooner, *Aust. J. Chem.* **15**, 40 (1962).
143. G. Haertling and R. L. Cook, *J. Am. Ceram. Soc.* **48**, 35 (1965).
144. H. A. Jones and H. S. Taylor, *J. Phys. Chem.* **27**, 623 (1923).
145. R. N. Pease and H. S. Taylor, *J. Am. Chem. Soc.* **43**, 2179 (1921).
146. N. J. Themelis and J. C. Yannopoulos, *Trans. Met. Soc. AIME* **236**, 414 (1966).
147. C. Vassilev, T. Nikolov, and M. Chimbulev, *Trans. Inst. Min. Met.* **77**, C36 (1968).
148. D. T. Hawkins and W. L. Worrell, *Met. Trans.* **1**, 271 (1970).
149. V. A. Lavrenko, V. S. Zenkov, V. L. Tikush, and I. V. Uvarova, *Russian Metallurgy* (*Engl. Transl. S.I.C. London*) **4**, 7 (1975).
150. C. Decroly and R. Winand, *Trans. Inst. Min. Met.* **77**, C134 (1968)
151. I. J. Bear and R. J. T. Caney, *Trans. Inst. Min. Met.* **85**, C139 (1976).
152. R. Rawlings and U. J. Ibok, *Trans. Inst. Min. Met.* **83**, C186 (1974).
153. J. O. Hougen, R. R. Reeves, and G. G. Manella, *Ind. Eng. Chem.* **48**, 318 (1956).
154. C. E. Guger and F. S. Manning, *Met. Trans.* **2**, 3083 (1971).
155. B. G. Baldwin, *J. Iron Steel Inst.* **179**, 30 (1955).
156. K. Otsuka and D. Kunii, *J. Chem. Eng. Jap.* **2**, 46 (1969).
157. R. A. Sharma, P. P. Bhatnagar, and T. J. Banerjee, *Sci. Ind. Res.* **16A**, 225 (1957).
158. H. K. Kohl and B. Marinček, *Arch. Eisenhüttenw.* **36**, 851 (1965).
159. H. K. Kohl and B. Marinček, *Arch. Eisenhüttenw.* **38**, 493 (1967).
160. Y. K. Rao, *Met. Trans.* **2**, 1439 (1971).
161. Y. Maru, Y. Kuramasu, Y. Awakura, and Y. Kondo, *Met. Trans.* **4**, 2591 (1973).
162. A. D. Kulkarni and W. L. Worrell, *Met. Trans.* **3**, 2363 (1972).
163. M. I. El-Guidny and W. G. Davenport, *Met. Trans.* **1**, 1729 (1970).
164. H. Y. Sohn and J. Szekely, *Chem. Eng. Sci.* **28**, 1783 (1973).
165. Y. K. Rao, *Chem. Eng. Sci.* **29**, 1435 (1974).
166. Y. K. Rao and Y. K. Chuang, *Chem. Eng. Sci.* **29**, 1933 (1974).
167. Y. K. Rao and Y. K. Chuang, *Met. Trans.* **7B**, 495 (1976).
168. T. E. Dancy, Howe Memorial Lecture presented at 106th AIME Annual Meeting, 1977, Atlanta, Ga.
169. R. Wild, *Chem. Proc. Eng.* **50**, 55 (1969).
170. R. Lawrence, Jr., in *Alternative Route to Steel*, Proceedings of the Annual Meeting of the British Iron and Steel Institute, (1971), p. 43.
171. C. W. Sanzenbacher and D. C. Meissner, *Can. Min. Metall. Bull.* **69**, 120 (1976).
172. P. P. Borthayre, paper presented at Latin American Iron and Steel Congress, 1974, Bogota, Columbia.
173. H.-D. Pantke, paper presented at the Latin American Direct Reduction Seminar, 1973, Mexico City, Mexico.
174. W. L. Davis, Jr., J. Feinman, and J. H. Gross, paper presented at the Latin American Direct Reduction Seminar, 1973, Mexico City, Mexico.
175. H. A. Kulberg, paper presented at the 135th Ironmaking Conference, AIME, 1976, St. Louis.
176. K. Meyer, G. Heitmann, and W. Janke, paper presented at the 24th Ironmaking Conference, AIME, 1965, Pittsburgh.

177. G. Meyer and R. Wetzel, paper presented at 30th Ironmaking Conference, AIME, 1971, Pittsburgh.
178. V. J. Azbe, *Rock Products* **47**(9), 68 (1944).
179. H. R. S. Britton, S. J. Gregg, and G. W. Winsor, *Trans. Faraday Soc.* **48**, 63 (1952).
180. E. P. Hyatt, I. B. Cutler, and M. E. Wadsworth, *J. Am. Ceram. Soc.* **41**, 70 (1958).
181. G. Narsimhan, *Chem. Eng. Sci.* **16**, 7 (1961).
182. T. R. Ingraham and P. Marier, *Can. J. Chem. Eng.* **41**, 170 (1963).
183. A. W. D. Hills, *Chem. Eng. Sci.* **23**, 297 (1968).
184. E. T. Turkdogan, R. G. Olson, H. A. Wriedt, and L. S. Darken, *Trans. SME, AIME* **254**, 9 (1973).
185. Z. Asaki, Y. Fukunaka, T. Nagase, and Y. Kondo, *Met. Trans.* **5**, 381 (1974).
186. C. N. Satterfield and F. Feakes, *AIChE J.* **5**, 115 (1959).
187. E. H. Baker, *J. Chem. Soc.* p. 464 (1962).
188. A. W. D. Hills, *Trans. Inst. Mining Met.* **76**, 241 (1967).
189. F. Thummler and W. Thomma, *Met. Rev.* **12** (115), 69 (1967).
190. J. Frenkel, *J. Phys.* (*USSR*) **9**, 385 (1945).
190a. R. L. Coble, *J. Am. Ceram. Soc.* **41**, 55 (1958).
190b. D. L. Johnson, *J. Appl. Phys.* **40**, 192 (1969).
190c. W. D. Kingery, *J. Appl. Phys.* **30**, 301 (1959).
191. I. B. Cutler, *High Temperature Oxides*, Vol. 3, A. M. Alper, ed., Academic Press, New York (1970).
192. R. L. Coble and J. E. Burke, *Prog. Ceram. Sci.* **3**, 197–253 (1963).
193. S. Chapman and T. G. Cowling, *The Mathematical Theory of Nonuniform Gases*, Cambridge University Press, Cambridge (1939).
194. J. O. Hirschfelder, C. F. Curtiss, and R. B. Bird, *Molecular Theory of Gases and Liquids*, John Wiley and Sons, New York (1954).
195. P. Grieveson and E. T. Turkdogan, *J. Phys. Chem.* **68**, 1547–1551 (1964).
196. J. P. Stark, *Acta Met.* **14**, 228–229 (1966).
197. L. S. Darken, *Trans. AIME* **175**, 184–194 (1948).
198. R. B. Bird, W. F. Stewart, and E. N. Lightfoot, *Transport Phenomena*, John Wiley and Sons, New York (1960).
199. H. Towers and J. Chipman, *Trans. AIME* **209**, 769–773 (1957).
200. T. B. King and P. Koros, *Trans. AIME* **224**, 299–306 (1962).
201. T. Saito and Y. Kawai, Sci. Rept. Tohuku University, Series A5, (1953), pp. 460–468.
202. P. Grieveson and E. T. Turkdogan, *Trans. Met. Soc. AIME* **230**, 1609–1614 (1964).
203. K. Mori and K. Suzuki, *Trans. Iron Steel Inst. Japan* **9**, 409–412 (1969).
204. D. P. Agarwal and D. R. Gaskell, *Met. Trans.* **6B**, 263–267 (1975).
205. L. Yang, S. Kado, and G. Derge, *Physical Chemistry of Process Metallurgy*, G. R. St. Pierre, ed., Interscience Publishers, New York (1961), pp. 535–541.
206. S. W. Strauss, *Nucl. Sci. Eng.* **8**, 362–363 (1960).
207. P. Kozakevitch, *Surface Phenomena of Metals*, Monog. 28, Society of Chemical Industry, London (1968), pp. 223–245.
208. K. Mohma and H. Suto, *Trans. Japan Inst. Metals* **2**, 148–153 (1961).
209. J. H. Swisher and E. T. Turkdogan, *Trans. Met. Soc. AIME* **239**, 602–610 (1967).
210. G. R. Belton, *Met. Trans.* **7B**, 35–42 (1976).
211. B. M. W. Trapnell, *Chemisorption*, Butterworths, London (1955).
212. T. B. King, *Physical Chemistry of Melts*, Inst. Min. Met., London (1953), pp. 35–41.
213. A. Adachi, K. Ogino, and T. Suetaki, *Tech. Rept. Osaka Univ.* **14**, 713–719 (1964).
214. P. Kozakevitch, in *Liquids: Structure, Properties, Solid Interactions*, T. J. Hughel, ed., Elsevier Publishing Co., Amsterdam (1965), pp. 243–280.
215. G. Cavalier, *Proc. Natl. Phys. Lab.* **2**, 4D (1958).
216. E. T. Turkdogan and P. M. Bills, *Am. Ceram. Soc. Bull.* **39**, 682–687 (1960).

217. P. Kozakevitch, in *Physical Chemistry of Process Metallurgy*, G. R. St. Pierre, ed., Interscience Publishers, New York (1961), pp. 97–116.
218. P. Kozakevitch, *Rev. Met.* **47**, 201–210 (1950).
219. P. Kozakevitch, *Rev. Met.* **51**, 569–587 (1954).
220. C. F. Cooper and J. A. Kitchener, *J. Iron Steel Inst.* **193**, 48–55 (1959).
221. C. F. Cooper and C. L. McCabe, Res. Rept. Metals Research Laboratory, Carnegie Inst. Tech., Pittsburgh, Pa. (Sept. 1959).
222. J. H. Swisher and C. L. McCabe, *Trans. Met. Soc. AIME* **230**, 1669–1675 (1964).
223. H. Eyring, *J. Chem. Phys.* **3**, 107–115 (1935).
224. S. Glasstone, K. J. Laidler, and H. Eyring, *The Theory of Rate Processes*, McGraw-Hill, New York (1941).
225. L. S. Darken and E. T. Turkdogan, in *Heterogeneous Kinetics at Elevated Temperatures*, G. R. Belton and W. L. Worrell, eds., Plenum, New York (1970), pp. 25–95.
226. P. Kozakevitch and G. Urbain, *Rev. Met.* **60**, 143–156 (1963).
227. R. D. Pehlke and J. F. Elliott, *Trans. Met. Soc. AIME* **227**, 844–855 (1963).
228. H. Schenck, M. G. Frohberg, and H. Heinemann, *Arch. Eisenhüttenw.* **33**, 593–600 (1962).
229. E. T. Turkdogan and P. Grieveson, *J. Electrochem. Soc.* **114**, 59–64 (1967).
230. T. Fuwa, S. Ban-ya, and T. Shinohara, *Tetsu to Hagane*, **54**, S436 (1968).
231. K. Mori and K. Suzuki, *Trans. Iron Steel Inst. Japan* **10**, 232–238 (1970).
232. K. Narita, S. Koyama, T. Makino, and M. Okamura, *Trans. Iron Steel Inst. Japan* **12**, 444–453 (1972).
233. M. Inouye and T. Choh, *Trans. Iron Steel Inst. Japan* **8**, 134–145 (1968); **12**, 189–196 (1972).
234. R. G. Mowers and R. D. Pehlke, *Met. Trans.* **1**, 51–56 (1970).
235. E. T. Turkdogan and J. V. Vinters, *Met. Trans.* **3**, 1561–1574 (1972).
236. L. A. Baker, N. A. Warner, and A. E. Jenkins, *Trans. Met. Soc. AIME* **239**, 857–864 (1967).
237. R. J. Fruehan and L. J. Martonik, *Met. Trans.* **5**, 1027–1032 (1974).
238. D. R. Sain and G. R. Belton, *Met. Trans.* **7B**, 235–244 (1976).
239. A. Forster and F. D. Richardson, *Trans. Inst. Min. Metall.* **84**, C116–122 (1975).
240. E. T. Turkdogan, P. Grieveson, and J. F. Beisler, *Trans. Met. Soc. AIME* **227**, 1265–1274 (1963).
241. T. B. King and S. Ramachandran, in *Physical Chemistry of Steelmaking*, J. F. Elliott, ed., John Wiley and Sons, New York (1958), pp. 125–133.
242. A. Nilas and M. G. Frohberg, *Arch. Eisenhüttenw.* **41**, 951–956 (1970).
243. K. M. Goldman, G. Derge, and W. O. Philbrook, *Trans. AIME* **200**, 534–540 (1954).
244. E. T. Turkdogan and M. L. Pearce, *Trans. Met. Soc. AIME* **227**, 940–949 (1963).
245. F. D. Richardson, *Physical Chemistry of Melts in Metallurgy*, Academic Press, New York (1974).
246. E. T. Turkdogan and P. Grieveson, *Trans. Met. Soc. AIME* **224**, 316–323 (1962).
247. R. G. Ward, *J. Iron Steel Inst. (London)* **201**, 11–15 (1963).
248. E. S. Machlin, *Trans. Met. Soc. AIME* **218**, 314–326 (1960).
249. E. D. Tarapore and J. W. Evans, *Met. Trans.* **7B**, 343–351 (1976).
250. R. G. Ward and T. D. Aurini, *J. Iron Steel Inst. (London)* **204**, 920–923 (1966).
251. R. J. Fruehan and E. T. Turkdogan, *Met. Trans.* **2**, 895–902 (1971).
252. G. R. Belton, R. J., Fruehan, and E. T. Turkdogan, *Met. Trans.* **3**, 596–598 (1972).
253. D. C. Boyd, W. C. Phelps, and M. T. Hepworth, *Met. Trans.* **6B**, 87–93 (1975).
254. E. T. Turkdogan, P. Grieveson, and L. S. Darken, *J. Phys. Chem.* **67**, 1647–1654 (1963).
255. P. A. Distin and S. G. Whiteway, *Can. Met. Q.* **9**, 419–426 (1970).

256. G. J. W. Kor and E. T. Turkdogan, *Met. Trans.* **6B**, 411–418 (1975).
257. E. T. Turkdogan, *Trans. Met. Soc. AIME* **230**, 740–750 (1964).
258. D. E. Rosner, *Inst. J. Heat Mass Transfer* **10**, 1267–1279 (1967); D. E. Rosner and M. Epstein, *Inst. J. Heat Mass Transfer* **13**, 1393–1414 (1970).
259. A. W. D. Hills and J. Szekely, *J. Chem. Eng. Sci.* **19**, 79–81 (1964).
260. E. T. Turkdogan and K. C. Mills, *Trans. Met. Soc. AIME* **230**, 750–753 (1964).
261. L. S. Nelson and H. W. Levine, in *Heterogeneous Kinetics at Elevated Temperatures*, G. R. Belton and W. L. Worrell, eds., Plenum, New York (1970), pp. 503–517.
262. D. G. C. Robertson and B. B. Staples, in *Process Engineering of Pyrometallurgy*, M. J. Jones, ed., Inst. Min. Metall., London (1974), pp. 51–59.
263. W. F. Porter, F. D. Richardson, and K. N. Subramanian, in *Heat and Mass Transfer in Process Metallurgy*, A. W. D. Hills, ed., Inst. Min. Metall., London (1967), pp. 79–114.
264. K. N. Subramanian and F. D. Richardson, *J. Iron Steel Inst. (London)* **206**, 576–583 (1968).
265. J. K. Brimacombe and F. D. Richardson, *Trans. Inst. Min. Metall.* **80**, C140–151 (1971); **82**, C63–72 (1973).
266. J. T. Davies, *Turbulence Phenomena*, Academic Press, New york (1972).
267. W. O. Philbrook and L. D. Kirkbride, *Trans. AIME* **206**, 351–356 (1956).
268. T. E. Brower and B. M. Larsen, *Trans. AIME* **172**, 137, 164 (1947).
269. E. T. Turkdogan, *Chem. Eng. Sci.* **21**, 1133–1144 (1966).
270. D. Poggi, R. Minto, and W. G. Davenport, *J. Metals, AIME* **21**, 40–45 (1969).
271. P. Kozakevitch and P. Leroy, *Rev. Met.* **51**, 203–209 (1954).
272. P. Kozakevitch *et al.*, *Congrés International sur les Aciers a' L'Oxygéne*, Le Touqet: Inst. Res. Sidérurgie, 1963, pp. 248–263.
273. P. Kozakevitch, *Open Hearth Proc. AIME* **52**, 64–75 (1969).
274. H. W. Meyer, W. F. Porter, G. C. Smith, and J. Szekely, *J. Metals, AIME* **20**, 35–42 (1968).
275. B. Trentini, *Trans. Met. Soc. AIME* **242**, 2377–2388 (1968).
276. D. J. Price, in *Process Engineering of Pyrometallurgy*, M. J. Jones, ed., Inst. Min. Metall., (1974), pp. 8–15.
277. S. Okano *et al.*, Intern. Conf. Science and Technology of Iron and Steel, Tokyo, 1970, Part I, pp. 227–231.
278. K. N. Subramanian and N. J. Themelis, *J. Metals, AIME* **24**, 33–38 (1972).
279. R. M. Davies and G. I. Taylor, *Proc. Royal Soc.* **A200**, 375–390 (1950).
280. M. H. I. Baird and J. F. Davidson, *Chem. Eng. Sci.* **17**, 87–93 (1962).
281. P. H. Calderbank, *Chem. Eng.* pp. CE209–233 (1967).
282. F. Yoshida and K. Akita, *AIChE J.* **11**, 9–13 (1965).
283. I. Leibson, E. G. Holcomb, A. G. Cacoso, and J. J. Jacmic, *AIChE J.* **2**, 296–306 (1956).
284. A. C. Lochiel and P. H. Calderbank, *Chem. Eng. Sci.* **19**, 485–503 (1964).
285. R. I. Guthrie and A. V. Bradshaw, *Trans. Met. Soc. AIME* **245**, 2285–2292 (1969).
286. V. G. Levich, *Physicochemical Hydrodynamics*, Prentice-Hall, Englewood Cliffs, N.J. (1962).
287. S. Hu and R. C. Kintner, *AIChE J.* **1**, 42–48 (1955).
288. F. D. Richardson, *Iron Coal* **183**, 1105–1116 (1961).
289. R. J. Fruehan, *Ironmaking Steelmaking* **3**, 33–36 (1976).
290. E. T. Turkdogan, *Trans. Inst. Min. Met.* **83**, C67–81 (1974); *Open Hearth Proc. AIME* **58**, 405–425 (1975).
291. J. C. Fulton and S. Ramachandran, *Electr. Furnace Steel Proc. AIME* **30**, 43–50 (1972).
292. R. J. Choulet, F. S. Death, and R. N. Dokken, *Can. Met. Q.* **10**, 129–136 (1971).
293. R. J. Fruehan, *Ironmaking Steelmaking* **3**, 153–158 (1976).

294. J. J. Byerley, G. L. Rempel, and N. Takebe, *Met. Trans.* **5**, 2501–2506 (1974).
295. U. Kuxmann and T. Z. Benecke, *Erzbergb. Metallhuett. Wes.* **19**, 215–221 (1966).
296. N. J. Themelis and P. R. Schmidt, *Trans. Met. Soc. AIME* **239**, 1313–1318 (1967).
297. A. V. Bradshaw, *Vide* **23**, 376–415 (1968).
298. P. A. Distin, G. D. Hallett, and F. D. Richardson, *J. Iron Steel Inst. (London)* **206**, 821–833 (1968).
299. A. V. Bradshaw and F. D. Richardson, *Vacuum Degassing of Steel*, Special Report 92, The Iron and Steel Institute, London (1965), pp. 24–44.
300. J. H. Flux, *Vacuum Degassing of Steel*, Special Report 92, The Iron and Steel Institute, London (1965), pp. 1–23.
301. R. G. Olsson and E. T. Turkdogan, *J. Iron Steel Inst. (London)* **211**, 1–8 (1973).
302. H. S. Philbrick, *Vacuum Metallurgy Conference 1963*, American Vacuum Soc., Boston (1964), p. 314.
303. E. T. Turkdogan, *J. Iron Steel Inst. (London)* **210**, 21–36 (1972).
304. M. Volmer and A. Weber, *Z. Phys. Chem.* **119**, 277–301 (1926).
305. R. Becker and W. Doring, *Ann. Phys.* **24**(5), 719–752 (1935),
306. E. R. Buckle, *Proc. Royal Soc. (London)*, **A261**, 189–196 (1961).
307. M. L. Turpin and J. F. Elliott, *J. Iron Steel Inst. (London)*, **204**, 217–225 (1966).
308. L. von Bogdandy *et al.*, *Arch. Eisenhüttenw.* **32**, 451–460 (1961).
309. G. K. Sigworth and J. F. Elliott, *Met. Trans.* **4**, 105–113 (1973); *Can. Met. Q.* **11**, 337–349 (1972).
310. G. Forward and J. F. Elliott, *Met. Trans.* **1**, 2889–2897 (1970).
311. E. T. Turkdogan, *J. Iron Steel Inst. (London)* **204**, 914–919 (1966).
312. R. K. Iyengar and W. O. Philbrook, *Met. Trans.* **3**, 1823–1830 (1972).
313. N. Lindskog and H. Sandberg, *Scand. J. Metall.* **2**, 71–78 (1973).
314. T. Emi, *Scand. J. Metall.* **4**, 1–8 (1975).
315. E. T. Turkdogan, in *Sulfide Inclusions in Steel*, J. J. de Barbadillo and E. Snape, eds., American Society of Metals, Metals Park, Ohio (1975), pp. 1–22.
316. C. Wagner, *Z. Elektrochem.* **65**, 581–591 (1961).
317. J. C. Billington and F. D. Richardson, *Trans. Inst. Min. Met.* **65**, 273–297 (1955–1956).
318. M. Nagamori, P. J. Mackey, and P. Tarassoff, *Met. Trans.* **6B**, 295–301 (1975).
319. E. Scheil, *Z. Metallk.* **34**, 70–72 (1942).
320. E. T. Turkdogan, *Trans. Met. Soc. AIME* **233**, 2100–2112 (1965).
321. E. T. Turkdogan and R. A. Grange, *J. Iron Steel Inst. (London)* **208**, 482–494 (1970).
322. D. Burns and J. Beech, *Ironmaking Steelmaking*, **1**, 239–250 (1974).
323. G. J. W. Kor and E. T. Turkdogan, *Met. Trans.* **3**, 1269–1278 (1972).
324. J. C. Yarwood, M. C. Flemings, and J. F. Elliott, *Met. Trans.* **2**, 2573–2581 (1971).
325. W. G. Pfann, *Zone Melting*, Wiley, New York (1966).
326. M. C. Flemings, *Solidification Processing*, McGraw-Hill, New York (1974).
327. M. Zief and W. R. Wilcox, *Fractional Solidification*, M. Dekker, New York (1967).
328. H. Schildknecht, *Zone Melting*, Academic Press, New York (1966).
329. A. Lawley, Zone refining, in *Techniques of Metals Research*, Vol. 1, Pt. 2, 1968.

5

Melting and Solidification in Metals Processing

Julian Szekely

5.1. Introduction

Melting and solidification phenomena processes play an important role in many metals processing operations. The melting of steel scrap in the basic oxygen furnace (BOF) the melting of copper anodes in an Asarco furnace, ingot solidification, and continuous casting of steel, copper, and aluminum may be quoted as examples.

Our purpose here is to describe the methods that are available for calculating the rates at which these melting and solidification processes take place. When examined from the viewpoint of transport theory, melting, solidification, and vapor deposition fall in the general category of "heat transfer with a change in phase." In order for such a phase change to take place in a one-component system the following conditions have to be satisfied:

1. The thermodynamic criteria which dictate the phase change must be met, i.e., the system has to be brought to the melting temperature, condensation temperature, etc.
2. The latent heat of fusion (volatilization, etc.) must be supplied or removed. This supply or removal of thermal energy may take place by convection, conduction, radiation, or by some combination of these mechanisms.

Julian Szekely • Department of Materials Science and Engineering, Massachusetts Institute of Technology, Cambridge, Massachusetts

These considerations have to be modified somewhat for multicomponent systems, where phase change does not occur at a given temperature, but rather over a temperature range which depends on composition.

It follows that in order to calculate the rate at which phase change takes place, information has to be available on the temperature (or temperature range) over which the phase change occurs; moreover, we have to calculate the rate at which the latent heat is being supplied or removed.

In melting and solidification processes there is a continuous movement of the phase boundary (e.g., growth of solid crystals from a melt, etc.) and for this reason those problems are often called *moving boundary problems.*

In the following we shall discuss how these phase change problems may be described quantitatively through a mathematical formulation, emphasizing, wherever possible, the physical basis of the mathematical statements. This discussion will then be followed by a description of the mathematical techniques available for the solution of the differential equations that result from the formulation.

Regarding organization of this chapter, Section 5.2 will be devoted to the formulation of melting and solidification problems involving one-component systems; the behavior of multicomponent systems will be discussed in Section 5.3, and in Section 5.4 we shall discuss the methods available for the solution of these moving boundary problems.

5.2. Melting and Solidification Problems Involving One-Component Systems

5.2.1. The Formulation of Solidification Problems

The consideration involved in the formulation of melting or solidification problems for elements (or pure compounds) are perhaps best illustrated through a simple example:

Let us consider a large container in which a molten substance is being held initially at a uniform temperature $T_{m,i}$. At time $= 0$ the temperature at the bottom of the container (corresponding to the $y = 0$ plane) is reduced to $T_{s,0}$ which is below the freezing point of the melt. As a result, for time larger than zero a solidified layer is formed, the thickness of which increases with time. A schematic representation of the system is shown in Figure 5.2-1 where the planar surface separating the solid and the molten regions is designated by the coordinate position

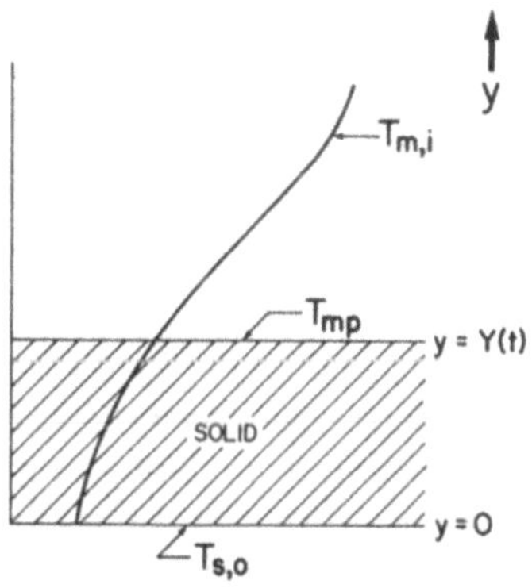

Figure 5.2-1. Sketch of a solidifying system.

$y = Y(t)$. The temperature profiles, also shown in the figure, clearly indicate that heat is being transferred from the melt through the solid to the cooled surface, which corresponds to the $y = 0$ plane.

If we assume, for the moment, that the domain occupied by the molten region is large and that the thermal properties are independent of temperature, then the mathematical statement of the problem is readily written down by establishing heat balances across infinitesimal volume elements in the solid and in the molten regions. Thus we have

$$\alpha_s \frac{\partial^2 T_s}{\partial y^2} = \frac{\partial T_s}{\partial t} \qquad \text{for } 0 \leq y \leq Y(t) \text{ (for the solid)} \qquad (5.2\text{-}1)$$

and

$$\alpha_m \frac{\partial^2 T_m}{\partial y^2} = \frac{\partial T_m}{\partial t} \qquad \text{for } Y(t) \leq y \leq \infty \text{ (for the melt)} \qquad (5.2\text{-}2)$$

where

α = thermal diffusivity, i.e., $\alpha = \dfrac{\text{thermal conductivity}}{(\text{specific heat}) \times \text{density}}$,

T = temperature,

t = time,

and the subscripts s and m refer to the solid and the molten phases, respectively.

We note here that equations (5.2-1) and (5.2-2) are the well-known Fourier's equations for unsteady-state heat conduction, written for the solid and the molten phases, respectively. Before the problem can be completely specified the boundary conditions have to be stated also. The system of equations (5.2-1)–(5.2-2) requires six boundary conditions.

Some of these boundary conditions are readily stated, viz.:

$$T_s = T_{s,0} \qquad \text{at } y = 0 \text{ for } t > 0 \tag{5.2-3}$$

$$T_m = T_{m,i} \qquad \text{for all } y, \text{ at } t = 0 \tag{5.2-4}$$

$$T_m \to T_{m,i} \qquad \text{as } y \to \infty \tag{5.2-5}$$

Here equation (5.2-3) expresses the constraint that the plane corresponding to $y = 0$ is maintained at a temperature $T_{s,0}$ for all times. The stipulation that all the system is molten initially at a uniform temperature $T_{m,i}$ is expressed in equation (5.2-4), whereas equation (5.2-5) is a bounding constraint, that when y is large the temperature in the melt retains its original value.

The remaining boundary conditions are somewhat more complicated because they refer to the melt–solid boundary, $y = Y(t)$, the position of which is time dependent. For a one-component system, the temperatures of the solid and the molten phases must equal at the melt–solid interface; thus we have

$$T_m = T_s = T_{mp} \qquad \text{at } y = Y(t) \qquad \begin{cases} (5.2\text{-}6) \\ (5.2\text{-}7) \end{cases}$$

The additional equation required is a relationship between the movement of the solidification front and the heat flux crossing the melt–solid boundary; this relationship may be developed through the establishment of a heat balance about the melt–solid interface. Thus we may write:

$$-\begin{bmatrix}\text{heat flux arriving}\\ \text{to the interface}\\ \text{from the melt}\end{bmatrix} + \begin{bmatrix}\text{heat flux leaving}\\ \text{the interface by}\\ \text{conduction through}\\ \text{the solid}\end{bmatrix} = \begin{bmatrix}\text{rate of heat}\\ \text{evolution due to}\\ \text{solidification}\end{bmatrix} \tag{5.2-8a}$$

The right-hand side of equation (5.2-8a) may also be expressed as

$$\begin{bmatrix}\text{rate of advancement of the}\\ \text{solidification boundary}\end{bmatrix} \times \begin{bmatrix}\text{latent heat of}\\ \text{solidification}\end{bmatrix}$$

If the melt is stagnant, then the heat flux arriving at the interface from the melt is due to conduction only; under these conditions equation (5.2-8a) may be written as

$$k_m \frac{\partial T_m}{\partial y} - k_s \frac{\partial T_s}{\partial y} = \rho_s \, \Delta H_s \frac{dY}{dt} \qquad \text{at } y = Y(t) \tag{5.2-8b}$$

where

ΔH_s = latent heat of solidification, and
ρ_s = density of the solid.

Finally, if we can express the physical fact that initially all the domain of interest was molten, that is,

$$Y(t)=0 \qquad \text{at } t=0 \tag{5.2-9}$$

then the system of equations (5.2-1)–(5.2-9) completes the statement of the problem.

It may be worthwhile to recapitulate what we have done here. We considered the solidification of a stagnant molten phase, of semiinfinite extent, which was initially at a uniform temperature. At time = 0, the surface corresponding to the $y=0$ plane was brought to a temperature below the freezing point of the medium. The problem was then formulated by writing the one-dimensional Fourier's equation for both the solid and the molten phases.

The *boundary conditions* employed were the customary expressions used in heat conduction problems, except for the fact that special boundary conditions were needed at the melt–solid interface. At this melt–solid interface, the position of which is time dependent, we had to specify that both the melt and the solid are at the freezing point; moreover, through the establishment of heat balance the rate of movement of the solidification front could be related to the net heat flow from the melt to the solid.

We note that in equation (5.2-1) the unsteady-state heat conduction equation remains the same for all one-dimensional solidification (or melting) problems, but the boundary conditions and the form of the heat transfer equation relating to the melt may differ, depending upon circumstances. Let us examine some specific cases.

5.2.1.1. Solidification from a Well-Stirred Melt of Infinite Extent

When solidification takes place from a well-stirred melt, the thermal mass of which is large, the temperature in the bulk of the melt remains constant and as sketched in Figure 5.2-2, and temperature gradients within the melt are confined to a narrow zone in the vicinity of the solidification front. The mathematical representation of such a system is readily accomplished if we consider that no conservation equation is required for the well-stirred melt (in view of its large heat capacity) and that the convective rate of heat transfer from the melt to the solid phase may be represented in terms of a heat transfer coefficient, h.

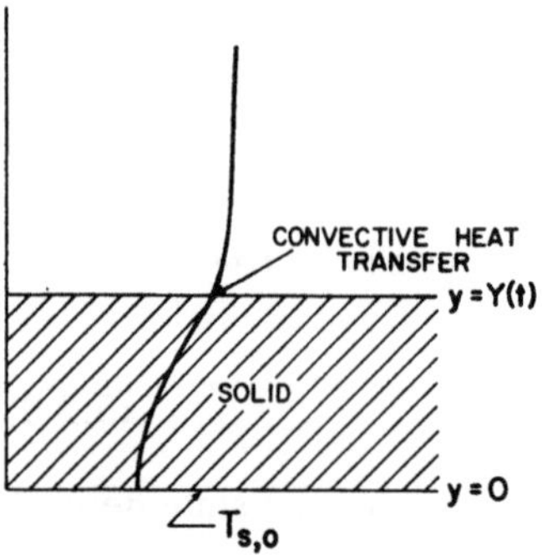

Figure 5.2-2. Sketch of the solidification front when there is convective heat transfer from the melt; clearly seen is the narrow region (loosely termed *stagnant film*) to which temperature gradients are confined in the melt.

Thus we have

$$\alpha_s \frac{\partial^2 T_s}{\partial y^2} = \frac{\partial T_s}{\partial t}, \qquad 0 \le y \le Y(t) \tag{5.2-1}$$

and the boundary conditions are written as

$$T_s = T_{s,0} \quad \text{at } y = 0 \quad \text{for } t > 0 \tag{5.2-3}$$

$$T_s = T_{mp} \quad \text{at } y = Y(t) \quad \text{for } t > 0 \tag{5.2-10}$$

and

$$h(T_{m,i} - T_{mp}) - k_s \frac{\partial T_s}{\partial y} = \rho_s \, \Delta H_s \frac{dY}{dt} \tag{5.2-11}$$

heat supplied by convection	heat removed by conduction	heat released due to solidification

We note that in equation (5.2-11) h is not a property value of the melt but rather depends on the properties of the system, such as the extent of agitation and the like.

Several further variations are possible on this general theme.

5.2.1.2. *Solidification from a Well-Stirred Melt of Finite Extent*

If the melt from which solidification takes place is of finite extent but still agitated, the sketch in Figure 5.2-2 has to be modified slightly to allow for the fact that the temperature in the bulk of the melt is a function of time in this specific case.

If the system is initially molten at a temperature $T_{m,i}$ and the depth of this melt is L, then the governing equations may be written as

$$\alpha_s \frac{\partial^2 T_s}{\partial y^2} = \frac{\partial T_s}{\partial t}, \qquad 0 \le y \le Y(t) \tag{5.2-1}$$

$$T_s = T_{s,0} \qquad \text{at } y = 0 \qquad \text{for } t > 0 \tag{5.2-3}$$

$$T_s = T_{mp} \qquad \text{at } y = Y(t) \tag{5.2-10}$$

$$h(T_m - T_{mp}) - k_s \frac{\partial T_s}{\partial y} = \rho_s \, \Delta H \frac{dY}{dt} \tag{5.2-11}$$

where T_m, the time-dependent melt temperature, is defined by

$$(L - Y)\rho_m C_p \frac{dT_m}{dt} = h(T_{mp} - T_m) \tag{5.2-12}$$

rate of heat loss from the agitated melt	convective heat transfer from the melt to the solidification crust

with

$$T_m = T_{m,i} \qquad \text{at } t = 0$$

5.2.1.3. The Convective Boundary Condition

In some cases it is unrealistic to assume that the temperature of the cooled surface is suddenly lowered to a specified value, as expressed by equation (5.2-3). Rather it may be more appropriate to postulate that this surface undergoes a convective heat exchange with a cooling medium; thus we have

$$-k \frac{\partial T_s}{\partial y} = h_c(T_s - T_c) \qquad \text{at } y = 0 \tag{5.2-13}$$

heat flux conducted to the outer surface of the solidified crust	heat flux owing to convective heat exchange with the surroundings

where

h_c = convective heat transfer coefficient between the outer surface of the solid layer, and
T_c = temperature of the cooling medium (assumed constant).

It is noted that $T_s \to T_c$ when h_c is large.

A particular case of the "convective boundary condition," which may frequently arise in practical systems, is that convective heat transfer from the cooled surface is augmented by thermal radiation. Under these conditions equation (5.2-12) has to be expressed in the following form:

$$-k\frac{\partial T'}{\partial y} = h_c(T'_s - T'_c) + \varepsilon\sigma(T'^4_s - T'^4_c) \tag{5.2-14}$$

where

ε = emissivity,
σ = Boltzmann's constant, and
T' = denotes absolute temperature.

Before proceeding further, some comment should be made regarding the physical appropriateness of the formulation developed in this section.

The postulate of a one-dimensional heat flow field considered in the examples is reasonable when the depth of the domain (melt or solid) is small compared to the other dimensions of the system, and at locations that are removed from the *other surfaces*. This point is illustrated in Figure 5.2-3 which shows the initial stages of the solidification of an ingot. It is seen that the assumption of one-dimensional heat flow would not be appropriate in the corners and at the wall in the vicinity of the top surface.

We note, furthermore, that in the vast majority of practical applications of solidification phenomena involving large-scale systems natural convection is likely to exist. Under these conditions, the assumption of a stagnant melt is unreasonable and the postulate of a "well-stirred" molten phase should be closer to the truth.

5.2.2. Solidification Problems in Spherical and in Cylindrical Coordinates

The principles discussed in the preceding section are readily used for representing solidification problems in spherical or cylindrical

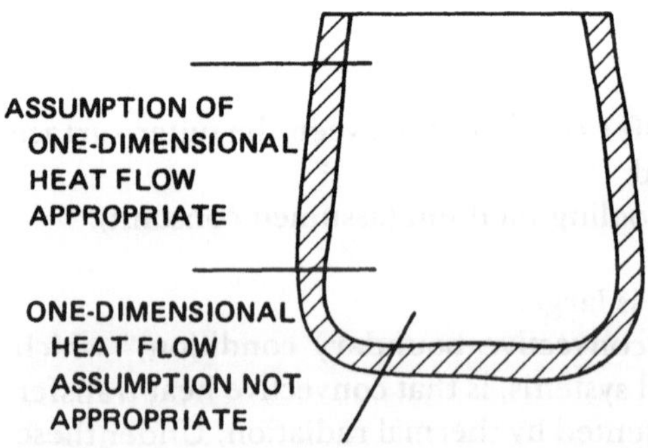

Figure 5.2-3. Sketch of the initial stages in the solidification of an ingot.

coordinates. Thus in the statement of these problems we have to establish a heat balance in each of the phases which leads to Fourier's equation; moreover, the boundary conditions at the phase boundary have to express the conservation of heat (i.e., the appropriate equivalent of equations (5.2-8) or (5.2-11) together with the stipulation that at the phase boundary the melt and solid temperatures must be the same.

To illustrate the procedure, let us consider the buildup of a solidified crust on a long, cylindrical rod of radius R_0, initially at a uniform temperature well below that of the freezing point of the melt, immersed in a well-stirred melt, the extent of which is large compared to the radius of the rod. If we assume that the material from which the rod is made is different from that of the melt and that these are mutually insoluble in each other, then we have the following:

Heat balance on the rod

$$\frac{\partial T_R}{\partial t}=\alpha_R\left(\frac{\partial^2 T_R}{\partial r^2}+\frac{1}{r}\frac{\partial T_R}{\partial r}\right), \qquad R_0 \le r \le 0 \qquad (5.2\text{-}15)$$

Heat balance on the solidified crust

$$\frac{\partial T_s}{\partial t}=\alpha_s\left[\frac{\partial^2 T_s}{\partial r^2}+\frac{1}{r}\frac{\partial T_s}{\partial r}\right], \qquad R(f) \le r \le 0 \qquad (5.2\text{-}16)$$

where subscripts R and s refer to the immersed rod and the solidified crust, respectively.

Since the melt itself is considered to be well stirred, no heat balance equation is required for the molten domain, but rather the heat flow from the melt to the rod has to appear in the boundary conditions.

The boundary conditions are written as follows:

Initial rod temperature specified

$$T_R = T_{R,0} \qquad \text{at } t=0 \qquad (5.2\text{-}17)$$

Heat balance at the rod–solidified crust interface

$$k_R\frac{\partial T_R}{\partial r}=k_s\frac{\partial T_s}{\partial r} \qquad \text{at } r=R_0 \qquad (5.2\text{-}18)$$

Symmetry

$$\frac{\partial T_R}{\partial r}=0 \qquad \text{at } r=0 \qquad (5.2\text{-}19)$$

Continuity of temperature at the rod–solidified shell interface

$$T_R = T_s \qquad \text{at } r = R_0 \tag{5.2-20}$$

Temperature of the solid must equal the melting point at the melt–solid interface

$$T_s = T_{mp} \qquad \text{at } r = R(t) \tag{5.2-21}$$

Heat balance at the melt–solid interface, the cylindrical equivalent of equation (5.2-11)

$$h_m(T_m - T_{mp}) - k_s \frac{\partial T_s}{\partial r} = \rho_s \, \Delta H_s \frac{dR}{dt} \tag{5.2-22}$$

And finally we have:

$$R(t) = R_0 \qquad \text{at } t = 0 \tag{5.2-23}$$

which completes the statement of the problem.

The equivalent expressions for solidification in spherical coordinates are readily written down by using a similar procedure. Thus if we consider the growth of a solidified shell on an initially cold sphere of radius $\tilde{R}_0$ immersed in a well-stirred melt, we have the following:

Heat balance on the sphere

$$\frac{\partial T_R}{\partial t} = \alpha_R \left(\frac{\partial^2 T_R}{\partial r^2} + \frac{2}{r} \frac{\partial T_R}{\partial r} \right) \qquad \tilde{R}_0 \leq r \leq 0 \tag{5.2-24}$$

Heat balance on the solidified shell

$$\frac{\partial T_s}{\partial t} = \alpha_s \frac{\partial^2 T_s}{\partial r^2} \frac{2}{r} \frac{\partial T_s}{\partial r} \qquad R(t) \leq r \leq \tilde{R}_0 \tag{5.2-25}$$

Initial temperature specified

$$T_R = T_{R,0} \qquad \text{at } t = 0 \tag{5.2-26}$$

Symmetry

$$\frac{\partial T_R}{\partial r} = 0 \qquad \text{at } r = 0 \tag{5.2-27}$$

Heat balance at the interface separating the sphere and the solidified shell

$$k_R \frac{\partial T_R}{\partial r} = k_s \frac{\partial T_s}{\partial r} \qquad \text{at } r = R_0 \tag{5.2-28}$$

Temperature of solid is at the melting point at the melt–shell interface

$$T_s = T_{mp} \qquad \text{at } r = R(t) \tag{5.2-29}$$

As before, the initial condition, specifying a zero initial thickness of the solidified shell is given as:

$$R(t) = 0 \qquad \text{at } t = 0 \tag{5.2-30}$$

Thus equations (5.2-24)–(5.2-30) represent the statement of the freezing problem in spherical coordinates.

The techniques available for the solution of the problems stated in Sections 5.2-1 and 5.2-2 will be discussed subsequently. However, some comments may be made purely on physical grounds, regarding the general behavior of the solidification profiles that one might expect.

In the examples described through equations (5.2-1)–(5.2-9) solidification took place from a quiescent melt, while the cooled surface was maintained at a constant temperature. Under these conditions one would expect a monotonous growth of the solidified crust with the actual growth rate decreasing with time, because heat has to be conducted through a progressively thicker solidified shell. Such a behavior is sketched in Figure 5.2-4a, which may be contrasted with that expected for solidification from a well-stirred melt of infinite (i.e., large) extent. Here we recall from equation (5.2-11) that the growth rate of the solidified crust is given by the following expression:

$$h(T_{m,i} - T_{mp}) - k_s \frac{\partial T_s}{\partial y} = \rho_s \, \Delta H_s \frac{dY}{dt} \tag{5.2-11}$$

Inspection of equation (5.2-11) shows that Y increases with time only, when the absolute value of the second term on the left-hand side is larger than the convective heat transfer term. Since the convective heat

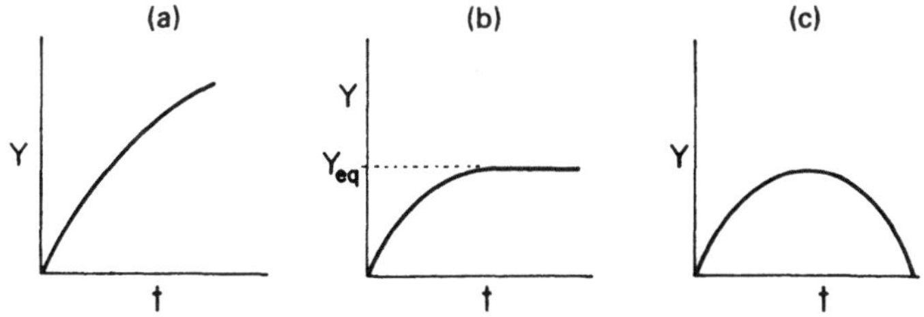

Figure 5.2-4. Sketch of the development of the solidification boundary for (a) a semi-infinite quiescent melt where the cold surface is maintained at a constant temperature; (b) a large, well-stirred melt where the cold surface is maintained at a constant temperature; and (c) a small cold object immersed in a large melt.

transfer term remains constant and the absolute value of the conductive term decreases with time (because heat has to be conducted through a progressively thicker solidified layer), eventually a stage is reached when the terms on the left-hand side of equation (5.2-11) exactly balance each other. Then no further growth occurs. The thickness corresponding to such a situation is Y_{eq} as sketched in Figure 5.2-4b.

We note here that if solidification were to take place from a well-stirred melt of *finite extent* such as discussed in Section 5.2.1.2, then we would expect a continuous growth of the solidified layer, because the $(T_m - T_{mp})$ term in equation (5.2-11) decreases with time, since both sensible and latent heat is being extracted from the system.

As sketched in Figure 5.2-4c, the growth of solidified shells on finite cylindrical or spherical bodies immersed in large, well-stirred melts is rather different. Under these conditions (provided the immersed body is sufficiently below the freezing temperature of the melt) a solidified crust is formed initially, but this crust *remelts* as the finite body is being heated up.

We must stress that the heat balance equations at the melt–solid boundary of the general form

$$\begin{bmatrix}\text{rate of heat transfer}\\ \text{(conductive or convective)}\\ \text{from the melt to the solid}\end{bmatrix} - \begin{bmatrix}\text{rate of heat conduction}\\ \text{into the solidified shell}\end{bmatrix}$$

$$= \left[\text{density} \times \begin{pmatrix}\text{latent heat of}\\ \text{solidification}\end{pmatrix} \times \begin{pmatrix}\text{rate of advancement of}\\ \text{the melt–solid interface}\end{pmatrix}\right] \tag{5.2-31}$$

predicate that the melt–solid interface may either advance (i.e., solidification takes place) or recede (i.e., melting takes place), depending on the relative magnitudes of the two terms appearing on the left-hand side of equation (5.2-31).

It follows that the expressions developed here for solidification problems apply equally well to melting.

5.2.3. Formulation of Melting Problems

As an illustration of the similarity between melting and solidification problems, let us consider the melting of a solid cylindrical specimen held in a cylindrical container, as sketched in Figure 5.2-5. Initially, the whole metal block is at a uniform temperature, $T_{s,i}$, but from time = 0 onward the upper surface, corresponding to the $y = L$ plane, receives radiative heat transfer from a source (e.g., flame) held at a temperature, T_E. Let us assume, moreover, for the sake of simplicity, that the heat flow

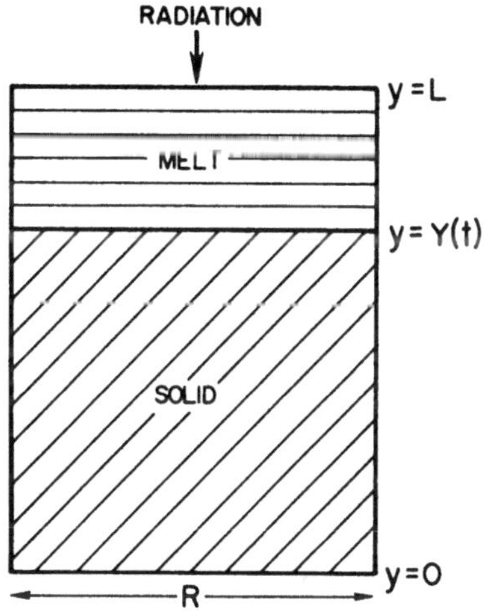

Figure 5.2-5. Sketch of the one-dimensional melting problem represented by equations (5.2-32)–(5.2-40).

problem is one dimensional and that bottom surface corresponding to the $y = 0$ plane is isolated.

Under these conditions, the problem is readily stated by writing the unsteady-state conduction equations for the solid and the molten phases. Thus we have†

$$\alpha_m \frac{\partial^2 T'_m}{\partial y^2} = \frac{\partial T'_m}{\partial t}, \qquad Y(t) \leq y \leq L \tag{5.2-32}$$

and

$$\alpha_s \frac{\partial^2 T'_s}{\partial y^2} = \frac{\partial T'_s}{\partial t}, \qquad 0 \leq y \leq Y(t) \tag{5.2-33}$$

The boundary conditions are written as

$$T'_s = T'_{s,i} \qquad \text{at } t = 0, \qquad 0 \leq y \leq L \tag{5.2-34}$$

$$-k_m \frac{\partial T'_m}{\partial y} = \varepsilon\sigma(T'^4_E - T'^4_m), \qquad y = L, \qquad t > 0 \tag{5.2-35}$$

$$T'_m = T'_s = T'_{mp}, \qquad y = Y(t) \tag{5.2-36–37}$$

$$k_m \frac{\partial T'_m}{\partial y} - k_s \frac{\partial T'_s}{\partial y} = \rho_m \, \Delta H_s \frac{dY}{dt}, \qquad y = Y \tag{5.2-38}$$

$$k_s \frac{\partial T'_s}{\partial y} \simeq 0, \qquad y = 0 \tag{5.2-39}$$

$$Y = L \qquad \text{at } t = 0 \tag{5.2-40}$$

† The T' denotes absolute temperatures.

Inspection of equations readily shows that these are very similar to the equations of a solidification problem in an identical geometry. These considerations are easily generalized to other geometries and will not be pursued further.

5.2.3.1. *Melting–Ablation*

The sketch in Figure 5.2-5 is not an entirely realistic representation of the majority of melting processes. Often when a solid body melts, the molten phase does not retain the original shape of the body (as in the case depicted in Figure 5.2-5) but rather the melt is allowed to drain from the system as shown in Figure 5.2-6. In metallurgical practice, problems of this type occur in the smelting of copper concentrates in a reverbatory furnace, in the melting of copper in the Asarco furnace, and in the melting of steel scrap or prereduced pellets in electric arc furnaces.

In the majority of cases the thermal resistance of the molten film is negligibly small so that the formulation of one-dimensional ablation problems is quite straightforward. Let us consider the melting ablation of a large slab, sketched in Figure 5.2-7; we shall assume that the solid is semi-infinite, extending from $y=0$ to $y=\infty$. Initially, the solid is at a uniform temperature T_0, which is below the melting point; for times larger than zero, the surface corresponding to the $y=0$ plane receives a heat flux q_y. Over an initial time period, say $0 \leq t \leq t_m$, this heat flow raises the temperature at the outer surface of the solid to the melting point. It follows that for times larger than t_m, melting of the solid takes place and since the molten phase is removed (ablates) the solid surface gradually recedes toward $y \to \infty$.

For such a one-dimensional system the problem is then readily formulated by writing the one-dimensional unsteady-state heat conduction equation

$$\alpha_s \frac{\partial^2 T_s}{\partial y^2} = \frac{\partial T_s}{\partial t} \tag{5.2-41}$$

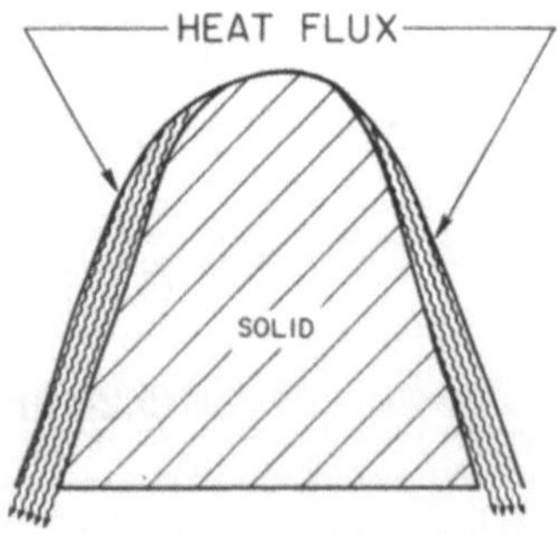

Figure 5.2-6. A draining molten film in the melting of a solid body.

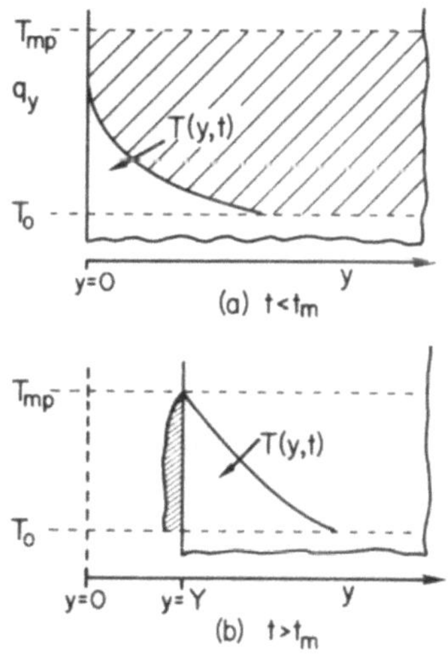

Figure 5.2-7. Sketch of the temperature distribution in an ablating slab during the (a) preheat period, and (b) actual ablation period.

For $0 \leq t \leq t_m$, i.e., the preheating stage, the boundary conditions are the following:

$$T_s = T_0 \qquad \text{at } t = 0 \tag{5.2-42}$$

$$q_y = k\frac{\partial T}{\partial y} \qquad \text{at } y = 0 \tag{5.2-43}$$

$$T \rightarrow T_0 \qquad \text{at } y \rightarrow \infty \tag{5.2-44}$$

For $t \geq t_m$, i.e., the melting–ablation period

$$T_s = T_{mp} \qquad \text{at } y = Y(t) \tag{5.2-45}$$

where, as seen in Figure 5.2-7, $Y(t)$ designates the (moving) position of the melting boundary.

$$q_y = k\frac{\partial T_s}{\partial y} = \rho \, \Delta H \frac{dY}{dt} \qquad \text{at } y = Y(t) \tag{5.2-46}$$

Inspection of equations (5.2-41)–(5.2-46) indicates that one-dimensional ablation problems tend to be simpler than the similar melting or solidification problems. This is due to the fact that in ablation it is assumed that the thermal resistance of the molten film is negligible so that we have to concern ourselves with the heat conduction equation in the solid phase only.

Another factor which affords simplification is that q_y, the heat flux falling on the melting surface, is likely to be constant during the melting period, provided the temperature of the heat source does not vary with time.

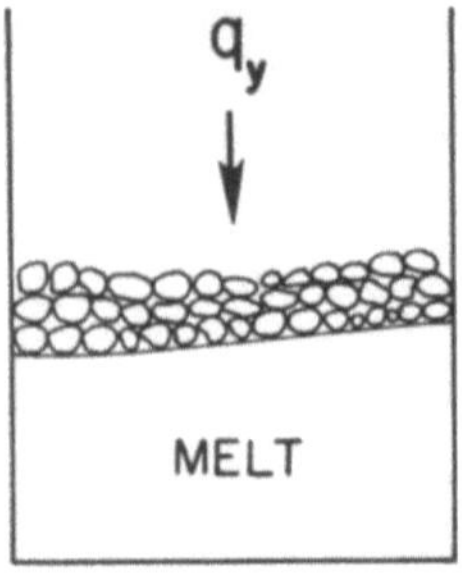

Figure 5.2-8. Sketch of the melting of a bed of granular solids receiving a heat flux from above, and in contact with its own melt.

If the slab were heated by convection, then

$$q_y = h_c(T_E - T_{mp}) \tag{5.2-17}$$

whereas if the slab were heated by a radiant source, we have

$$q_y = \varepsilon\sigma(T_E'^4 - T_{mp}'^4) \tag{5.2-18}$$

where T_E is the temperature of the heat source (e.g., the flame) and the prime in equation (5.2-18) indicates temperatures on the absolute scale.

We note here that two-dimensional ablation problems (e.g., the proper treatment of the system sketched in Figure 5.2-6) would be much more difficult to handle because of the continuous change of the shape of the ablation specimen.

The rigorous treatment of the melting–ablation problem sketched in Figure 5.2-8, depicting the melting of a granular bed of solids in contact with its melt, would also be quite difficult, because the percolation of the molten material into the as yet unmelted porous solid matrix would represent an additional heat transport mechanism which is difficult to assess, although problems of this type are of considerable importance in scrap melting and the melting of prereduced pellets. An approximate representation of such a melting process is possible, however, through the statement of overall heat balances.

5.3. Melting and Solidification of Multicomponent Systems

5.3.1. Introduction

In the melting and solidification of multicomponent systems, phase change occurs over a temperature (and composition) range within which molten and the solid phases may coexist. This behavior, which is well-known from thermodynamic studies, does introduce some conceptual difficulties although the mathematical handling of many multicomponent phase change problems need not be any more complex than the cases discussed in the preceding section.

This point is illustrated in Figure 5.3-1 which shows the well-known phase diagram of the iron–carbon system. The reader will recall from thermodynamic studies that if a melt containing, say, 1 wt % carbon and 99 wt % iron, initially at say, 1550° C, were gradually cooled, solidification would commence at about 1480° C, and the initial solid phase formed would contain much less carbon than 1 wt %, say, of the order of 0.3 wt %. As the system is cooled further, the fraction of solids present would increase until ultimately the whole melt is completely solidified. In contrast to the behavior of pure (one-component) systems the *phase change occurs over a temperature range* and ideally the modeling of the system would have to include an appropriate allowance for the composition and the temperature changes (together with the diffusional phenomena).

However, in many practical situations it is permissible to gloss over these complicating factors and make a simple allowance for the fact that in the solidification of multicomponent systems the latent heat of

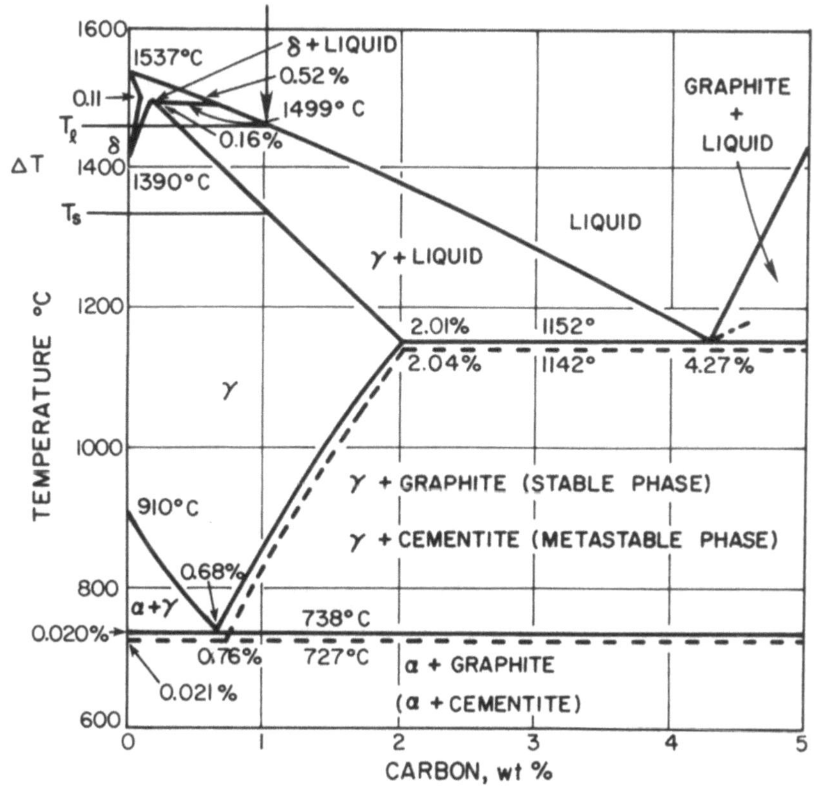

Figure 5.3-1. The iron–carbon phase diagram.

solidification is released over a temperature range rather than at a fixed melting point. Yet in other cases the exact nature of the composition dependence of the melting temperature has to be known accurately (e.g., scrap melting in the BOF). Alternatively, our specific objective may be to establish the composition of the two-phase region which separates the completely molten and the completely solidified regions (macrosegregation, continuous casting, etc.). Under these conditions the problem is indeed quite complex.

In the following we shall present illustrations of both these approaches.

5.3.2. Formulation of Solidification Problems through the Use of an Effective Specific Heat

Let us consider the solidification of a multicomponent melt where substantial fluid motion is absent and where our principal interest is to calculate the time-dependent temperature profiles in the solid phase and the time required for the complete solidification of the system.

For a one-dimensional problem Fourier's equation may be conveniently recast in terms of the enthalpy, H_T which now denotes the sum of the sensible and the latent heat of the material. Thus we have

$$\frac{\partial}{\partial y}\left(k\frac{\partial T}{\partial y}\right)=\rho\frac{\partial H_T}{\partial t} \tag{5.3-1}$$

Equation (5.3-1) applies to the whole domain, including the molten, the solid, and the two-phase regions. Allowance is made for a temperature-dependent thermal conductivity, moreover, by assigning an appropriate temperature dependence to H_T, the enthalpy; proper account may be made for the latent heat released on solidification.

Figure 5.3-2 shows a plot of H_T against the temperature. The steep portion of this plot for $T_s \leq T \leq T_m$ (i.e., within the melting range) reflects the latent heat of solidification. An alternative way of writing equation (5.3-1) is to assign an *effective specific* heat to the molten, two-phase, and solid regions. Thus we have

$$\frac{\partial}{\partial y}\left(k\frac{\partial T}{\partial y}\right)=\rho\left(\frac{dH_T}{dT}\right)\frac{\partial T}{\partial t} \tag{5.3-2}$$

where

$$\frac{dH_T}{dT}=Cp(T), \qquad \text{the effective specific heat}$$

The effective specific heat as a function of temperature is sketched in

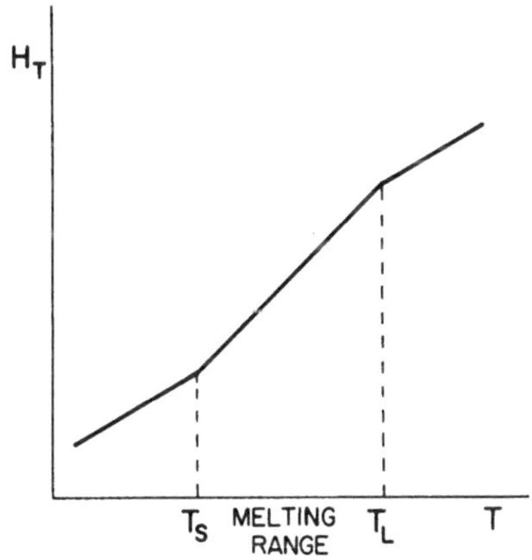

Figure 5.3-2. Sketch of the enthalpy H_T as a function of temperature; note the steep portion of this plot within the melting range.

Figure 5.3-3. Here the shaded area denotes the latent heat of solidification.

The formulation of solidification (or melting) problems through the use of equations (5.3-1) or (5.3-2) or their multidimensional equivalents is very attractive because rather than having to deal with a moving boundary problem, we only have to solve the unsteady-state conduction equation with a time-dependent specific heat. This is a much easier task and numerous readily programmed subroutines are available for generating a computer solution of such problems.

As will be shown subsequently, the solution thus obtained provides the time-dependent temperature profiles within the system from which the position of the melt line or the solidification front may be deduced. More specifically, we note that regions below the solidus temperature will be unambiguously solid, and furthermore, the regions that are above the liquidus temperature will be clearly molten. The two-phase or mushy region (i.e., where the temperature is between the solidus and the liquidus) is rather less well defined. As shown subsequently, a more precise definition of the two-phase region requires a rather more sophisticated approach.

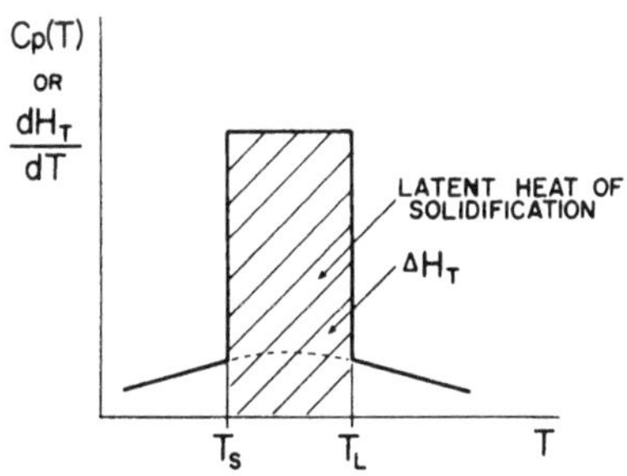

Figure 5.3-3. Sketch of $C_p(T)$ against the temperature; the shaded area denotes the latent heat of solidification.

5.3.3. The Modeling of Continuous Casting Systems

Figure 5.3-4 shows a schematic sketch of a continuous casting system. It is seen that molten metal is supplied at a controlled rate to a water-cooled mold where a solidified shell is formed at the mold–melt interface. The thickness of the solidified shell increases progressively with the vertical distance traveled and, upon exiting the mold, the shell is strong enough to contain the molten core.

A more detailed sketch of the solid–melt boundary is shown in Figure 5.3-4b where we can distinguish between the completely solidified shell and the two-phase or mushy zone. It is seen, furthermore, that part of the mushy zone adheres to the solidified shell, while the remainder is being swept away by the molten metal stream. This behavior, which has been proven experimentally, is quite noteworthy from the thermal and structural viewpoints. From the thermal viewpoint this means that in such cases part of the latent heat of solidification is released in the melt while the remainder is released in the mushy zone (or at the mushy zone–solid interface). The removal of part of the mushy zone has also important structural implications.

One way of formulating these problems is to combine the concepts developed in Section 5.2 with the concept of an effective specific heat discussed in the preceding section.

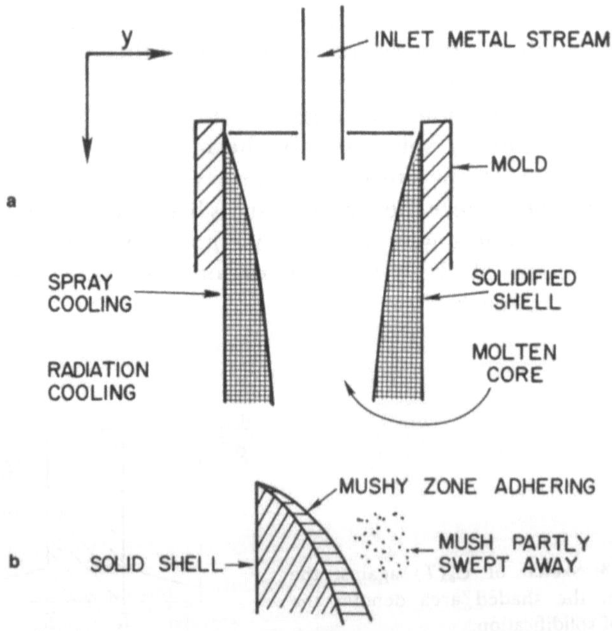

Figure 5.3-4. Schematic sketch of a continuous casting system.

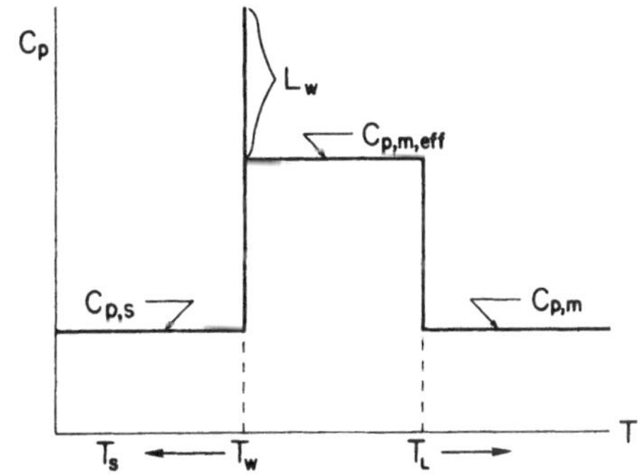

Figure 5.3-5. Sketch of $C_{p,m,eff}$ against temperature; also shown is L_W the latent heat of solidification given up at $T = T_w$.

Let us consider that when solidification takes place from an agitated melt, in general the surface of the solidified shell (which may include part of the mushy zone) in contact with the moving melt is at some temperature, T_w, which is an intermediate between the solidus, T_s, and the liquidus, T_L, temperatures. Thus

$$T_s \leq T_w \leq T_L$$

We shall postulate that from a thermal viewpoint, the numerical value of T_w (i.e., its position relative to T_s and T_L) designates the boundary separating the solidified shell and the moving melt. Let L_w designate the portion of the latent heat of solidification that is given up at this boundary and then an allowance may be made for the adsorption of the remainder of the latent heat of solidification by assigning an *effective specific heat* to the molten region within the temperature range, $T_w \leq T \leq T$.

This arrangement is shown schematically in Figure 5.3-5 and the relationship between $C_{p,m,e,ff}$ (effective specific heat), ΔH, and L_w is given by the following expression:

$$C_{p,m,e,ff} = C_{p,1} + \frac{\Delta H - L_w}{T_L - T_w} = C_{p,1} + \frac{\Delta H}{T_L - T_s} \tag{5.3-3}$$

We note, moreover, that L_w, ΔH, T_s, T_L, and T_w are further related through the phase diagram.

If we consider the simple case of a one-dimensional system, such as was sketched in Figure 5.3-4, and assume, for the sake of simplicity, that the system is perfectly mixed in the y direction, then for a casting velocity, U_c, the statement of the problem is readily given as follows:

Conservation of thermal energy in the solidified region, including the adhering mushy zone

$$\alpha_s \frac{\partial^2 T_s}{\partial y^2} = u_c \frac{\partial T_s}{\partial x}, \qquad 0 \leq y \leq Y(x) \tag{5.3-4}$$

Heat balance in the molten core

$$u_c C_{p,m,e,ff} \frac{\partial T_m}{\partial x} = h_c (T_m - T_w) \tag{5.3-5}$$

where h_c is the heat transfer coefficient between the molten core and the outer surface of the mushy zone adhering to the solid surface.

The rate of advancement of the solidification front may be expressed as

$$L_w \rho u_c \frac{dY}{dx} = k \left(\frac{\partial T_s}{\partial y} \right)_{y=Y} - h_c (T_m - T_w) \tag{5.3-6}$$

rate of liberation of latent heat — net rate of heat transfer from the melt–solid boundary

The remaining boundary conditions may be written as†

Inlet temperature of the melt specified

$$T_m = T_{m,i} \qquad \text{at } x = 0 \tag{5.3-7}$$

Heat transfer in the mold

$$k_s \frac{\partial T_s}{\partial y}\bigg|_{y=0} = h_m (T_s - T_{cw}), \qquad 0 \leq x \leq x_0$$

Heat transfer in the spray zone

$$k_s \frac{\partial T_s}{\partial y}\bigg|_{y=0} = h_{sp} (T_s - T_{sp}), \qquad x \leq x \leq x_1 \tag{5.3-8}$$

Heat transfer in the radiation zone

$$k_s \frac{\partial T_s}{\partial y}\bigg|_{y=0} = \varepsilon\sigma (T_s'^4 - T_E'^4), \qquad x > x_1$$

and finally,

$$\frac{\partial T_m}{\partial y} = 0 \qquad \text{at } y = \frac{w}{2} \tag{5.3-9}$$

† Here, T_{cw} and T_{sp} denote the cooling water and spray temperatures respectively, and h_m and h_{sp} are the mold and spray heat transfer coefficients.

The system of equations (5.3-4)–(5.3-9) may be solved numerically. We note here that numerous formulations of varying degrees of sophistication have been developed for representing heat transfer in continuous casting systems, in the event the growth of the solidified shell is more or less adequately described by the majority of these. The principal reason for this apparent agreement between measurements and predictions, based on a variety of models, is due to the fact that after a solidified shell of sufficient thickness has been formed, the conduction of heat through this shell represents the principal resistance to heat flow. This aspect of the problem is readily illustrated.

We note, however, that more recent work in the modeling of continuous casting systems has been concerned with the cleanliness and structure of the product cast. These factors are markedly influenced by both the movement in the molten core and by the fraction of the mushy zone that is swept away by the metal stream.

As noted in the introductory comments to Section 5.3, for multicomponent systems the latent heat of solidification is not released at a uniquely defined temperature corresponding to the melting point of a pure substance, but rather, phase change occurs over a range of temperatures as defined by the phase diagram. In the preceding sections we discussed the effect of these phenomena on solidification problems; let us now turn our attention to the melting process in a multicomponent system.

5.3.4. Scrap Melting in the Basic Oxygen Furnace

When steel scrap is being melted in the basic oxygen furnace, low carbon steel is being brought into contact with a melt the initial temperature of which is below that of the melting point of pure iron; however, the initial carbon content of the melt is high enough that the "effective melting point" of the solid metal is lowered sufficiently that melting can take place. This situation may be idealized somewhat, if we consider the melting of a slab extending from $y = -Y_0$ to $y = Y_0$ and which has an initially uniform temperature, $T_{s,i}$. Let this slab, consisting of pure iron, be immersed in an agitated melt of iron and carbon. Let us postulate, furthermore, that both the temperature and the carbon content of the melt are time dependent, say $T_B(t)$ and $C_B(t)$, as sketched in Figure 5.3-6.

The problem may be formally stated by writing the unsteady-state heat conduction equation for the slab, while representing the movement of the melt line in terms of a time-dependent effective melting point which has to be determined from the phase diagram.

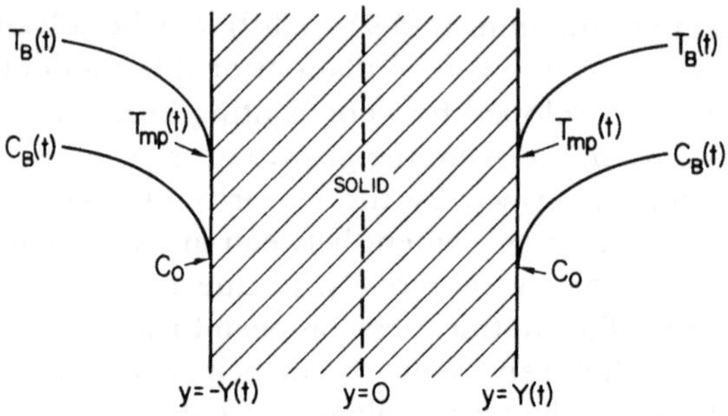

Figure 5.3-6. Sketch of the temperature and the concentration gradients for the scrap melting problem.

Upon allowing for symmetry about the plane $y = 0$, we have

Unsteady-state conduction equation within the solid slab of half thickness

$$\alpha \frac{\partial^2 T_s}{\partial y^2} = \frac{\partial T_s}{\partial t}, \qquad 0 \leq y \leq Y(t) \tag{5.3-10}$$

The boundary conditions are written as

Initial temperature specified

$$T_s = T_{s,i} \qquad \text{at } t = 0 \tag{5.3-11}$$

The effective melting temperature at the melt–solid boundary is time dependent—the exact relationship to be determined subsequently

$$T = T_{mp}(t) \qquad \text{at } y = Y(t) \tag{5.3-12}$$

and

$$k\frac{\partial T}{\partial y} = h[T_B(t) - T_{mp}(t)] + \rho\, \Delta H_s \frac{dY}{dt} \tag{5.3-13}$$

heat conduction into the slab	convective heat transfer from the melt to the slab	rate of advancement of the melt line

where h is the convective heat transfer coefficient between the melt and the slab.

The time-dependent bulk temperature of the melt $T_B(t)$ is assumed to be known and $T_{mp}(t)$ is calculated as follows:

We approximate the liquidus curve of the Fe–C phase diagram by the following expression:

$$T_{mp}(^{\circ}\text{C}) = 1530 - 0.83\times 10^{4} C_0, \qquad 0 \le C_0 \le 0.04 \qquad (5.3\text{-}14)$$

where C_0 is the weight fraction of carbon at $y = Y(t)$ and $T_{mp}(^{\circ}\text{C})$ is the liquidus temperature.

The quantity C_0 may now be evaluated by establishing a carbon balance at the solid–melt interface. Thus we have

$$C_0 = \frac{h_D[C_B(t) - C_0]}{(dY/dt) + h_D[C_B(t) - C_0]} \qquad (5.3\text{-}15)$$

where h_D is the mass transfer coefficient characterizing the transfer between the melt and the solid. The system of equations (5.3-10)–(5.3-15) may be solved numerically. The important characteristic of the solution is that the melting rate of a pure iron block, immersed in a molten iron–carbon alloy depends critically on the carbon content of the melt. It is stressed, furthermore, that such an iron block melts readily (because of the phase relationship) even when the temperature of the melt is substantially below the melting point of pure iron, provided the carbon content is high enough.

5.4. *Some Techniques of Solution and Computed Results for Melting and Solidification Problems*

5.4.1. *Introduction*

In the preceding section we discussed the formulation of melting and solidification problems for one-component and multicomponent systems. In a sense, this discussion was the most important part of the material presented here because the visualization of the key physical characteristics of a system and the translation of the resultant physical concepts into mathematical terms is the key step in any mathematical modeling procedure.

Once the equations are formulated their solution by analytical or numerical means may be undertaken, if needed, with the help of experts in these fields. In this section we shall present a very brief review of some of the techniques that are available for the solution of melting and solidification problems, together with some solutions in a graphical form which should be applicable to a broad range of practical problems.

5.4.2. Analytical Solutions of Melting and Solidification Problems

The complexity and the nonlinearity of moving boundary problems precludes the generation of closed-form analytical solutions, except for certain specialized circumstances.

Simple, closed-form, analytical solutions may be obtained for melting and solidification problems only when

1. the problem is one-dimensional spatially (i.e., heat flow in a slab),
2. the medium is infinite or semi-infinite,
3. the initial temperature distribution is uniform,
4. the thermal properties (specific heat, thermal conductivity, etc.) are independent of temperature, and
5. the phase boundary with the environment is maintained at a constant temperature throughout.

When all the above stipulations are met (at least approximately, as far as the physical reality is concerned) analytical solutions may be generated, as illustrated in the following.

Solidification of a Semi-infinite Melt, Initially at Its Melting Point

Let us consider a semi-infinite melt extending from $y=0$ to $y=\infty$ which is initially at its melting temperature, T_{mp}. At time $=0$, let the surface, corresponding to the $y=0$ plane, be brought to zero temperature (which is below the freezing point). As a consequence, as sketched in Figure 5.4-1, after some time a solid layer is formed, which moves progressively toward $y\to\infty$.

As discussed in Section 5.2, for a one-component system this problem is readily stated by writing the unsteady-state heat conduction equation, in the present case for the solid phase only, because the molten region was postulated to be at a uniform temperature (at the melting point) throughout.

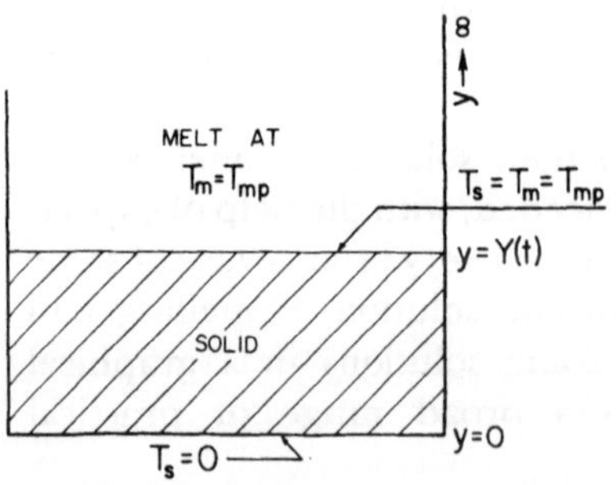

Figure 5.4-1. Sketch of the solidification of a semi-infinite melt, which is initially at its melting temperature.

Thus we have

$$\alpha\frac{\partial^2 T_s}{\partial y^2}=\frac{\partial T_s}{\partial t}, \qquad 0\leq y\leq Y(t) \tag{5.4-1}$$

and the boundary conditions are written as follows:

$$T_s=0 \qquad \text{at } y=0, \qquad t>0 \tag{5.4-2}$$

$$T=T_{mp}, \qquad y=Y(t) \tag{5.4-3}$$

and

$$-k_s\frac{\partial T_s}{\partial y}=\rho\,\Delta H_s\frac{\Delta Y}{dt} \tag{5.4-4}$$

with

$$Y(t)=0 \quad \text{at } t=0 \tag{5.4-5}$$

We seek the solution of the unsteady-state heat conduction problem in the solid region in the following form:

$$T_s=A+B\,\text{erf}\left[\frac{y}{2(\alpha_s t)^{1/2}}\right] \tag{5.4-6}$$

It is known that equation (5.4-5) is a solution of the unsteady-state heat conduction equation and the only remaining question is to select A and B such that the boundary conditions are satisfied.

The boundary condition contained in equation (5.4-2) requires that $A=0$; moreover, equation (4.5-3) can be satisfied only if $y/2(\alpha_s t)^{1/2}$ is a constant at $y=Y$.

In order to meet this stipulation let us assume that

$$Y=2\lambda\,(\alpha_s t)^{1/2} \tag{5.4-7}$$

where λ is an as yet unknown constant of proportionality. We note that this expression is consistent with the boundary condition contained in equation (5.4-5).

Then equation (5.4-6) may be written as

$$T_s=\frac{T_{mp}}{\text{erf}\,\lambda}\,\text{erf}\left[\frac{y}{2(\alpha_s t)^{1/2}}\right] \tag{5.4-8}$$

while λ may be evaluated by substituting from equations (5.4-7) and (5.4-8) to equation (5.4-4). Upon performing the operation indicated and some rearrangment, we have

$$\lambda e^{\lambda^2}\,\text{erf}\,\lambda=-\frac{C_p T_{mp}}{\Delta H \pi^{1/2}} \tag{5.4-9}$$

which may be solved for λ by trial and error.

This solution is readily generalized to situations where the surface temperature is maintained at a constant, nonzero value, say, $T_{s,0}$.

Upon applying a linear transformation, equations (5.4-8) and (5.4-9) may be written as

$$(T_s - T_{s,0}) = \frac{T_{mp} - T_{s,0}}{\operatorname{erf} \lambda} \operatorname{erf}\left[\frac{y}{2(\alpha_s t)^{1/2}}\right] \tag{5.4-10}$$

and

$$\lambda e^{\lambda^2} \operatorname{erf} \lambda = -\frac{C_p(T_{mp} - T_{s,0})}{\pi^{1/2} \Delta H_s} \tag{5.4-11}$$

Let us illustrate the use of this technique by working a simple example:

A water-cooled copper plate maintained at $T_{s,0} = 200°$ F is brought in contact with the surface of liquid steel at the melting temperature ($T_{mp} = 2700°$ F). Derive an expression for the rate of advancement of the solidification front.

$$\alpha_s = 0.44 \text{ ft}^2 \text{ hr}^{-1}$$

$$k_s = 20 \text{ Btu hr}^{-1} \text{ ft}^{-1} \text{ °F}^{-1}$$

$$C_{p,s} = 0.1 \text{ Btu lb}^{-1} \text{ °F}^{-1}$$

$$\Delta H = 110 \text{ Btu lb}^{-1}$$

Since the melt is at the freezing temperature we can use equation (5.4-11). Thus we have

$$\lambda e^{\lambda^2} \operatorname{erf} \lambda = -\frac{C_{p,s}(T_{mp} - T_{s,0})}{\pi^{1/2} \Delta H} \tag{5.4-11a}$$

Upon solving by trial and error for the above numerical values, we have

$$\lambda = 0.82$$

and the temperature distribution in the solid shell is given by

$$(T_s - 200) = 3320 \operatorname{erf} \frac{y}{2(0.44t)^{1/2}}$$

Once λ is known, the needed expression for the position of the solidification front is given by equation (5.4-7):

$$Y(t)[\text{ft}] = 1.09\bar{t}^{1/2}[\text{hr}]$$

Thus after one minute the thickness of the solidified layer is

$$Y = 1.09(1/60)^{1/2} \approx 0.14 \text{ ft} \approx 1.7 \text{ in.}$$

This technique may be used for the solution of more complex solidification problems, e.g., when a molten metal pool is in contact with a (cold) refractory surface, or for systems where temperature gradients exist in the molten and solid phases.

Numerous examples of these problems are discussed in the literature. We note that the constraints that have to be observed for generating analytical solutions for melting and solidification problems are unduly restrictive. Thus the technique cannot be used for solving problems when there exists *convection in the melt* or when the outer surface of the solid layer undergoes convective or radiative heat exchange with the surroundings.

These latter problems require the use of numerical methods.

5.4.3. Numerical Methods for the Solution of Melting and Solidification Problems

In recent years there has been an almost explosive growth in the literature devoted to numerical methods for the solution of partial differential equations and moving boundary problems. The detailed review of these techniques is way beyond the scope of this presentation where we shall confine our discussion to a particularly simple but useful technique and will make a brief mention of some computed results obtained through the use of alternative procedures.

5.4.3.1. *The Use of an Effective Specific Heat for Modeling Melting or Solidification Phenomena*

Let us consider a pure substance, extending, say from $y = 0$ to $y = L$, initially at a temperature, $T_{m,i}$, in the molten state as sketched in Figure 5.4-2. At time = 0, let the surface corresponding to the $y = 0$ plane start losing heat by convection to the environment which is maintained at a

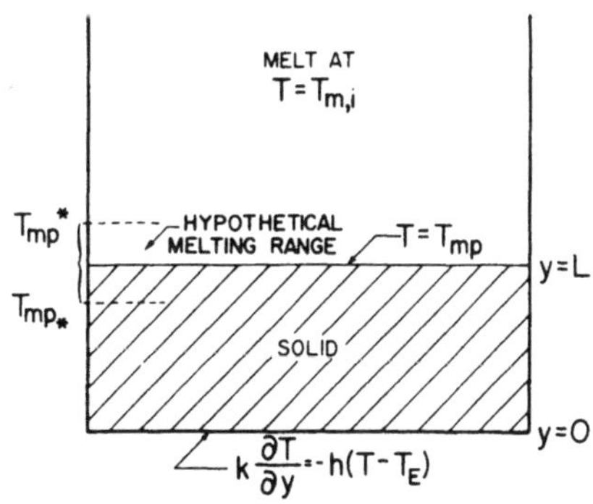

Figure 5.4-2. The illustration of the example given in Section 5.4.3.1.

temperature T_E, where $T_e < T_{mp}$. As a result a solidification front is formed which moves progressively toward the $y = L$ plane which we shall assume to be insulated.

This problem could be represented by writing the unsteady-state heat conduction equations for both the molten and the solid phases (neglecting convection of the melt for the moment).

As discussed in Section 5.2, we may, however, represent the problem by writing a single differential equation expressing the conservation of enthalpy which is valid over the whole domain. Thus we have

$$\frac{\partial}{\partial y}\left(k\frac{\partial T}{\partial y}\right) = \rho \frac{dH_T}{dT}\frac{\partial T}{\partial t}, \qquad 0 \leq y \leq L \tag{5.4-12}$$

Here dH_T/dT is the effective specific heat. Let us designate a melting range $T_{mp*} \leq T_{mp} \leq T^*_{mp}$; then outside this melting range, dH_T/dT corresponds to the specific heats in the molten and in the solid states, respectively, whereas within the melting range we have

$$\int_{T_{mp*}}^{T^*_{mp}} \frac{dH_T}{dT}\, dT = \Delta H_s \tag{5.4-13}$$

that is, the latent heat of solidification.

For a one-component system, this procedure represents a simplification of the physical reality which need not be too serious, however, provided the melting range chosen is narrow. For a multicomponent system this procedure would, of course, be exact with T_{mp*} and T^*_{mp} corresponding to the liquidus and the solidus temperatures, respectively. A plot of dH_T/dT against the temperature for such a case has been given in Figure 5.3-3; however, for certain computational purposes it may be more convenient to avoid the sudden step changes at $T = T_{mp*}$ and $T = T^*_{mp}$. Under such conditions a smooth curve may be drawn for $dH_T/dT = f(T)$ within the range $T_{mp*} \leq T \leq T^*_{mp}$, which still

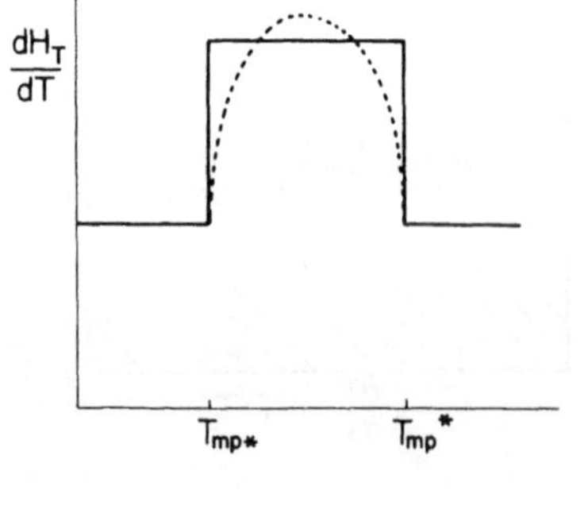

Figure 5.4-3. Sketch of the effective specific heat plotted against the temperature within the melting range. The dashed line denoted the smoothed curve.

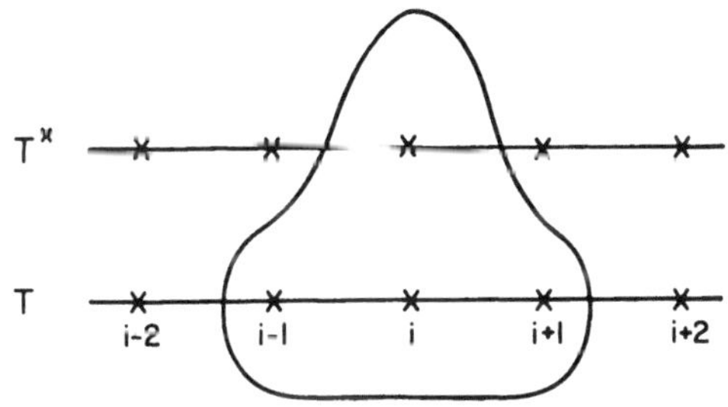

Figure 5.4-4. Illustration of the use of equation (5.4-14).

satisfies the constraint posed by equation (5.4-14) as sketched in Figure 5.4-3 with the dashed line.

We may now proceed by putting equation (5.4-12) in a finite difference form. If i, $i-1$, and $i+1$ denote adjacent spatial grid points, whereas T^* designates the temperature at a subsequent time step (i.e., at $t+\delta t$), then, upon using a central differencing formula, we have

$$T^* = T + \frac{k\delta t}{(dH_T/dT)\delta y^2}(T_{(i+1)} - 2T_i + T_{(i-1)}) \qquad (5.4\text{-}14)$$

As sketched in Figure 5.4-4, equation (5.4-14) provides a simple formula for calculating the value of the temperature at grid point i at a subsequent time step from the values at adjacent grid points (i.e., $i-1$, i, and $i+1$) and the preceding time step. Such a calculation is readily performed on digital computers.

It should be emphasized that this simple finite differencing procedure, involving an explicit, so-called *marching technique*, is just one of the many numerical techniques that may be deployed for solving partial differential equations. In many cases the simple marching technique would be very inefficient and more sophisticated implicit techniques should be used. References to these will be given at the end of this chapter.

We note that the use of an effective specific heat is probably the computationally most straight forward way of solving melting and solidification problems. The finite difference approach to the moving boundary problem proper may lead to computational instabilities unless care is taken in the development of an appropriate numerical approach.

5.4.3.2. *Integral Profile Techniques*

The use of integral profile methods and particularly the much more sophisticated development of this technique, *finite element methods*, has gained considerable popularity in recent years. A good description of this technique is available in the references listed under Suggested

Reading. Here we shall confine ourselves to presenting a brief outline of this procedure, together with a useful set of computed results for continuous casting.

Let us consider the growth of a solidified shell from a well-stirred melt which is at a temperature $T_{m,i}$, when the cold face of the solid crust is maintained at a temperature $T_{s,0}$. Following our earlier discussion, this problem is readily formulated:

$$\alpha \frac{\partial^2 T_s}{\partial y^2} = \frac{\partial T_s}{\partial t}, \qquad 0 \leq y \leq Y(t) \tag{5.4-15}$$

with

$$T_s = T_{mp}, \qquad y = Y(t) \tag{5.4-16}$$

$$T_s = T_{s,0}, \qquad y = 0 \tag{5.4-17}$$

and

$$-k\frac{\partial T_s}{\partial y} + h(T_{m,i} - T_{mp}) = \rho\, \Delta H_s \frac{dY}{dt} \tag{5.4-18}$$

Let us integrate both sides of equation (5.4-15) between the limits 0 and Y. Thus we have

$$\alpha \int_0^Y \frac{\partial^2 T_s}{\partial y^2}\, dy = \int_0^Y \frac{\partial T_s}{\partial t}\, dy \tag{5.4-19}$$

that is,

$$\alpha \left[\frac{\partial T_s}{\partial y}\bigg|_Y - \frac{\partial T_s}{\partial y}\bigg|_0 \right] = \frac{d}{dt} \int_0^Y T\, dy \tag{5.4-20}$$

We may now proceed by postulating that T is some function of the dimensionless distance (y/Y), i.e.,

$$T = T_{mp}\left[a_0 + a_1\left(\frac{y}{Y}\right) + a_2\left(\frac{y}{Y}\right)^2 + a_3\left(\frac{y}{Y}\right)^3 \right] \tag{5.4-21}$$

The constants a_0, a_1, etc., are not known at this stage, but may be determined from the boundary conditions. Once these constants are known, or there exists a functional relationship between them and Y, we may substitute into equation (5.4-20) which yields a first-order, ordinary differential equation for Y in terms of t. This differential equation is readily integrated.

Hills used this technique for obtaining useful general solutions for a one-dimensional continuous casting problem. Upon defining

x = distance from the mold inlet,
u_c = casting speed, and
h = the mold heat transfer coefficient,

Hills presented his results in terms of the following dimensionless parameters:

Dimensionless distance in direction of motion

$$\zeta = \frac{xh^2}{u_c p C_p k}$$

Dimensionless thickness of solidified layer

$$\tilde{Y} = \frac{hY(x)}{k}$$

Dimensionless latent heat + superheat

$$\Delta H_T = \frac{\Delta H_T}{C_p T_{mp}}$$

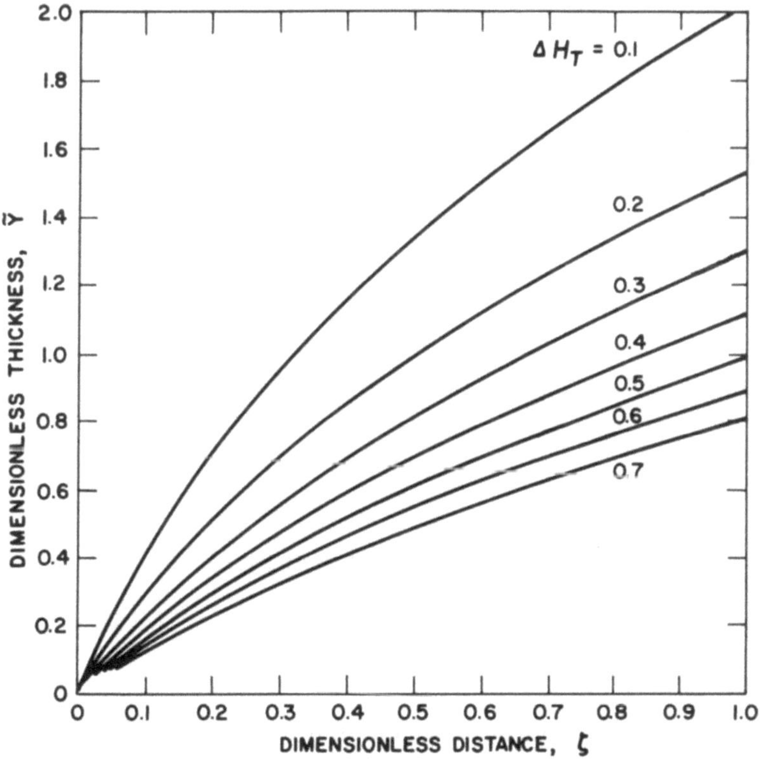

Figure 5.4-5. Plot of the dimensionless thickness Y against the dimensionless distances ζ.

The computed values of $\tilde{Y}$ are shown as a function of ζ and $\Delta\tilde{H}_T$ in Figure 5.4-5. Notwithstanding the simplifying assumptions made in the computation of Figure 5.4-5 (in particular the constant mold heat transfer coefficient!), these results were found to be in reasonable agreement with actual measurements. It has to be noted, however, that the use of Figure 5.4-5 requires that an estimate be made of the mold heat transfer coefficient. Nonetheless, this procedure is perhaps the most convenient available for obtaining a quick estimate of the shell thickness emerging from the mold of a continuous casting machine, as illustrated by the following example:

Example. Use Figure 5.4-5 to estimate the thickness of the solidified skin on a 70–30 brass slab, 1-in. × 15-in., as it leaves a 10-in. deep graphite mold.

Data

superheat = 100° F
melting point = 1750° F
latent heat (ΔH) = −88.2 Btu lb^{-1}
specific heat (C_p) = 0.09 Btu lb^{-1} °F^{-1}
thermal conductivity of solid (k) = 70 Btu hr^{-1} ft^{-1} °F^{-1}
density (ρ) = 531 lb ft^{-3}
casting speed (u_c) = 4 ft min^{-1} = 240 ft hr^{-1}
distance traveled through mold (x) = $\frac{10}{12}$ = 0.834 ft
heat transfer coefficient (h) = 200 Btu hr^{-1} ft^{-2} °F^{-1}
dimensionless latent heat

$$\tilde{\Delta H}_T = \frac{\Delta H + C_p(T_{mp} - T_i)}{C_p T_{mp}} = \frac{88.3 + (0.09)(100)}{(0.09)(1750)} = 0.617$$

dimensionless distance traveled

$$\zeta = \frac{xh^2}{u_v \rho C_p k} = \frac{0.834 \times 200^2}{240 \times 531 \times 0.09 \times 70} = 0.042$$

For $\tilde{\Delta H}_T = 0.617$ and $\zeta = 0.083$, we obtain from Figure 5.4-5

$$\tilde{Y} = \frac{hY}{k} \simeq 0.13$$

Therefore

$$Y = \frac{0.13 \times 70}{300} = 0.015 \sim 0.2 \text{ in.}$$

5.5. Concluding Remarks

In this brief treatment we sought to provide an introduction to the formulation and solution of melting and solidification problems in metals processing. As shown in Section 5.2, in the formulation of these problems for one-component systems, we follow procedures similar to those used in the statement of heat conduction, except for the fact that allowance must be made for the heat release or heat absorption at the melt–solid boundary and for the movement of this boundary.

In principle, the modeling of multicomponent systems is a rather more complex task, because the heat flow problem is inherently coupled with diffusion. In some cases, however, when we do not wish to represent the position of the phase boundary precisely, these systems may be modeled by using an effective specific heat.

The available solution techniques that were discussed range from quite simple analytical expressions, appropriate to idealized situations, to methods that entail machine computation. A large number of numerical methods are available for the solution of moving boundary (phase change) problems, a fact which renders the solution of these a more or less routine task.

To the author's mind, the key in all these problems is to recognize the basic physical situation, and then, through the statement of the appropriate conservation equation, to arrive at the governing equations. The practising metallurgist may then seek help from others in his bid for generating numerical or analytical solutions. However, the basic problem must be identified and formulated by those who are familiar with the physical system; the process metallurgist is in a unique position to perform this task.

Suggested Reading

Books

M. C. Flemings, *Solidification Processing*, McGraw-Hill, New York (1974); good general description of solidification with emphasis on properties.

L. I. Rubenstein, *The Stefan Problem*, The American Mathematical Society, Providence, R.I. (1971); highly mathematical, but very comprehensive treatment of both the analytical and the computational aspects of solidification problems.

H. S. Carlsaw and J. S. Jaeger, *Conduction of Heat in Solids*, Chap. 11, Oxford University Press, New York (1959); an excellent source of analytical solutions for solidification problems.

J. Szekely and N. J. Themelix, *Rate Phenomena in Process Metallurgy*, Chap. 10, John Wiley, New York (1971); many metallurgical problems.

T. Goodman, in *Advances in Heat Transfer*, Vol. I, T. F. Irvine and J. P. Harnett, eds., Academic Press, New York (1964); good description of integral profile methods.

J. Crank, *The Mathematics of Diffusion*, Oxford University Press, New York (1956); good discussion of numerical analysis as applied to partial differential equations.

A. R. Mitchell, *Computational Methods in Partial Differential Equations*, John Wiley, London (1969).

D. H. Norrie and G. deVries, *The Finite Element Method*, Academic Press, New York (1973); a useful treatment.

Journal Articles in Selected Areas

Continuous Casting

Symposium on Continuous Casting, Chicago, 1973, AIME, New York (1973).
A. W. D. Hills, *J. Iron Steel Inst. (London)* **203**, 18 (1965).
J. K. Brimacombe and F. Weinberg, *J. Iron Steel Inst. (London)* **211**, 24 (1973).
E. A. Mizikar, *Trans. Met. Soc. AIME* **239**, 1747 (1967).
S. Asai and J. Szekely, *Ironmaking Steelmaking* **3**, 205 (1975).

Scrap Melting

J. Szekely, Y. K. Chuang, and J. W. Hlinka, *Met. Trans.* **3**, 2825 (1972).
K. Mori and J. Nomura, *Tetsu to Hagane* **55**(5), (1969).

Dissolution of Deoxidants

R. I. L. Guthrie, H. Heinein, and L. Gourtaoyanis, in *Symposium on the Physical Chemistry of the Production of Alloy Additives*, AIME, New York (1973).

Index

MIX
Papier aus verantwortungsvollen Quellen
Paper from responsible sources
FSC® C105338

If you have any concerns about our products,
you can contact us on
ProductSafety@springernature.com

In case Publisher is established outside the EU,
the EU authorized representative is:
Springer Nature Customer Service Center GmbH
Europaplatz 3, 69115 Heidelberg, Germany

Printed by Libri Plureos GmbH
in Hamburg, Germany